THE POWER OF PLAGUES

Second Edition

T0178924

THE POWER OF PLAGUES

Second Edition

Irwin W. Sherman
University of California at San Diego

ASM PRESS

Washington, DC

Library of Congress Cataloging-in-Publication Data

Names: Sherman, Irwin W., author.

Title: The power of plagues / Irwin W. Sherman.

Description: 2nd edition. | Washington, DC : ASM Press, [2017] | Includes bibliographical references and index.

Identifiers: LCCN 2017005669 (print) | LCCN 2017006703 (ebook) | ISBN 9781683670001 (hardcover : alk. paper) | ISBN 9781683670018 (ebook) | doi:10.1128/9781683670018

Subjects: LCSH: Communicable diseases–History. | Epidemics–History. | Diseases and history.

Classification: LCC RA643 .S55 2017 (print) | LCC RA643 (ebook) | DDC 616.9–dc23

LC record available at https://lccn.loc.gov/2017005669

SKY10060567_112023

Address editorial correspondence to

ASM Press, 1752 N St., N.W.,
Washington, DC 20036-2904, USA

Send orders to ASM Press, P.O. Box 605, Herndon, VA 20172, USA

Phone: 800-546-2416; 703-661-1593
Fax: 703-661-1501
E-mail: books@asmusa.org
Online: http://www.asmscience.org

Cover: "La peste en Mandchourie" *Le Petit Journal* n° 1057 du 19 Fev. 1911
Design: Lou Moriconi (http://lmoriconi.com/)

Table of Contents

Preface to the Second Edition

Plagues, the historian Asa Briggs observed, "are a dramatic unfolding of events; they are stories of discovery, reaction, conflict, illness and resolution." They test "the efficiency and resilience of local and administrative structures" and "expose the relentlessly political, social and moral shortcomings...rumors, suspicions, and at times violent...conflicts." This book was written to make the science of epidemic diseases—plagues—accessible and understandable. It is a guide through the maze of contagious diseases, their past importance, the means by which we came to understand them, the methods and practices for control and eradication, and failing this how they may affect our future.

My objective in writing the *Power of Plagues* has been to provide an understanding of epidemic diseases and their impact on lives past, present and future. As a biologist I wanted to examine infectious diseases that have been and continue to afflict humankind. In this 2nd edition I describe the nature and evolution of a selected group of diseases and then aim to show how past experiences can prepare us for future encounters with them. It is also a status report of where we are today with infectious diseases that are emerging and re-emerging to inflict harm not only to the individual but to larger segments of the world's populations. This edition covers modern disease outbreaks, such as anthrax, cholera, Ebola disease, HIV, influenza, Lyme disease, malaria, tuberculosis, and Zika disease, as well as the threat of antimicrobial resistance.

For decades it has been a commonly held belief that all epidemic diseases in the developed world could be eliminated by vaccines, as was the case with smallpox. If not, then we deluded ourselves that new and more powerful chemotherapeutic agents and antibiotics would be developed and these could be relied on to cure an emerging new disease. Some of us were convinced that our water was safe to drink and our food could be eaten with little fear of contracting a disease and that the transmitters of disease—mosquitoes, flies, lice, ticks and fleas—could easily be eliminated by the spraying of deadly insecticides. If by chance a few of our neighbors became infected, some held the illusion that the infected individuals would be treated quickly and effectively so that a severe disease outbreak could be averted.

These views, however, have come to be challenged: drug-resistant tuberculosis emerged as a worldwide threat, and there were outbreaks of hantavirus and SARS (severe acute respiratory syndrome) in 2012 and 2003, respectively. In the summer of 1999 an outbreak of West Nile virus (WNV) caused illness in 62 people and 7 died. This took place in New York City, not Africa. Then in 2001 there was a terrorist attack of anthrax that killed 5 and sickened 17 in New York City, Florida, and Washington, DC. A particularly lethal bird influenza—one that killed millions of animals in a dozen Asian countries—caused alarm in 2004. This virus, which rarely infects humans, spread rapidly in the population and killed 42 people. If this "new" flu had been able to be transmitted from person-to-person by coughing, sneezing, or even a handshake, then a natural bioterrorist attack would have occurred. An outbreak of Ebola disease in Africa in 2014 and an outbreak of Zika disease in Brazil and the Caribbean in 2015 caused public health authorities to sound alarms that these might be forerunners of a pandemic. Luckily, the doomsday scenarios were not realized. But in the future, might a pandemic occur? And if there is such a catastrophe, how will we deal with it?

This book has been conceived as a conversation about how we came to understand the nature of severe outbreaks of epidemic disease—plagues. It tells about the microbe hunters who were able to identify and characterize the infectious disease agent, its mode of transmission, and how control was effected and health restored. It tells of the ways in which plagues and culture interact to shape values, traditions, and the institutions of Western civilization.

As with the previous edition, I have not taken a chronological approach in the examination of the plagues that have afflicted humans. Chapters have been written so that they are more-or-less independent and as a result they need not be read in a proscribed sequence. However, in some instances readers may attain a somewhat better understanding when the chapters on principles and protection (Chapters 1, 10, and 11) as well as the Appendix on Cells and Viruses are read early on. Some readers will be disappointed that their "favorite" plague has not been included in these pages. To have added many more epidemic diseases would have made the book much longer and encyclopedic—something I wanted to avoid. Rather, the particular plagues included have been selected for their value in teaching us important lessons. The style of the *Power of Plagues* is such that readers without any background in the sciences should easily be able to understand its message. This book is intended to promote an understanding of infectious disease agents by a sober and scientific analysis and is not a collection of horror stories to provoke fear and loathing. Learning about how infectious diseases have shaped our past has proven to be an exciting and enlightening experience for me. My hope is that readers of this book will also find that to be true.

1
The Nature of Plagues

doi:10.1128/9781683670018.ch1

Disease can be a personal affair. Peter Turner, a World War II veteran, was a commander of the Pennsylvania Division of the American Legion. In the summer of 1976, Turner, a tall, well-built 65-year-old, decked out in full military regalia, attended the American Legion convention in Philadelphia. As a commander, Turner stayed at the Bellevue-Stratford Hotel, headquarters for the meeting. Two days after the convention Turner fell ill with a high fever, chills, headache, and muscle aches and pains. He dismissed the symptoms as nothing more serious than a "summer cold." His diagnosis proved to be wrong. A few days later he had a dry cough, chest pains, shortness of breath, vomiting, and diarrhea. Within a week his lungs filled with fluid and pus, and he experienced confusion, disorientation, hallucinations, and loss of memory. Of 221 fellow Legionnaires who became ill, Commander Turner and 33 others died from pneumonia. The size and severity of the outbreak, called Legionnaires' disease, quickly gained public attention, and federal, state, and local health authorities launched an extensive investigation to determine the cause of this "new" disease. There was widespread fear that Legionnaires' disease was an early warning of an epidemic. Though no person-to-person spread could be documented, few people attended the funerals or visited with the families of the deceased veterans.

Statistical studies of Legionnaires' disease revealed that all who had become ill spent a significantly longer period of time in the lobby of the Bellevue-Stratford Hotel than those who remained healthy. Air was implicated as the probable pathway of spread of the disease, and the most popular theory was that infection resulted from aspiration of bacteria (called *Legionella*) in aerosolized water either from cooling towers or evaporative condensers. Unlike infections caused by inhalation, in aspiration secretions in the mouth get past the choking reflex and, instead of going into the esophagus and stomach, mistakenly enter the lungs. Protective mechanisms that normally prevent aspiration are defective in individuals who are older, in smokers,

Figure 1.1 (Left) *Woman with Dead Child*. Kathe Kollwitz etching. 1903. National Gallery of Art, Washington, D.C.

and in those who have lung disease. The Legionnaires were near-perfect candidates for contracting the disease.

After the outbreak, the hotel, which had been the choice of conventions such as that held by the Legionnaires as well as those of Hollywood stars such as John Wayne, Grace Kelly, and Elvis Presley, was shunned by guests. The hotel closed down and was empty for almost 3 years, during which time there was talk of tearing the building down. After tens of millions of dollars in renovation, however, there was a new owner, and after reopening in 1989, today it is the Hyatt at The Bellevue.

Since the Philadelphia outbreak, there have been numerous reports of Legionnaires' disease. For example, in 1985 in Stafford District Hospital in Stafford, England, there were 175 cases and 28 deaths; in 1999 in Bovekarspel, Holland, a hot tub was responsible for 318 cases and 32 deaths; in 2001, a hospital in Murcia, Spain, reported 800 cases; in 2005 at the Seven Oaks Home for the Aged in Toronto, Canada, 127 were sickened and 21 died; and in 2015 in a housing development in the South Bronx, NY, 128 were infected with *Legionella* and 13 died. It is estimated that in the United States there are 8,000 to 18,000 cases of legionellosis a year that require hospitalization, and worldwide the numbers are even greater.

A few years after the outbreak of Legionnaires' disease in Philadelphia, another "new" disease appeared. Mary Benton, a graduate student and English composition teaching assistant at UCLA, knew something was amiss as she prepared for Monday's class. She had spent the previous day happily celebrating her 24th birthday, but by evening she was doubling over in pain every time she went to the bathroom. Mary figured she probably had an infection or was suffering from overeating. Mary, who was previously healthy and active, became concerned as her symptoms worsened. By the time she saw her physician, she had nausea, chills, diarrhea, headache, and a sore throat. Her temperature was 104.7°F, her heart rate 178 beats/min, and she had a red rash, initially on her thighs, but it had become diffuse over her face, abdomen, and arms. Her blood pressure had fallen to 84/50 mm Hg, she had conjunctivitis in both eyes, and her chest X-ray was normal, but a pelvic examination revealed a brownish discharge. Though her doctors administered antibiotics, oxygen, and intravenous fluids, her condition deteriorated over the next 48 h. She died of multiorgan failure: low blood pressure, hepatitis, renal insufficiency, and internal blood clots. Laboratory tests provided clues to the cause of death. Cultures made from her blood, urine, and stools were negative, but the vaginal sample contained the bacterium *Staphylococcus aureus*. The "new" disease that felled Mary Benton was named toxic shock syndrome, or TSS. The source of Mary's infection, and whether it might be spread through the population as a sexually transmitted disease (STD), raised many concerns. TSS continued to appear for the next 10 years among previously healthy young women residing in several states. As with Mary Benton, each case began with vomiting and high fever, followed by light-headedness and fainting; the throat felt sore, and the muscles ached. A day later there appeared a sunburn-like rash, and the eyes became bloodshot. Within 3 to 4 days the victims suffered confusion, fatigue, weakness, thirst, and a rapid pulse; the skin became cool and moist; and breathing became rapid. These symptoms were followed by a sudden drop in blood pressure; if it remained low enough for a long enough period, circulatory collapse produced shock.

TSS was a gender-specific disease. From 1979 to 1996, it affected 5,296 women, median age 22, with a peak death rate of 4%. TSS, however, was not an STD. Ultimately it was linked to the use of certain types of tampons, especially those containing cross-linked carboxymethyl cellulose with polyester foam, which provided a favorable environment for the toxin-producing *S. aureus*. Elevated vaginal temperature and neutral pH, both of which occur during menses, were enhanced by the use of these super-absorbent tampons. In addition, tampons obstruct the flow of menstrual blood and may cause reflux of blood and bacteria into the vagina. By the late 1980s, when these tampon brands were removed from the market, the number of deaths from TSS declined dramatically.

The effects of disease at the personal level can be tragic (Fig. 1.1), but when illness occurs in many people, it may produce another emotion—fear—for now that disease might spread rapidly, causing death, as well as inflaming the popular imagination. The 2003 outbreak of SARS (severe acute respiratory syndrome) had all the scary elements of a plague—panic, curtailed travel and commerce, and economic collapse. It began in February 2003 when a 64-year-old Chinese physician who was working in a hospital in Guandong Province in southern China traveled to Hong Kong to attend a wedding and became ill. He had a fever, a dry cough, a sore throat, and a headache. Unconcerned, he felt well enough to go sightseeing and to shop with his brother-in-law in Hong Kong; during that day, however, his condition worsened and he found that he had difficulty breathing. Seeking medical attention at a nearby hospital, he was taken immediately to the ICU (intensive care unit) and given antibiotics, anti-inflammatory drugs, and oxygen. These were to no avail, and several hours later he suffered respiratory failure and died. The brother-in-law, who was in contact with him for only 10 h, suffered from the same symptoms 3 days later and was hospitalized. Again, all measures failed, and he died 3 weeks after being hospitalized.

Laboratory tests for the physician (patient 1) and his brother-in-law (patient 2) were negative for Legionnaires' disease, tuberculosis, and influenza. A third case of this severe respiratory syndrome occurred in a female nurse who had seen the physician in the ICU, and the fourth case was a 72-year-old Chinese-Canadian businessman who had returned to Hong Kong for a family reunion. He stayed overnight in the same hotel and on the same floor as the physician. (He would ultimately carry SARS to Canada when he returned home.) Patient 5 was the nurse who attended the brother-in-law, and patients 6, 7, 8, and 9 were either visitors to the hospital or nurses who had attended patient 4. Patient 10 shared the same hospital room with patient 4 for 5 days. In less than a month 10 patients had SARS, with 6 (patients 3, 4, 6, 8, 9, and 10) surviving and 4 (patients 1, 2, 5, and 7) dying. Over the next 4 months the SARS survivors sowed the seeds of infection that led to more than 8,000 cases and 800 deaths in 27 countries, representing every continent except Antarctica.

On February 1, 2016, the World Health Organization (WHO), after recording a surge in the number of babies born with microcephaly—an abnormally small head—sounded the alarm that Zika virus was a threat to pregnant women and could cause serious harm to their fetuses. Six months later, on August 1, 2016, the *Los Angeles Times* reported that there were 1,638 confirmed cases of microcephaly and other neurological defects in Brazil as a consequence of the Zika virus. Worldwide, 64 countries and territories have reported to the WHO evidence of mosquito-borne transmission of Zika. There has been a steady march of the Zika virus across the Americas—an epidemic—and that is because the vector, the thoroughly "domesticated" *Aedes* mosquito, stays close to people and is present primarily in the Southwest and Southeast United States, as well as the Caribbean, Central and South America, and Europe. Indeed, by October 2016, according to the CDC, there were 3,936 cases in the continental U.S. and 25,955 cases in the U.S. territories of Puerto Rico, the U.S. Virgin Islands, and American Samoa. The number of cases of microcephaly may reach hundreds. The CDC director, Thomas Frieden, in an understatement, warned that without a vaccine "this is an emergency that we need to address."

Despite the recognition that disease, such as SARS, Legionnaires' disease, TSS, and Zika, may appear suddenly and with disastrous consequences, more often than not little notice has been given to the ways in which disease can and has shaped history. The influence of disease on history was often neglected because there appeared to be few hard-and-fast lessons to be learned from a reading of the past; sickness seemed to have no apparent impact except for catastrophic epidemics such as the bubonic plague, or it was outside our experience. We tend to live in an age in which diseases appear to have minimal effects—we are immunized as children, we treat illness with effective drugs and antibiotics, and we are well nourished. And so our impressions of how diseases can affect human affairs have been blunted. But this is an illusion: the sudden appearance of SARS, Legionnaires' disease, TSS, AIDS, and Zika are simply the most recent examples of how disease can affect society. Our world is much more vulnerable than it was in the past.

New and old diseases can erupt and spread throughout the world more quickly because of the increased and rapid movements of people and goods. Efficiencies in transportation allow people to travel to many more places, and almost nowhere is inaccessible. Today, few habitats are truly isolated or untouched by humans or our domesticated animals. We can move far and wide across the globe, and the vectors of disease can also travel great distances, and, aided by fast-moving ships, trains, and planes, they introduce previously remote diseases into our midst (such as West Nile virus and SARS, influenza and Zika). New diseases may be related to advances in technology: TSS resulted from the introduction of "improved" menstrual tampons that favored the growth of a lethal microbe, and Legionnaires' disease was the result of the growth and spread of another deadly "germ" through the hotel's air conditioning system.

This book chronicles the recurrent eruptions of plagues that marked the past (Fig. 1.2), influence the present, and surely threaten our future. The particular occurrence of a severe and debilitating outbreak of disease may be unanticipated and unforeseen, but despite the lack of predictability, there is a certainty: dangerous "new" diseases will occur.

Living Off Others

The "germs" that caused SARS, Legionnaires' disease, and TSS are parasites. To appreciate more fully the nature of these diseases as well as others and how they may be controlled, it helps to know a little more about parasites. No one likes to be called a parasite. The word suggests, at least to some, a repugnant alien creature that insinuates itself into us and cannot be shaken loose. Nothing could be further from the truth. Within the range of all that lives, some are unable to survive on their own, and they require another living being for their nourishment. These life-dependent entities are called parasites, from the Greek *parasitos*, meaning "one who eats at the table of another. "The business they practice, parasitism, is neither disgusting nor unusual. It is simply a means to an end: obtaining the resources needed for their growth and reproduction. We do the same—eating and breathing—in order to survive.

Parasitism is the intimate association of two different kinds of organisms (species) in which one benefits (the parasite) at the expense of the other (the host), and as a consequence of this, parasites often harm their hosts. The harm inflicted, with observable consequences, such as those seen in Commander Peter Turner and Mary Benton and those patients afflicted with SARS and Zika virus, is called "disease," literally "without comfort." Though parasites can be described by the one thing they are best known for—causing harm—they come in many different guises. Some may be composed of a fragment of genetic material wrapped in protein

Figure 1.2 *The Plague of Ashod* by Nicolas Poussin (1594-1665).
The painting probably represents bubonic plague since rats are shown on the plinth

(virus).* Others consist of a single cell* (bacteria, fungi, and protozoa), and some are made up of many cells (roundworms, flatworms, mosquitoes, flies, and ticks). Some parasites, such as tapeworms, hookworms, malaria, and HIV, as well as the Zika and Ebola virus, live inside the body, whereas others (ticks and chiggers) live on the surface. Parasites are invariably smaller in mass than their host. Consider the size of malaria, a microparasite, and hookworm, a macroparasite. Both produce anemia, or, as one advertisement for an iron supplement called the condition, "tired blood."

A malaria parasite lives within a red blood cell that is 1/5,000 of an inch in diameter. If only 10% of your blood cells were infected, the total mass of the malaria parasites would not occupy a thimble, and yet in a few days they could destroy enough of your red blood cells that the acute effects of blood loss could lead to death. In effect, you could die from an internal hemorrhage. Although the "vampire of the American South," the bloodsucking, thread-like hookworm, is only 0.5 in. in length and 0.05 in. in girth, if your intestine harbored 50 worms, you would lose a cupful of blood a day. Yet the entire mass of worms would weigh less than 5 hairs on your head.

Some parasites have complex life cycles and may have several hosts. In malaria the hosts are mosquitoes and humans; in blood fluke disease, the "curse of the pharaohs," the hosts are humans and snails; and in sleeping sickness the hosts are tsetse flies, game animals, and humans. All parasites—whether they are large or small—cause harm to their host, though not all kill their host outright. This is because resistance may develop in any population of hosts and not every potential host will be infected—some individuals may be immune or not susceptible due to a genetic abnormality or the absence of some critical dietary factor (vitamin deficiency).

To succeed in a hostile world where individual hosts are distinct and separate from one another, parasites need to disperse their offspring or infective stages to reach new hosts. To meet this requirement they produce lots of offspring, thereby increasing the odds that some of these will reach new hosts. It is a matter of numbers: more offspring will have a greater probability of reaching a host and setting up an infection. In this way the parasite enhances its chances for survival. Three cases will illustrate this: the red blood cell-destroying hookworms, malaria, and the white blood cell killer HIV.

When a malaria-infected mosquito feeds, it injects with its saliva perhaps a dozen of the thousands of parasites that are present in its salivary glands. Each malaria parasite invades a liver cell, and after a week each produces up to 10,000 offspring; in turn, every one of these infects a red blood cell. Within the infected red blood cell, a malaria parasite produces 10 to 20 additional infective forms to continue the destructive process. In little more than 2 weeks a person infected by a single malaria parasite will have produced >100,000 parasites, and 2 days later the blood will contain millions of malaria parasites.

Hookworms live attached to the lining of the small intestine, which they pierce with their razor-sharp teeth, allowing them to suck blood, as would a leech. Each female hookworm—no bigger than an eyelash—can live within the intestine for >10 years, producing each day >10,000 eggs. In her lifetime, this "Countess Dracula" can produce >36 million microscopic eggs.

The AIDS-causing virus, HIV, is a spherical particle so small that if 250,000 were lined up they would hardly be 1 in. in length. Each virus, however, has an incredible capacity to reproduce itself. After it invades a specific kind of white blood cell (the T-helper lymphocyte), where it replicates, a million viruses will be produced in a few short days. To gain some appreciation of the high reproductive capacity of this virus, we might think of the infecting HIV as a person standing on a barren stretch of beach; if we were to return to this beach a few days later, we would find it jammed and overcrowded with millions—a population explosion.

Any environment other than a living host is inimical to the health and welfare of the parasite. Some parasites have gotten around this with resistant stages such as spores, eggs, or cysts that enable them to move from one host to another in a fashion akin to "island hopping." Hookworms, tapeworms, blood flukes, and pinworms have eggs that are able to survive outside the body; the microscopic cysts of the roundworm *Trichinella* are able to resist the ordinarily lethal effects of the acids in our stomach to cause trichinosis, and now we are all too familiar with the possibility of a bioterrorist attack from anthrax (p. 416), which has resistant spores that allow it to spread by inhalation of "anthrax dust." The movement of a parasite from host to host—whether by direct or indirect means—is called transmission. When the transmission of parasites involves a living organism such as a fly, mosquito, tick, flea, louse, or snail, these "animate intermediaries" are called vectors. Transmission by a vector may be mechanical (e.g., the bite wound of a mosquito or fly) or developmental (e.g., parasites that grow and reproduce in snails in blood fluke disease, or in mosquitoes, as in malaria and yellow fever). Transmission of a parasite may also occur through contamination of eating utensils, drinking cups, food, needles, bedclothes, towels, or clothing, or in droplet secretions. In the 1976 outbreak of Legionnaires' disease in Philadelphia, transmission was not from person to person but through a fine mist of water in the air conditioning system, whereas in the case of SARS (and influenza), transmission is from person to person via droplet secretions from the nose and mouth.

Parasites and their free-living relatives come in a variety of sizes, shapes, and kinds (species). Bacteria, 1 to 5 micrometers (µm) in size, are prokaryotes* that can be free-living or parasitic. They may assume several body forms: rods (bacilli), spheres (cocci), or spiral. Protozoa, 5 to 15 µm in size, are one-celled eukaryotes[*] that can lead an independent existence (such as the freshwater *Amoeba* sp.) or be parasitic (such as the *Entamoeba* sp. that causes amebic dysentery or the cork-screw-shaped trypanosomes that cause African sleeping sickness). Bacteria and protozoa are too small to be seen with the unaided eye. The technological advance—the microscope—perfected in the 1600s allowed for their discovery, and so they are called microparasites. The ultimate microparasite is a virus—a small piece of nucleic acid (RNA or DNA) enclosed within a protein coat. A virus has no cell membrane, no cytoplasm, and no organelles; and because it has no metabolic machinery of its own, it requires a living cell to make more virus. Viruses are <0.1 µm in size; they cannot be seen even with the light microscope, but only with the electron microscope, which can magnify objects >10,000 times. Viruses, such as the agents of SARS, AIDS, Zika, Ebola, yellow fever, and the flu, are neither cells nor organisms.

Microparasites reproduce within their hosts and are sometimes referred to as infectious microbes, or, more commonly, "germs." Larger parasites, ones that can be seen without the use of a microscope, are referred to as macroparasites; they are composed of many cells. Those that most often cause diseases of humans or domestic animals are roundworms, such as the hookworm; flatworms, such as the blood fluke; blood-sucking insects, such as mosquitoes, flies, and lice; or arachnids, such as ticks. Macroparasites do not multiply within an infected individual (except in the case of larval stages in the intermediate hosts) but instead produce infective stages that usually pass out of the body of one host before transmission to another host.

"What's in a name? That which we call a rose by any other name would smell as sweet." When William Shakespeare penned these lines in *Romeo and Juliet*, he gave value to substance over name-calling. But being able to tell one microbe from another is more than having a proper name for a germ—it can have practical value. Imagine you have just returned from a trip and now suffer with a fever, headache, and joint pains, and worst of all you have a severe case of diarrhea. What a mess you are! When you see your physician, she tells you that the cause of your distress could be due to an infection with *Salmonella* or *Giardia* or *Entamoeba* or the influenza or SARS virus. Prescribing an antibiotic for diseases caused by a virus would do you no good, but for "food poisoning" caused by *Salmonella*, a bacterium, a course of antibiotic therapy might restore you to health. On the other hand, if your clinical symptoms were due to the presence of protozoan parasites such as *Giardia* or *Entamoeba*, they would not respond to antibiotics either, and other drugs would have to be prescribed to cure you. Determining the kind of parasite (or parasites) you harbor, therefore, will do more than provide the name of the offender; it will allow for the selective treatment of your illness.

*See: Cells and Their Structure in the Appendix

Plagues and Parasites

In antiquity, all disease outbreaks, irrespective of their cause, were called plagues; the word "plague" comes from the Latin *plaga*, meaning "to strike a blow that wounds." When a parasite invades a host, it establishes an infection and wounds the body (Fig. 1.2). Individuals who are infected and can spread the disease to others (such as SARS patient 4) are said to be contagious or infectious. Initially, Legionnaires' disease and TSS were thought to be contagious. Despite the obvious clinical signs of coughing, nausea, vomiting, and diarrhea, however, a person-to-person-transmissible agent was not found. In short, the victims of TSS and Legionnaires' disease were not infectious, in contrast to what we know in cases of influenza, SARS, and the common cold with a similar array of symptoms. Influenza and SARS are different kinds of diseases of the upper respiratory system: the flu is contagious 24 h before symptoms appear, has a short (2-to-4-day) incubation period, and requires hospitalization infrequently; whereas SARS has a longer (3-to-10-day) incubation period, the patient is infectious only after symptoms appear, and the infection requires that the victim be hospitalized.

Infectiousness, however, may persist even after disease symptoms have disappeared; such infectious but asymptomatic individuals are called carriers. The most famous of these carriers was the woman called "Typhoid Mary," an Irish immigrant to the United States whose real name was Mary Mallon. In 1883 she began working as a cook for a wealthy New York banker, Charles Henry Warren, and his family. The Warren family rented their large house in Oyster Bay, Long Island, from a George Thompson. That summer, six of eleven people in the house came down with typhoid fever (caused by the "germ" *Salmonella typhi*), including Mrs. Warren, two daughters, two maids, and a gardener. Mr. Thompson, fearing he would be unable to rent his "diseased house" to others, hired George Soper, a sanitary engineer, to find the source of the epidemic. Soper's investigation soon led him to Mary Mallon, who had been hired as a cook just 3 weeks before the outbreak of typhoid in the Warren household. Mary had remained with the Warrens for only a month and had already taken another position when Soper found her. On June 15, 1907, Soper published his findings in the *Journal of the American Medical Association*: Mary was a healthy carrier of typhoid germs. Although she was unaffected by the disease (which causes headache, loss of energy, diarrhea, high fever, and, in a tenth of cases, death), she still could spread it. When Soper confronted Mary and told her she was spreading death and disease through her cooking, she responded by seizing a carving fork, rushing at him, and driving Soper off. Soper, however, was undaunted and convinced the New York City Health Department that Mary was a threat to the public's health. She was forcibly carried off to an isolation cottage at Riverside Hospital on Rikers Island in the Bronx. There, her feces were examined and found to contain the typhoid bacteria. Mary remained at the hospital, without her consent, for 3 years and then was allowed to go free as long as she remained in contact with the Health Department and did not engage in food preparation. She disappeared from the Health Department's view for a time but then took employment as a cook at the Sloane Maternity Hospital under an assumed name, Mrs. Brown.

During this time she spread typhoid to 25 doctors, nurses, and staff, 2 of whom died. She was sent again to Rikers Island, where she lived the rest of her life, 23 years, alone in a one-room cottage. During her career as a cook, "Typhoid Mary" probably infected many more than the 50 documented cases, and she surely caused more than 3 deaths. Mary Mallon was not the only human carrier of typhoid. In 1938 when she died, the New York City Health Department noted that there were 237 others living under their observation. She was the only one kept isolated for years, however, and one historian has ascribed this to prejudice toward the Irish and a non-compliant woman who could not accept that unseen and unfelt "bugs" could infect others. Mary Mallon told a newspaper: "I have never had typhoid in my life and have always been healthy. Why should I be banished like a leper and compelled to live in solitary confinement … ?"

Predicting Plagues

Recognizing the elements required for a parasite to spread in a population allows for better forecasting of the course a disease may take. Three factors are required for a parasite to spread from host to host: there must be infectious individuals, there must be susceptible individuals, and there must be a means for transmission between the two. Transmission may be by indirect contact involving vectors such as mosquitoes (in malaria and yellow fever) or flies (in sleeping sickness and river blindness) or ticks (in Lyme disease), or it may be by direct contact as it is with measles, influenza, SARS, and tuberculosis, where it is influenced by population density.

In the past, the sudden increase in the number of individuals in a population affected by a disease was called a plague. Today we frequently refer to such a disease outbreak as an epidemic, a word that comes from the Greek *epi*, meaning "among," and *demos*, "the people." Epidemiologists are disease forecasters who study the occurrence, spread, and control of a disease in a population, using statistical data and mathematical modeling to identify the causes and modes of disease transmission and to predict the likelihood of an epidemic, to identify the risk factors, and to help plan control programs such as quarantine and vaccination. When TSS broke out, epidemiologic studies linked the syndrome to the use of tampons, principally Rely tampons, and the recommendation was that the illness could be controlled in menstruating women by the removal of such tampons from the market. Acting on this advice, Procter & Gamble stopped marketing Rely tampons and the number of cases virtually disappeared.

For an infection to persist in a population, each infected individual on average must transmit the infection to at least one other individual. The number of individuals each infected person infects at the beginning of an epidemic is given by the notation R_0; this is the basic reproductive ratio of the disease, or, more simply, the multiplier of the disease. The multiplier helps to predict how fast a disease will spread through the population.

The value for R_0 can be visualized by considering the children's playground game of touch tag. In this game one person is chosen to be "it," and the objective of the game is for that player to touch another, who in turn also becomes "it." From then on each person touched helps to tag others. If no other player is tagged, the game is over, but if more than one other player becomes "it," then the number of touch taggers multiplies. Thus, if the infected individual (it) successfully transmits the disease (touches another), then the number of diseased individuals (touch taggers) multiplies. In this example the value for R_0 is the number of touch taggers that result from being in contact with "it."

The longer a person is infectious and the greater the number of contacts that the infectious individual has with those who are uninfected, the greater the value of R_0 and the faster the disease will spread. An increase in the population size or in the rate of transmission increases R_0, whereas an increase in parasite mortality or a decrease in transmission will reduce the spread of disease in a population. Thus, a change that increases the value of R_0 tends to increase the proportion of hosts infected (prevalence) as well as the burden (incidence) of a disease. Usually, as the size of the host population increases, so do disease prevalence and incidence.

If the value for R_0 is >1, then the "seeds" of the infection (i.e., the transmission stages) will lead to an ever-expanding spread of the disease—an epidemic or a plague—but in time, as the pool of susceptible individuals is consumed (like fuel in a fire), the epidemic may eventually burn itself out, leaving the population to await a slow replenishment of new susceptible hosts (providing additional fuel) through birth or immigration. Then a new epidemic may be triggered by the introduction of a new parasite or mutation, or there may be a slow oscillation in the number of infections, eventually leading to a persistent low level of disease. If R_0 is <1, though, then each infection produces <1 transmission stage and the parasite cannot establish itself.

The economic costs of the outbreak of SARS in 2003 were nearly $100 billion as a result of decreased travel and decreased investment in Southeast Asia. The University of California at Berkeley was so concerned about this epidemic that it put a ban on Asian students planning to enroll for the summer session. The question raised at the outset was: How long will the SARS outbreak last? Calculating the value of R_0 provided an answer. Analysis of ~200 cases during the first 10 weeks of the epidemic gave an R_0 value of 3.0, meaning that a single infectious case of SARS would infect about three others if control measures were not instituted. This value suggested a low to moderate rate of transmissibility and that hospitalization would block the spread of SARS. The prediction was borne out: transmission rates fell as a result of reductions in population contact rates and improved hospital infection control as well as more rapid hospitalization of suspected (but asymptomatic) individuals. By July of 2003 the R_0 value was much smaller than 1, and the ban on Asian students enrolling at the Berkeley campus of the University of California was lifted.

Epidemiologists know that host population density is critical in determining whether a parasite can become established and persist. The threshold value for disease establishment can be obtained by finding the population density for which $R_0 = 1$. In general, the size of the population needed to maintain an infection varies inversely with the transmission efficiency and directly with the death rate (virulence). Thus, virulent parasites, that is, those causing an increased number of deaths, require larger populations to be sustained, whereas parasites with reduced virulence may persist in smaller populations.

Measles, caused by a virus, provides an almost ideal pattern for studying the spread of a disease in a community. The virus is transmitted through the air as a fine mist released through coughing, sneezing, and talking. The virus-laden droplets reach the cells of the upper respiratory tract (nose and throat) and the eyes and then move on to the lower respiratory tract (lungs and bronchi). After infection, the virus multiplies for 2 to 4 days at these sites and then spreads to the lymph nodes, where another round of multiplication occurs. The released viruses invade white blood cells and are carried to all parts of the body using the bloodstream as a waterway. During this time the infected individual shows no signs of disease. But after an incubation period (8 to 12 days), there is fever, weakness, loss of appetite, coughing, a runny nose, and a tearing of the eyes. Virus replication is now in high gear. Up to this point the individual probably believes his or her suffering is a result of a cold or influenza, but when a telltale rash appears—first on the ears and forehead and then spreading over the face, neck, trunk, and to the feet—it is clearly neither influenza nor a common cold. Once a measles infection has begun, there is no treatment to halt the spread of the virus in the body.

Measles passes from one host to another without any intermediary; recovery from a single exposure produces lifelong immunity. As a consequence, measles commonly afflicts children, and for that reason it is called a "childhood disease." Although measles has been eradicated in the United States because of childhood immunization, it can be responsible for a death rate of ~30% in lesser-developed countries. It is one of the ten most frequent causes of death in the world today. One of the reasons that measles may disappear from a community is immunity that may be the result of natural recovery from an infection or immunization.

The spread of infection from an infected individual through the community can be thought of as a process of diffusion, in which the motions of the individuals are random and movement is from a higher concentration to a lower one. Therefore, factors affecting its spread include the size of the population, those communal activities that serve to bring the susceptible individuals in contact with infectious individuals, the countermeasures used (e.g., quarantine, hospitalization, and immunization), and seasonal patterns. For example, in northern temperate zones, measles spreads most frequently in the winter months because people tend to be confined indoors, while in Iceland, when the spring thaw is followed by a harvest, there are also summer peaks because of communal activities on the farm.

Epidemiologists have as one of their goals the formulation of a testable theory to project the course of future epidemics. It is possible to calculate the critical rate of sexual partner exchange that will allow an STD to spread through a population, i.e., when R_0 is >1. For HIV, with a duration of infectiousness of 0.5 year and a transmission probability of 0.2, the partner exchange value is 10 new partners per year. For other STDs, such as untreated syphilis and gonorrhea, with somewhat higher transmission probabilities, the values are 7 and 3, respectively. Despite the development of mathematical equations, predicting the spread of an epidemic can be as uncertain as forecasting when a hurricane, blizzard, or tornado will occur. Indeed, making predictions early in a disease outbreak by fitting simple curves can be misleading because it generally ignores interventions to reduce the contact rate and the probability of transmission. For SARS, fitting an exponential curve to data from Hong Kong obtained between February 21 and April 3, 2003, predicted 71,583 cases 60 days later, but using a linear plot, 2,410 cases were predicted. In fact, by May 30, 2003, according to the WHO, there were >8,200 cases worldwide and >800 deaths. By July 5, 2003, a headline in the *New York Times* declared "SARS contained, with no more cases in the last 20 days."

Other uncertainties in predictability may involve changes in travel patterns with contact and risk increased. Sociological changes may also affect the spread of disease—children in school may influence the spread of measles, as occurred in Iceland when villages grew into towns and cities. Quarantine of infected individuals has also been used as a control measure. Generally speaking, quarantine is ineffective, and more often than not it is put in place to reassure the concerned citizens that steps at control are being taken. As is noted above, though, there are other interventions that do affect the spread of disease by reducing the number of susceptible individuals. One of the more effective measures is immunization.

A Measles Outbreak

In the year 2015, for some, Disneyland wasn't the happiest place on Earth. It was in January of that year that a single measles-infected individual was able to spread the disease to 145 people in the United States and a dozen others in Canada and Mexico. Patient zero in the 3-month-old Disneyland outbreak was probably exposed to measles overseas and while contagious unknowingly visited the park. (The measles strain in the Disneyland outbreak was found to be identical to one that spread through the Philippines in 2014, where it sickened ~50,000 and killed 110. It is likely that patient zero acquired the virus there.)

Measles spreads from person to person by sneezing and coughing; the virus particles are hardy and can survive as long as 2 h on doorknobs, handrails, elevator buttons, and even in air. For the first 10 to 14 days after infection, there are no signs or symptoms. A mild to moderate fever, often accompanied by a persistent cough, runny nose, inflamed eyes (conjunctivitis), and sore throat, follows. This relatively mild illness may last 2 or 3 days. Over the next few days, the rash spreads down the arms and trunk, then over the thighs, lower legs, and feet. At the same time, fever rises sharply, often as high as 104 to 105.8°F (40 to 41°C). The rash gradually recedes, and usually lifelong immunity follows recovery. Complications, which may include diarrhea, blindness, inflammation of the brain, and pneumonia, occur in ~30% of cases. Between 1912 and 1916 there were 5,300 measles deaths per year in the United States. Yet all that changed in 1968 with the introduction of the measles vaccine; in the United States, measles was declared eliminated in 2000.

What, then, underlies the Disneyland outbreak?

On average, every measles-infected person is able to spread the disease to ten other people, i.e., its R_0 value is 10. With this multiplier, measles will spread explosively; indeed, with multiplication every 2 weeks and without any effective control (such as immunization), millions could become infected in a few months. It has been estimated that to eliminate measles (and whooping cough) ~95% of children under the age of 2 must be immunized. For disease elimination not everyone in the population need be immunized, but it is necessary to reduce the number of susceptible individuals below a critical point (called herd immunity).

An analysis of the Disneyland outbreak of measles shows that that those infected were unvaccinated. The researchers have calculated that the number of vaccinated individuals might have been as low as 50%. The outbreak that began in California was a reflection of the anti-vaccination movement, which had led some parents to believe the false claim that the vaccine for measles caused an increase in autism. (The "evidence" for this was based on just 12 children and has been thoroughly discredited by massive studies involving half a million children in Denmark and 2 million children in Sweden.) Then, too, some parents believe their children are being immunized too often and with too much vaccine because pharmaceutical companies are recklessly promoting vaccination in pursuit of profit. Other parents contend that the vaccine is in itself dangerous. It is not, as is evidenced in Orange County, where Disneyland is located: the outbreak sickened 35 people, including 14 children. And although a measles vaccine has been available worldwide for decades, according to the WHO, about 400 people a day died in 2013.

The response to the outbreak at Disneyland prompted the California Senate to pass a bill, SB 277, which required almost all California schoolchildren to be fully vaccinated in order to attend public or private school, regardless of their parents' personal or religious beliefs. In signing the bill, Governor Edmund G. (Jerry) Brown wrote: "While it is true that no medical intervention is without risk, the evidence shows that immunization powerfully benefits and protects the community."

The Evolution of Plagues

"A recurrent problem for all parasites … is how to get from one host to another in a world in which such hosts are never contiguous entities," wrote the historian William McNeill. He went on: "Prolonged interaction between human host and infectious organisms, carried on across many generations and among suitably numerous populations on each side, creates a pattern of mutual adaptation to survive. A disease organism that kills its host quickly creates a crisis for itself since a new host must somehow be found often enough and soon enough, to keep its chain of generations going." Based on this view, it would seem obvious that the longer the host lives, the greater the possibility for the parasite to grow, reproduce, and disperse its infective stages to new hosts. The conventional wisdom, therefore, is that the most successful parasites are those that cause the least harm to the host, and over time virulent parasites would tend to become benign.

At first glance, it would appear that the progress of the disease myxomatosis in Australia supports this evolutionary perspective. The story of myxomatosis begins in 1839, when the Austin family migrated from England to Australia. Over time they became rich from sheep farming. To reestablish their English environment, the Austins imported furniture, goods, and a variety of animals. In 1859 a ship came from England to Australia with rabbits. Since the rabbits had no natural predators, they multiplied rapidly, destroying plants and the native animals. The Austins began to wage war on the rabbits. By 1865, >20,000 rabbits were killed on the Austin estate. And still the rabbits continued to spread, traveling as much as 70 miles per year. Control measures such as fences, barbed wire, ditches, and the like did not work. A viral disease of wild rabbits from South America, called myxomatosis and lethal to domestic rabbits, was introduced into Australia in the 1950s to act as a biological control agent. The vector for the myxoma virus is a mosquito. In 1950, 99% of the rabbits died of myxomatosis. Several years later the virus killed only 90%, and it declined in lethality with subsequent outbreaks. It was also found that the viruses from the later epidemics were less virulent than the earlier forms and that these less virulent forms were much better at being transmitted by mosquitoes—the rabbits lived longer and the number of infected rabbits was higher with milder disease. One may conclude that the virus had evolved toward benign coexistence with the rabbit host. McNeill, impressed by the results of the introduction of the myxoma virus into Australia, wrote: "from an ecological point of view ... many of the most lethal disease-causing organisms are poorly adjusted to their role as parasites ... and are in the early stages of biological adaptation to their human host; though one must not assume that prolonged co-existence necessarily leads toward mutual harmlessness. Through a process of mutual accommodation between host and parasite ... they arrive at a mutually tolerable arrangement ... (and based on myxomatosis) ... some 120-150 years are needed for a human population to stabilize their response to drastic new infections." There is, however, reason to question McNeill's conclusions.

A recent reexamination of myxomatosis in Australia shows that the mortality of the rabbits, after the decrease in the virulence of the virus and the increase in rabbit resistance, was comparable to the mortality of most vector-borne diseases of humans, such as malaria. In other words, the virus was hardly becoming benign. Further, the decrease in virulence observed over the first 10 years of the study did not continue, but reversed. It appears that myxomatosis is not an example of benign evolution.

An alternative to the contention that parasites evolve toward a harmless state is that natural selection favors an intermediate level of virulence. This intermediate level is the result of a trade-off between parasite transmission and parasite-induced death. Since the value for R_0 increases with the transmission rate as well as the duration of the host's infectiousness, an increase in transmission would reduce the duration of infection, and then selection may favor intermediate virulence. And because R_0 depends directly on the density of susceptible hosts, if the number of susceptible individuals is great, then a parasite may benefit from an increased rate of transmission even if this kills the host sooner and prevents transmission at a later time. If susceptible hosts are not abundant, however, then the parasite that causes less harm to the host (i.e., is less virulent) may be favored since that would allow the host to live longer, thereby providing more time for the production of transmission stages. The hypothesis that virulence is always favored when hosts are plentiful and is reduced when there are fewer hosts neglects the fact that a feedback exists in the host-parasite interaction: a change in parasite virulence impacts the density of the host population, and this in turn alters the pressures of natural selection on the parasite population, and so on. Thus, although parasite virulence generally tends to decline over evolutionary time, it never becomes entirely benign, and in the process the parasite population becomes more efficient in regulating the size of the susceptible host population.

The view that parasites evolve toward becoming benign suggests that parasites are inefficient if they reproduce so extensively that they leave behind millions of progeny in an ill or dead host. Indeed, some biologists have contended that enhanced virulence is the mark of an ill-adapted parasite or of one recently acquired by the host. This is not true. The number of parasite progeny lost is not of evolutionary significance; rather, it is the number of offspring that pass on their genes to succeeding generations that determines evolutionary success. Natural selection does not favor the best outcome for the greatest number of individuals over the greatest amount of time, but instead favors those characteristics that increase the passing-on of a specific set of genes. Consider a particular species of weed that is growing in your garden. The production of 1,000 seeds that yields only 100 new weed plants might be considered wasteful in terms of seed death and the amount of energy the weed put into seed production, but if the surviving seeds ultimately yield more weed plants in succeeding generations, then that weed species is more efficient in terms of evolutionary success. Parasites are like weeds. They have a high biotic potential, and those that leave the greatest number of offspring in succeeding generations are the winners, evolutionarily speaking. Evolutionary fitness, be it for a parasite, human, bird, or bee, is a measure of the success of the individual in passing on its genes into future generations through survival and reproduction. When the fitness of the host is reduced by a parasite, there is harm, illness, and an increased tendency toward death. Host resistance is the counterbalance to virulence or the degree of harm imposed on the host by the presence of the parasite. If host resistance is lowered, a disease may be more pathogenic although the parasite's inherent virulence may be unchanged. How negatively a host will be affected, i.e., how severe or how pathogenic is the disease, is thus determined by two components: virulence and host resistance. In addition, virulence is not so much a matter of a particular mutation but rather how that mutation is filtered through the process of natural selection; it is through natural selection that the final outcome may be a lethal outbreak or a mild disease, and, of course, when a new pathogen emerges, R_0 must be a number >1.

Since parasite survival requires reaching and infecting new hosts, effective dispersal mechanisms may require that the host become sick: sneezing, coughing, and diarrhea may assist in parasite transmission. The conventional wisdom is that it takes a prolonged period of time for virulence to evolve; the evolution of parasite virulence, however, may be quite rapid (on the order of months) and need not take years, as was the case with the myxoma virus. The basis for this is that a parasite may go through hundreds of generations during the single lifetime of its host. Then, too, because of competition between different parasites living in a single host, it might be advantageous for one kind of parasite to multiply as rapidly as it can before the host dies from the other infectious species. Succinctly, the victorious parasite is the one that most ruthlessly exploits the pool of resources (food) provided by the host and produces more offspring, thus increasing its chances to reach and infect new hosts.

If parasite dispersal depends on the mobility of the host as well as host survival, then severe damage inflicted on the host by enhanced virulence could endanger the life of the parasite.

Consider, for example, the common cold. It would be very much in the interest of the cold virus to avoid making you very sick, since the sicker you become, the more likely you are to stay at home and in bed; this would reduce the number of contacts you would have with other potential hosts, thereby reducing the possibilities for virus transmission by direct contact. Similarly, the development of diarrhea in a person with the disease cholera or *Salmonella* infection (which causes "food poisoning") facilitates the dispersal of these intestinal microbes via fecally contaminated water and food, and in the absence of diarrhea parasite transmission would be reduced.

AIDS is a consequence of an increase in the virulence of HIV. The enhancement in HIV virulence is believed to have resulted from accelerated transmission rates due to changes in human sexual behavior: the increased numbers of sexual partners was so effective in spreading the virus that human survival became less important than survival of the parasite. As the various kinds of plagues are considered in greater detail in subsequent chapters, recognition of the evolutionary basis for virulence may suggest strategies for public health programs. Clean water may thus favor a reduction in the virulence of waterborne intestinal parasites (such as cholera), and clean needle exchange and condom use would both reduce transmission and lessen HIV virulence. But some contend that this indirect mechanism may be too weak and too slow to reduce virulence substantially, and that a better approach could be direct selection by targeting the virulence factor itself. For example, immunization that produces immunity against the toxin produced by the diphtheria microbe also results in a decline in virulence. Future efforts will determine which strategy is the better means for effective "germ" control to improve the public's health.

doi: 10.1128/9781683670018.ch2

2
Plagues, the Price of Being Sedentary

In Stanley Kubrick's classic film *2001: A Space Odyssey*, Richard Strauss's music ("Thus Spake Zarathustra") provides a haunting and frightening background to the sequence of scenes that represent the dawn of humanity. The sun rises on a barren African savannah. A band of squat, hairy ape-men appear; they eat grass. Though herds of tapirs graze close by, the ape-men ignore them, since the means and the tools necessary to attack or kill the tapirs have not yet been developed. These ape-men are vegetarians who forage for roots and edible plants. On the dawn of the second day, the ape-men are seen huddled around a water hole; the landscape is littered with bones. One, the leader of the group, picks up a bone, smashes the skeleton of an antelope, and then the bone is used to kill a tapir (Fig. 2.1). Shortly thereafter, the raw pieces of tapir flesh are eaten and shared by other hairy apelike creatures, members of the clan. At the dawn of the third day, the meat-eating, tool-using man-apes drive off a neighboring band of apelike creatures. Bone tools used for killing animal prey are now used to threaten and drive off rival tribes. In slow motion, accompanied by the slowly building tones of Strauss's music, the leader of the man-apes flings his weapon, a fragmented piece of bone, into the air. It spins upward, twisting and turning, end over end. There is a jump cut of 4 million years into the future, and the bone dissolves into a white, orbiting space satellite. Kubrick's science fiction film has been described as a countdown to tomorrow, a visual masterpiece and a compelling drama of human evolution. Absent from the film is an examination of how the enlightened roving bands of early apelike humans settled down and become increasingly disease-ridden. Here is that part of the story.

Figure 2.1 (Left) Hollywood's view of *Australopithecus* as seen in the movie *2001: A Space Odyssey*. Turner Entertainment Co. Licensed by Warner Bros Entertainment Inc. All rights reserved, Alamy Stock Photo

Becoming Human, Becoming Parasitized

It is now generally accepted that Africa was the cradle of humanity. The earliest evidence of hominids, that is, animals ancestral to modern humans and not closely related to other monkeys and apes, is found in Africa. The evidence for this comes from unearthed bones and teeth (fossils). The fossil record shows that one of our oldest ancestors—called *Australopithecus*—lived in Africa about 4.2 million to 3.8 million years ago. These early hominids, which split from the ape lineage (and were discovered in Kenya in1994), are named *A. anamensis*, and to judge from the structure of the teeth and the position of the opening where the spinal cord enters the skull, one can conclude that they were apelike humans, not apes. Our ancestor *Australopithecus* spent time in trees and basically behaved similarly to chimpanzees, or so we believe, since fossils provide no record of behavior. Whether *A. anamensis* walked on two feet is uncertain, but evidence for erect, upright posture in *Australopithecus* comes from bones discovered in Ethiopia and Tanzania that are 3.8 million to 3.0 million years old, from a species named *A. afarensis*. One of these finds, a small female, was discovered and named Lucy by Donald C. Johanson of the University of California at Berkeley. The limb structure and the way the hip joint and pelvis articulate make it clear that Lucy walked on two legs (Fig. 2.2). This was dramatically shown by Mary Leakey and her team, who discovered three sets of fossilized footprints left in wet volcanic ash some 3.2 million years ago. *A. afarensis* weighed about 75 lb and was not very "brainy," its brain being no larger than the brains of living African great apes. When *A. afarensis* descended from the trees and stood upright with two feet firmly planted on the ground, not only did it affect posture, but it dramatically changed lifestyles and diets, and disease patterns began to be altered.

Descent from the trees to the ground placed the australopithecines into a new environment, an ecological niche that was very different from the forest canopy. This freed them from some diseases but allowed for the acquisition of new ones. For example, in the treetops australopithecines would have been bitten by mosquitoes that carried parasites acquired from other animals living in the canopy, but at ground level they would be exposed to other airborne bloodsuckers such as ticks and flies, or they would come in contact with different food sources and contaminated water. Their teeth were small and underdeveloped, as in modern human beings, and the canines, highly developed in existing ape species, were small like ours. We can infer from their teeth that these australopithecines probably chewed fruits, seeds, pods, roots, and tubers. Since no stone tools have been found associated with the fossils, it is believed that *A. afarensis* did not make or use durable tools or understand the use of fire. They were opportunistic scavengers or vegetarians. The life span of an australopithecine has been estimated to have been between 18 and 23 years.

Figure 2.2 *Australopithecus* reconstruction of Mr. and Mrs. Lucy.
Courtesy of Ken Mowbry, American Museum of Natural History

Beginning about 3 million years ago, the climate in Africa changed from tropical warm and wet to a more temperate cool and dry one, and as a consequence, the dense woodlands were replaced by more open grassy habitats, a savannah. This climate change presented a challenging environment for the woodland-dwelling australopithecines. Although we do not know whether the climate change triggered it, at about this same time, ~2.5 million to 1.8 million years ago, there appear in the fossil record several different kinds (species) of hominids, with two or three coexisting species in eastern and southern Africa. One of these species was *A. boisei*, a small-brained vegetarian, and the other was *Homo habilis* (Fig. 2.3b). The name *Homo habilis*, or "handy man," is based on the fact that altered stones and animal remains have been found with the fossil bones. *H. habilis* was more than a scavenger and a gatherer. *H. habilis* was also a hunter who made and used stone tools: simple stone flakes, scrapers and "choppers" that were chipped from larger stones (Fig. 2.3a). (These stone tools, first found in Africa's Olduvai Gorge, are called Oldowan tools.) The fashioning of tools suggests a great leap in human intelligence and begins the technological changes that would forever mark *Homo* as a tool maker and a tool user. *H. habilis* used the flake tools to cut up the carcasses of the animals that were killed; these were

Figure 2.3a Oldowan tools used by *Homo habilis,* Courtesy Didier Descouens, CC-BY-SA 4.0

transported to a home base where the meat was fed upon. *H. habilis*, with a somewhat larger brain, was "smarter" than *A. afarensis*, but the fossil finds tell us nothing of the numbers of individuals, whether there was division of labor among males and females, or anything about their behavior. We speculate, however, that there were 50 to 60 individuals in a group living in an area of 200 to 600 square miles. We imagine that *H. habilis* lived at the edge of shallow lakes and in crude rock shelters.

There is no fossil record of the parasites that afflicted *H. habilis* since their soft bodies have disintegrated over time, but we do know that with meat eating came an increase in parasitism. As these nomadic hunters encountered new prey, they met new parasites and new vectors of parasites. The result was zoonosis; that is, animal infections were transmitted to humans. What were these zoonotic infections? We surmise that the parasites of *H. habilis* were those acquired from the wild animals that were killed and scavenged. The butchered meat might have had parasites such as the bacteria anthrax and tetanus, the roundworm that causes trichinosis, and a variety of intestinal tapeworms. *H. habilis* would probably have been bitten by mosquitoes, ticks, mites, and tsetse flies, and probably also had head lice. *H. habilis* also may have suffered from viral diseases such as the mosquito-transmitted yellow fever, as well as non-vector-borne viruses that cause hepatitis, herpes, and colds, and he may have had spirochete infections such as yaws. It is doubtful, but *H. habilis* could also have been infected with the parasites that cause sleeping sickness, malaria, and leprosy. They certainly must have been infected with filaria, pinworms, and blood flukes,

Figure 2.3b Diorama in the Nairobi National Museum of *Homo habilis*,
CC-BY 2.0, https://www.flickr.com/photos/ninara/17147417090/; CC-BY 2.0 license

Figure 2.3c *Acheulean tools used by Homo erectus,* Courtesy Didier Descouens

Figure 2.3d Diorama of *H. ergaster* the "African equivalent" to fossils of *H. erectus*. Alamy Stock Photo.

but probably did not have typhus, mumps, measles, influenza, tuberculosis, cholera, chickenpox, diphtheria, or gonorrhea. At the time when *H. habilis* roamed the African savannah, the human population was quite small, consisting of about 100,000 individuals, and we expect that rates of human-to-human transmission of parasites were low.

Roughly 1.6 million years ago (or ~1.8 million years ago in Africa), *H. habilis* was replaced by *H. erectus* (meaning "erect man") (Fig. 2.3d). (In Africa, *H. erectus* is equivalent to fossils that have been named *H. ergaster*.) *H. erectus* was close to modern humans in body size, and its skull capacity suggests that it was somewhat larger-brained than *H. habilis*—but still with barely half the capacity of modern humans. *H. erectus* had smaller cheek teeth, suggesting that they were omnivores; they were smaller-faced; they developed a culture characterized by living in caves; and they hunted game animals using bifaced flake stone tools fashioned into "hand axes." This stone tool technology (called Acheulean tools) (Fig. 2.3c) allowed *H. erectus* to process more completely the harder parts of animals and plants by grinding, crushing, splitting, and cutting up these before eating. As such, these stone tools represented a technological advance and served as extensions of the hands and teeth to break down food before digestion. *H. erectus* was able to start fires and made use of fire to cook the food. The *H. erectus* population was now somewhat less than a million. According to the "long journey" hypothesis, about 2 million years ago the *H. erectus* populations began to move out of Africa via the Middle East, but climate and geography prevented them from turning west, and so they took a more southerly route into present-day China and Indonesia. Then they turned north and moved west again across the more central parts of Europe and Asia. The earliest fossil remains of *H. erectus* were found in Indonesia (Java) by Eugene Dubois in 1891 and so were named Java Man; 2 decades later, when Davidson Black found similar fossils in caves in China, they became known as Peking Man. The cave sites in China are ~500,000 years old, and the last of these were abandoned ~230,000 years ago.

When the populations of *H. erectus* left Africa, some of their parasites went with them, but only those that could be transmitted directly from person to person. Those vectors that remained restricted to Africa, such as the species of mosquito, snail, and fly that transmit diseases such as filariasis, blood fluke disease, and sleeping sickness, respectively, did not follow the migratory path. Indeed, even today they remain diseases that are characteristic of Africa. But as *H. erectus* encountered new environments with new kinds of animals, they were subjected to sources of new parasites; with an increase in the number of humans living in more-restricted geographic environments, the probability for large-scale infections was enhanced.

Tool making and tool use, as well as human cooperation, made hunting possible. Together they contributed to further increases in the size of the human population, and over time *H. erectus* evolved into humans closely resembling us. A half-million years ago the human populations of Africa and those in Europe and Asia began to diverge from one another. Some 200,000 years ago the fossil record shows individuals who were larger-brained; those with a more graceful face, found in southwestern Europe and dated to 40,000 years ago, were called Cro-Magnon man. In a fit of hubris, Carolus Linnaeus gave them the scientific name *Homo sapiens*, literally "wise man." *H. sapiens* not only used the stone technology of *H. erectus* but also made tools from bone and antlers. They made artistic carvings and cave paintings, kept records on bone and stone, played music on simple wind instruments, adorned themselves with jewelry, and buried their dead in ritual ceremonies; their living sites were highly organized and stratified, and they hunted and fished in groups. The intermittent technological advances in tool making seen with *H. habilis* and *H. erectus* were constantly refined. Clearly, they were our immediate ancestors. At this time the human population numbered about a million individuals.

In western Europe, human skeletons were found first in the Neander Valley of Germany in 1856; they were called Neanderthals. Subsequently, Neanderthal fossils were found in the Middle East and parts of western Asia. They date from between 190,000 and 29,000 years ago. Some archeologists have classified them as a separate species,

Figure 2.4a Neanderthal man in profile; Neanderthal woman cleaning a reindeer skin, (both) Wellcome Library, London (CC-BY 4.0)

Homo neanderthalensis. The Neanderthals, who previously have been depicted as brutish cavemen (Fig. 2.4a and b), have been "modernized" on the basis of a brain size slightly greater than our own, and they left evidence showing that they cared for the sick and performed ritual burials (Fig. 2.4c). Their stone tools, however, were cruder than those of Cro-Magnon man. Within a few thousand years the Cro-Magnons, with their superior weapons and other advanced cultural practices, had completely displaced the Neanderthals. In Africa there are skeletal remains that are more modern than those of Neanderthals, dating back 200,000 years. Thus, nearly a quarter of a million years ago, populations of *H. sapiens* lived in Africa, Asia, and Europe. Their migration into the Americas took place ~35,000 to 12,000 years ago when they crossed the land bridge at the Bering Strait from Asia into the Americas. Forty thousand years ago humans moved into Australia. Thus, over the past 5 million years, new hominid species have emerged, coexisted, competed, and colonized new environments, in some instances succeeding, in others becoming extinct. The fossil record is of necessity incomplete, and much will be learned from future anthropological digs, but what is certain is that

Figure 2.4b A 1953 B- grade movie poster representing a monster-like Neanderthal man. Courtesy of Popcorn Posters.

Figure 2.4c Neanderthal Family, Reconstruction. Ian Tattersall, American Museum of Natural History

we did not arrive by a straight-line descent from the apes; we are not the single topmost limb in the hominid evolutionary tree, but simply one of its many branches.

Hunter-gatherers, unable to preserve and store fruits, vegetables, and meat, were forced to roam over large distances in search of wild edible plants and to hunt down game animals and find sources of drinking water. Moving from place to place, these nomadic bands were not surrounded by heaps of rotting meat or feces, and exposure to parasite-infested waters was limited. Though the hunter-gatherers did come together in groups, the size of their populations was small, and so diseases of crowds requiring human-to-human transmission were absent. Based on what we know about modern hunter-gatherer societies like those in present-day New Guinea, the Australian aborigines, and the Kalahari bushmen, we believe our hunter-gatherer ancestors were a relatively healthy lot. Gradually, however, conditions would change as the size of human populations increased and people adopted sedentary habits—living for extended periods of time in permanent or semipermanent settlements. This would, over time, dramatically increase the incidence of human disease.

The Road to Plagues:
More Humans, More Disease

Today we speak of the problems associated with the population bomb—the unbri-dled growth of humans—that threatens our very existence. This growth in human populations cannot be calculated with any certainty until the middle of the 18th cen-tury, but we can make some educated guesses. Three hundred thousand years ago there were 1 million; 25,000 years ago that number had grown to 3 million; and 10,000 years ago the estimated human population was 5 million. By A.D. 1 it was 300 million. The phenomenal growth spurt in the human population coincides with the initiation of agriculture and the domestication of animals, which is generally dated to 8000 B.C. Between 8000 B.C. and A.D. 750, the population of the world increased 160 times to 800 million. Not only was the human population increasing, so too was overcrowding. For example, in 8000 B.C. human density was 0.2 people per square mile, but by 4000 B.C. it was 4 people per square mile.

What is the basis for this growth in the human population? The English clergyman Thomas Malthus (1766-1834) wrote *An Essay on the Principle of Population* in 1798, in which he stated that a population that is unchecked increases in geometric fashion. Malthus assumed that there would be a uniform rate of doubling, and this is of course naive, because it leads to impossibly large numbers. (By way of example, if you doubled a penny every day over a month, the final amount would be >$20 million.) It has been said that explosions are not made by force alone, but by a force that exceeds restraint. As Malthus correctly observed, there are factors that will eventually bring population growth to a halt; for example, restraint could result from the fact that the food supply increases only arithmetically. The consequences of unrestrained population growth, in Malthus's words, would lead to "misery and vice," or, in today's vernacular, to starvation, disease, and war. These would tend to act as "natural restraints" on population growth. Thus, the Malthusian model suggested that a natural population has an optimal density.

If we were to make a graph plotting the human population on an arithmetic scale from 500,000 years ago to the present, we would find that the resulting curve sug-gests that the population remained close to the baseline from the remote past to about 500 years ago, and then it surged abruptly as a result of the scientific-industrial revolution (Fig. 2.5). More instructive, however, would be to plot the same data for a longer time period using a logarithmic scale, since this allows for more of the data points to be placed in a smaller space. This log-log plot reveals that the human pop-ulation has moved upward in a stepwise fashion, and that there were three surges: those reflecting the development of tool making or the cultural revolution, followed by the agricultural, and finally by the scientific-industrial revolution. What were the

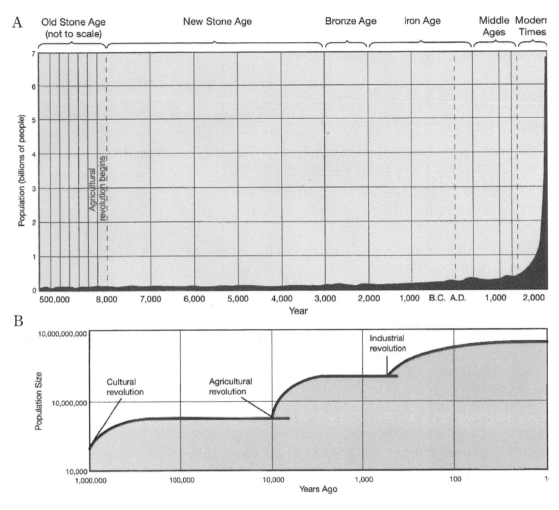

Figure 2.5 A. Growth of the human population for the last 500,000 years. If the Old Stone Age (Paleolithic) were in scale it would reach 18 feet to the left. B. Log-log plot of the human population over the last million years.

checks on human growth rates that limited population size so that at equilibrium (the flat part of the logarithmic "curve") there was a zero rate of change and the number of deaths equaled the number of births? Two kinds of checks occurred to set the upper limit (or the set point) for population growth: external or environmental factors (including limited food, space, or other resources) and self-regulating factors (such as fewer births, deliberate killing of offspring, or an increased death rate due to accidents or more-virulent parasites). For Malthus, disease and warfare as well as "moral restraint" (birth control) acted as "natural restraints"—the Four Horsemen of the Apocalypse: Disease, Famine, War, and the Pale Rider, Death. Indeed, it has been estimated that prior to the introduction of agriculture the earth could have supported a population of between 5 million and 10 million people who were engaged in hunting and gathering. Agriculture changed the environmental restraint so that the set point, or upper limit, of population size was increased.

The Effect of Agriculture

Human history took off 50,000 years ago in what Jared Diamond, professor of geography and physiology at the University of California at Los Angeles School of Medicine, called "the Great Leap Forward." Fifty thousand years ago *H. sapiens* used standardized stone tools that could be used for cutting, scraping, and grinding, as well as pieces of bone that could be fashioned into fishhooks and spears, needles, awls, harpoons, and eventually bows and arrows. These tools could also be used as weapons, and now humans could begin to hunt down and kill their animal prey at a distance. Not only did these early humans use the meat of animals for their nourishment, but they began to clothe themselves in the skins of these animals. Through the invention of rope it was possible to make snares and nets so that birds and fishes could also become part of the diet. All this attests to the fact that between 100,000 and 50,000 years ago there was a significant change in the cognitive capacity of the human brain without a significant change in its size. Coupled with this change in brain organization was the anatomical improvement of the voice box; now humans could not only speak but also begin to develop language.

For 99% of human existence we were hunter-gatherers, so why some 10,000 years ago did we settle down to become farmers? This change from hunting and gathering to farming has been termed the agricultural revolution, the time when humans domesticated plants and animals and exerted control over food production. Although the term "revolution" would seem to indicate that it appeared suddenly and dramatically, this was certainly not the case. The human control of food production was not discovered or invented, nor was it a conscious choice made by our ancestors "to farm" or "not to farm," since at the time there would have been no farmers to serve as role models. No, domestication of plants and animals evolved as a consequence of human choice made without any awareness of its future long-term consequences.

Development of techniques and practices for agriculture and animal husbandry progressed step by step in sequential fashion. They were not developed over a short time, and not all the wild animals and wild plants that would eventually be domesticated in a particular region would be domesticated at the same time. Indeed, it probably took thousands of years to shift the human diet from wild foods alone to foods both cultivated and wild. The reason for this time lag is that food production evolved as a result of the accumulation of many separate choices, and there were trade-offs, especially in the allocation of time and effort.

Consider for the moment that you are a hunter-gatherer who has accumulated enough wisdom and technology to set up a small garden. Some of the choices you would be faced with are: Which plants should I grow? How much time should I spend planting instead of hunting or scavenging? What are the benefits of tending the garden over going out to hunt and gather wild plants? Perhaps your most important consideration might be which of the two, hunting or gardening, will save you from starvation in the future. It has been speculated by Jared Diamond that "all other things being equal, people seek to maximize their return of calories, protein or other specific food categories by foraging in a way that yields the most return with the greatest certainty in the least time for the least effort. Simultaneously they seek to minimize the risk of starving. ... One suggested function of the first gardeners 11,000 years ago was to provide a reliable reserve larder as insurance in case wild food supplies failed." Although the factors that contributed to the shift from hunting and gathering to farming still remain controversial with regard to their relative importance, one thing is certain: once there was a shift from nomadic hunting and gathering to more-sedentary food production, there could be no turning back.

At the end of the Pleistocene era (~11,000 years ago), the climate began to change: the glaciers had receded; the climate became milder and drier; and many large mammals had become extinct in Europe, Asia, and Africa. This led to a decline in the abundance of wild game, and hence the life of the hunter-gatherer became more precarious: to obtain the same amount of food as in the past would require the expenditure of greater amounts of time and energy. The reduction in the number and kind of game animals was coupled with a change in climate that favored plants with the potential for domestication, particularly the cereal grains. Now there were larger and larger areas with wild cereals, and these could be harvested with little difficulty, using stone and bone tools fashioned into sickles with flint blades. In addition to the technologies for harvesting cereals such as wild barley and wheat, the technologies needed for processing and storing these grains came into being. These technologies, which appear crude and simple by today's standards, allowed for the first unconscious steps of plant domestication to take place. Tools included baskets to carry the grain from the field to the home base; mortars and pestles to remove the husks and to pulverize the grain; a technique of heating the grain to allow storage without sprouting; and the construction of underground storage pits, some of which were made waterproof by plastering. Coincident with farming, the density of the human population increased. It is not clear whether the rise in density of human populations led to the domestication of plants and animals or vice versa, but it is certainly true that with the availability of more and more calories it was possible to feed more and more people. Diamond, in his book *Guns, Germs, and Steel*, has observed that the adoption of food production is an autocatalytic process, a positive feedback, in which a gradual rise in population densities required that people obtain more food, and in turn those who took the steps to produce it were rewarded. Once "farming" began, people could become more and more sedentary—they settled down. In turn, birth spacing could be shortened, resulting in more births and larger families, who required still more food, and so on.

Farming populations became better nourished thanks to an increase in the availability of the number of edible calories per square mile, and eventually farmers replaced the nomadic groups of hunter-gatherers by converting them to engage in the practice of farming or by displacing them by the sheer force of greater numbers.

The life of nomadic hunter-gatherers was such that population levels were well below the maximum limit that would be imposed by their reproductive biology and the availability of food. What then limited their increase? The inability of the hunter-gatherer mother to carry more than a single child along with her normal baggage, coupled with her inability to nurse more than one child at a time, limited the practical interval between births to four years. It is likely that hunter-gatherers effectively spaced their children by means of lactation amenorrhea, sexual abstinence, infanticide, and spontaneous abortion. In contrast, once humans settled down, they were freed from the encumbrances imposed on the hunters and gatherers who had to carry their children around, so that now they could have as many children as they could bear and raise. Consequently, the birth interval for the "farmer" was reduced to two years. Agriculture also encouraged higher birth rates because additional children provided cheap labor. Further, farming had another advantage over hunting and gathering: more calories could be produced per unit land area and time expended. While 200 square miles could support 50 to 60 hunter-gatherers, more than 10,000 "farmers" could be supported on this same land area. The higher birth rate of the food producers, together with their ability to feed more people per square mile, allowed these "farmers" to achieve much higher population densities than those who were engaged in hunting and gathering.

Once agriculture and animal husbandry yielded a surplus of food that could be stored, there was a need for some members of the sedentary population to guard it. This was of course impractical for the nomadic hunter-gatherers since they would have to both hunt and gather and at the same time protect their bounty from others. But the availability of surplus stored food allowed members of the settled population to specialize: some became guards, and others were armed and served as soldiers who could group together to steal food from others. Perhaps this foreshadowed the origins of war. And when food production did not require every member of the settled community to directly work the land, then it became possible for some members of the group to engage in other activities. Once humans formed agriculturally based societies, those who chose to participate in such group activity may have been at a competitive advantage over those who did not. The latter likely became outcasts. Hence the agricultural revolution may have provided selective advantages to those who were in a controlled group, and superstition and organized religious practices were effective means for control by promoting group cohesion. It may be not be accidental that the first known and highly organized religions arose coincident with the agricultural revolution.

Some of the surplus food could be used to feed those "who provide religious justification for wars of conquest, artisans such as metalworkers who develop swords, guns and other technologies; and scribes, who preserve far more information than can be remembered accurately." In time, political stratification would develop: heading the settled community would be the elite, consisting of hereditary chiefs (or kings) and bureaucrats. Under the appropriate circumstances these complex political units, which governed "the settled," could also be mustered into formidable armies of conquest. Stored food and the land upon which it was grown became valuable resources that could be taxed, and surpluses could be traded for other goods; commerce and banking began to emerge. Thus, with larger populations family and inheritance schemes result, class structures with elaborate religious practices emerge, and writing is invented. Through agriculture and its prospect for increased food production, there was a population expansion that favored technological advances, as well as the development of cities (urbanization) and the rise of civilizations.

Another consequence of humans settling down was an increase in the amount of human disease. Agriculture by itself did not create new infections; it simply accentuated those that were already present or it converted an occasional event into a major health hazard. This was largely due to the fact that transmission of infectious agents becomes easier as individuals are crowded together; the practice of using human excrement ("night soil") or animal feces (manure) as fertilizer allows for the transmission of infective stages; and finally, the closer association with domestic animals allows for their diseases to be transmitted to humans.

The Lethal Gifts of Agriculture

Permanent settlements developed independently in several parts of the world, including the Middle East, China, and the Americas. But those that have been best studied are found in the so-called Fertile Crescent, a region bounded by the Tigris and Euphrates Rivers and curving around the Mediterranean and the Nile Valley to include present-day Syria, Lebanon, Israel, Egypt, Turkey, Jordan, Iran, and Iraq. The oldest village known, just outside present-day Jericho in Israel, may have sprung up around a shrine used by roving bands of hunters and gatherers. By 10,500 years ago it had evolved into a small farming village. At first this settlement and others like it were simply collections of villages on the banks of natural streams, but soon they were able to spread out via networks of irrigation canals. The surplus of food and the practice of irrigation contributed to larger and larger concentrations of people, allowing some people to quit farming and to become full-time artisans, priests, or members of other professions. Meanwhile, the farmers who provided the food for these ever-enlarging villages continued to live on their outskirts. And 5,500 years ago the first undisputed city—a place where farmers do not live—Uruk in Mesopotamia (present-day Iraq), was established.

We have little precise information about the parasitic diseases that afflicted our ancestors more than 10,000 years ago. To be sure, they had their parasites, but we believe their impact may not have been quite as severe as in the period that stretches from the present back to the time of the earliest cities. The reason for this is that the pattern and the impact of disease depend on several factors: the population density, the character and quality of the water supply, food and shelter, the frequency of contact among individuals, human contact with animals, and the climate. Once human populations were concentrated into larger and larger communities and their numbers increased, however, then the enhanced potential for infectious disease organisms to be transmitted could (and did) affect the size and well-being of the population. Farming and domestication also reduce the biological variability of animals and crop plants, leading to purebred strains, making local populations of plants and animals more and more uniform. As a consequence, any upset in the balance could decimate an entire population, whether that was a crop or a flock or a herd. (A modern example of this is the 1845-1851 potato famine in Ireland.) Agriculture necessitated artificial flooding of land, which was naturally devoid of adequate amounts of water, enabling longer growing seasons; it also required tilling or plowing and replenishing the soil with fertilizer. The cheapest and most available of fertilizers is human waste or animal feces. Intensive agricultural practices, requiring crop protection from weeds and pests, demand some kind of control measures. Tilling the soil necessitates some kind of work force: animal power, human power, or machines. Disruptions in the availability of these could lead to disaster. With purebred strains (monocultures), disruption becomes that much easier. Living in villages or cities—that is, in permanent settlements—involves the risk of parasite invasion. Those infected with intestinal parasites can through their feces more easily transmit disease to others, and where the water supply becomes contaminated through the use of either night soil or fecally contaminated streams, the spread of disease can be great indeed. A single contaminated water source serving a large population can be a much greater threat than several sources supplying smaller bands of hunters and gatherers. Irrigation practices thus created a favorable environment for transmission of parasites—moisture was abundant; there was a liquid medium in which parasites and/or their cysts and eggs could persist; and the water could also be used for drinking, bathing, washing of clothes, and waste disposal.

The disruptive effects of an epidemic disease are more than simply the loss of individual lives. Often the survivors are demoralized, they lose faith in inherited customs, and if it affects the working age group, it can lead to a material as well as a spiritual decline. As a consequence, the cohesion of the community may collapse and it may become susceptible to invasion from neighbors. Once disease is widespread in an agricultural community, it can produce a listless and debilitated peasantry, handicapped for sustained work in the fields, for digging irrigation canals, and for resisting military attack or throwing off alien political domination. All this may allow for economic exploitation.

The smaller population size of hunters and gatherers makes it seem probable that person- to-person "civilized" infectious diseases, such as measles, influenza, smallpox, and polio, could not have established themselves, because these are density-dependent diseases requiring a critical number of individuals for transmission. Although there is no hard literary or archeological evidence, it does seem reasonable to suggest, as did William McNeill, that "the major civilized regions of the Old World each developed its own peculiar mix of infectious, person-to-person diseases between the time when cities first arose (about 3000 BC) and about 500 BC. Such diseases and disease-resistant populations were biologically dangerous to neighbors unaccustomed to so formidable an array of infections. This fact made territorial expansion of civilized populations much easier than would otherwise have been the case."

The Accident That Caused Societal Differences

Imagine for a moment that you are living in the year 1492 and you have just graduated from the University of Padua in Italy with a degree in medicine. Before you are able to set up your practice, you receive a letter from an old friend, Giuseppe Diamonte, who writes from Spain that King Ferdinand and Queen Isabella are about to provide funds for the discovery of a new route to India; the expedition is to be under the command of a fellow Italian, Christopher Columbus, and he is in need of a naturalist to collect plants and animals and to act as the ship's doctor. Giuseppe writes that he has recommended you for the position. You are enthusiastic, so you board a ship in Venice, land in Barcelona, and travel by horseback to Palos, Spain. The journey across the Atlantic Ocean begins on August 3. Upon arrival in "India" (in actual fact the present-day Dominican Republic) on October 12, you are astounded to be greeted by a band of near-naked "Indians" who have paddled their canoes to greet the ship and its crew. No iron tools or oceangoing vessels can be seen, the village consists of a scattering of huts, and there is little that could be considered a city. The natives have no writing, and what agriculture there is is on an entirely different scale from that with which you were familiar in Europe. Ten weeks earlier you left iron tools and weapons, agriculture, oceangoing ships, large cities, horses, carts and carriages, writing, money, banking, painting, sculpture, cathedrals, palaces, buildings of brick and stone, established religion and ritual, and music. You are perplexed. You ask yourself: Why has the rate of technological and political development been so much faster in Europe-Asia than in the Americas? In short, why were the Americas technologically a few thousand years behind Europe? Why were stone tools, comparable to those used by Europeans and Asians 10,000 years earlier, being used?

Fast forward to the future. In *Guns, Germs, and Steel*, Diamond argues persuasively that it was not biological differences but geography that was the decisive element. It was differences not in the braininess or genetics of the human populations but in the plant and animal resources available on a particular continent—an accident of geography—that made the difference. Diamond believes that the fortuitous accident began in the Fertile Crescent, which contained a suitable array of plants and animals, called "founders," and that these formed the basis for domestication. What were the founder plants? Those locally available in Southwest Asia/Fertile Crescent that would not serve as the basis for domestication were plants with a large amount of indigestible material such as bark or that were poisonous, low in nutritional value, or tedious to prepare and gather. The desirable attributes of plants that make them suitable for domestication include having a larger proportion of edible parts (large seeds) and a lower proportion of woody, inedible parts; being easy to harvest en masse (with a sickle); a seasonal nature; being easy to grind, easy to sow, and easily stored; and being high in yield and high in calories. Plants with these characteristics were selected for domestication. They fall into four categories: grasses (wheat, barley, oats, millet, and rice), legumes (peas and beans), fruit and nut trees (olives, figs, dates, pomegranates, grapes, apples, pears, and cherries), and fiber crops (flax, hemp, and cotton).

Of the 148 big wild terrestrial plant-eating animals (herbivores)—those suitable for domestication—only 14 were able to serve as founders. What were these founder animals that could be domesticated? In Europe and Asia in about 4000 B.C. it was the "Big Five"—sheep, goats, pigs, cows, and horses. In East Asia the cow was replaced by the yak, water buffalo, and gaur. On the other hand, although in the Americas there were mountain sheep and goats, llamas, bison, peccaries, and tapirs and in Australia there were kangaroos and in Africa zebras, buffaloes, giraffes, gazelle, antelope, elephants, and rhinoceroses, all of the latter were unsuitable for domestication. Domestication involves more than taming and requires a special suite of animal characters: social species that occupy territories and animals that are herbivores. A potentially domesticated animal species must also have the right reflexes—it must be predictable and not panic easily, and it must not be ferocious or nasty in disposition. It must grow quickly, and it must be able to breed in captivity. The appropriate domesticated animals were the cow, goat, horse, sheep, donkey, yak, and camel. Once these animals were domesticated, what benefits did they provide? Food in the form of meat and milk, clothing and fiber from wool and hides, manure for use as a fertilizer, and animal power for land transport of goods and people, as well as for plowing fields. Indeed, before there were domesticated beasts of burden, the only means for moving goods and people across the land was on another person's back! Domesticated horses, goats, camels, and cows were hitched to wagons to move humans and their possessions, and reindeer and dogs were used to pull sleds across the snow. Horse-drawn chariots revolutionized warfare, and after the invention of saddles and stirrups, it became possible for marauding Huns on horseback to strike fear into the legions of Rome.

There was a downside to animal domestication. Domesticated animals could be the source of human disease. As human populations settled down, they created heaps of waste—middens of animal bones, garbage, and feces. These served as the breeding grounds for and a source of microparasites; they also attracted insects that could act as vectors of disease, as well as wild birds and rodents carrying their own parasites and potential new sources of human disease. With each domesticated species of animal came the possible human exposure to new disease agents—parasites. For example, the numbers of diseases acquired from domestic animals (zoonotically) has been estimated to be: dogs, 65; cattle, 45; sheep and goats, 46; pigs, 42; horses, 35; rats, 32; and poultry, 26. Specifically, the human measles virus has its counterpart in the distemper virus of dogs and rinderpest in cattle. Smallpox has its closest relatives in the virus of cows and poxviruses in pigs and fowl, and human tuberculosis is a cousin of bovine tuberculosis. More recent examples of the "jump" from one animal species to another include HIV, in which a chimpanzee virus became humanized; monkey pox transmitted to humans by the bite of pet prairie dogs; SARS from civet cats; and Ebola from bats.

With the clearing of forests, the planting of crops, and destruction of wild game animals, new ecological niches were created for insects and scavenging rodents. Mosquitoes and flies that once fed on game animals now found a new source of blood: humans. These "bloodsuckers" could act as vectors for malaria, yellow fever, and African sleeping sickness. Ditches, irrigated fields, and pottery vessels could also serve as breeding grounds for insects and snails, facilitating the transmission of blood fluke disease, yellow fever, malaria, elephantiasis, and river blindness.

The crowd diseases of humans such as smallpox, measles, pertussis (whooping cough), tuberculosis, and influenza were initially derived from very similar ancestral infections of domesticated animals. At first those who hunted, farmed, and domesticated animals fell prey to the parasites they acquired, and some died, but in time resistance to these new diseases developed. When such a partially immune people came in contact with others who had had no such experience, a devastating epidemic could occur. It was these contagious diseases (caused by a wide variety of worms and "germs") that would ultimately play a decisive role in the European conquests of native Americans, Africans, and Pacific Islanders; determine the outcomes of wars; loom large in the economic growth and prosperity of nations; and contribute to slavery and colonialism.

doi: 10.1128/9781683670018.ch3

3
Six Plagues of Antiquity

As humans changed their lifestyles, their relationship with infectious diseases came to be altered. For 2 million years these human populations consisted of small groups of hunter-gatherers with limited contact with other such groups, and there were no domesticated animals. Such a population structure, with little or no exposure to new sources of infection and where parasite survival and transmission were minimized, led to a situation in which epidemic diseases were virtually nonexistent. Indeed, only those diseases with very high transmission rates that induced little or no immunity, as well as macroparasitic diseases that did not involve vectors for transmission and sexually transmitted diseases, were able to establish themselves in the groups of hunter-gatherers. Although some vector-borne diseases, such as malaria and yellow fever, may have been present at this stage of human history, it was only after human populations settled down and adopted an agricultural life, or continued a nomadic existence that depended on the husbandry of large herds of animals, that conditions favored the emergence of epidemic diseases (plagues). Historically, plagues (Fig. 3.1) came to be recorded only in our recent past, a time when we became farmers.

By 8000 B.C. the human population was settled in villages—first in the valleys of the Tigris and Euphrates Rivers in Mesopotamia and then along the Nile in Egypt, the Indus in India, and the Yellow River in China. Agriculture provided increased amounts of food for the people, but it also contributed to the conditions that would result in a decline in human health. It was the agricultural revolution, with the cultivation of crops and animal husbandry, that provided the driving force for the growth of cities (urbanization). Urban life also enhanced the transmission of certain diseases through the air and water; by direct contact; and by vectors such as snails, mosquitoes, and flies. The diseases of antiquity (5000 B.C. to A.D. 700) were characterized by parasites with long-lived transmission stages (e.g., eggs) as well as those involving

Figure 3.1 (Left) *Plague in an Ancient City* (detail) circa 1652-1654 by Michiel Sweerts (1624-1664).

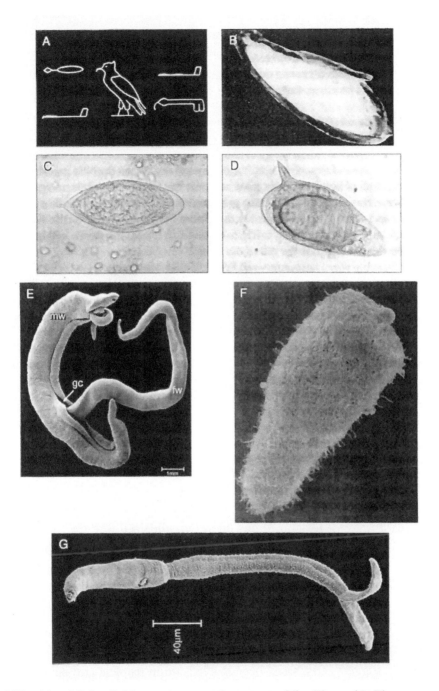

Figure 3.2 The blood fluke *Schistosoma,* causative agent of the Pharoah's Plague.
A. Hieroglyphic; B. Calcified egg from a mummy; C. *Schistosoma haematobium* egg as
seen with light microscope; D. Schistosoma mansoni egg with miracidium inside; E. adults
in copula, as seen with a scanning electron microscope (from David Halton); F. ciliated
miracidium as seen with the scanning electron microscope (courtesy of Vaughan South-
gate); and G. cercaria, as seen with a scanning electron microscope (from David Halton).
mw, male worm; fw, female worm; gc, gynecophoric canal

person-to-person contact. Thus, most became established only when a persistent low level of infectious individuals could be maintained, i.e., were endemic; this required populations greater than a few hundred thousand.

The Pharaohs' Plague

A look back

The Assyrian and Babylonian literature, as well as the Egyptian papyrus from Kahun written in about 1900 B.C., describe a disease that causes blood to appear in the urine (hematuria). And near the Louvre Museum in Paris there is a stone from ancient Egypt that says, "Anyone who moves this boundary stone will be covered with bloody urine." Hematuria was described by the father of Arabian medicine, Avicenna (930-1037), in his *Canon of Medicine*, but the condition, called a-a-a, was recognized much earlier and is mentioned in the Ebers papyrus, dated ~1500 B.C. and named after Georg Ebers, who in 1862 found the paper in a tomb in Thebes, Egypt. There is even a hieroglyphic sign showing a penis dripping fluid, and this too may be blood (Page 44). Such a sign was not considered to be connected with disease, but as a mark of puberty in the male child. Many remedies for this disease are described in the Ebers papyrus, suggesting that this condition was widespread. And in a relief of the tomb of Ptah-Hetep I and Mehou of the Sixth Dynasty at Sakarrah there are figures of fishermen and bargemen with enlarged abdomens—surely representing the pathology of chronic snail fever or blood fluke disease. In 1910, Marc Armand Ruffer examined several Egyptian mummies from the Twentieth Dynasty (1200-1000 B.C.) and found the calcified eggs of the blood fluke in the kidneys of several mummies (Fig. 3.2B). Fossil snails capable of transmitting blood fluke disease have been found in the well water of Jericho. It has been hypothesized that the water was infested with infected snails, resulting in a high level of disease. Too debilitated by disease to defend the city or repair the decaying walls, the people of Jericho were easily defeated by Joshua's army. Joshua, though unaware of the cause of the disease that contributed to his success, but wanting to prevent its spread, destroyed Jericho and proclaimed a curse on anyone who would rebuild it. The city remained deserted for 500 years. Centuries of recurring drought destroyed the snails, and thus the city has remained free of disease to this day. All this suggests that snail fever has existed in tropical and subtropical parts of the world, but especially in Egypt, since ancient times.

Soon after humans settled down and food production began in the Fertile Crescent (~8000 B.C.), it spread to other parts of Eurasia and North Africa. The plants (as well as the domesticated animals) that formed the basis for agriculture in the valleys between the Tigris and Euphrates Rivers were also cultivated in the Nile Valley of Egypt, and this triggered the rise of Egyptian civilization. These same agricultural practices, though, sustained by the Nile, also sowed the seeds of Egypt's decline.

The land of the pharaohs flourished for 27 centuries, and its accomplishments, even today, are truly impressive. When Herodotus, the Greek historian, made a tour of Egypt in 400 B.C., he wrote of "wonders more in number than those of any other land." And he went on to say that "when the Nile inundates the land all of Egypt becomes a sea and only the towns remain above water. Anyone traveling from Naucratis to Memphis sails right alongside the pyramids, and when the waters recede they leave behind a layer of fertile silt—'black land'—the Egyptians call it, to distinguish it from the sterile 'red land' of the deserts. Egypt is the gift of the river."

In some ways Nature favored Egypt because, unlike Mesopotamia, which stood on an open plain and was unprotected from marauding tribes, the deserts that bordered the Nile discouraged invasion, and so the people lived in relative security. The villages shared the river and merged into cities. To tap the bounty of the Nile required the cooperation and organization of the people, with social and political structures developing therefrom. All power was invested in the pharaohs, who were both kings and gods. Below the pharaoh was a vast bureaucracy that rested on the shoulders of the workers and the peasantry. Egypt's people, however, who built enduring stone monuments for their pharaohs, were racked with a debilitating disease, snail fever. And although medical science began in Egypt, the doctors and surgeons could not keep this disease at bay. There is a reason for this: the early civilizations of Egypt and those of the Fertile Crescent (Sumer, Assyria, and Babylon) were based on agriculture, and this agriculture required irrigation and/or natural flooding by the rivers. Irrigation farming, especially in the tropics, created conditions favorable for the transmission of snail fever caused by the blood fluke. Blood fluke disease—the plague of the pharaohs—is not a fatal disease, as is malaria or yellow fever; it is, however, a corrosive disease. And although there may have been a time when the natural flooding of the Nile River made snail fever a seasonal problem, once there was irrigation it became a year-round problem, since infections could be acquired from the standing water in the irrigation channels. Consequently, as William McNeill wrote in his book *Plagues and Peoples*,

> there was a listless and debilitated peasantry handicapped ... for the ... demanding task of resisting military attack or throwing off alien political domination and economic exploitation. Lassitude and chronic malaise ... induced by parasitic infections was conducive to successful invasion by the only kind of large-bodied predators human beings have to fear: their own kind, armed and organized for war and political conquest.

McNeill also suggested that the rule of the pharaohs may have been due to the power of the snail and the blood fluke and malaria—the classic plagues of Egypt—which debilitated the populace.

And so it was that snail fever did its work. By 660 B.C., Egypt became subject to internal political dissension and to attack by their iron-armed neighbors (the Assyrians), and their civilization, based on agriculture and copper weapons, began to collapse. The Persians overran Egypt in 525 B.C. The cause of snail fever, the disease that set the Egyptian civilization on its inexorable downward spiral, was unknown to the ancient Egyptians because the transmission stages of the parasite (eggs, miracidia, and cercaria) are microscopic; in addition, the adult worms themselves are tiny and live within the small blood vessels, and so they were unnoticed for thousands of years.

Search for the destroyer

Blood fluke disease, also known as snail fever and endemic hematuria, involves feces or urine, water, snails, and a flatworm. The first Europeans to experience the disease on any scale appear to have been the soldiers of Napoleon's army during the invasion of Egypt (1799-1801). The symptoms of the disease, bloody urine, were rife among the soldiers. Baron Jean Larey, a military surgeon, noted its high frequency in the men; he believed, however, that the excessive heat during the long marches was the cause. The connection between hematuria and a parasite did not occur until 1851. In that year Theodor Bilharz, a German physician working in Egypt, while carrying out an autopsy on a young man, made a startling discovery: worms were found in the blood vessels, a location never before encountered (Fig. 3.2E). He named the worm *Distomum* (meaning "two mouths") *haematobium* (from the Greek words *haema*, meaning "blood," and *bios*, meaning "to live in"). In 1858 the name was changed to *Schistosoma* (from the Greek words *schisto*, meaning "split," and *soma*, meaning "body"). Today, blood fluke disease is called schistosomiasis or bilharzia, the latter in honor of Bilharz's discovery. (During World War I, British soldiers found it easier to call the disease "Bill Harris.")

In 1851, Bilharz reported that he had seen microscopic eggs with a pointed spine in the female worm (Fig. 3.2C), and in the following year he observed these eggs in the bladder; within the egg he observed a small, motile embryo. He also found that the eggs would hatch to release a small ciliated larva (Fig. 3.2F) that swam around for about an hour and then disintegrated. This work was confirmed in 1863 when John Harley, a London physician, examined a patient with hematuria who had previously lived in the Cape of Good Hope in South Africa. Examining the blood-tinged urine in a drop of water under the microscope, he found schistosome eggs, and several of these hatched to give progeny that swam by using their cilia (Fig. 3.2G). But there remained a puzzle: how was the infection transmitted? Bilharz and others were aware that flukes closely related to *Schistosoma* had intermediate stages in snails, but when Harley examined snails from a region where schistosomiasis was prevalent, he found no evidence of larval stages. Despite this failure, the suspicion remained that humans acquired the infection either by eating infected snails or by drinking water containing the ciliated larva called miracidia. In 1870, Spencer Cobbold, working in London, obtained eggs from a young girl living in the Cape of Good Hope and found that although the eggs would not hatch in urine, they did so in fresh or brackish water. Then, in about 1904, Japanese physicians found that a related blood fluke, named *Schistosoma japonicum*, could also infect humans, but this species had eggs without a spine. In 1905, Patrick Manson discovered another type of schistosome egg, one with a spine on its side (Fig. 3.2D); this was in the feces of an Englishman who had lived in the West Indies but had never visited Africa; this new type was duly named *Schistosoma mansoni*. Now there were three known species of human-infecting blood flukes.

The life cycle and mode of transmission of the schistosome to a human were first demonstrated between 1908 and 1910 in Japan. Fujinama and Nakamura found that when the tails of mice were immersed in water from rice fields known to have a high incidence of bilharzia, they became infected with *S. japonicum*. Shortly thereafter it was possible to show that the miracidia were able to penetrate freshwater snails in the rice paddies, and Ogata found that a tailed larva (called a cercaria) emerged from infected snails and could directly penetrate the skin of mice (Fig. 3.2G). This suggested that species other than *S. japonicum* might have a similar life cycle. At the outbreak of World War I the British became concerned about the potential deleterious effects of schistosomiasis on their troops in Egypt. In 1915 the British War Office sent Robert Leiper to Cairo "to investigate bilharzia … and advise as to preventive measures to be adopted." Leiper collected freshwater snails, identified them, and determined whether they were infected, either by allowing the snails to release cercaria or by dissecting the snails to find other larval stages (called sporocysts). Within weeks he and his team identified the snails *Bulinus* and *Biomphalaria* as the vectors. (Because the snail vector is critical to transmission, schistosomiasis is called "snail fever.") Leiper went on to show, by placing the tails of the mice in cercaria-infested water, that the skin of mice could be directly invaded by the cercaria. This suggested that the infection was acquired by bathing in infested water. But could the infection also be acquired by ingestion? Because Leiper was able to show that when cercaria were placed in dilute hydrochloric acid (similar in concentration to that found in the human stomach) they were killed, this route of infection seemed most unlikely. Leiper was also able to show that the adult *S. mansoni* and *S. haematobium* were different from one another and that cercaria hatched from *Biomphalaria* produced eggs with lateral spines whereas those from *Bulinus* produced eggs with a terminal spine. The pathology of the two species was also found to differ: *S. mansoni* remained in the liver and laid its eggs there, whereas *S. haematobium* early in its development left the liver for the veins surrounding the bladder. Thus, 65 years after the discovery of the adult worm by Bilharz, the life cycle of snail fever was known. The discharged eggs, on reaching freshwater, release a swimming larva, the miracidum. Miracidia are short-lived, but if they encounter a suitable snail, they penetrate the soft tissues (usually the foot), migrate to the liver, and change in form (sporocyst); and for 6 to 7 weeks, by asexual reproduction, the numbers of parasites increase. During this time the snail sheds thousands of fork-tailed cercaria, which can swim and directly penetrate human skin, and in 5 to 8 weeks they develop into adult worms.

Snail fever, the disease

Schistosomes differ from other flukes (trematodes) in that the sexes are separate and they inhabit the blood vessels. The adult worms are ~10 mm in length, and the stouter males have a groove running lengthwise, called the gynecophoric canal, where the female normally resides (Fig. 3.2E). It is this groove in the male that is the basis for the worm's generic name *Schistosoma*, meaning "split body." Both males and females have two suckers at the head end of the worm, and the more anterior one surrounds the mouth. (Bilharz mistakenly took the two suckers for two mouths, and thus he called the worm *Distomum*, "two mouths.") The schistosome adults, in sexual union, live in blood vessels (veins) close to the bladder and small intestine. Mating occurs in the gynecophoric canal, and then the paired worms move "upstream" into smaller veins, where the female worm deposits the fertilized eggs. The pathology of schistosomiasis is due not to the adult worms themselves but to the eggs. Each day hundreds of embryo-containing eggs move across the walls of the veins into the bladder or intestine, aided by the host's inflammatory response, and in the process eggs become enclosed in a small tumor called a granuloma. It is the passage of eggs through the bladder wall that results in bleeding and gives the telltale sign of hematuria. Once in the bladder or intestine, the egg becomes freed of the granuloma and is eliminated from the body either with the urine or in the feces.

More than two-thirds of the eggs, however, fail to work their way out of the body and are washed back in the veins, and by means of the bloodstream they scatter throughout the body, where they accumulate in various organs. Accumulation is greatest in the liver and spleen. The piling-up of eggs blocks the normal blood flow, and this leads to tissue death. The egg also acts as an irritating foreign substance that the body attempts to wall off by surrounding it with a fibrous capsule. The egg-laden liver eventually becomes filled with scar tissue. In the bladder blood fluke, *S. haematobium*, the scarred areas block the migration of eggs through the lower bowel tissues, and more eggs are swept back into other sites. The earliest signs of infection, fever, chills, sweating, headache, and cough, occur within 1 or 2 months. Six months to a year later the accumulation of eggs produces organ enlargement, especially the liver and spleen; the enlarged and cirrhotic liver causes

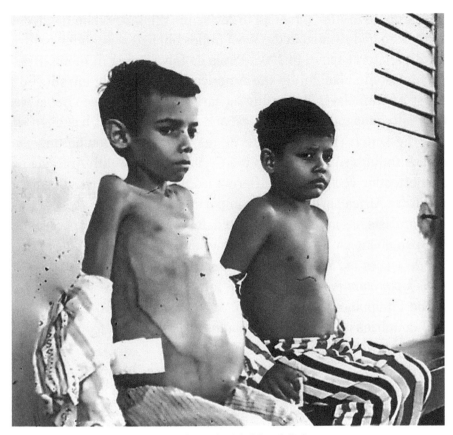

Figure 3.3 Two young boys infected with blood flukes

the abdomen to become bloated, appetite diminishes, blood loss leads to anemia, and there is dysentery (Fig. 3.3).

Schistosomiasis is an arithmetic disease: the severity of its symptoms and cumulative damage are directly related to the number of worms present, and the latter depends on the degree of exposure. In heavy cases there may be hundreds of worms, and the adults may live for 20 or 30 years. Clearly, with time and increased invasion by cercaria, a person becomes more and more debilitated. Yet over the centuries the adult inhabitants of areas where the disease is endemic, such as Africa, developed some measure of immunity largely as a result of continuous exposure; Europeans and Americans with no such immunity suffer more-severe symptoms as a result of higher burdens of worms.

Where snail fever is found
Wars and human migrations carried the blood flukes of the East African lakes to the Nile River, and from there it was distributed along the trade routes to most of the continents of the world. Although in 1902 Manson believed schistosomiasis to

be a disease unique to Africa, he had to revise his thinking when he discovered an Englishman, who had resided in the West Indies but had never been to Africa, passing eggs with a lateral spine. In 1908, Piraja da Silva, living in Bahia, Brazil, wrote that the schistosome common in the Americas was probably introduced from Africa by West African slaves beginning as early as 1550. Indeed, Bahia was one of the ports of entry for the African slaves, and more recently it has been suggested that, under the Dutch (1630-1654), Recife may also have been an important slave entry point. Although snails native to the Caribbean and South America have been found to be effective vectors (and different from the snail species in Africa), snails introduced from Africa have also been important in transmission.

Schistosomiasis has not been eliminated. It is estimated that at present there are 240 million people infected with schistosomes and 700 million people at risk. Ninety percent of the cases are found in sub-Saharan Africa, resulting in >200,000 deaths annually. *S. japonicum* is found in Southeast Asia and the western Pacific, as well as China, the Philippines, and Indonesia. *S. haematobium* and *S. mansoni* are both found in 43 countries in Africa, but the latter species is also found in the Americas (Brazil, Suriname, Venezuela, and the Caribbean).

People who come to the freshwater pools to work, bathe, drink, wash clothes, and swim may also use the water for elimination of their body wastes. Individuals may be infected and reinfected almost daily as they paddle through the cercaria-infested waters that they have come to use as their outdoor toilets. Schistosomiasis remains one of Africa's greatest tragedies. The highest incidence occurs in African children. In some instances, technology has expanded the numbers of cases. Indeed, every new irrigation scheme and each new dam may pose a new threat.

The Aswan High Dam of Egypt, begun in 1960 with Soviet financing and engineering and requiring 30,000 Egyptians toiling around the clock, was completed in 1971. The High Dam, by controlling the level of water in Lake Nasser, has brought electricity to many parts of Egypt as well as making four crops per year possible through year-round irrigation, but it has also created conditions favorable for the schistosome-carrying snails. Before High Dam construction there was already perennial irrigation in the Nile Delta and the prevalence of schistosomiasis was 60%, whereas in the 500 miles of river between Cairo and Aswan when there was annual flooding the prevalence was 5%. Some 4 years after the dam was completed, the average prevalence between Cairo and Aswan increased 7-fold (35%; range, 19 to 75%).

Schistosomiasis is generally a disease associated with agriculture, but it has also been a military problem ever since the days of Napoleon. During World War II, when U.S. troops stormed ashore on the Pacific island of Leyte in October 1944, they were unaware that in addition to being attacked by Japanese bullets they were also being invaded by the cercaria of *S. japonicum*. By January 1945 the first cases were diagnosed, and in the end 1,700 men were put out of action at a cost of 300,000 fighting man-days and $3 million. Five years later, 50,000 Chinese communist soldiers prepared for an invasion of Taiwan, but schistosomiasis became so widespread among the troops that the campaign was abandoned and the island was retained by Chiang Kai-Shek's forces.

Where did schistosomiasis originate? It probably first occurred in animals living in the rain forests and lakes of East Africa and then spread together with its vector snails along the Nile and out into the Middle East and Asia via the trade routes. (Blood flukes occur in birds and mammals other than humans. Indeed, "swimmer's itch," or cercarial dermatitis, is found in lakes and along the seashore in Michigan, Minnesota, Wisconsin, New Jersey, and New England, as well as in other parts of the world, and is caused by cercaria [Fig. 3.2G] that normally infect aquatic birds and mammals. The skin rash and pustules are the result of their failure to continue their migration past human skin.)

Snail fever today

Diagnosis of schistosomiasis is made by the examination of stools and urine under the light microscope and the finding of eggs. Sometimes this visual test is supplemented by biopsy and immunologic methods. Preventive measures include education of the population regarding the means for preventing transmission, the treatment of infected persons, and the control of the snail vector using molluscicides. In some cases (e.g., growing rice), avoidance of contact may be impossible. Human exposure can be reduced, however, by the provision of a safe water supply for bathing and washing as well as by the sanitary disposal of human wastes. Other measures may be lining of irrigation channels with cement to discourage snails; intermittent irrigation of rice paddies to disrupt the life cycle; or storage of water away from snails for 2 or 3 days, a time that exceeds the survival time of the miracidia.

The earliest treatment for the disease, developed in 1918, required the intravenous administration of an antimony compound (tartar emetic). In 1929 intramuscular injections of another antimony compound, stibophen, were used, but the cure rates were not as good as with tartar emetic, and both drugs showed severe toxic reactions and sometimes resulted in death. Later, an oral drug, niridazole, was introduced (1964), but it wasn't until the 1970s that a truly effective drug with low toxicity was developed: praziquantel (trade name: Biltricide). There is no preventative vaccine or drug.

The development of new drugs for the treatment of schistosomiasis can be a long and expensive undertaking. Further, determination of drug efficacy and safety may require extensive animal and human testing. One such heroic effort is worthy of mention. Claude Barlow, an American physician, volunteered for a chemotherapy trial and exposed his abdomen to 224 cercaria, and when cercarial dermatitis developed, he knew he had been infected. He came down with severe schistosomiasis, which required many intravenous injections of tartar emetic for cure. In his old age Barlow wrote: "even today I shudder every time I see a hypodermic needle."

Today, with hindsight, it is easy to understand why those living in ancient Egypt were unable to control schistosomiasis, but why, 2,000 years later, are we still failing? There are certainly economic constraints such as the cost of pesticides to kill the snails and the cost of drugs, as well as the necessary infrastructure for providing clean water and sanitary disposal of wastes, but in the final analysis it is the habits of the human population that are of critical importance to the elimination of this disease. Since there is no animal reservoir, humans are required for the perpetuation of the disease. As long as infected individuals continue to urinate and defecate in the same waters where the vector snail lives, and to expose their bare skin, there will be blood fluke disease.

The Plague of Athens

The valleys of the Nile and Tigris-Euphrates spawned the civilizations of Egypt and Mesopotamia. The flat fertile valleys made it easier to subject the populations to a single ruler and to oversee individuals so that each had a prescribed role in the society. As a consequence, there emerged in the Fertile Crescent and Egypt a unified system of kingdoms. The different geography of Greece gave rise to another kind of sociopolitical system. Greece, the southernmost extremity of the Balkan landmass, is composed of limestone mountains separated by deep valleys and is cut almost in two by the narrow Corinthian Gulf. The southwestern peninsula was called Peloponnesia, and the northeastern part was called Attica. East of the mainland are the islands of the Aegean Sea; to the south is Crete. In Greece, because the small populations were separated from one another by mountains and the sea, central control was impossible and individuals were not compelled to become specialists but rather masters in a range of accomplishments of hand and mind. Greece, smaller in area than Florida, was never able to support more than a few million inhabitants, but it has played an enormous role in the history of Western civilization. We see some of this today in their monuments, sculptures, paintings, and writings, and we speak of the "glory that was Greece." Indeed, the prestige of the Greeks in the arts and their ideas on medicine, astronomy, and geography were accepted with unquestioning faith until the 17th century, when a new scientific spirit of experiment and inquiry came into being.

The land and the climate of Greece were unsuitable for farming grains, and as a result, the economy rested on the large-scale movement of goods by ship: the Greeks planted vines and olive trees and produced wine and oil, and these were exchanged for grains and other less valuable commodities. Indeed, an acre of land with vines and olive trees could yield a quantity of wine and oil that could be exchanged for an amount of grain requiring many acres for its production. The outlying communities on the Mediterranean and Black Seas that provided these less valuable commodities were called by the Greeks "barbarians," from the Greek word *barbaros*, meaning "foreign" or "uncivilized." It was the barbarian societies that provided grain, metals, timber, and slaves in exchange for oil and wine and so contributed to the emergence of the Greek civilization.

The Greeks had an unshakable belief in the worth of the individual. While to the east there were absolute monarchies, the Greeks evolved a democratic society where each individual was respected and counted. What developed were states consisting of a city and its surrounding lands whose inhabitants, citizens (literally "those living in the city"), were valued. (Not all individuals, however, were citizens; slaves and women did not have the same rights.) The city-states of Greece, insulated from the barbarians, consisted of urban centers dedicated to commercial transactions with limited local farming; as they prospered, the population grew. The inhabitants of the city-states were healthy but naive to the diseases that were endemic in the Middle East, and this was to prove decisive in the Peloponnesian War.

Before the Greeks of historical times (~750 B.C.), civilizations had flourished in Mycenae (1600 to 1200 B.C.) and before that in Crete (the Minoans). Later, the Dorians invaded from the north, destroying the Mycenaean cities and society, and this conquest ushered in for 450 years (1200 to 750 B.C.) what is called the Dark Age. At the close of the Dark Age there emerged two powerful city-states that were essentially military garrisons governed by a commander and his captains. These were Athens (in Attica) and Sparta (on the Peloponnesus). Sparta was settled by the Dorians, and Athens gave refuge to Mycenaeans. Each city-state represented opposing philosophies: stern military discipline in Sparta versus intellectual and political freedom in Athens. As the population of Athens grew too large for the limited space of Attica, the Athenians began sailing out of their port of Piraeus to colonize the Aegean islands and the western coast of Asia Minor (in what is today Turkey). These Greek colonies were called Ionia.

Belief in liberty and freedom made Greek city-states resist domination by others. And since each state had its own habits, rules, and government, the loyalties of the citizens were to a particular city-state, the polis. (People engaged in the civic life of the polis give us the term "politics.") As a consequence, war between Athens and Sparta was inevitable. In 431 B.C. the Peloponnesian War began between these two city-states, and it lasted for 27 years. The cause of the war was probably economic, although of this we cannot be certain, but what is known is that its outcome was determined by disease.

The Greeks, apparently against all odds, managed to defeat the numerically far superior Persian forces in two battles, in 490 B.C., at Marathon, and again in 480 B.C., in the great naval battle of Salamis. This led a number of Greek city-states to join together with Athens in a sea league both to punish the Persians and to obtain recompense for the cost of the war. In time, however, Athens turned this league into an instrument of its own imperial power, appropriating the funds of the league for the creation of monuments of imperial splendor (notably, the Parthenon). This naturally provided a focal point for the jealousies and rivalries of the various city-states, especially Sparta.

Corinth was a commercial and colonial power and an ally of Sparta. Because of this, its interests were in competition with those of Athens. Athens began to interfere in the affairs of Corinth. Corinth naturally objected and threatened Athens, so in retaliation Athens began an embargo of Corinth and other city-states on the Isthmus of Corinth. This crippling embargo caused Corinth to urge Sparta to declare war on Athens. The Athenians had a great fleet but a poorly trained army, whereas the Spartans were an effective military power on the land but lacked a fleet, and so were not a sea power. Pericles, the leader of the Athenians, decided to rely mainly on Athenian naval supremacy. His strategy was to bring all the people in Attica into the city, abandon the outlying countryside to destruction by the Spartans, and rely on the navy to supply the city with food and other necessities that would be carried through the fortified corridor from the port of Piraeus into the city itself. The hope was that Sparta would eventually be worn out and frustrated. Sparta did invade Attica from the north, and the Athenians gathered themselves and remained secure within the walled fortifications of their city. But when the Spartans destroyed the olive and grape orchards in the outlying countryside, the source of Athens' wealth came into jeopardy. Further, as the large numbers of peasants from the countryside sought refuge within Athens, the city became overcrowded. In 430 B.C. disaster struck. An epidemic that started in Ethiopia moved into Egypt, and from there it was brought by ship to Piraeus. The epidemic raged for about 2 years and killed about a fourth of the Athenians, including, in 429 B.C., Pericles.

In 1994-1995 a mass grave was uncovered prior to construction of a subway station just outside Athens' ancient cemetery. There were some 90 skeletons, 10 belonging to children; the grave may have contained as many as 150 people. The skeletons in the graves were placed helter-skelter with no soil between them, and the bodies were placed in the pit within a day or two, suggesting burial in a state of panic. The grave was dated to between 430 and 426 B.C. It is believed these are the remains of Athenians killed by the plague. The historian Thucydides, himself a survivor of the plague, wrote:

> The bodies of the dying were heaped one on top of the other, and half-dead creatures could be seen staggering about in the streets or flocking around the foun-

tains in their desire for water. ... The catastrophe was so overwhelming that men, not knowing what would happen next to them, became indifferent to every rule of religion or of law. All the funeral ceremonies which used to be observed were now disorganized, and they buried the dead as best they could. Many people, lacking the necessary means of burial because so many deaths had already occurred in their households, adopted the most shameless methods. They would arrive first at a funeral pyre that had been made by others, put their own dead upon it and set it alight; or, finding another pyre, they would throw the corpse that they were carrying on top of the other one and go away. ... Seeing how quick and abrupt were the changes of fortune which came to the rich who suddenly died and to those who had previously been penniless ... people began openly to venture on acts of self indulgence which before they used to keep in the dark. ... No fear of god or law of man had a restraining influence. As for the gods, it seemed to be the same thing whether one worshiped them or not, when one saw the good and bad dying indiscriminately.

What was this devastating plague? Despite Thucydides' detailed description, the precise identity of the disease is not known. It was clearly not the bubonic plague, for the characteristic symptom of the bubo (swelling of the lymph nodes in the region of the groin and armpits) is not found in Thucydides' description. Other suggested candidates are measles, typhus, Ebola, mumps, and even toxic shock syndrome. The case for typhus seems strongest both epidemiologically—the age group is similar—and from the standpoint of the symptoms. Typhus is characterized by fever, pustules, and a rash of the extremities; it is known as a "doctors' disease" from its frequent incidence among caregivers. But the fit is not exact. The rash in Thucydides' description does not precisely match that of typhus, nor does the state of mental confusion.

The plague of Athens demoralized the citizenry, destroyed the fighting power of the Athenian navy, and prevented the launching of an attack against Sparta. Though the war dragged on for many more years, the spirit of Athens had already been broken, and by 404 B.C. defeat was complete. Sparta deprived Athens of her navy, and her land defenses were razed to the ground. The plague of Athens clearly changed the course of history.

The Roman Fever

Of the antiquity of human malaria there is no doubt. Enlarged spleens, presumably due to malaria, have been found in Egyptian mummies more than 3,000 years old, and the Ebers papyrus (1570 B.C.) mentions fevers. More recently, evidence of malaria has been detected in lung and skin samples from mummies dating from 3204 B.C. to 1304 B.C. Clay tablets from the library of Ashurbanipal (2000 B.C.), king of Assyria, mention enlarged spleens, headaches, as well as periodic chills and fever, indicating that more than 4,000 years ago the region between the Tigris and Euphrates Rivers was already malarious. Malaria probably came to Europe from Africa via the Nile Valley or resulted from closer contact between Europeans and the people of Asia Minor. Early Greek poems from the end of the 6th century B.C. describe intermittent fevers, and Homer's *Iliad* (ca. 750 B.C.) mentions malaria, as do the writings of Aristophanes (445-385 B.C.), Plato (427-387 B.C.), and Sophocles (496-406 B.C.). The Greek physician Hippocrates (460-370 B.C.) discussed in his *Book of Epidemics* the two kinds of malaria, one with recurrent fevers every third day (benign tertian) and another with fevers on the fourth day (quartan), which are today called *Plasmodium vivax* and *Plasmodium malariae*. He also noted that those living near marshes had enlarged spleens, but he never speculated as to the relationship. Indeed, Hippocrates believed that the intermittent fevers were the result of an imbalance in the body's fluids (bile, blood, and phlegm) brought about by drinking stagnant marsh water. Hippocrates recognized that at harvest time, when Sirius the Dog Star was dominant in the night sky, fever and misery would soon follow. From the historical descriptions, malaria was obviously present in Greece but probably had little influence on military campaigns. Only once in his account of the Peloponnesian War did Thucydides refer to an illness suggestive of malaria, which affected the Athenian army encamped on marshy grounds while besieging Syracuse.

Hippocrates did not describe the deadly malignant tertian malaria (*Plasmodium falciparum*), and so we suspect that it did not exist or was rare. It has been speculated that although this kind of malaria was periodically brought into southern Europe from North Africa and Asia Minor, such infections were infrequent due to the absence of a suitable mosquito vector in the Mediterranean. With greater agricultural activities, including deforestation and soil erosion, conditions arose that favored the establishment of a suitable habitat on the shores of Europe for several Asian and North African species of *Anopheles* mosquitoes. Over time, the competence of the mosquito to transmit the disease increased, and so the highly virulent *P. falciparum* came to be established in Europe by the second century A.D., and from that time onward it plagued the Romans.

The great age of Greek expansion lasted 200 years (750 to 550 B.C.). Greek colonization took place along the shores of the Aegean and Black Seas, west into Sicily and southern Italy, and by 750 B.C. colonies had been established on the west coast of Italy as far north as the Bay of Naples. The Ionian Greeks from Asia Minor sailed and traded farther west, along the coast of the Mediterranean, reaching present-day Spain. The Greeks also traded with inland villages, moving up the Rhone River into Gaul and even as far north as England and Ireland. It was the practice of the Greeks to keep to themselves and to remain apart from the indigenous peoples; and so, as extensions of the homeland, the colonies were one of the means by which Greek civilization was spread to other parts of the world. The flourishing trade with the colonies also became the means by which infectious diseases could be transmitted to the naive populations in distant lands.

In ~2000 B.C. the Latins, a group to which the Romans belonged, and of Indo-European origin, possibly with forebears in central Asia, migrated first into central Europe and then to the northernmost part of Italy. By 1000 B.C. they had settled on the ~700-square-mile volcanic Latium plain, bounded on the north by the Tiber River. The soil was rocky but fertile, and the Latins prospered as farmers. Then, in ~800 B.C., the Etruscans coming out of Asia Minor landed on the coast north of the Tiber, from whence they moved inland, and by 600 B.C. they dominated all of Italy from the Alps in the north to Salerno in the south; here, though, they were halted by the already established Greek colonies. The Etruscans, a highly civilized people who were traders and merchants, brought to the Romans their first contact with the eastern Mediterranean. The Romans made allies of or subdued the other tribes of the Latium plain, including the Etruscans. By the beginning of the 4th century B.C. Rome was the leading city in central Italy. In time that city became an empire that would last 500 years. By 350 B.C. the Romans had moved southward, reaching the Greek settlements at the foot of the peninsula, and in 275 B.C., when the Greeks were defeated, Rome became the master of the entire Italian peninsula. Later, there would be other conflicts and other victories over Carthage (264-241 B.C., 218-201 B.C., and 149-146 B.C.), the Macedonian Empire of Alexander (197 B.C.), the Seleucid Empire in Syria (190 B.C.), and the Ptolemaic Empire in Egypt (31 B.C.). By 55 B.C. the Romans had invaded Britain.

Rome lived off its imports. Cargo shipped from the provinces was unloaded at the seaport at Ostia and carried up the Tiber River to the city. The Roman Empire, with its center in Rome, developed an ever-extended series of colonies, and by the year A.D. 100 there was a vast trade network that included India, China, and the northern parts of Africa and the Middle East. The regular movement of goods and people to and from Rome also made for the spread of infections. Thus, over time the chances of the Mediterranean population contracting an unfamiliar infection became greater and greater.

There is no evidence that malaria was a public health problem in Italy among the ancient Etruscans, but there is clear evidence of malaria being devastating to the Roman Empire. Indeed, in ancient Rome temples were dedicated to the goddess Febris, who is described as an old hag with a prominent belly and swollen veins. The medical literature of the time also contains accurate descriptions of malaria, and there are references to marshes as the source of the disease. The disease was so prevalent in the marshland of the Roman Campagna, near Ostia, that the condition was called the Roman fever, and eventually it was given the Italian name *mal' aria*, literally "bad air," because it was believed that this recurrent fever occurring during the sickly summer season was due to vapors emanating from the marshes. For almost 2,000 years Rome was the home of the Roman fever.

Although malaria was uneven in its distribution and its endemicity fluctuated cyclically in the Roman Empire, epidemics of malaria occurred in Rome and the Campagna every 5 to 8 years. The Pontine Marshes southwest of Rome were a lethal source of malaria. As one writer put it, "the Pontine ... creates fear and horror. Before entering it you cover your neck and face well before the swarms of large blood-sucking insects are waiting for you in this great heat of summer, between the shade of the leaves, like animals thinking intently about their prey. ... Here you find a green zone, putrid, nauseating where thousands of insects move around, where thousands of horrible marsh plants grow under a suffocating sun." Not even the Romans, who were able to conquer most of the Western world, were able to master the Pontine marshes. It has been estimated that even if malaria occurred in only one-sixth of the Campagna it could devastate the agricultural economy. Indeed, it required the efforts of the fascist dictator Benito Mussolini in the late 1920s to drain and fill these swamps, making them habitable and agriculturally productive. Colonies on the coast of Italy also failed because of malaria, and in some districts of Rome, where the urban population may have reached 750,000 to 1 million people, the death rate could be quite high. In some particularly unhealthy places life expectancy was only 20 years, whereas in places where malaria was absent life expectancy could be as high as 40 or 50. In the Roman Empire malaria was a disease of children, but severe illness was also found among immigrants without acquired immunity. After the establishment of Christianity, the Roman fever plagued pilgrims visiting the holy city. Some have claimed that foreign invaders of Rome—particularly the French and Germans—were more effectively repelled by the deadly fevers of the Pontine marshes and the Campagna than any man-made weapons. Indeed, Alaric died from malaria during the siege of Rome in the summer of A.D. 410, and Attila the Hun's failure to march on Rome in A.D. 452 was partly due to the threat of this disease as well as a famine in Italy. And in A.D. 1155, Frederick Barbarossa and his army were so decimated by malaria at Rome that they were forced to retreat across the Alps.

Plagues and the Rise of Christianity

Plagues of various sorts were not unfamiliar in Roman history. An epidemic—probably smallpox—struck the city in A.D. 65. It was brought, or so we believe, by Roman troops who had been campaigning in Mesopotamia. Mortality was heavy, and in some cases half the population died. A new round of smallpox infections began in A.D. 251 to 266, and it was reported that 5,000 a day died at its height. By this time measles may also have become established among Mediterranean populations. No accurate estimates of population losses can be given, but they must have been high.

The early Romans owed their loyalty to their pagan gods. Their religion was one of form and ritual rather than of spiritual observance. In essence, the Romans had a contractual relationship with their gods: if you do something for me, I will do something for you. At first their religion was animism: gods represented the spirits in water, rocks, fire, trees, beasts, sun, moon, stars, and lightning. The spirits were amoral, and they either helped or harmed the worshippers according to the manner in which they were treated. In effect, the role of religion was the appeasement of the multitude of gods so that the worshippers would receive some kind of benefit. From the Etruscans the Romans borrowed elaborate religious ceremonies and gods with a human form, a form that could be represented as painted images or carved statues. And it was the Etruscans who gave the Romans their earliest contact with the gods and goddesses of Greece, many of whom were absorbed into the Roman religion. Cults developed to worship specific gods at specific times (called holy days or, later, holidays), and those who presided over the cult rituals were called priests, but these individuals were neither moral nor spiritual. In time the emperor himself became a god, and so loyalty had to be sworn to him.

Then, in the 1st century A.D., there appeared a new kind of religion, preached by Jesus of Nazareth (the Christ) and his disciples. Although Jesus preached for less than 3 years in what is now Palestine, his many disciples traveled throughout the Roman Empire spreading the word of the Christian religion. Jesus' preaching took the form of parables and miraculous healings; he encouraged the poor and the oppressed, spoke of forgiveness, and proselytized that there should be detachment from wealth and property and that the outcasts and sick of society should be given special care. Because the Roman Empire consisted of many cities, Christianity became an urban movement. (Indeed, because of this, those living outside the city were called rustic or, in Latin, *paganus*, from which the word "pagan," meaning non-Christian, is derived.) Jesus gathered around him a community of followers who regarded themselves as God's people, and they went forth establishing a missionary movement. The early Christians had a moral ideal: they separated themselves from pagan idolatry and they espoused universal salvation. Because Jesus' disciples preached the coming of a new king, it appeared to the Romans that there might be a revolution in the making. But at the outset the Romans simply regarded Jesus as a minor political rebel whose followers could be used as convenient scapegoats. As a result, the Christians were blamed for all types of disasters, including plagues, inflation, fires, and even barbarian incursions. The reasons for this were many: the Christians did not worship the emperor, they did not observe the pagan ritual acts, they insisted that they alone possessed God's truth, and Christ's teachings were critical of the established order. Christ's omnipotence was also believed to be demonstrable by those who had survived a debilitating or deadly disease.

The rise and consolidation of Christianity may have also been affected by disease. The expectations of the poor Romans were that with Christ's second coming they would be freed from their rich masters. Christianity, unlike paganism, preached care of the sick as a recognized religious duty. Those who were nursed back to health felt gratitude and commitment to the faith, and this served to strengthen Christian churches at a time when other institutions were failing. Another positive feature of Christianity was that the teaching of the faith made life meaningful even in the face of sudden death since it was perceived as a release from an individual's suffering. The capacity of Christian doctrine to cope with the psychic shock of epidemic disease made it attractive for the populations of the Roman Empire. Paganism, on the other hand, was less effective in dealing with the randomness of death. In time the Romans came to accept the Christian view. Rome became the headquarters of Christianity, and in A.D. 337, with the conversion of Emperor Constantine, Christianity became the church of the empire.

The Antonine Plague

The Roman Empire expanded its frontiers until A.D. 161. But from that time onward its defenses began to crumble. Toward the end of the 1st century A.D. a warlike people riding on horseback from Mongolia, the Huns, began a westward movement. The Huns brought to the Roman Empire new infections, but others such as the Roman fever also rebuffed them. Marcus Aurelius Antoninus was born in A.D. 121, and throughout his reign as emperor (140-180) he was engaged in defensive wars on the northern and eastern borders of the Roman Empire. In A.D. 164 the Roman legions under the command of Avidus Claudius were sent to Mesopotamia to repel an invasion by the Parthians, and in this they succeeded. But the troops returned with a devastating plague that spread throughout the countryside and reached Rome by 166. This epidemic of Antoninus, the Antonine plague, spread to other parts of Europe, causing so many deaths that cartloads of bodies were removed from Rome and other cities. Within the city of Rome, Emperor Antoninus made administrative reforms; he concerned himself with famine as well as plague in the empire. A devotee of stoicism, he ruthlessly persecuted the Christians, believing them to be a threat to imperial power. In 161, when the Huns had reached the northeast border of Italy, Antoninus was forced to contemplate battle, but fear and disorganization delayed a direct confrontation with the Germanic tribes on the Rhine-Danube frontier until 169. When he and his legions moved into the northern frontier with the objective of securing the empire's northwesterly boundaries (as far as the Vistula River), a plague broke out among the troops; it raged until 180 and affected not only the Roman legions but also the Huns. As the plague ravaged his army, Antoninus elected to retreat to Rome but was never to reach his destination. In Vienna, on the seventh day of his illness, May 17, 180, he died from this plague. The plague returned again in 189, and though it was less widespread than the first epidemic, at its peak there were more than 2,000 deaths a day in the city of Rome.

The Antonine plague (A.D. 164 to 189) is also associated with the physician Galen of Pergamum (A.D. 129-216), whose ideas dominated medicine until the 16th century. Galen's hero was Hippocrates (460-377 B.C.), who had laid down the principles of medicine in Greece. Galen was first appointed surgeon to the gladiators in Asia Minor and then moved to Rome, where he practiced medicine. Though Galen was a skilled anatomist, an experimentalist, and a searcher for new drugs, when faced with the plague he fled Rome. He was, however, recalled by Emperor Marcus Aurelius to Rome, where he died. But before his death he left a description of the plague's symptoms: high fever, inflammation of the mouth and throat, thirst, diarrhea, and a telltale sign: pustules on the skin that appeared after 9 days. Even today precisely what this plague was remains a mystery, but most historians suspect that this was the first record of a smallpox epidemic. Some believe either that smallpox moved into the Roman Empire with the legions returning from Mesopotamia or else that the Huns carried it with them from Mongolia and then on to Rome.

The Cyprian Plague

In A.D. 250, Cyprian, the Christian archbishop of Carthage, described a disease that appeared to have originated in Ethiopia, then moved into Egypt, and eventually came to the Roman colonies of North Africa: vomiting, diarrhea, gangrene of the hands and feet, a burning fever, and a sore throat. The Cyprian plague became a pandemic, advancing quickly through direct person-to-person contact as well as by contaminated clothing. It is suspected that the disease was either smallpox or measles. Mortality is said to have been high, with the number of deaths exceeding those who survived. The plague of Cyprian lasted 16 years, causing panic among the people; those who fled to the surrounding countryside served as "seeds" for initiating fresh outbreaks. The land they left lay fallow. Despite this, the Roman Empire survived the devastation wrought by the Cyprian plague and was even able to overcome subsequent invasions by the Huns. But by 275 the Roman legions were forced to retreat from the Danube and the Rhine to the city of Rome. The situation was so precarious that the emperor decided to fortify Rome itself to protect against this plague. This also proved to be ineffectual.

The plague of Cyprian strengthened Christianity. Cyprian wrote:

> Many of us are dying in this mortality, that is many of us are being free from the world. This mortality is a bane to Jews and pagans and enemies of Christ; to the servants of God it is a salutary departure. … Without any discrimination the just are dying with the unjust. … The just are called to refreshment, the unjust are carried off to torture; protection is more quickly given to the faithful; the punishment to the faithless. … This plague and pestilence which seems horrible and deadly, searches out the justice of each and every one.

This ability of Christianity to deal with the horrors and hardships of a plague made church doctrine an attractive alternative to the stoic and pagan philosophies, which were impersonal, uncompassionate, and ineffectual in explaining the randomness of death due to disease, and so served to strengthen its hold on the Roman peoples. The attraction of Christianity for the people of Rome not only altered their current religious and cultural practices but also influenced future social and political development.

St. Sebastian, often painted as a naked youth wearing a crown, tied to a tree and with his body pierced by arrows, is the patron saint of archers,

Figure 3.4 St. Sebastian in a painting by Andrea Mategna (1490) in Ca d' Oro, Venice , Courtesy of Wikiart.org

athletes, and soldiers as well as the protector from plague (Fig. 3.4). His answers to prayers for protection from the plague, first in Rome and later in Milan (1575) and Lisbon (1599), were the cause for his elevation to sainthood. Documentation and myth along with the fate of St. Sebastian have been woven together in *The Golden Legend*, a text from the Middle Ages written in 1275 by the archbishop of Genoa, Jacobus de Voragine (1229-1298). It is written that Sebastian was born of a wealthy family in Narbonne in 257 and was educated in Milan. He became an officer in the Imperial Roman Army during the time of Emperor Diocletian (284-305) and was secretly converted to Christianity. Diocletian's name is often associated with the last and the most terrible persecutions of the early Christian Church. During Diocletian's persecutions, Sebastian visited his fellow Christians in the prisons, giving them food and comfort. But in 286, when Diocletian learned of his works in converting others to the faith, he ordered Sebastian to be shot to death by his archers. Sebastian was tied to a tree, shot with arrows, and left for dead. The arrows did not kill him, however, and a Christian widow, Irene, tended his wounds. Upon his recovery he continued to preach, and after he confronted the Emperor Diocletian to denounce his cruelty, the enraged emperor ordered that he be beaten to death by the blows of a club. The destructive effects of the plague (particularly bubonic plague) caused people to compare their being struck down by death to an attack by an army of archers, and so they prayed for salvation from a divine being; it is claimed that their prayers were answered by St. Sebastian. So, in time, Sebastian the saint of archers became the people's protector from the plague.

Over the next 3 centuries Rome slowly collapsed under pressure from the Germanic tribes (Goths and Vandals) as well as recurrent outbreaks of mysterious plagues such as that of Antonine, Cyprian, and others. Gradually, the Four Horsemen of the Apocolypse—Famine, Disease, War, and Death—led to a disintegration of the Roman Empire. And when the Germanic peoples moved into Italy and Gaul, crossed the Pyrenees into Spain, and even reached North Africa, they too became subject to this plague; by 480 the Vandals themselves were so sickly that they were unable to resist invasion by the Moors.

The Justinian Plague

The historian Procopius of Caesarea wrote in *The Persian Wars* that "it embraced the entire world and blighted the lives of all men." He described the symptoms of the disease: great swellings, or buboes, appeared in the groin and armpit as the lymph nodes enlarged; there was delirium and frantic restlessness; and some became comatose. There was general hysteria and panic. If the buboes were lanced, there was the possibility of survival, but most died in 5 days. It raged in the city of Constantinople (present-day Istanbul), the capital of the Roman Empire in the East, for months, and so numerous were the corpses that they were cast into hollow towers of some incomplete fortifications." Procopius' eyewitness account claimed that 300,000 died, but this is probably an exaggeration. There is, however, no question that all of Europe suffered from this plague, and 5 years later it had reached Ireland, and by that time it had killed close to a million people.

This disease arrived in A.D. 541 and raged intermittently in Europe, North Africa, and the Middle East until 757. Called the plague of Justinian, it was probably bubonic plague that came to the Mediterranean from an original focus in northeast India or via central Africa. From Lower Egypt the disease reached the harbor town of Pelusium in 540, from where it spread to Alexandria and then by ship to Constantinople, the capital of Justinian's empire. It is speculated that Justinian's General Belisarius was unable to accomplish the goal of Justinian, that is, the reestablishment of the Roman Empire in all its glory, because of the outbreak of plague.

The plague-induced losses in the population reduced the taxes and services available to the government. Lacking in manpower and money, Justinian was unable to send aid to his beleaguered troops. The survivors of the plague were damaged both physically and mentally and had diminished self-confidence. As disasters continued to befall the empire, the Roman and Persian forces were unable to offer more than token resistance to the Moslem armies that swarmed out of Arabia in 634.

Bubonic plague recurred every 3 or 4 years for decades, and its effects lasted well into the 7th century. It is estimated that by the year 600 the plague of Justinian—the first pandemic of bubonic plague—had reduced the population by 100 million people, almost 50% of the population of Western Europe. McNeill has observed that the plague of Justinian resulted in "the perceptible shift away from the Mediterranean as the preeminent center of European civilization and the increase in the importance of more northerly lands." The Justinian plague and the subsequent outbreaks of plague over 2 more centuries marked the end of the classical world—the Greek and Roman civilizations—and ushered in the so-called Dark Ages. Plague so diminished trade in the Mediterranean that it left most countries with only a bartering economy, cities withered, feudalism grew, religion became more fatalistic, and Europe turned inward upon itself.

doi: 10.1128/9781683670018.ch4

4
An Ancient Plague, the Black Death

During the last 2,500 years, three great plague pandemics have resulted in social and economic upheavals unmatched by armed conflicts or any other infectious disease. In Constantinople, the capital of the Roman Empire in the East, it was the first plague pandemic (A.D. 542-543) that surely contributed to Justinian's failure to restore imperial unity. In the year 1346 the second pandemic began, and by the time it disappeared in 1353, the population of Europe and the Middle East had been reduced from 100 million to 80 million people (Fig. 4.1). This devastating pandemic, known as the Black Death, the Great Dying, or the Great Pestilence, put an end to the rise in the human population that had begun in 5000 B.C., and it took more than 150 years for the population to return to its former size. Some believe this catastrophic crash in population to be Malthus's prophecy come true, while others, such as the historian David Herlihy, consider the Black Death to be not a catastrophe promoted by "positive checks" (i.e., disease, war, and famine) but an exogenous factor that served to break a Malthusian stalemate. That is, despite fluctuations in population size, relatively stable population levels were maintained over prolonged periods of time due to "preventive checks" (i.e., changes in inheritance practices, delay in the age of marriage, and birth controls). The Black Death did more than break the Malthusian stalemate; it allowed Europeans to restructure their society along very different paths.

Although those living in the medieval period recognized that plague was a contagious disease spread from person to person, its cause was not identified. Indeed, most believed it to be "a vicious property of the air" itself. The Black Death is most associated with Florence, one of the great cities of Europe at the time, and because it felt the full impact of the epidemic, it is sometimes called the Plague of Florence. Giovanni Boccaccio (1313-1375), who lived in Florence during the plague, described what he witnessed:

Figure 4.1 (Left) *The Plague* by Felix Jenewein (1900) shows a mother carrying a coffin with her child (Courtesy Wellcome Library, London, CC-BY 4.0) towards the beginning of spring ... the doleful effects of the pestilence began to be horribly apparent by symptoms ... an issue of blood from the nose was a manifest sign of inevitable death; but in men and women alike it first betrayed itself by the emergence of certain tumors in the groin or armpits, some of which grew as large as a common apple, others as an egg ... from the two said parts of the body it soon began to propagate and spread itself in all directions; after which the form changed, black spots ... making their appearance in many cases on the arm or thigh or elsewhere, now few and large, then minute and numerous ... almost all within three days from the appearance of said symptoms, sooner or later, died ... such as were left alive inclining almost all of them to shun and abhor all contact with the sick and all that belonged to them, thinking thereby to make their own health secure ... they banded together and formed communities in houses where there were no sick, and lived a separate and secluded life, which they regulated with the utmost care ... eating and drinking moderately of the finest wines, holding converse with none but one another ... diverting their minds with music and, other delights as they could devise.

The contagious nature of plague led to the belief that the only way security could be achieved was in total isolation of the sick. But despite its being clearly recognized that the enemy was the sick, it was the lack of appreciation that microbes were the cause of infectious diseases that led to the institution of crude and generally ineffectual public health measures. In 1374 the Venetian Republic required that all ships and their crew, passengers, and cargo had to remain on board for 40 days while tied up at the dock; this gave rise to the term "quarantine" (from the Italian word *quaranta*, meaning "forty"). Soon all European ports restricted entry by quarantine, but the disease continued unabated. "Cordons sanitaires" (literally rings of armed soldiers ordered to guard against the fugitives of disease) restricted the movement of people and may have reduced the spread of plague, but oftentimes the infected individuals were shut up in their homes with the uninfected members of the family, leading to higher mortality. More-effective measures included the burning of clothing and bedding and the burying of the dead in shallow unmarked graves sprinkled with lye. The public, unable to identify the real source of the plague, used "outsiders" as scapegoats. The Black Death also led to societal and religious changes: feudal structure began to break down, the laboring class became more mobile, merchants and craftsmen became more powerful, and guild structures were strengthened. There was also a decline in papal authority, and people lost faith in a Christian church that was powerless to stem the tide of death. The horrors of the plague during this time are depicted in the 1562 painting *Triumph of Death* (Fig. 4.2). by Pieter Bruegel (1525-1569) and are graphically described in the introduction to Boccaccio's classic collection of short stories *The Decameron*

Figure 4.2 *Triumph of Death* by Pieter Brueghel (1562), Courtesy Wellcome Library, London, CC-BY 4.0

and in Albert Camus' book *La Peste* (*The Plague*). It is even depicted in popular movies such as *The Seventh Seal* by Ingmar Bergman, *Monty Python and the Holy Grail*, and the movie based on the book *Arrowsmith* by Sinclair Lewis.

Though the Black Death was undoubtedly the most dramatic outbreak of plague ever visited upon Europe, it did not disappear altogether. Between 1347 and 1722 plague epidemics occurred in Europe at infrequent intervals. In England the epidemics occurred at 2- to 5-year intervals between 1361 and 1480. In 1656-1657, 60% of the population of Genoa died, half of Milan in 1630, and 30% in Marseilles in 1720. In the Great Plague of 1665, described in the diary of Samuel Pepys (and fictionalized in Daniel Defoe's *Journal of the Plague Year*), at least 70,000 Londoners died (out of a population of 450,000).

The third plague pandemic began in the 1860s in the war-torn Yunnan region of China. Troop movements from the war in that area allowed it to spread to the southern coast of China. Plague-infected rodents, now assisted by modern steamships and railways, quickly spread the disease to the rest of the world. It is estimated that in the three pandemics plague killed more than 200 million people.

A Look Back

At the beginning of the 12th century the European population grew quickly after achieving relatively stable numbers during the early Middle Ages. One of the reasons for the unchanging numbers of people was poor harvests and famine, but with the introduction of new crops, windmills, waterwheels, horse collars, and the moldboard plow, agricultural production increased. Money instead of barter began to be used in trade, and with this prosperity new towns grew, and so did the population. But by the 13th century there was ushered in a sustained period of cold winters and rainy summers. Agriculture could not keep up with the population rise, and over the next century famines occurred every few years. The result was poverty and misery, especially in the crowded and filthy cities.

The second pandemic occurred from 1346 to 1353 and probably originated from the steppes of central Asia, where it was epidemic among small mammals such as marmots and gerbils. Plague spread westward along the trade routes to enter Europe. An outbreak occurred along the Volga River, from where it spread west to the Don River and then down to the Black Sea. By 1347 there were outbreaks in the port cities of Kaffa (present-day Feodosia), Constantinople (present-day Istanbul), and Genoa, and as plague continued to move by ship, it spread via North Africa to Spain by 1348. By this time it had already reached central Europe, including France, Germany, Switzerland, Austria, and even as far as Great Britain. From there it traveled to Scandinavia aboard a ship from London that docked in Bergen in 1349, carrying crew, wool cargo, and plague. By 1351 it was in Poland, and when in 1352 it reached Russia, plague had completed its circuit and a deadly noose had been secured around Europe. In its path the Black Death killed an estimated 17 million to 28 million Europeans, representing ~30 to 40% of the population at the time.

In 1358, Agnolo di Tura, a citizen of Siena, described the situation:

> Father abandoned child; wife, husband; one brother, another; for this illness seemed to strike through the breath and sight. And so they died. And no one could be found to bury the dead for money or for friendship … and in many places in Siena great pits were dug and piled deep with huge heaps of dead. … And I, Agnolo di Tura, called the Fat, buried my five children with my own hands and so did many others likewise. And there were also so many dead throughout the city who were so sparsely covered with earth that the dogs dragged them forth and devoured their bodies.

The short-term effect of plague was shock and fear. Panic and fright broke the continuity of the existing economic structure and disrupted the routines of work and service. People were terrified, and they deserted the cities and towns. In consequence, communities were thrown into chaos. But the Black Death also had its positive aspects: it contributed to technological advances through the invention of labor-saving devices, as well as a reevaluation of Galenic medicine, and with so many dead the value of labor rose sharply, resulting in greater prosperity among the working survivors.

Public Health

More than 2,000 years ago—long before there was Pasteur's germ theory of disease—the Greek physician Hippocrates of Cos warned that for certain diseases (particularly plague) it was "dangerous to associate with those afflicted." And in A.D. 549, during an outbreak of plague in Constantinople, the emperor Justinian enacted laws calling for the delay and isolation of travelers coming from regions where there was evidence of this disease. Therefore, in spite of a lack of appreciation that microbes could cause disease, those living in medieval Europe (and dying from the Black Death) recognized that the disease to which they were exposed was contagious and that the only way to preserve the public health would be to isolate the sick totally. When plague reached Europe in 1347, ports on the Adriatic and Mediterranean Seas were the first to deny entry to ships coming from pestilential areas, especially Turkey, the Middle East, and Africa. As early as 1348, Florence, one of the most plague-ridden cities, issued a restriction called quarantine on travelers and goods. The Venetian republic formally excluded infected and suspected ships in 1348, and in 1377 the first official quarantine station was established in Dubrovnik. There persons and ships were isolated on a nearby island for 30 days to await signs of illness or continued good health. Later, other port cities established quarantine stations on shore or on neighboring islands. But even with quarantine, and with 90% of the passengers dying aboard ship, the populations of the port cities were decimated by plague. Quarantine was sometimes employed within the confines of the city itself, and oftentimes the sick were shut up in their homes with the uninfected members of the family. Under these conditions, household quarantine did not prevent the spread of the disease; rather, it resulted in higher mortality.

Plague undermined confidence in local church leaders, and many people left on pilgrimages to seek salvation elsewhere. Pilgrimages increased the number of travelers on the highways, and these strangers provoked hostility; at the same time they were also the "seeds" for spreading the infection. To restrict the movements of such strangers, cordons sanitaires were established. Towns closed their gates to travelers. These measures circumscribed plague to certain towns while protecting others. Restricting the long-distance movement of people and baggage from infected cities may also have helped to reduce the spread of the disease.

Discrimination

Restrictions such as cordons sanitaires and quarantine frequently were expanded to limit personal freedom, and to identify the culprits. The medieval public wanted to know who was to blame. They found in their community likely sources—strangers, lepers, beggars, the poor, prostitutes, and the Jews. Plague aided in the spread of anti-Semitism, a practice already begun when Jews were slaughtered beginning with the First Crusade in 1096. In some Muslim countries it was the Christians who were blamed; in both Christian and Muslim countries, however, the Jews were blamed. Although Jews, as the largest cultural minority, had been free to practice their religion by both Roman and canon law, once plague claimed the lives of the political leaders of a town, such freedom was soon abrogated. Early in 1348 the rumor arose that Jews were spreading the disease by poisoning the Christian wells. Though this appeared absurd to the kings and popes, and Pope Clement VI of Avignon (France) tried to discredit the charge by issuing a bull calling the accusations "unthinkable," Jews continued to be blamed and persecuted. By the fall of 1348 formal accusations were brought against Jews, and when court officials extracted confessions, the rumors began to be believed. "Well poisoning" by Jews spread from Avignon to Strasbourg and then to all of Germany and Poland. Riots occurred in Strasbourg, and when the members of the local government tried to protect the Jews, they were thrown out of office. With a new government in place, more than 900 Jews, about half the entire community, were rounded up, and on February 14, 1349, they were burned on the grounds of the Jewish cemetery. At Freiburg all Jews were placed in a large wooden building and burned to death. In northern Italy legislation was enacted to identify the suspect groups—they chose a color, yellow—and Jews were required to wear a yellow Star of David on their clothing; prostitutes were also forced to have yellow labels. (It is ironic that yellow was traditionally used to identify lepers and criminals.) And because Jews were prohibited from owning land, they had become merchants, bankers, financiers, moneylenders, and pawnbrokers. Although they had the knowledge for mercantile trade, they were still discriminated against, especially in Germany and France. Killing of the Jewish moneylenders and bankers had another benefit—no loans to repay. The king of Poland, Casimir, had a Jewish mistress and needed the expertise of the Jews, and so they were invited to settle in Poland; as a consequence, in western Russia and Poland, Jews became a significant part of the population after the Black Death. (Their numbers would once again be reduced 3 centuries later by pogroms and gas chambers, latter-day instruments of anti-Semitism).

Church

In cities of 50,000 people, more than 500 died each day. Priests who gave last rites had a very high mortality rate, and there was a loss of faith in the clergy because they seemed so powerless to prevent death or the spread of death from disease. There was also a decline in papal authority. The Church passed the responsibility for plague on to God, suggesting that this was Judgment Day, that people had sinned and so nothing could be done to reduce the suffering. All of the monks of a monastery near Avignon and another near Marseilles succumbed.

Fear of plague also led to a greater consciousness of religion, especially the "magical religion" embodied in the cults of healer saints. These patron saints, who not only knew suffering but also had the power to heal the sick and provide comfort, usurped the veneration of God alone and set the stage for long and divisive debates over the nature of religion. One of these healer saints was St. Sebastian, and another was St. Roch (Fig. 4.3). Although it is not known whether the life of St. Roch is truth or fiction, his story comes to us in written form in 1414 from the Venetian humanist and scholar Francisco Diedo. Roch was born in Montpellier, France, in 1295, and he is said to have had a birthmark in the form of a red cross on his breast. As a young man he gave to the poor and embarked on pilgrimages to Rome. Wherever he traveled there was plague, and he was able to cure the afflicted by placing his hands on the buboes. He eventually contracted the disease,

Figure 4.3 St. Roch, the patron saint of those suffering from plague. The original hangs in the Galleria Dell' Academia, Venice, Italy, Courtesy Wikipedia.com

was expelled from Montpellier, and died in 1327. St. Roch is often depicted as a pilgrim carrying a purse and a staff and pointing to a bubo on his inner thigh.

Another threat to the Church was a pilgrim movement of another sort. The Brethren of the Flagellants began in eastern Europe, but its strongest bases were in Germany and France. The flagellants, sometimes as many as a thousand, marched in procession through the town; they were led by a master carrying a banner of purple velvet and were dressed in cloth of gold. Masked and dressed in dark clothing emblazoned with a red cross, each flagellant carried a whip made of leather thongs and tipped with metal studs, with which they beat their backs and chests. (This is graphically shown in Goya's painting of the flagellants as well as in Ingmar Bergman's movie *The Seventh Seal*.) The flagellants were a counterculture to the Church, and they claimed divine authorization for their mission, the alleviation of the plague. Pope Clement VI initially encouraged the flagellant movement, and up until 1349 the flagellants had their way in recruiting other pilgrims, but when the pope saw he could not control it, he issued a bull denouncing the movement and its practices. Eventually the movement ceased for reasons not fully understood. In its time, however, the flagellant movement did some good: it brought about a spiritual revival, sinners confessed and robbers returned stolen property, hope was raised (albeit temporarily), and it provided a theatrical diversion. But in the final analysis, the movement did more harm than good. Jews became the special victims of the flagellants, and their persecutions were the forerunners of the pogroms. In Frankfurt in 1349 the flagellants rushed into the Jewish quarter and incited the people to engage in wholesale slaughter. In Brussels the mere announcement of their arrival triggered a massacre of 600 Jews. Death from the Black Death itself coupled with that due to virulent anti-Semitism virtually wiped out Jewish communities in many parts of Europe and also led to permanent shifts in their populations to Poland and Lithuania.

Medicine

Medicine was also affected. Medieval society had four kinds of medical practitioners: academic physicians, who knew theory but did not care for the sick; surgeons, who learned their trade as apprentices and who were the principal caregivers of the sick; barbers, who did bloodletting and minor surgery; and those who practiced folk medicine, mostly women. Academic physicians were generally older men who relied on the teachings of Galen, who believed, as did Hippocrates before him, that disease was a result of an imbalance in the four humors of the body (blood, phlegm, and black and yellow bile). But since plague appeared to have little to do with humoral changes, confidence in the academic physicians diminished. When these holders of the chairs of medicine at the great European universities died, the newer and younger appointees could move into other clinical areas, such as anatomy. Surgeons, who wore a costume with a beak containing perfume or spices, a cloak of waxy leather, eye lenses, and a wand with incense (Fig. 4.4), and who cared for the sick, died at higher rates than did the other medical practitioners. Because of this, their role in curing disease was little valued. Indeed, the stench of death was so great during the plague years that to "purify" the air, the perfume Eau de Cologne was invented in Germany and named after the city of Cologne. Today, the perfume is known as "4711," the address of the household where it was first made. New prestige fell to the barbers, and bloodletting and surgery (Fig. 4.5) became an integral part of their practice rather than barbering alone. This also led to an emphasis on studies of human anatomy in health and disease, and the Galenic system, which had no clear theory of contagion, declined in importance. Slowly, very slowly, change occurred both in the thought and in the practice of medicine, and it was the Black Death that instigated that change.

One of the first physicians to advance a theory of contagion was Giovanni Fracastoro (1483-1533), in his book *On Contagion and Contagious Diseases*. Infectious disease, Fracastoro wrote, could be transmitted by semenaria ("germs") in three

Figure 4.4 Dr. Pestis, the plague doctor in costume, Courtesy Wikipedia.com

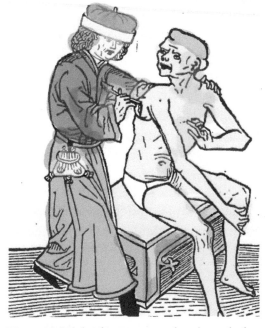

Figure 4.5 A barber-surgeon lancing a bubo. Woodcut

ways: by direct contact, through carriers such as dirty linen, and through airborne transmission. His theory was put into practice when as physician to Pope Paul III he recommended the transfer of the Council of Trent from Trent to Bologna as a response to plague. Other physicians, however, did not subscribe to Fracastoro's theory, and soon the practices were displaced by misguided suggestions until they were revived in the 19th century by Louis Pasteur, Robert Koch, and their associates.

Education

As the death toll from the plague increased, the numbers of learned individuals decreased. This affected the universities, where lawyers, physicians, and clerics were trained. As the numbers of university students declined, so too did the number of universities. Before 1348 all of Europe had 30 universities, but by the time the plague ended 5 of these had been wiped out completely. The institution of cordons sanitaires as well as other restrictions on travel prevented students from enrolling at distant universities, and so local universities were established. Cambridge University in England acquired four new colleges, each founded by a bequest of a rich and pious patron, and all benefiting poor local scholars and clerics. Similarly, new universities were established throughout Europe in cities such as Vienna, Prague, and Heidelberg, so that it no longer became necessary to travel to Bologna or Paris for an education. This not only diminished the dominance of certain centers of learning but also led to curricular reform, and instruction began to be carried out in the vernacular tongue.

Economy and Social Order

The immediate effect of the Black Death was paralysis. Trade ceased. The economic effect of the Black Death was inflation and a sharp rise in the cost of food. The Hundred Years' War between England and France was halted by a truce in 1349 because of the lack of able-bodied men. In depopulated Europe, even soldiers received higher wages. Another innovation to compensate for depopulated armies was the development of improved firearms; fewer soldiers, but with better and more-destructive weapons, could then successfully engage the enemy in battle.

The nature of farming was altered, too. In medieval times, farming involved serfs working the lands of their lords. The serfs and their families occupied a small village consisting of several small, whitewashed, two-room huts made of reeds and clay-like sod with a thatched roof. In one room lived the family, and in the neighboring room lived their animals. The floor was dirt, and it might be covered with leaves or rushes, and in the center were flat stones on which the fire was placed. There was no chimney or windows, and the smoke of the fire escaped through a hole in the roof or through the open door. Furniture, such as it was, consisted of a table, a few stools, a storage chest, and wooden pallets that served as beds. Adjacent to each hut was a half acre of land used by the serf and his family for the garden, chicken coop, and pigpen; the nearby stream had ducks and geese. Manure and night soil (human feces) were used as fertilizer, and plow animals such as horses or oxen were shared among several villages.

Medieval villages were organized around large fields that were used for growing grains such as wheat or barley or oats. The land was cultivated by the serf and his family but owned by the lord of the land (landlord), who usually lived in a large, fortified manor house. The serfs labored for the landlord, and their duties were mostly unspecified and subject to the whims of the landlord. The landlord also had almost complete legal power over these poor and illiterate peasants. This feudal system of serfs and landlords declined coincident with the substitution of money for manual services. In effect, salary began to replace labor duties, and rents were paid to the landlord. Since during the Black Death so many peasants died, if the lord wanted to farm his land, he had to either obtain more laborers (usually at a dearer cost) or rent out his land to the survivors from neighboring villages or from the now-deserted towns. The nobility, in league with the landlords, tried to enforce service without payment of the laborers, and over time this engendered great hostility among them as their numbers continued to decline due to disease. In England, when the crown instituted a series of poll taxes in 1381, there was an uprising (called the Peasants' Revolt). The revolt failed and repressive measures were instituted, but in the end the landlords had to capitulate because they recognized that without a labor force they would receive no income from the land. The landlords devised another scheme: they would continue to own the land, but now estate managers or stewards were hired to manage the fields and collect the rents. The stewards now lived in a grand house, and the rent-paying peasants now were tenant farmers on the lord's land. But as disease and death continued to deplete the available labor pool, the tenant farmers had to recruit and pay other peasants or landless city folk. In this way, as the landlord's fields came to be worked by non-landowning and rent-paying tenant farmers, the feudal system slowly began to change in character. To compensate for their loss in income (due to higher wages), the landlords continued to acquire more and more land and the tenant farmers in turn began to use less labor-intensive farming practices. This they could accomplish by the invention of the moldboard plow and through the conversion of farmland into pastureland. In England, in particular, pasturelands were devoted to sheep farming, which became so profitable that in some regions growing of crops was completely replaced by the raising of sheep for wool.

Mills that once were used for grinding wheat and barley could now be diverted to the spinning of cloth, operating of the bellows of furnaces, and sawing of wood. In England sheep husbandry exceeded all other crops and wool became the basis of prosperity. But the tenant farmers were continually besieged by ever-increasing rents, as well as by taxes imposed by the crown on the landowning lords. This is immortalized in the nursery rhyme "Baa Baa Black Sheep":

> Baa baa black sheep, have you any wool?
> Yes, merry, have I, three bags full:
> One for my master, one for my dame,
> And one for the little boy that lives in the lane.

The hardworking peasant had to give one-third of his income to the king, his "master," and another third to the landlord (his "dame"); he was left with only one-third for himself ("the little boy").

The tenant farmers who had raised sheep for wool became richer and more powerful than their predecessors. By taking advantage of the anarchy caused by the land-grabbing and warring aristocracy (culminating in the Wars of the Roses, 1455-1485), the sheep raisers were able to buy up larger and larger parcels of land from the estates of the ruined lords, becoming in the process wool barons. The smaller labor force in the cities led to their having a stronger negotiating position, and so there was a need to pay higher wages; as a consequence, there was an improvement in the standard of living. Many individuals, however, fled the cities and moved to the country, and there the survivors were induced by landowners to harvest the crops by the provision of higher wages. The standard of living among the urban and rural populations improved, and the laboring classes became more mobile. The Black Death was thus a contributing factor in altering the social and economic structure in England, and by the 16th century landlord and serf ceased to exist, although feudalism did linger for several more centuries in continental Europe.

Governments that sought to control wages sowed the seeds of discontent both in the city and on the farm. This led to what is called factor substitution—cheap land and capital were substituted for the more expensive labor. Peasants would not accept a lease unless the landlord provided additional capital in the form of oxen and seeds. In the city, tools and machines (capital) were substituted for human labor. As plague continued to deplete the numbers of guild members as well as the pool from which new members could be drawn, the guilds had to take on new apprentices without family connections to a particular trade. Shorter years of apprenticeship, rapid turnover in guild members, and increased recruitment of new members led to a decline in the quality of the product. Sometimes the higher labor costs led to technological innovation in the form of labor-saving devices. One such was the invention of movable type and the printing press. Scribes who had been employed to copy sheets of manuscript and then to assemble these into a bound volume could not keep up with the demand for their services with an emerging and increasingly more literate population; a way out of this labor-intensive method of bookmaking was to find a cheaper solution. Johannes Gutenberg's invention in 1453 of printing using movable metal type was one way that this was accomplished.

Another was to employ an economy of scale in sea- and oceangoing transport vessels. Bigger ships with smaller crews could remain at sea for longer periods of time and would be able to sail directly from port to port, but this would require better ship construction, improvements in navigational instruments, and new business enterprises such as maritime insurance to protect the investment in cargo and the ship. As a consequence, merchants such as bankers and craftsmen became more powerful. The new economy became more diversified, there was a more intensive use of capital, technological innovations became more and more important, and there was a greater redistribution of wealth. In time the aristocracy found it had to yield power to the masses. The social and economic fabric of Europe began to be altered, and it was the Black Death that instigated such change.

Finding the Killer

As plague raged through medieval Europe, it became increasingly apparent that this disease was contagious. Even if Fracastoro's idea of "seeds of contagion" was accepted, however, at this time there would still be no means to identify precisely the causative agent. Identifying the "seed" would not only require a technological innovation; it would also require a change in the concept of infectious (contagious) diseases. Three centuries after Fracastoro's theory of contagion, the concept that disease could result from the invasion of the body by microbes or germs—the germ theory of disease—was established. And with a 17th-century technological innovation, the microscope, it was actually possible for humans to see germs! Two schools of thought, one in France under the leadership of Louis Pasteur (1882-1895) and the other in Germany, led by Robert Koch (1843-1910), were responsible for firmly grounding the germ theory, and for all of their lives these two microbe hunters remained fierce competitors (see p. 418).

As plague ravaged China during the third pandemic, Pasteur dispatched Alexandre Yersin (1863-1943), a Swiss-born member of the French medical colonial corps, to Hong Kong to study and attempt to isolate the germ of plague. Yersin arrived when the epidemic was in full force and was given a small table in a dark corridor next to a patients' room, where he could leave his microscope, notebook, stains, pipettes, and a few cages with guinea pigs, mice, and rats. At first Yersin was not permitted access to the morgue, but through connivance and bribery he was finally able to visit the morgue for a few minutes to examine a sailor who had just died

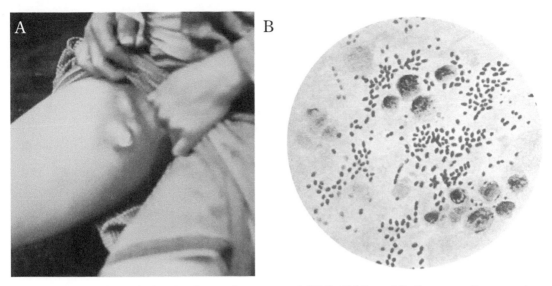

Figure 4.6 A. Bubo of bubonic plague (courtesy of CDC, 1993) and B. the causative agent, *Yersinia pestis,* stained and seen with light microscope (courtesy CDC)

from plague. Yersin punctured the bubo on the dead soldier's thigh (Fig. 4.6A) with a sterile needle and removed some fluid; he then examined the fluid under the microscope, inoculated a few guinea pigs, and sent the remainder to the Pasteur Institute in Paris. On June 24, 1894, he wrote Pasteur that the fluid was full of rod-shaped bacilli (Fig. 4.6B) that stained poorly with Gram stain; i.e., they were Gram negative. (Gram stain consists of a tincture of crystal violet. After bacteria are killed and washed, some lose the dye whereas others retain the purple color; the former are called Gram-negative and the latter Gram-positive bacilli.) Yersin also wrote: "without question this is the microbe of plague." A few days later Yersin found that the guinea pigs he had injected with the fluid that he had aspirated from the bubo had died, and that their bodies swarmed with the same bacilli. Yersin was intrigued by the large number of dead rats in the streets of Hong Kong, as well as in the hospital corridors and the morgue. When he examined some of these dead rats, he found the same bacilli to be present. He correctly concluded that plague infects both rats and humans. At about the same time, Koch, convinced that plague was caused by a microbe, sent his associate Shibasaburo Kitasato and a large number of assistants as well as abundant equipment to find the plague germ. Kitasato did culture a bacterium from the finger of a sailor who had died from plague, but it was Gram positive. Further, Kitasato was never able to prove that this bacillus could produce plague in humans or other animals. Bubonic plague, caused by a rod-shaped, Gram-negative bacterium, was named *Yersinia pestis* after its discoverer, Alexandre Yersin (although *Pasteurella pestis* was the name used before 1970).

Finding the Vector

Yersin discovered the plague microbe, but he did not find the means whereby the disease could be transmitted. He asked himself: was it airborne, or did it have to do with rat feces or dried rat urine? The answer to the question came from Paul-Louis Simond (1858-1947), a French army physician who was sent by Pasteur to Vietnam and India to follow up on Yersin's observations. Simond noted that not only were there large numbers of dead and dying rats in the streets and buildings, but that 20 laborers in a wool factory who had been cleaning the floor of dead rats had died of plague, while none of the other factory workers who had no contact with rats became ill. He began to suspect that there must be an intermediary between a dead rat and a human, and that the intermediary might be the rat flea *Xenopsylla cheopis* (Fig. 4.7). Simond found that healthy rats groomed themselves and had few fleas, but that sick rats unable to groom their fur had many; and when the rats died, the fleas moved off onto other healthy rats

Figure 4.7 Flea as seen with the scanning electron microscope. Courtesy CDC/Janice Haney Carr

or onto humans. To prove this he did an experiment: a sick rat was placed at the bottom of a jar, and above this he suspended a healthy rat in a wire mesh cage; although the healthy rat had no direct contact with the plague-infected one, it did become infected by exposure to its fleas (which he determined could jump as high as 4 in. without any difficulty). As a control Simond placed a sick rat without fleas together with healthy rats in a jar. None of the healthy ones became sick, but when he introduced fleas into the jar, the healthy rats developed plague and died. On June 2, 1898, he wrote Pasteur that the problem of plague transmission had been solved.

The most common scenario for the origins of the Black Death, i.e., the second pandemic, is that its source was microbes left over from the first pandemic (the Justinian plague) that had moved eastward and remained endemic for 7 centuries in voles, marmots, gerbils, and the highly susceptible black rats (*Rattus rattus*) of the arid plateau of central Asia (roughly corresponding today to Turkestan). Plague-infected rats moved westward along the caravan routes between Asia and the Mediterranean known as the Silk Road, and in this way plague traveled from central Asia, around the Caspian Sea, to the Crimea. There the rats boarded ships and moved from port to port and country to country, spreading plague to the human populations living in filthy, rat-infested cities. This has been described poetically in Robert Browning's poem "The Pied Piper," which is based on a legend that on June 26, 1284, the German city of Hamelin became infested with rats. A Pied Piper comes to Hamelin and agrees to rid the city of rats for payment of a large sum of money. He is able to enchant the rats by playing his flute, leading them to the river, where they drown. But when the city fathers refuse payment, the Pied Piper leads the children to a cave, where they disappear. Versions of the tale were gathered by the Grimm brothers (1812), and this plague-inspired tale has over time come to have a moral interpretation: evil will befall those who do not carry out their promises.

Simond's hypothesis was that the transmission of plague was from the black rat (*R. rattus*) via the rat flea (*X. cheopis*) to humans. This has been the accepted scenario for decades. Recently, however, this has been called into question. First, there is a complete lack of evidence of the involvement of rats and rat fleas in the historical epidemics in Europe; and second, the speed of transmission of the epidemics was very different: while the medieval epidemics spread extremely rapidly, the modern epidemics spread rather slowly. Although rats may have been important in warmer countries such as China and India, in northern Europe there are no historical references to rats. For example, there is no mention of rats during the plague epidemics in London during the 1600s, and Samuel Pepys, who in his diary described trifling events in great detail, makes no mention of sick or dead rats. It is the view of Hufthammer and Walløe that there were in fact very few rats in northern Europe at the time, and collections of ancient rat bones suggest that they were patchily distributed. It has been proposed that the Black Death cannot be attributed to a singular introduction of *Y. pestis* but to repeated climate-driven reintroductions from a reservoir of gerbil populations in Asia into Europe via caravans. It is hypothesized that since rats were uncommon in medieval Europe it is unlikely that they were responsible for the dissemination of human plague during the second pandemic. Instead, it is suggested that the mode of transmission of *Y. pestis* during the time of the Black Death was not, as shown by Simond, via the black rat and the rat flea *X. cheopis* (during the third pandemic), but rather directly from human to human by the human flea *Pulex irritans* and the human body louse *Pediculus humanus humanus*. Hufthammer and Walløe add that "they were present in all European countries and in sufficient numbers to be real candidates during ancient human epidemics" and that they were present in people's clothing and bedding in the Middle Ages and early modern times. Further, it has been shown experimentally that *P. irritans* is capable of transmitting plague from a dying human plague victim to guinea pigs and rats and that *Y. pestis* can remain in soil in burrows dug by small mammals.

According to this hypothesis, plague was carried over long distances by people in their clothing and in the wool and other goods they transported. It was caravans passing through Asian plague foci that were responsible for transporting plague between Asia and Europe; camels are known to become infected relatively easy from infected fleas in plague foci and can transmit the disease to humans. After a plague infection established itself in a caravan (in its animals, in the traders or in in fleas in its cargo) the disease could have spread to other caravans during the time when goods and animals were redistributed and transported across Eurasian trade routes.

This hypothesis, the authors claim, unlike the rat model, can account for the rapid spread of plague epidemics and also explains why all members of one household in a town might have become plague victims while neighboring households escaped. The maritime import of plague is most evident during the second pandemic in the isolation of ships before unloading their cargo, and the black rat may have played a role in maintaining plague outbreaks on ships as well as importing plague into harbors, but its role as a potential plague reservoir in Europe can be questioned.

DNA sequences obtained from skeletons in Germany spanning the 14th to 17th centuries suggest that plague outbreaks in Europe did not arrive from the East but rather that a virulent strain of *Y. pestis* remained (or was reimported) for several centuries in Europe, while over time it worked its way back to Asia, causing outbreaks and killing millions in 19th-century China, including the one that reappeared in 1994 in epidemic form in countries including Malawi, Mozambique, and India, and that afflicts Madagascar today.

The Disease of Plague

At present, most human cases of the plague are of the bubonic form, which results from the bite of a flea, usually a flea that previously fed on an infected rodent. The bacteria spread to the lymph nodes (armpits and neck but frequently the area of the groin), which drain the site of the bite, and these swollen and tender lymph nodes give the classic sign of bubonic plague, the bubo. Three days after the buboes appear, a high fever develops, the individual becomes delirious, and hemorrhages in the skin result in black splotches. Some contend that these dark spots on the skin gave the disease the name Black Death, whereas others believe "black" is simply a mistranslation of *pestis atra*, meaning not "black" but rather a "terrible" or "deadly" disease.

The buboes continue to enlarge, sometimes reaching the size of a hen's egg, and when these buboes burst, there is agonizing pain. Death can come 2 to 4 days after the onset of symptoms. In some cases, however, the bacteria enter the bloodstream. This second form of the disease, which may occur without the development of buboes, is called septicemic plague. Septicemic plague is characterized by fever, chills, headache, malaise, massive hemorrhaging, and death. Septicemic plague has a higher mortality than does bubonic plague. In still other instances, the bacteria move via the bloodstream to the alveolar spaces of the lungs, leading to a suppurating pneumonia or pneumonic plague. Pneumonic plague is the most feared form of the disease and is the only form of the disease that allows for human-to-human transmission. It is characterized by watery and sometimes bloody sputum containing live bacteria. Coughing and spitting produce airborne droplets laden with the highly infectious bacteria, and by inhalation others may become infected. Pneumonic plague is the rapidly fatal form of the disease, with a fatality rate of 100% and death occurring within 24 h of exposure.

Some have suggested that the nursery rhyme "Ring around the Rosie" refers to bubonic plague in 17th-century England.

> Ring around the rosie,
> A pocket full of posies,
> Achoo! Achoo!
> We all fall down.

The rosies are the pink body rash, posies the perfumed bunches of flowers used to ward off the stench of death, "achoo" is the coughing and sneezing, and death is signified by "we all fall down." The first time this rhyme was suggested to be plague related was in 1961, however, and because there is no evidence of a version prior to 1881, it is unlikely that the rhyme actually relates to the plague.

Y. pestis is one of the most pathogenic bacteria: the lethal dose that will kill 50% of exposed mice is a single bacterium that is injected intravenously. Typically, *Y. pestis* is spread from rodent to rodent by fleas, but it can also survive for a few days in a decaying corpse and may persist in a frozen body.

The reasons for the high degree of pathogenicity of *Y. pestis* remain unclear, however, though some virulence factors have been identified. The genes of *Y. pestis* are located on one chromosome and a virulence plasmid (a circular DNA molecule). Located on this plasmid are a set of genes that are activated by body temperature and millimolar concentrations of calcium, conditions found in the mammalian host. These genes code for the T3SS (type III secretion system), containing an "injectosome," a hypodermic-like structure, and a translocon that forms a pore across the host cell membrane. The T3SS functions to inject multiple toxic proteins, named Yops (*Yersinia* outer proteins), directly into the host cell cytoplasm. Once inside the cell, the Yops trigger a preprogrammed chain reaction that results in cell death. Yops also inhibit phagocytosis by macrophages and block cytokine function (see p. 97). Together this prevents the bacterium from being ingested, thereby avoiding the immune mechanisms of the host; another protein is able to degrade fibrin, a material found in the blood clot, allowing the *Yersinia* to move throughout the body by allowing it to escape from the clot at the site of the flea bite.

While the above description refers to the disease in humans, the disease in fleas also has a distinctive pattern. Small mammals such as urban and sylvatic (or wood) rats, as well as squirrels, prairie dogs, rabbits, voles, coyotes, and domestic cats, are the principal hosts for *Y. pestis*. More than 80 different species of fleas are involved as plague vectors. Fleas are blood-sucking insects (Fig. 4.7), and when a flea bites a plague-infected host (at the bacteremic/septicemic stage), it ingests the rod-shaped bacteria; these multiply in the blood clot in the proventriculus (foregut) of the flea. This bacteria-laden clot obstructs the flea's bloodsucking apparatus, and as a consequence the flea is unable to pump blood into the midgut, where it would be digested. As a result, the flea becomes hungrier, and in this ravenous state the flea bites the host repeatedly; with each bite it regurgitates plague bacteria into the wound. In this way infection is initiated. *Y. pestis* can also be pathogenic for the flea, and fleas with their foregut blocked rapidly starve to death. If the mammalian host dies, its body cools down, and the fleas respond by moving off the host to seek another live warm-blooded host. If there is an extensive die-off of rodents, however, then the fleas move on to less preferred hosts such as humans, and so an epidemic may begin.

The Plague-Causing Bacterium, *Yersinia pestis*

Y. pestis has been subdivided into three varieties—Antiqua, Medievalis, and Orientalis. Epidemiological and historical records support the hypothesis that Antiqua, presently resident in Africa, is descended from bacteria that caused the first pandemic, whereas Medievalis, resident in central Asia, is descended from bacilli that caused the second pandemic; those of the third pandemic, and currently widespread, are all Orientalis. It is believed that *Y. pestis* probably evolved during the last 1,500 to 20,000 years because of changes in social and economic factors that were themselves the result of a dramatic increase in the size of the human population, which was coincident with the development of agriculture. This triggered the evolution of virulent *Y. pestis* from the enteric, food-borne, avirulent pathogen *Y. pseudotuberculosis*. This required several genetic changes: a gene whose product is involved in the storage of hemin resulted in blockage of the flea proventriculus and enhanced flea-mediated transmission, and other gene products (phospholipase D and plasminogen activator) facilitated blood dissemination in the mammalian body and allowed for the infection of a variety of hosts by fleas.

Plague Today

The mortality and morbidity from plague has been significantly reduced in the 21st century. The disease, however, has not been eradicated. On November 2, 2007, Eric York, age 37, a wildlife biologist, was found dead in his home on the South Rim of the Grand Canyon. He died from plague after performing an autopsy on a dead cougar on October 27. This was the first case reported in Arizona since 2000. A year earlier a woman in Los Angeles was reported in the *Los Angeles Times* Online to have caught the Black Death. She had been taken to the hospital on April 13 suffering with characteristic swollen inguinal lymph nodes (buboes), fever, and other telltale symptoms of plague. It was suspected (but not proven) that the woman caught the disease after being bitten by infected fleas in her home.

Time magazine reported on October 30, 2015, that a 16-year-old girl from Oregon who had been on a hiking trip was in intensive care after being hospitalized a week earlier with bubonic plague. It was the 16th reported case that year, and so far, according to the Centers for Disease Control and Prevention, there already had been 4 deaths. In 2015 most of the bubonic plague cases were from Arizona, California, Colorado, Georgia, New Mexico, Utah, as well as Oregon. Two cases were linked to exposure in or near Yosemite National Park.

Between 1900 and 2012, 1,006 confirmed or probable human plague cases occurred in the United States. More than 80% of U.S. plague cases have been the bubonic form. In recent decades an average of 7 human plague cases have been reported each year (range, 1 to 17 cases per year). Plague has occurred in people of all ages (infants up to age 96), though 50% of cases occur in people ages 12 to 45. It occurs in both men and women, though historically is slightly more common among men, probably because of increased outdoor activities that put them at higher risk.

Worldwide between 2010 and 2015 there were 3,248 cases of plague reported and 584 deaths. Following the reappearance of plague in the 1990s in India, the World Health Organization considered plague as a reemerging disease. Over the years there has been a major shift of cases from Asia to Africa, with >90% of all cases occurring in Madagascar, Tanzania, Mozambique, Malawi, Uganda, and the Democratic Republic of the Congo (DRC). Most cases are of the bubonic variety; cases of pneumonic plague, however, were reported from the DRC in 2006 and Uganda in 2007. Madagascar is the most seriously affected country, with an overall fatality rate of 23% associated with pneumonic plague. In the United States plague has been encountered when houses have encroached on rural areas that harbor plague-infected rodents (sylvatic plague), such as prairie dogs, ground squirrels, mice, and rats. Infection may also be acquired from domestic pets (cats and dogs) that have come in contact with prairie dogs and can transmit disease to humans via fleas or in the pneumonic form. In the United States there has been an increase in the number of human cases of pneumonic plague, especially among veterinarians who have been exposed to infected cats.

Control of plague can be effected by surveying wild populations for infections, monitoring die-offs in the rodent population, making plague-infested areas known to the public, reducing the appeal of residential areas to rodents, and treating rodent burrows with cabaryl dust or bait stations to kill fleas. In some cases, rodenticides have been used.

Because of sylvatic plague and enzootic infections (i.e., outbreaks of a disease in animals other than humans), total eradication cannot be achieved.

Though human disease is rare, a feverish patient who has been exposed to rodents or flea bites in an area where plague is endemic should be considered a possible plague victim. The potential for the spread of pneumonic plague has been increased by air travel. Passengers who have fever, cough, or chills and come from areas where plague is endemic should be placed in isolation and, if necessary, treated. The diagnosis of plague has remained virtually unchanged since the days of Yersin: Gram staining and culture of bubo aspirates or sputum. The bacteria can also be grown in the laboratory on blood agar.

Unless specific treatment is given, the condition of a plague-infected individual deteriorates rapidly, and death can follow in 3 to 5 days. Untreated plague has a mortality rate of >50%. Streptomycin, tetracyclines, and sulfonamides are the standard treatments, and gentamicin and fluoroquinolones are alternatives when these other antibiotics are unavailable. In May of 2015 the U.S. Food and Drug Administration approved moxifloxacin (Avelox) to treat patients with plague. Earlier treatment was effected by using a serum Yersin had prepared in horses after injecting them with the plague bacilli he had isolated in Hong Kong in 1894.

Vaccines

The first vaccine for plague was developed by Waldemar Haffkine (1860-1930). Haffkine, the son of a Jewish schoolmaster, was born and educated in Odessa, Russia. While studying at Odessa University, Haffkine came under the influence of Elie Metchnikoff (a future Nobel Prize recipient). Haffkine's efforts to combat anti-Semitism by joining the Jewish Self-Defense League resulted in his arrest by the Russian authorities, but he was released after Metchnikoff intervened. Upon the completion of his degree, Haffkine attempted to find a post at the university, but it was denied him because of his refusal to be baptized. In 1889 he emigrated to Paris, where he joined Metchnikoff at the Pasteur Institute to work on a vaccine for cholera. When plague broke out in India in 1896, Haffkine moved to Bombay, where in 1897 he prepared an effective vaccine using dead bacilli. In 1902, however, 19 people out of the thousands who had received the vaccine died; this created a distrust of Haffkine's vaccine for plague. It was later discovered that the deaths had been due to another doctor's mistake in contaminating the vaccine with tetanus bacilli; Haffkine was blamed, however, and his career never recovered. He worked at other posts in India and retired to France in 1914. Although he visited Odessa in 1927, he found he could not adapt to the anti-Semitism that had been wrought by the Russian Revolution, and so he returned to France until his death.

Currently, there is no licensed vaccine available against plague. A formalin-killed whole-cell vaccine against *Y. pestis* was in use in the United States until 1999, when it was discontinued. This vaccine provided protection against bubonic plague, but there was good evidence that the vaccine provided little protection against primary pneumonic plague, and adverse side effects were known to occur. A live attenuated strain of *Y. pestis* (EV76) has also been used. It protected against bubonic plague efficiently and induced high titers of antibodies, but it did not confer long-lasting immunity and there were mild to severe side reactions. The immunogenicity and virulence of the EV76 vaccine preparations used in different countries were found to be highly variable, most likely because of the genetic drift of the bacteria used. Indeed, high genomic plasticity is observed in *Y. pestis*, which results from frequent chromosomal rearrangements between the numerous copies of insertion sequences present in its genome.

Recently, recombinant protein-based vaccine candidates have been developed. Two of these vaccines, RypVax and rF1V, have been shown to be superior in clinical trials. These vaccines rely mainly on a combination of two protein antigens, F1 and LcrV, and are strong inducers of an antibody response. These vaccines, however, provide poor and inconsistent protection in African green monkeys, probably due to a weak cellular immune response that is essential for good protection against plague.

Coda

Plague cannot be eradicated since it is present in wildlife rodent reservoirs. It has been shown that warmer springs and wetter summers in Kazakhstan increase the prevalence of plague in its main host, the great gerbil. Such conditions were present during the first and second pandemics—conditions that might become more common in the future with global warming. A plague outbreak may cause widespread panic, as occurred in India in 1994 when a relatively small outbreak, with 50 deaths, was reported in the city of Surat. This led to a nationwide collapse in tourism and trade, with an estimated loss of $600 million.

It should not be overlooked that plague has been weaponized throughout history, from the catapulting of plague-infected corpses over city walls in the Crimean port city of Kaffa in 1346-1347, to the Russians hurling cadavers of plague victims into the ranks of the enemy in 1710 during the battle against Swedish forces in Reval, to the Japanese dropping infected fleas from airplanes during World War II, to a refined aerosol formulation developed by the Soviet Union. The use of an aerosol released by terrorists in a confined space could result in significant mortality and widespread panic. Although plague cannot match the so-called "big three" diseases (malaria, HIV/AIDS, and tuberculosis) in the numbers of current cases, it far exceeds them in pathogenicity and rapid spread under the right conditions. Plague should not be seen as an historical curiosity. It is an infectious disease we cannot afford to ignore.

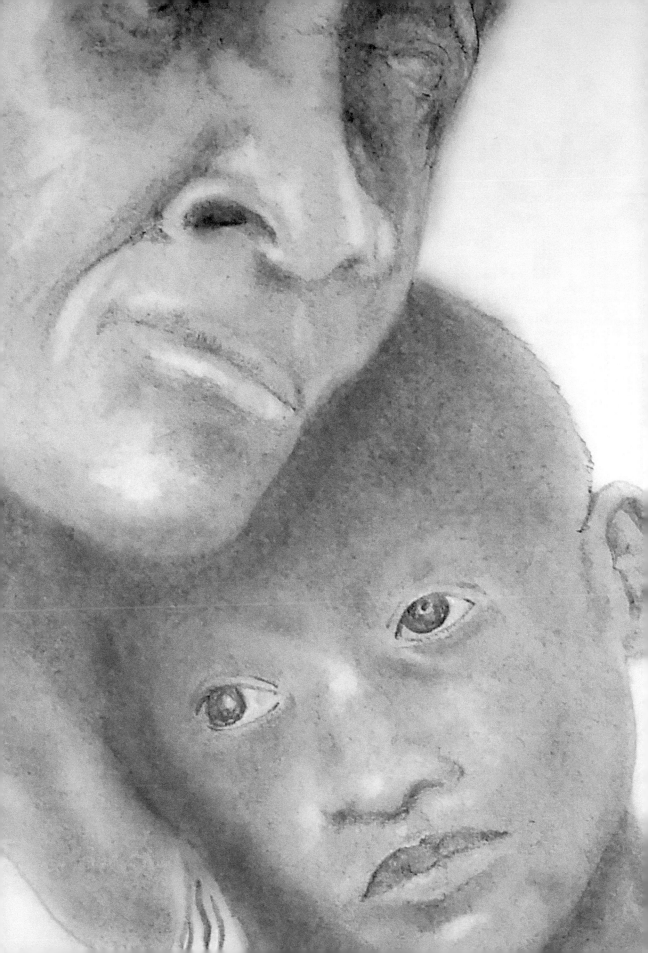

doi: 10.1128/9781683670018.ch5

5
A 21st Century Plague

Flesh and muscle melt from the bones of the sick in packed hospital wards and lonely bush kraals. Corpses stack up in morgues until those on top crush the identity from the faces underneath. Raw earth mounds scar the landscape, grave after grave, without name or number. Bereft children grieve for parents lost in their prime, for siblings scattered to the winds. The victims don't cry out. Doctors and obituaries do not give the killer a name. Families recoil in shame. Leaders shirk responsibility. The stubborn silence heralds victory for the disease: denial cannot keep (it) at bay.

So wrote Johanna McGeary as she described the plague that stalks the continent of Africa: AIDS (Fig. 5.1).

AIDS, acquired immune deficiency syndrome, called by some the plague of the 21st century, is a deadly global disease for which there is no vaccine or cure. The magnitude of the AIDS epidemic can best be appreciated by looking at the increase in the number of individuals infected. In 1981, when the disease first appeared in the United States, there were 12 cases; by 1994 the numbers of cases was 400,000; and in 2014 it was estimated there were >1 million people infected. Worldwide, the Joint United Nations Programme on HIV and AIDS (UNAIDS) estimates that in 2014 >36.9 million people were infected, and in that year alone 1.2 million people died of AIDS-related illnesses. The AIDS epidemic has killed nearly 700,000 people in the United States, and since the epidemic began worldwide, 25.3 million people have died from AIDS-related illnesses. With preventive measures and drug treatment, the vast majority of these deaths would not have occurred.

Figure 5.1 (Left) A sketch by the author of Lorraine aged 11, who has AIDS, comforted by her grandmother. (Based on a photograph by James Nachtwey—VIII Photo Agency and used with permission.)

A Look Back

AIDS began as most epidemics do, with a small number of cases, a smaller number of deaths, and then an almost geometric increase in the numbers of cases and deaths. In 1981, when AIDS was first described, an estimated 10,000 people worldwide were infected by the virus that causes AIDS—the virus called HIV. The discovery came about in an interesting way.

In 1981 the Centers for Disease Control (CDC) began receiving an increased number of requests for pentamidine, a drug used to treat an unusual type of pneumonia. Many of these requests came from Dr. Michael Gottlieb at the University of California at Los Angeles, who began to see an increasing number of patients with this unusual pneumonia caused by a fungus, *Pneumocystis jirovecii*. (There are also viral and bacterial pneumonias, all causing fever and lung congestion that may lead to respiratory failure.) The pneumonia caused by *Pneumocystis* is usually found in children with leukemia or adults with lymphomas or those on immunosuppressive drugs. Gottlieb found, however, that these were not his kind of patients. Instead, they were young, male, and homosexual. In New York City and San Francisco there were similar infections, as well as the appearance of a rare cancer of the skin and blood vessels that was signaled by a red-purple blotching of the skin, called Kaposi's sarcoma. This sarcoma had been described in Italian and Jewish men from around the Mediterranean, and it was most commonly found in older individuals; but in New York City, San Francisco, and Los Angeles the sarcoma was common and was present in young, gay males. At this time the syndrome—a collection of characteristic disease symptoms—was called GRID, for gay-related immune deficiency.

Two years later, in 1983, two laboratories, one in France headed by Luc Montagnier and one at the National Institutes of Health in the United States headed by Robert Gallo, identified the causative agent of GRID: a virus named by Gallo HTLV-III (human T-cell lymphotropic virus) and by Montagnier LAV (lymphadenopathy virus). In June 1985, Montagnier and Gallo held a joint news conference to announce that HTLV-III and LAV were likely one and the same virus. Despite this, the French and U.S. governments engaged in a bitter dispute over priority. Two years later U.S. president Ronald Reagan and French prime minister Jacques Chirac announced an agreement in which Montagnier and Gallo would be recognized as codiscoverers of the virus, renamed human immunodeficiency virus, or HIV, and the disease complex it produced was called acquired immune deficiency syndrome, or AIDS. Montagnier was awarded the Nobel Prize for Physiology or Medicine in 2008 for the "discovery of human immunodeficiency virus." Gallo was not mentioned.

HIV Discovered

Scientists as problem solvers constantly ask questions of Nature. George Gaylord Simpson, the paleontologist, put it another way: "Cats may be killed by curiosity but science would die without it." More often than not, scientific questions are framed in very specific terms so that the query results in a precise answer. Biologists, as they explore the nature of living things, may ask three kinds of questions: structural, functional, and evolutionary. A structural question might be: what is this "thing" made of and how is it put together? A functional question would be: how does this "thing" work? And an evolutionary question would be: how did this "thing" come about? Few biologists start out attempting to answer all three types of questions with a single exploration of the "thing"; rather, it is hoped that through the efforts of several scientists asking many questions, doing lots of experiments, and obtaining lots of information (data), the answers to all these different kinds of questions will be found. In short, scientists in general, and biologists in particular, expect that discoveries from different exploratory paths will converge so that a more complete understanding of the "thing" emerges. That is indeed what happened with the "thing" we call HIV/AIDS.

The story of the discovery of HIV begins more than a century ago (1884), with the development of a porcelain filter by Charles Chamberland, who was working in the laboratory of Louis Pasteur. Chamberland's filter had very small pores, so small that it was possible by using this filter, Chamberland wrote, "to have one's own pure spring water at home, simply by passing the water through the filter and removing the microbes." A few years later (1898) Dmitri Ivanowski took the mottled, dying leaves from a tobacco plant with mosaic disease, ground them up, and filtered the extract through a Chamberland filter. When the recovered sap was introduced into healthy plants, it produced the same symptoms of disease as did the unfiltered sap. Although Ivanowski believed this result could be explained by the presence of a toxin elaborated by bacteria dissolved in the filtered sap, or by the bacteria on the tobacco plant passing through the pores of the filter, he had in fact discovered the virus that causes tobacco mosaic disease. Indeed, since bacteria could not possibly pass through the pores of the porcelain filter, the infectious agent had to be smaller than a bacterium; today, we know that these agents of disease, smaller than bacteria, are viruses.

In 1911 a sick chicken was brought to a young physician, Peyton Rous (1879-1970), who was working at the Rockefeller Institute in New York City. The chicken had a large and rather disgusting tumor in its breast muscle. Rous wondered what could cause such a tumor (called a sarcoma), and so he took the tumor tissue, ground it with sterile sand, suspended it in a salt solution, shook it, centrifuged it to remove the sand and large particles, and filtered it through a Chamberland filter. The sap (or filtrate) was used to inoculate chickens, and within several weeks some developed sarcomas. Rous examined the filtrate and the sarcoma under a light microscope and found that neither contained bacteria; he concluded that he had discovered an infectious agent capable of causing tumors and that the infective agent was smaller than a bacterium. Clearly, he had found a virus that caused a cancer. (Viruses such as Ivanowski's tobacco mosaic virus and the virus that produced "spontaneous" chicken tumors, named the Rous sarcoma virus [RSV], would be seen only after the development of the electron microscope in the 1940s.) Rous had shown that a chicken cancer could result from a virus infection, but he failed to find similar virus-causing cancers in mice or humans. He received no support for his beliefs from other scientists; the sarcoma story remained a biological curiosity. As a result, Rous turned his attention to other aspects of pathology and did not return to the theme of cancer until 1934. (Rous's 1911 work was ultimately recognized in 1966 by his being awarded the Nobel Prize, just 4 years before his death at age 91.)

In the 1950s it was discovered that under certain conditions viruses could cause cancers such as leukemia and other tumors in mice. At about this time it also became possible to grow a variety of eukaryotic cells (see Appendix) in laboratory dishes (in vitro), and with these tissue cultures the effects of viruses could be studied in isolated cells rather than in mice or chickens. Indeed, tissue-cultured cells could be transformed into cancerous ones simply by adding virus to the laboratory-grown cells. At this time it was also found that viruses were of two kinds, those containing DNA and those containing RNA.

Renato Dulbecco (1914-2012), who began to study DNA viruses in tissue culture cells in the late 1950s, found that although some viruses caused tumors, on occasion virus could not be detected in the infected cell because its genetic material (DNA) had become inserted or integrated into the cell's chromosomes. In other words, the virus had been incorporated into the host cell genes and behaved as if it were a part of the cell's genetic apparatus. The virus DNA was now a part of the cell's heredity!

Dulbecco worked with DNA viruses, and this made it easier to visualize how the DNA of a virus could be integrated into the DNA of the host cell. RSV, however, is an RNA-containing virus, and it was not obvious how the genetic information of RSV could become a part of the tumor cell's heredity. Howard Temin (1934-1994) attended Swarthmore College, majoring in biology, and then went on to Caltech (California Institute of Technology) to do graduate work with Dulbecco; his doctoral thesis was on RSV. Temin's experiments, carried out from 1960 to 1964 at the University of Wisconsin in a laboratory located "in the basement, with a sump in … the tissue culture lab and with steam pipes for the entire building … in the biochemistry lab," convinced him that when the RSV nucleic acid was incorporated into the host cell, it acted as a "provirus." Under appropriate conditions the provirus would trigger the cell to become cancerous.

A fellow graduate of Swarthmore College, David Baltimore (b. 1938), did his doctoral work at MIT (Massachusetts Institute of Technology) and at the Rockefeller Institute, where he studied the virus-specific enzymes of RSV. His first independent position was at the Salk Institute in La Jolla with Dulbecco, studying RSV in tissue culture. When he returned to MIT as a faculty member, he continued to study the RSV enzymes. In 1970, Temin and Baltimore simultaneously showed that a specific enzyme, reverse transcriptase, in RSV was able to make a DNA copy from the virus's RNA. They independently went on to show that the replication of RNA viruses involves the transfer of information from the viral DNA copy, and that the viral DNA was integrated into the chromosomes of the transformed cancer cells. Later, other investigators were able to show that purified DNA from the transformed cancer cell, when introduced into normal cells, caused the production of new RNA tumor viruses. Clearly, RSV was a cancer-causing virus. For their discoveries Dulbecco, Temin, and Baltimore shared the 1975 Nobel Prize for Physiology or Medicine.

Viruses are unable to replicate themselves without taking over the machinery of a living cell, and in this sense they are the ultimate parasites. The material containing the viral instructions for replication, that is, its genes, may be composed of one of two kinds of nucleic acid, DNA (deoxyribonucleic acid) or RNA (ribonucleic acid). The viral nucleic acid is packaged within a protein wrapper called the core; this, in turn, is encased in an outer virus coat or capsid; and the outermost layer is called the envelope. In the case of HIV, the genetic material is in the form of RNA, not DNA, and thus in order for this virus to use the machinery of the host cell (which can copy only from DNA) it must subvert the cell's machinery to copying viral RNA into DNA. As a consequence, in HIV the information flow is RNA→DNA→RNA→protein, and because the flow of information appears to be the reverse of what one typically finds in cells (DNA→RNA→protein), these viruses (which include RSV) are called retroviruses.

The Genes of HIV

HIV is a virus—an entity that one biologist defined as "a bit of bad news wrapped in protein." A virus such as HIV (Fig. 5.2) is spherical in shape, resembling a 20-sided soccer ball, but it is a very small soccer ball, measuring only ~100 nm in diameter. Put another way, if you lined up HIV particles in single file, 1 in. would contain 250,000. Because of their small size, viruses cannot be seen with the light microscope, but they can be seen with the electron microscope, which magnifies objects 100,000 times.

The genes (in the form of RNA) of HIV code for at least nine proteins (Fig. 5.3). These proteins are divided into three classes: structural, regulatory, and accessory. The major structural proteins are Gag (group-specific antigen), Pol (polymerase), and Env (envelope). The *gag* gene encodes a long polyprotein that is cleaved by

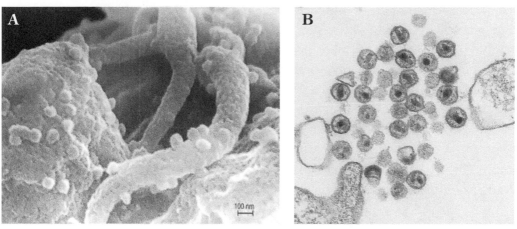

Figure 5.2A A scanning electron microscope image of HIV budding from the surface of a lymphocyte. (Courtesy of CDC/C. Goldsmith/P. Feorino/ E.L. Palmer/W.R. McManus, 1989). Figure 5.2B Appearance of HIV as seen with the transmission microscope. (Courtesy of CDC/Maureen Metcalf/Tom Hodge, 2011)

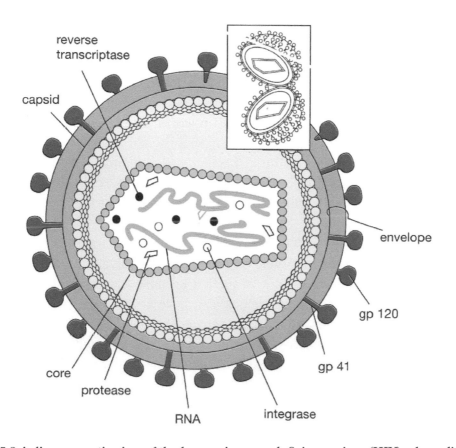

Figure 5.3 A diagrammatic view of the human immunodeficiency virus (HIV) when sliced in half

a viral protein-digesting enzyme, or protease, to form the capsid. The *pol*-encoded enzymes are also initially synthesized as part of a large polyprotein precursor; the individual *pol*-encoded enzymes, protease, reverse transcriptase, and integrase, are also cleaved from the polyprotein by the viral protease. The envelope glyco-proteins are also synthesized as a polyprotein precursor. Unlike the Gag and Pol precursors, which are cleaved by the viral protease, the envelope precursor, known as gp160, is processed by a host cell protease during virus trafficking to the host cell surface; gp160 processing results in the generation of the surface envelope glycoprotein, gp120, and a transmembrane glycoprotein, gp41. (The regulatory proteins are Tat and Rev, and the accessory proteins are Vpu, Vpr, Vif, and Nef.)

HIV Targets Immune Cells

To understand how HIV is able to cause AIDS, it is necessary to digress briefly to discuss the immune system. Blood consists of cells suspended in a yellowish, salt- and protein-containing fluid called plasma (after the blood has clotted, the fluid is called serum). There are two kinds of cells in our blood: red blood cells and white blood cells. The red cells contain hemoglobin (responsible for the red color of blood) and are involved in the transport of oxygen to, and the removal of carbon dioxide from, the tissues. The white blood cells have a different function; they are a part of the body's defense system and play a key role in immunity In our bodies there about a trillion white cells, and they are of five sorts. There are three kinds of gran-ule-bearing white cells called eosinophils, basophils, and neutrophils, and there are two kinds that lack granules in the cytoplasm, called lymphocytes and monocytes (macrophages) (see Fig. 10.3). The lymphocytes and macrophages are produced in the bone marrow but are found in regional centers such as the spleen and lymph nodes, as well as in the blood.

Lymphocytes are divided into two different types called T and B. The B lympho-cytes make antibody either on their own or by being activated by a T helper cell, called T4. The T4 lymphocytes are so named because they have on their surface a receptor molecule, CD4. Macrophages also have CD4 on their surface. T cells, unlike B cells, do not make antibody, but they are involved in what is referred to as cell-mediated immunity. Cell-mediated immunity is the immunity that is responsi-ble for transplant rejection, and for delayed hypersensitivity reactions such as occur after exposure to a bee sting. The T cells can communicate with one another using chemicals called chemokines, soluble chemical messengers that attract or activate other white cells, especially T and B lymphocytes and macrophages. To be activated, these white cells must have on their surface a receptor (analogous to a docking sta-tion) for the chemokine. One chemokine receptor, called CCR5, is an entry cofactor for the invasion of T cells by HIV, and it also activates neutrophils. Another chemokine receptor, called CXCR4, is an entry factor for the invasion of macrophages by HIV and activates monocytes, lymphocytes, basophils, and eosinophils.

Now let us look more closely at how HIV and immune cells interact with one another (Fig. 5.4). HIV, a roughly spherical particle 1/60 the size of a red blood cell, consists of an outermost layer, the envelope, and within there is an associated matrix enclosing a capsid, which itself encloses two copies of the single-stranded RNA genome and several enzymes. The glycoproteins of the HIV envelope resemble lollipops: the "stick" is called gp41 and the "candy ball" is gp120. The gp120 contains the determinants that interact with the CD4 receptor and coreceptor CCR5/CXCR4, while gp41 not only anchors the gp120/gp41 complex in the membrane but also contains domains that are critical for the membrane fusion reaction between viral and host cell lipid bilayers during virus entry.

The two viral proteins gp40/gp120 act in concert to anchor the HIV to CD4 on the surface of the T cell. Within an hour of docking the gp120 changes its shape so that it can bind to the chemokine receptors, CCR5/CXCR4, and this allows the virus to fuse with and gain entry into the cell. Precisely how virus-host cell membrane fusion occurs is not well understood, but once HIV enters the cell, the capsid breaks down and the viral RNA and reverse transcriptase are released. The latter begins to synthesize viral DNA from the viral RNA; the proviral DNA moves to the nucleus of the cell, where another viral enzyme, called integrase, inserts the proviral DNA into the DNA of the cell. When this infected cell begins to make proteins, it synthesizes (via viral mRNA) a long string of polyprotein. After this, a viral enzyme, protease, cuts the polyprotein into smaller proteins that serve a variety of viral functions: some become structural elements (envelope and capsid) while others become enzymes such as reverse transcriptase, integrase, and protease. Once this process (which takes about 15 h) has been completed, the newly assembled virus particles move to the inner surface of the host cell's membrane, where they mature. Then the infectious viral particles are released into the bloodstream by budding from the surface of the HIV-infected cell. The entire process from virus binding to virus release takes about 48 h, and during that time a single virus particle can produce several hundred thousand new infectious particles. An infected individual can produce 10.3 billion virus particles in a single day. At present there are ~30 approved therapeutic drugs (see p. 105) targeting the various steps in the replication cycle: (i) entry, (ii) reverse transcription, (iii) integration, (iv) maturation, and (v) protein cutting. Treatment with these antiretroviral drugs (see p. 105) may serve to control (but does not eradicate) the infection.

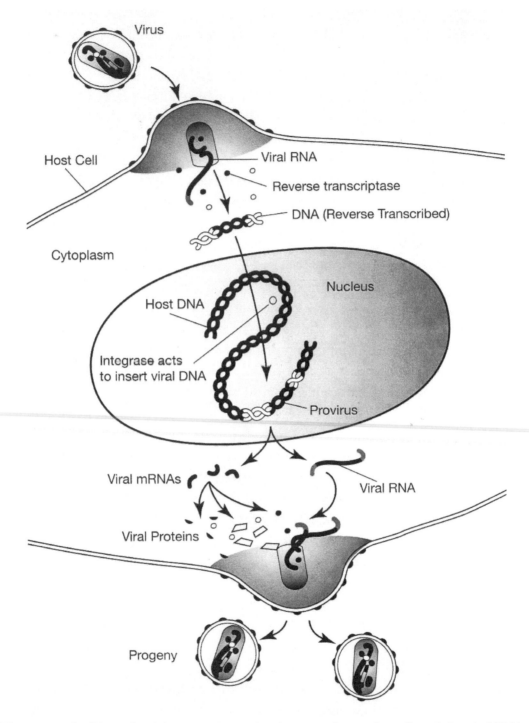

Figure 5.4 The life cycle of the retrovirus. The virus attaches to the cell, and the viral RNA is reverse-transcribed into DNA by the viral transcriptase and integrates itself (as a provirus) into the host DNA (using the integrase). There it gets transcribed into mRNAs coding for viral proteins (integrase, transcriptase, protease) and viral RNA genes. The viral proteins and the virus genomic RNA are assembled into virus progeny.

HIV and AIDS

One of the characteristics of an HIV infection is depletion of T4 cells (Fig. 5.5). It is not known how the virus depletes T4 cells other than by the direct lysis, or explosion, of the host cell. In a healthy individual there are ~1,000 T4 cells/mm³ of blood, and in an HIV-infected individual the number declines by about 40 to 80 T4 cells/year. When the count reaches 400 to 800/mm³, the first opportunistic infections appear. Opportunistic infections are those that under ordinary circumstances cause no disease but in individuals with a weakened immune system are able to take the opportunity afforded them by the body's crippled defenses to cause clinical disease. The first opportunistic diseases usually appear 1 to 3 months after infection and are

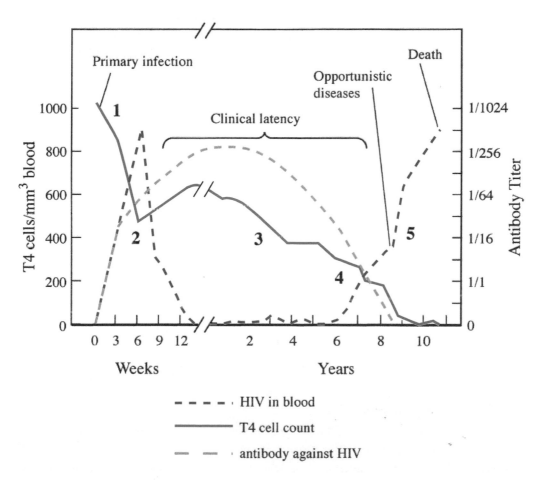

Figure 5.5 Clinical characteristics of an HIV infection. At 1, virus production occurs and this results in fever, fatigue, diarrhea and body aches. At 2, antibody is produced, and the number of virus particles in the blood begins to decline. At 3, the T4 lymphocytes decline. At 4, antibody levels decline dramatically, and T cell function is lost. At 5, virus particle production is re-established (Source CDC)

annoying infections of the skin and mucous membranes, such as thrush—painful sores in the mouth (though sometimes on the vulva and prepuce) due to the fungus *Candida albicans*—and shingles, a virus of the peripheral nerves, due to herpes zoster (chicken pox virus). In addition, there is severe athlete's foot and white patches on the tongue caused by Epstein-Barr virus. Once these symptoms appear, the person is said to have ARC—AIDS-related complex. Such individuals may also have swollen lymph glands, excessive weight loss (10 to 15% body weight), profuse night sweats, fevers, persistent cough, and diarrhea. Some individuals may be infected but not show ARC for years, and yet such individuals are infectious.

Once the T4 count drops to <200 cells/mm^3, the individual is said to have AIDS. At this T-cell level the AIDS-defining opportunistic infections are *Pneumocystis* pneumonia, cryptococcal meningitis, and toxoplasmosis, which together account for 50 to 75% of deaths. When the T4-cell count drops below 100 T4 cells/mm^3, several other opportunistic infections result, including cytomegalovirus (CMV)—a herpes virus leading to blindness and damage to the lungs and digestive tract—*Mycobacterium avium* (bird tuberculosis), and chronic cold sores due to herpes simplex. In the immunodeficient individual—and that's what an AIDS patient is—these diseases are debilitating and deadly, and together they cause 90% of the deaths. In addition, some other conditions may occur: encephalopathy may lead to hallucinations, dementia, or motor loss; and cancers of blood cells, called lymphomas, may develop.

Failure To Control by Antibody or Vaccine

Soon after Montagnier and Gallo identified HIV, a specific and sensitive test for antibodies to HIV was developed. With this diagnostic test it is possible to screen the serum of large numbers of individuals and determine who is infected. (Such individuals are called serum positive or seropositive.) This simple test also has had immediate and profound effects for public health, since blood supplies in the United States and other countries could now be screened for HIV. By 1985, through the use of screening, an HIV-free blood supply was ensured, thereby preventing millions of potential transfusion-related infections. In addition, the antibody test could be used in epidemiology to determine the global scope and evolution of the disease. The availability of the antibody test has allowed for the identification of individuals before they show clinical signs of the disease, and it permits a more accurate description of the true clinical course of HIV infections. The FDA approved a rapid HIV antibody test that can be performed outside the laboratory and provide results in about 20 min. Those with positive results should have a confirmatory test, however, and those who are seronegative should recognize that although they are presently without evidence of infection they could in time become positive, and therefore a repeat of the test would be necessary to ensure that the individual is HIV free.

Although antibodies to HIV can be found in the plasma of millions of people worldwide, these individuals continue to be infected and may transmit the virus to others. Clearly, although HIV elicits a strong antibody response, these antibodies neither neutralize the virus nor clear it from the body, and there is no protection against reinfection. Why? The antibodies to HIV are unable to reach the critical receptor sites and do not block the conformational changes in gp120 that are necessary for HIV attachment and entry. Further, HIV is able to persist for long periods of time in nondividing T cells, so that even if an effective antibody were produced, its large size would not allow it to enter the infected cell and kill the virus. In addition to its ability to hide within resting T cells, HIV is able to cripple the immune system by depleting the helper T cells not only by outright destruction (through lysis and viral release) but also by decreasing the rate of production of T helper cells. The result can be the appearance of deadly opportunistic infections, which may include candidiasis of bronchi, invasive cervical cancer, coccidioidomycosis, cryptococcosis, cryptosporidiosis; CMV disease (particularly CMV retinitis), encephalopathy, herpes simplex, histoplasmosis, isosporiasis (an intestinal disease caused by a parasite), Kaposi's sarcoma, lymphoma, *M. avium* complex, tuberculosis, *Pneumocystis* pneumonia, pneumonia, degeneration of the white matter of the brain, *Salmonella* septicemia, and toxoplasmosis of the brain.

AZT, the First Antiretroviral Drug

When Jerome Horwitz (1919-2012), an organic chemist at the Michigan Cancer Center at Wayne State University in Detroit, opened the March 21, 1987, issue of the *New York Times* and read the headline "U.S. Approves Drug to Prolong Lives of AIDS Patients," he was stunned. The reason for Horwitz's astonishment was that the "approved drug"— azidothymidine, or AZT—was a compound he had spent a decade developing as a cure for leukemia; in the 1987 article, however, AZT was named as an antiviral drug made by Burroughs Wellcome (now GlaxoSmithKline) under the brand name Retrovir.

According to the article, AZT had been tested on AIDS-infected individuals in the United States as early as July 1985, with a pivotal clinical trial beginning in February 1986. During the clinical trial 19 deaths occurred among 137 patients taking the placebo, compared to a single death among the 144 AIDS patients taking AZT. By the time of its approval by the FDA, AZT had been made available to >5,000 patients with AIDS. AZT reached the market with remarkable speed, proceeding from laboratory to clinical trials and FDA approval in less than two and a half years. Horwitz's contribution to the development of AZT was never mentioned.

In the early 1960s, with the war on cancer in full swing (and before there was any thought of an AIDS epidemic), many scientists pulled drugs randomly off the shelf to test whether they had any potential as anticancer agents, but Horwitz found this approach intellectually unsatisfying. Instead, he decided on a more rational approach. Knowing that dividing cells, especially cancer cells, require nucleosides (for the synthesis of DNA), he created what he called "fraudulent nucleosides" that were similar to the real thing. These "fakes," he theorized, would gum up the cancer cell's replication machinery and halt the cell's rapid division. The theory was fine; when AZT (3' -azido-2' ,3' -dideoxythymidine) was synthesized in 1964, however, and failed to help mice with leukemia, Horwitz "dumped it on the junkpile," wrote up his failure, and moved on. He was so disappointed in AZT that he didn't even apply for a patent.

AZT collected dust on the shelf until the mid-1980s, when there was a public awareness of the growing number of deaths from AIDS. This prompted a search for a treatment.

At Burroughs Wellcome, as in many pharmaceutical company laboratories, the research goal was to find inhibitors for particular biological targets. Researchers at Wellcome had experience with drugs against herpesviruses and other viruses, and they began testing compounds that might be effective against retroviruses. AZT showed promise in laboratory animals, and in collaboration with the National Cancer Institute (NCI) and in discussion with the FDA, Wellcome sought permission to begin the testing of AZT in humans. On July 3, 1985, the first AIDS patient received AZT at the NCI, and the result was encouraging. Further studies in 19 patients with AIDS showed that the drug harmed the bone marrow at high doses but was not seriously toxic; more importantly, many of the AZT-treated individuals improved. Wellcome applied for and received a patent for AZT, and on March 20, 1987, the FDA approved AZT for the treatment of HIV infections. Marketed as Retrovir (and also known generically as zidovudine) by Wellcome, it was priced at $8,000 a year. By 1992 AZT had sales of $400 million. Horwitz, without a patent for AZT, received not a penny, although Wellcome did donate $100,000, not even enough for an endowed professorship, to the Michigan Cancer Center in Horwitz's name. Horwitz remained at Wayne State, developed two anticancer drugs, and patented them, and in 2003, Wayne State University licensed them to a pharmaceutical company. The company paid a hefty licensing fee, and at age 86 Horwitz received the first royalty check of his career. Horwitz died in 2012 at age 93.

George Hitchings (1905-1998) began in 1942 to study analogs to inhibit the growth of malignant cells at the Burroughs Wellcome laboratories in Tuckahoe, NY. The underlying principle was that even if there was not an absolute difference in the metabolism of the malignant cell, there were probably enough differences to allow for differential inhibition. In 1944, Gertrude Elion (1918-1999) joined Hitchings at Wellcome. Although at the time none of the enzymes or the steps in the formation of nucleic acids were known and the deciphering of the double helix structure of DNA by Watson and Crick was a decade away, Hitchings and Elion began studying the growth-inhibitory properties of analogs of pyrimidines and purines. And 44 years later Hitchings and Elion shared the Nobel Prize for Physiology or Medicine for "discoveries of important principles for drug development."

Gertrude Elion was born in New York City to parents who immigrated to the United States from Russia and Lithuania. At age 15 she was motivated to embark on a research career when her beloved grandfather died from stomach cancer. Elion's parents were poor, so she attended the free Hunter College with a major in chemistry. Graduating in 1937, and with the worldwide economic depression, she was unable to go directly to graduate school and worked as a teacher and laboratory technician. Two years later, with some money saved and with the help of her parents, she enrolled for graduate study at New York University and was the only female in the chemistry classes. During the time she was enrolled for the M.S. degree, she taught chemistry in New York secondary schools by day and worked on research at night, graduating in 1941. With World War II ongoing, jobs began to open up for women in chemistry. Her first chemistry job, however, was far from glamorous: testing the acidity of pickles and the color of mayonnaise for the A&P grocery chain; later she was hired by Johnson & Johnson to synthesize sulfa drugs, but the lab closed after 6 months. In 1944 she was hired by Hitchings and began to study for the Ph.D. at the Brooklyn Polytechnic Institute, taking evening courses; when she was told, however, that she would have to become a full-time student to complete the Ph.D., she abandoned that goal. She remained at Wellcome for the rest of her career, and by the time she retired 39 years later, she was the holder of 45 patents, had 23 honorary degrees and a long list of honors, and was the head of the department of experimental therapeutics. Despite receiving the Nobel Prize, she never completed the requirements for a Ph.D.!

Elion's initial job was to make purine compounds that would antagonize the growth of the bacterium *Lactobacillus*. To do this she went to the library, looked up the methods for synthesis found in the old German literature, and made the compounds. In 1948 she had in hand a purine analog that inhibited the growth of *Lactobacillus* and showed in vitro activity against the DNA-containing vaccinia (smallpox) virus; it was, however, too toxic for animals and so was abandoned. By 1951, Elion had made and tested >100 purine analogs in the *Lactobacillus* screen and discovered that 2 of them—6-mercaptopurine (6-MP) and 6-thioguanosine (6-TG)—were also particularly effective against rodent tumors and mouse leukemia. The finding that 6-MP could produce complete remission of acute leukemia in children (although most relapsed later) led to its approval by the FDA in 1953. With combination therapy, using three or four drugs to produce and consolidate remission, plus several years of maintenance therapy with 6-MP and methotrexate, 80% of children with acute leukemia were cured. In 1959 the compound azathioprine (marketed as Imuran) was found to blunt transplant rejection, especially in kidney transplantation. And in 1968 Elion turned her attention to antivirals. Diamino arabinoside was found to be active against herpes simplex and vaccinia virus, and in 1970 the guanine analog acycloguanosine, or acyclovir, was found to be effective against herpes, varicella, and Epstein-Barr virus, though not CMV. Acyclovir, marketed as Zovirax, is not toxic to mammalian cells; it interferes with the replication of herpesvirus and only herpesvirus, proving that these drugs can be selective. In time the discoveries of acyclovir, 6-MP, and 6-TG would lead to the development of AZT by Elion's colleagues at Wellcome.

Beyond AZT

Before the mid-1990s there were few antiretroviral treatments for HIV, and the clinical management of those with AIDS consisted largely of prophylaxis for common opportunistic infections. In 1987 AZT was the first antiretroviral used in the treatment of AIDS. AZT acts by targeting the HIV reverse transcriptase and belongs in the category of nucleoside reverse transcriptase inhibitors (NRTIs). Although AZT blocks the activity of the HIV reverse transcriptase and has 100-to-1 affinity for HIV over human cells, when used at high doses it does limit the polymerase needed by healthy cells for cell division. And at higher concentrations it depresses the production of red and white blood cells in the bone marrow.

Following AZT, other NRTIs were developed: abacavir, tenofovir (an analog of AMP), lamivudine, didanosine (ddI), zalcitabine, stavudine, and emtricitabine. Treatment of HIV was further revolutionized in the 1990s with the introduction of these NRTIs, as well as the nonnucleoside reverse transcriptase inhibitors efavirenz, nevirapine, and rilpivirine and the protease inhibitors lopinavir, indinavir, nelfinavir, tipranavir, saquinavir, darunavir, amprenavir, fosamprenavir, atazanavir, and ritonavir. This allowed for FDA approval of regimens consisting of combinations of antiretroviral drugs, called highly active antiretroviral therapy, or HAART. (HAART is initiated when the T-lymphocyte count declines to $350/mm^3$ and when there are between 10,000 and 100,000 HIV particles per ml of plasma.) As early as 1995 the HIV integrase was recognized as a potential antiretroviral target with the discovery of susceptibility by oligonucleotides, synthetic peptides, and polyphenols, but the milestone was the publication in 2000 of a diketoacid inhibitor that led to FDA approval of the first integrase inhibitor, raltegravir (approved in 2007); following this was the clinical development of elvitegravir (approved in 2012) and dolutegravir (approved in 2013). A CCR5 antagonist, maraviroc, is also available for HIV therapy.

Currently there are 15 million people taking antiretroviral drugs, and it is estimated that 30 million deaths may have been averted between 2000 and 2014 by HAART. HAART can suppress viral replication for decades; it alone, however, cannot eliminate all HIV particles. Nevertheless, HAART has changed HIV from an inexorably fatal condition into a chronic disease in which the patient enjoys a longer life expectancy. Indeed, in 2001 the life expectancy of an individual with HIV was 36 years, while in 2014, thanks to HAART, it was 55 years!

In 2001 HIV treatment required eight or more antiretroviral pills a day at a cost of $10,000 per year, but by 2014 a variety of single-daily-tablet regimens containing a combination of antiretroviral drugs (tenofovir/emtricitabine/efavirenz or tenofovir/emtricitabine/rilpivirine or tenofovir/emtricitabine/elvitegravir or abacavir/lamivudine/dolutegravir) were introduced at an annual cost of $100.

As with most viruses, the AIDS-causing virus is hard to kill without doing any damage to the host. And even with combination drug therapy, resistant viruses may emerge. Treatment with an antiretroviral drug can be a two-edged sword: it may benefit the individual, but if it does not reduce infectiousness, it might not significantly benefit the community. It has been calculated that the number of virus particles must be reduced to <50 per ml of serum for the individual to lose the capacity to be infectious. (It has been shown, however, that even with successful reduction a viral reservoir may remain in the body, and that if antiretroviral therapy is discontinued or if there is the emergence of resistance, the virus may rebound.)

Despite decades of attempts to produce a vaccine for HIV, there is none. The National Institutes of Health explains that there is no vaccine for HIV/AIDS because HIV has "unique ways of evading the immune system, and the human body seems incapable of mounting an effective immune response against it." A recently published study by Carnathan et al. in the *Proceedings of the National Academy of Sciences* using monkeys further clarifies the problem. The researchers evaluated five different strategies for immunizing 36 rhesus monkeys against the simian immunodeficiency virus (SIV). After being given an initial shot of one of the five different vaccines, each of the monkeys received booster shots at 16 weeks and then again at 32 weeks. Next, the monkeys were exposed to a low dose of SIV. In general, the researchers found that none of the vaccines prevented an SIV infection. Oddly, all the immunized monkeys had detectable levels of circulating "killer" T cells, but these cells did not prevent infection. "The possibility that certain immunization regimens designed to protect against HIV infection and AIDS result in increased risk of virus transmission is not just a theoretical concern, because three recent large-scale clinical trials … have shown a trend toward higher infection rates in vaccinated individuals than in placebo recipients," noted the authors.

"Catching" HIV

HIV is an infectious disease, but not a very contagious one (R_0 = 2 to 5). It is not spread very easily by fomites (inanimate objects) since the virus is killed by high temperature, by detergents, and by 10% chlorine bleach. It is not a vector-borne disease; indeed, if it were transmitted by bloodsucking mosquitoes or flies, it would not be a young person's disease. HIV is principally found in the secretions and body fluids of infected individuals. The most infectious sources are blood, breast milk, vaginal secretions, and semen, where there are sufficient concentrations of virus to cause infection. Viruses are found both free and inside T lymphocytes. A tear or lesion in the skin or mucous membranes must occur for infected fluid to initiate an infection. Infection may also result from injection of infected blood into the body, organ transplants, and ingestion of infected breast milk; virus can also cross the placenta.

There are three means of HIV transmission. The first is unprotected sexual intercourse. The most dangerous kind is anal intercourse because the rectal area is susceptible to rips and tears, therefore permitting entry of virus into the bloodstream of the recipient. Lower transmission occurs from an infected female to a male because the amount of vaginal secretion is much less than that of semen. HIV, however, can be transmitted via oral sex (with cunnilingus being less efficient than fellatio), through the sharing of HIV-contaminated blood and blood products, and from infected mother to child due to small ruptures in the placenta or via breast milk. The explosive spread of AIDS in some parts of the world is, in part, a reflection of the increased number of partners involved in the exchange of HIV-laden fluids.

Control of AIDS

In the absence of a vaccine and an effective cure by drug treatment, what can be done to avoid infection and reduce the spread of HIV? There are three tactics:
(i) practicing abstinence; (ii) having a monogamous relationship with a previously celibate or uninfected individual; and (iii) changing risky behavior, especially for high-risk groups such as those with many sex partners, those engaging in anal intercourse, those using contaminated needles, and those who shun the use of condoms.

The power of risk-behavioral changes to control the spread of AIDS has been clearly shown in a theoretical transmission model using data from the San Francisco Young Men's Health Study. The model addressed three questions. (i) If an HIV vaccine were available, what proportion of young gay men in the community would have to be vaccinated in order to eradicate HIV? (ii) How effective would such a vaccine have to be in order to ensure eradication? (iii) What effect would changes in sexual risk behavior have if a mass vaccination program were instituted? In this model the value for R_0 (the number of secondary cases produced when one infectious case is introduced into an uninfected population; see p. 12) was estimated to be between 2 and 5 using sexual risk behavior data such as the number of receptive anal sex partners and the use of condoms. (At the time of the study, the infection rate in young gay men in San Francisco was 18%.) The model predicted with $R_0 = 2$ that 80% of those vaccinated would have to become immune to ensure eradication; i.e., susceptibility to infection would have to be reduced by 95% and immunity would have to persist with a half-life of 35 years. If $R_0 = 5$, then the minimum efficacy of the vaccine would have to be 80% and coverage 100% to achieve HIV eradication. Since participants in the study indicated they would not participate in a vaccine trial if it were only 50% effective, however, it is possible that a mass vaccination program could actually increase the severity of the epidemic. In the absence of a change in risk behavior, and with $R_0 = 5$, a vaccine that was 60% efficacious could not eradicate the epidemic, and even with a vaccine of 80% efficacy, 100% coverage would be needed. If risk behavior were halved, on the other hand, eradication would be possible with a vaccine that was only 60% efficacious. The theoretical model predicted that without a mass vaccination campaign and solely a reduction in risk behavior HIV could be eradicated. For example, if $R_0 = 2$, the risk behavior levels would have to be decreased by 50%, and for $R_0 = 5$, the decrease would have to be 80%. This model calculates that even when a highly effective vaccine does become available (and to date none exists) coverage will have to be very high to achieve eradication, and in San Francisco it is extremely unlikely that HIV eradication could occur without significant reductions in risk behavior combined with an effective vaccine. This model predicts that the eradication of HIV will require simultaneous deployment of efficacious prophylactic vaccines and behavioral interventions.

Although early in the AIDS epidemic the virus was spread via blood products (i.e., preparation of factor VIII for hemophiliacs) and whole blood used during transfusions, since 1985 it has been possible to test for the virus. Consequently, through screening of the blood supply, blood and blood products are unlikely sources of transmission. Health care workers may be at a higher risk, but only if contact is made between the body fluids of an infected individual and the bloodstream of the recipient.

"Safer sex" describes those practices that minimize the transfer of bodily secretions that contain the virus. Using a condom during oral, anal, or vaginal sex can reduce the transmission of the virus, but the condom must be used correctly. It needs to be placed on the penis after it is erect, leaving room at the tip for the ejaculate. Withdrawal should occur before the penis becomes flaccid, and the condom should never be reused. Latex condoms are perishable and can deteriorate with time and heat. Water-based lubricants such as K-Y jelly, not petroleum-based ones such as Vaseline, should be used.

A popular misconception is that condoms have a high failure rate and so there is little benefit in using them to protect against AIDS. The truth is that condoms are very effective in blocking HIV transmission, but they must be used consistently and correctly. Most of the statistics showing high failure rates relate to the use of condoms for birth control, and the failure rate relates to the number of births when using condoms. But these failures may be due to inconsistent or improper use of condoms and not an inherent flaw in the condom itself. At least one study showed that most failures come from a small minority of people who did not, on closer examination, follow the proper condom use steps outlined above.

Those who avoid the use of condoms sometimes justify this reckless behavior by claiming that HIV is so small that it can pass through the holes in a condom. These individuals know little about HIV or the molecular characteristics of the latex rubber in a condom. First of all, most virus is inside T4 cells, and these cells are larger than sperm, which do not pass through a condom; otherwise, condoms would be useless for contraception. Even the free virus does not pass through the condom wall because it contains millions of atoms, much larger than even the molecules in water, which do not pass through the intact condom wall either. Finally, some people are concerned about condoms breaking during use. Condom failure is very rare; condoms are rigorously tested during manufacture, and where ripping does occur, it is commonly the result of improper storage and use of petroleum-based lubricants. The best evidence for protection by condoms comes from a study of couples in Europe in which one member was HIV infected and the other was not. Among 123 couples who reported consistent condom use, none of the uninfected individuals became infected, whereas among 122 couples who used condoms inconsistently, 12 of the uninfected partners became infected. "Safer sex," including condom use, minimizes risk but does not eliminate it entirely. In a similar fashion, while driving a car we may manage to reduce the risk of accidental injury or death by wearing seat belts and driving carefully, but we can't eliminate threats to our existence completely.

HIV and the African Connection

There are three types of HIV: HIV-1, discovered in 1983, has spread rapidly from many foci; HIV-2, discovered in 1985, has spread very slowly and is confined mostly to countries of West Africa such as Mali, Mauritania, Nigeria, and Sierra Leone; and HIV-0, discovered in 1990, is found only in Cameroon and Gabon and has barely spread. HIV-like viruses occur in cows, lions, horses, sheep, goats, and simians (monkeys), but in most of these natural hosts they cause little or no immunodeficiency; once they infect domestic equines and house cats, though, they do cause disease. The simian viruses, called SIV, have been infecting primates for as long as 16 million years. Discoveries in the 1980s in chimpanzees, *Pan troglodytes*, in the Gabon rain forest showed a similarity between the virus of the chimpanzee and HIV: it was called SIVcpz. Serum taken from infected chimpanzees or serum produced against the SIVcpz by injection reacted with HIV-1, and vice versa. SIVcpz is transmitted among chimps sexually and appears to be harmless, perhaps because it has been coevolving with monkeys for a very long time.

The transmission of a disease from animals to humans is called a zoonosis. In the case of HIV, an extremely rare zoonosis later established itself as a human-to-human infection. SIVcpz from *Pan* entered the human population, or so we believe, through forest people who engage in the bush meat trade, that is, hunters who, using primitive butchery methods, dismember chimpanzees and other apes; sell and eat the meat, either raw or improperly cooked; and in the process of butchery come in contact with chimpanzee body fluids such as saliva and blood. In this way the hunters exposed themselves to the risk of a zoonotically transmitted disease. This species jump of virus from chimpanzee to human probably occurred in the 1920s in the Kinshasa region of the Democratic Republic of the Congo (DRC). Once the virus had entered the human population in Africa, it was transmitted sexually. The area around Kinshasa is full of transport links such as roads, railways, and rivers. It also had a growing sex trade around this time. The high population of migrants and sex trade workers might explain how HIV spread along these routes, first to Brazzaville in 1937. By 1980 half of all infections in the DRC were in locations outside of Kinshasa, reflecting the growing epidemic.

There are four main groups of HIV strains (M, N, O, and P), each with a slightly different makeup. This supports the hunter theory, because every time SIV passed from a chimpanzee to a human it would have developed in a slightly different way within the human body and produced a slightly different strain. The most-studied HIV is HIV-1 of the M group, the strain that has spread throughout the world and is responsible for the vast majority of HIV infections. During the 1950s and 1960s colonial rule in Africa ended, there were civil wars, vaccination programs involved the reuse of needles and syringes, there was the growth of large cities, there was a sexual revolution, travel within Africa and between Africa and the rest of the world increased, and with this transmission also increased. By the 1960s a subtype of HIV-1 M, called the "B" subtype, took hold in the Caribbean when many Haitian professionals who were working in the DRC returned to Haiti. A recent genetic analysis has revealed that this HIV-1 strain then traveled to the United States. Upon arriving in New York City around 1970 or 1971, it circulated and diversified for 5 years before it dispersed across the country, being transmitted between gays with multiple partners, in intravenous drug users, and in the global blood market. By the late 1970s—years before it was officially recognized by the medical community in 1981—thousands of people were infected. From New York City, HIV-1 eventually moved to Western Europe, Australia, Japan, South America, and other places. The HIV-1 subtype B is now the most geographically spread subtype, with some 78 million infections.

If the hunters of the African forest had HIV, and it caused an AIDS-like disease, why did it go unnoticed for so many years? Probably because it remained isolated in a small population. Further, within Africa a disease of this sort might not be noticed as being special and different from the wasting that occurs with malnutrition, or infections such as malaria, yellow fever, and hookworm. The emergence of AIDS is probably the result of a change in the modes and frequency of transmission of HIV. This, in turn, is related to changes in human behavior—intravenous drug use, sex with many partners, population movements, prostitution, and anal intercourse. With greater efficiency of transmission, the virulent forms of HIV emerged.

The Sooty Mangabey Connection: HIV-2

The closest relative of HIV-2 is the SIV from the sooty mangabey monkey called SIVsm—indeed, there is 80 to 90% genetic homology. HIV-2 probably arose in West Africa in the rain forests of Benin. Antibodies to HIV-2 and SIVsm show these two viruses to be indistinguishable. In West Africa 10% of prostitutes test positive for both SIVsm and HIV-2, but in the United States and Central Africa there is little reaction with SIVsm. HIV-2 is endemic in West Africa, where there appears to be no clinical epidemic; HIV-2 is less pathogenic, and individuals with HIV-2 are less at risk for developing AIDS than those infected with HIV-1.

If HIV-2 originally came from the sooty mangabey SIV, where in turn did the sooty mangabey get its SIV? Perhaps from large cats harboring feline immunodeficiency virus (FIV), in which it was a benign disease. (In the domesticated cat, where it is transmitted in the saliva, it causes an AIDS-like disease if the cats also harbor feline leukemia virus.)

The Social Context of AIDS

HIV contributes to social vulnerability, reduces life expectancy, limits productivity, and stifles economic growth. AIDS exacerbates and prolongs poverty and increases malnutrition. HIV infections demand a greater proportion of the already meager annual income of those living in Asia, Africa, and Eastern Europe, thereby reducing access to food and health care. AIDS has diminished the number of available teachers, and in this way it impacts the educational system. Globally, HIV is one of the five leading causes of death. The modern plague of AIDS is a forceful reminder that the global impact of infectious disease is yet to be blunted.

doi: 10.1128/9781683670018.ch6

6
Typhus, A Fever Plague

Those attending a New York Philharmonic concert and listening to the cannons, the ringing of church bells, and the melody of "La Marseillaise" in Tchaikovsky's *1812 Overture* probably knew that it was composed in 1880 to mark the opening of the Moscow Exhibition of 1882 and the consecration of the Cathedral of Christ the Savior, built to give thanks for the Russian victory over Napoleon in 1812. They may have also known from the program notes that the first performance was given on August 20, 1882, to celebrate the 70th anniversary of Russia's victory over Napoleon and that it was a great success. Although even today audiences continue to be impressed by the power of the work, Tchaikovsky felt little enthusiasm for it, and had it not been for a very lucrative commission, he probably would never have composed the overture. Everyone at the performance of the Philharmonic, conducted by Leonard Bernstein, was aware that the defeat of Napoleon, set to music, dealt with the impact of war on the people, the bravery of the Russian soldiers, and the role of the Russian winter, and that it celebrated the glory of the people of Russia. Few in the audience, however, would have known that the Russians had another ally—typhus fever—and that it was this disease that contributed in a significant way to Napoleon's downfall (Fig. 6.1).

Typhus fever is sometimes called "war fever" because frequently it is a companion to hostilities. War involves overcrowding and intermixing of populations, resources are diverted, and often famine and malnutrition increase. These so-called enabling factors in turn lead to decreases in personal hygiene and medical care, and frequently a breakdown of the social structure. During wartime individuals are subjected to increased stress, they become more susceptible to new diseases, and endemic diseases may become more severe. An invading army may introduce a disease into the defending army as well as the surrounding communities, or the endemic diseases of the community may become a source of infection for the invaders. For example, in 1741 the Austrian army

Figure 6.1 (Left) Napoleon's troops in Vilna after the Russian Campaign in 1812. Engraving by Eugene Le Roux after A. Raffet. Courtesy Wellcome Library, London, CC-BY 4.0.

surrendered Prague to the French because 30,000 soldiers had died of typhus and the Austrians were unable to defend the city. Following the end of World War I, the returning armies carried typhus with them. Between 1917 and 1923 there were 20 million to 30 million cases and 3 million deaths in Eastern Europe. In 1915 an epidemic of typhus occurred in Serbia and nearly all the country's 400 doctors contracted the disease; more than a quarter died. In the wake of World War II a few million people died of typhus. Between the 1950s and 1980s large epidemics of typhus became less frequent, and its geographic distribution declined because of improvements in living standards. During the 1990s, however, the disease reemerged in places where sanitary conditions were poor. More than 45,000 cases of epidemic typhus occurred in Burundi in association with a civil war during the 1990s. Outbreaks occurred in the rural highlands of Peru and Africa (Uganda, Ethiopia, Nigeria, and Rwanda). Sporadic cases of epidemic typhus have been reported from several northeastern states of the United States as well as California, where there were between 26 and 70 cases between 2011 and 2013, and 30 in 2015.

In 1935, Hans Zinsser, the Harvard Medical School physician and bacteriologist, wrote in his classic work *Rats, Lice and History*: "Typhus had come to be the inevitable and expected companion of war and revolution; no encampment, no campaigning army, and no besieged city escaped it."

A Look Back

The origin of this plague, called "prison fever" or" ship fever" as well as "war fever," remains obscure, but it is generally believed that typhus was a disease of rats, transmitted by lice. Subsequently it became a human infection. In 429 B.C., during the Peloponnesian War, Pericles, the 65-year-old leader of the Athenians, came down with a mysterious illness. His condition included bouts of high fever, chest pains, vomiting, diarrhea, fetid breath, and a bumpy red rash. His illness occurred at the same time as the plague of Athens, and although some believe the plague of Athens was due to smallpox, it is entirely possible that Pericles suffered with and died from typhus. Indeed, the word "typhus" comes from a Greek word, *typhos*, meaning "hazy" or "smoky," and was used by the Greek physician Hippocrates to describe the confused state and stupor of the victim.

The earliest severe epidemic of typhus was in Spain in 1489-1490, when the forces of Queen Isabella and King Ferdinand were at war with the Moors over the possession of Grenada. The disease was described as "a malignant spotted fever" and was assumed to have been introduced by soldiers who came to Grenada from the island of Cyprus. In 1490, of the 20,000 soldiers who were unaccounted for, 17,000 had died of typhus and 3,000 were killed by the Moors. A second outbreak of typhus occurred during the time of the civil war in Grenada in 1557; it raged unchecked for 13 years.

Perhaps the most dramatic event involving typhus, recounted in Tolstoy's *War and Peace* and memorialized in the *1812 Overture*, was the defeat of Napoleon in Russia. In the spring of 1812, Napoleon was at the height of his power and glory. His Grand Army had experienced almost 20 years of unbroken successes. The Napoleonic Empire spread eastward to the Russian frontier and Austria, and in the north, west, and south it was bounded by the North Sea, the Atlantic Ocean, and the Mediterranean Sea. Great Britain was Napoleon's most formidable enemy, and the bulk of her wealth stemmed from trade, especially with India. Napoleon wanted to control India, which was at that time a colony of Great Britain, but the naval superiority of Britain prevented the French from intercepting the trading ships from India. Further, Napoleon could not invade and conquer Britain because the English Channel formed a barrier his armies could not cross. Therefore, Napoleon's military strategy was to take an indirect approach: conquer India, block the trade routes between India and Britain, and force Britain to capitulate. But cutting off the trade routes to India had failed once before when Lord Horatio Nelson defeated the French navy in 1798 at Aboukir Bay. It thus appeared that the only way for Napoleon to conquer India and the East would be through the formation of an alliance. He decided that Russia was strategically located for a land invasion of India, and he sought to make her an ally through a treaty of cooperation. In 1807, Napoleon had defeated the Russians at Tilsit, and the two countries had entered into a treaty, but cooperation soon failed and friction arose over Poland, which the Russians wanted to retain. There were other rifts as well, and so Napoleon decided to go for a dramatic success—the complete and utter defeat of Russia, and the saving of Poland. Napoleon believed that Russia, as a consequence of defeat, would become a submissive partner and would permit France to have easy entry to India by land. From 1811 onward Napoleon planned the invasion of Russia, and in March 1812 he signed agreements with Prussia and Austria to provide troops. Before long the force of his Grand Army numbered 500,000 men and 1,100 cannons, whereas Russia had fewer than 250,000 men.

During the summer of 1812 all went well, and the Grand Army moved quickly to the border between Poland and Prussia. In 4 days, without any opposition from Russia, the troops were camped in Vilna, Poland. Poland was filthy and dirty, and the roads were so rudimentary that the column straggled, while supplies of food were slow in arriving. The French troops foraged for food in the countryside, pillaging village after village. The Polish villagers were infested with fleas, bedbugs, and lice, and their homes were so filthy that one could both see and hear the cockroaches as they swarmed during the day and night. The exceptionally dry summer that year also restricted the amount of water available for bathing, and because of the hostility of the Poles, the men slept together in close groups. Lice in the infested hovels crept everywhere, clinging to the seams of the soldiers' clothing; nits and head lice infested their hair. Within a few days in Vilna the soldiers developed high fevers and pink rashes broke out on their bodies. By the time Napoleon's army left Vilna, the troops had contracted typhus, a disease endemic in Poland and Russia. Preventive measures could not be instituted, however, since at the time the cause of the disease and its manner of spread were not known.

The French army marched eastward toward Moscow, unopposed. Indeed, the Russian strategy was to give up land to the invading army, thus drawing the French deeper and deeper into Russia. By the third week of July, at Vitebsk, 80,000 men had died, and yet the French were still 150 miles from the Russian frontier and 300 miles from Moscow. As the march to Moscow continued, more and more of Napoleon's men became sicker; they not only suffered from typhus but lacked provisions as well. By mid-August, at Smolensk, the 174,000 men in the army of Napoleon were still 200 miles from Moscow. Smolensk fell on August 17, and the Battle of Valutino followed shortly thereafter, with 6,000 French casualties. But the Russians continued to retreat. Napoleon could not face withdrawal and humiliation, and so he pressed on rather than letting his army rest. On August 25 the march resumed, and the typhus raged on; Napoleon had now lost more than 100,000 troops. With nearly half of his central column fallen, the strike force was now reduced to 160,000 men. The two armies finally met at Borodino on September 5, 1812. Napoleon had 130,000 seasoned troops, and General Kutusov, the Russian commander in chief, had 120,000 troops. Napoleon did not attack, possibly because he was in great pain from a bladder infection and a feverish cold. For 2 critical days he delayed, but after the combatants met on September 7, there were 30,000 French and 50,000 Russian soldiers dead. It was a standoff. The Russians then withdrew beyond Moscow, and the French moved onward to Moscow. The French army suffered another 10,000 casualties due to sickness, and by the time they arrived in Moscow, only 90,000 men were in the central force. Eight out of 10 soldiers of Napoleon's army had fallen on the way to Moscow.

By the time Napoleon entered Moscow, most of its citizens had taken their belongings and evacuated the city; the streets were empty. Fires, deliberately set to deprive the French of food and shelter, had destroyed three-quarters of the city, and typhus continued to ravage the army. A month later Napoleon commanded his soldiers to leave Moscow, since there was little they could use and because it would be impossible to remain there without shelter as winter approached. By the time the army left, another 10,000 men had been lost to disease and wounds. Still the Russians refused to surrender, and so Napoleon and his troops began their march back to Poland. The first snows fell on November 3, and then the temperature dropped and it became bitter cold. Many of the troops froze to death. The cold and snow became more intense, and by the time the Grand Army reached Smolensk, to meet with the reserves, Napoleon realized that his troops had been wasted with typhus; discipline was breaking down, and food was lacking. When the French army departed the city of Smolensk, another 20,000 were sick in hospitals and living in ruined houses. By the time the French army reached Vilna on December 8, another 15,000 had died and the effective force consisted of only 20,000 men. Vilna was already crowded with the sick, and typhus was rampant. One observer described it: "Men lay on rotten straw soaked with their own excrement, without medical attention, and so hungry that they gnawed on leather or even human flesh." The French army moved on by December 10, leaving the sick and wounded in Vilna. Typhus remained rampant through the countryside for the entire winter. Dung was burned in the streets because it was believed that smoke would drive off the disease. It didn't. At the end of December about 20,000 men had arrived in Vilna, but by the following June fewer than 3,000 were alive. Word was sent to Napoleon on December 12 that the army no longer existed. By the time the French army had crossed into Germany, there were fewer than 4,000 men, and of these only 1,000 were ever healthy enough to be fit for duty.

It is estimated that during the Russian Campaign 400,000 French soldiers died from illness, exposure, or wounds; as many as 220,000 may have died from disease alone. The Russians also suffered from dysentery, malnutrition, and typhus, and it is estimated that at least 100,000 Russian soldiers died from either disease or battle wounds. So ended Napoleon's fantasy—the conquest of Russia and India—defeated by cold, hunger, illness, the Russians, and his own physical deterioration. At the Battle of Berezina, on November 29, 1812, Marshal Ney of Napoleon's army wrote to his wife, "General Famine and General Winter, rather than Russian bullets have conquered the Grand Army." He left out General Typhus, the unseen ally of the Russians.

Why did typhus become such an explosive epidemic during the 1812 campaign? There were many factors. The infected peasants served as a source of infection, and their lice were effective vectors; the soldiers were stressed by lack of food and water; the scorched earth policy of the Russians stressed the soldiers further, and this created movement of lice-infested and typhus-infected refugees; the soldiers were forced to live in crowded conditions for safety and warmth, and they had to sleep on the cold ground; clothing could not be washed or changed. It was a prescription for a military disaster.

Typhus in People

What is typhus? Before the discovery of the causative agent, typhus was believed to be the result of foul-smelling air. Indeed, the presence of typhus in prisons gave the disease one of its other names, "jail fever." Because bad smells came from the filthy

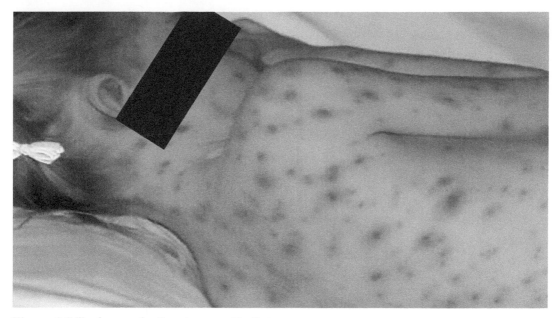

Figure 6.2 Typhus rash. Courtesy medbullets.com.

and unwashed prisoners who were crowded together in jails, it was thought that "bad air" was the source of infection. English judges to this day usually wear a small bunch of sweet-smelling flowers (called nosegays) to ward off the evil odors, thereby preventing them from contracting "jail fever." Typhus is often characterized by the sudden appearance of headaches, chills, a high fever, coughing, severe muscular pain, and a blotchy rash that appears on the 5th to 6th day after infection, initially on the upper trunk and then the entire body, except for, usually, the face, palms, and soles of the feet (Fig. 6.2). Delirium, confusion, and ultimately death occur. Untreated, the disease may last up to 3 weeks, with a mortality rate of 30 to 70% under epidemic conditions. Humans who recover are immune thereafter to further infections.

Finding the Killer

In 1909, Charles Nicolle (1866-1936), the director of the Pasteur Institute in Tunis, was assigned the task of doing something about typhus, which was decimating the local population. He first transmitted the disease to guinea pigs by injecting blood from an infected human, thereby proving that typhus persisted in the blood of an infected individual and remained infectious. He learned how the disease was transmitted quite by accident, however. Tunis had many patients with typhus, both within the hospital and on the streets; Nicolle also observed that the clothing of typhus patients was infectious. On occasion he found that people who worked in the hospital laundry also became infected, but once the patients were admitted to the hospital, and after they had a hot bath and were dressed in hospital clothing, they were no longer infectious. Nicolle deduced that there must be a vector in the clothes and underwear of the patients, and he suspected that the vector was a louse (Fig. 6.3). He injected a chimpanzee with blood from a typhus patient and allowed the chimpanzee to come down with typhus; lice were then obtained from this infected chimpanzee, and these were transferred to a healthy monkey. It came down with typhus. Nicolle recognized that the typhus agent was highly infectious and could be absorbed through the eye; so if a sufficient amount of dry fecal material from a louse were accidentally placed on the tip of the finger, it could be rubbed into the eye and cause infection. In a similar way, he observed, a person with louse droppings on the skin or clothes can become infectious.

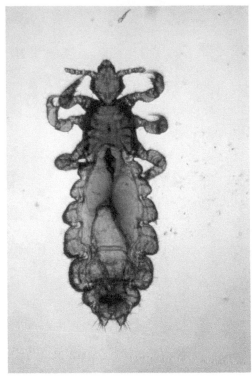

Figure 6.3 *Pediculus humanus humanus* (body louse). Courtesy CDC/ Dr. Dennis Juranek, 1979.

Nicolle also solved another mystery: he noticed that typhus tended to occur in individuals who were unable to clean themselves of their body lice and lice-infested clothing—in particular criminals lodged in the filthy jails—and that "prison fever" could spread from felon to judge, not by foul-smelling air but through infected lice. In September 1909 he wrote that it was obvious that typhus was transmitted by lice. Nicolle received the 1928 Nobel Prize for his work on the transmission of typhus; the discovery of the causative agent of typhus, however, was left to others.

Typhus is not the same as typhoid fever, which is a waterborne disease caused by a bacillus, *Salmonella*. Today we know the causative agent of the disease typhus to be a bacteria-like organism, *Rickettsia prowazekii*, related to the organisms that produce Rocky Mountain spotted fever, *Rickettsia rickettsii*, described by H. T. Ricketts in 1906, with a tick vector and a fatality rate of 20%. Rickettsias are small, oblong, Gram-negative bacteria-like organisms (Fig. 6.5) that are intracellular parasites. Rickettsias range in size from 0.3 to 0.6 μm and therefore are of intermediate size between viruses and bacteria.

Howard Ricketts (1871-1910) described the infectious agents responsible for both Rocky Mountain spotted fever and typhus. Ricketts graduated from Northwestern

University Medical School, and before taking up a position as an assistant professor of pathology at the University of Chicago, he traveled to the Pasteur Institute in Paris. There he became a bona fide microbe hunter. In the spring of 1906, Ricketts went to Montana to study "spotted fever," and while examining stained specimens of blood from infected monkeys and guinea pigs, he found "a bipolar bacillus." He saw the same bacillus in humans suffering from the disease; however, despite his training in standard techniques of bacteriology, such as culturing in nutrient broth and on blood agar, all attempts on his part to grow these bacilli failed. Ricketts did note that "spotted fever" was associated with people who had tick bites, and upon examining the salivary glands and gut of ticks that had fed on infected guinea pigs, he found the telltale microbe. But when ticks were fed on uninfected guinea pigs, no such microbes were seen. He concluded that this "spotted fever" of the Rocky Mountains was transmitted by the bite of ticks that carried the "bipolar bacilli."

In 1906 an outbreak of typhus in Mexico caught Ricketts's attention, and while there (and unaware of the work of Nicolle in Tunis), he showed that lice could transmit the disease, that typhus could be transmitted to monkeys, and that after recovery the individual was immune. Ricketts, though, was unable to follow up on these leads; he contracted typhus and died. (Not until 1938 did H. R. Cox discover that rickettsias are obligate intracellular parasites that require a living cell; he found that they could be grown in embryonated chicken eggs, making it possible to produce large quantities of rickettsias to be used as a phenol-killed vaccine.) A few years after Ricketts's death (1913), a Czech born in Bohemia, Stanislaus von Prowazek (1875-1915), carried on the work of Ricketts at the Bernhard Nocht Institute in Hamburg, Germany. In collaboration with Henrique da Rocha Lima, a Brazilian, he studied typhus in Serbia and Turkey and found rickettsias in lice that had fed on typhus-infected patients. In 1915, while working in a German hospital facing an outbreak of typhus among Russian prisoners of war, Prowazek contracted typhus and died. Da Rocha Lima described the causative agent of typhus in 1916 and named the organism *Rickettsia prowazekii*, after his fallen colleague and Ricketts.

A Lousy Business

It is the human louse—including the body louse *Pediculus humanus corporis* (Fig. 6.3) and the head louse *P. humanus capitis*, as well as the crab louse *Phthirus pubis*—that is involved in rickettsial transmission of typhus from one human to another by means of fecal contamination of the bite wound.

In days gone by, when people bathed and changed clothes less frequently and there were no insecticides, everyone from the royal family to the lowly peasant had lice. Archbishop Thomas Becket (b. 1118) was murdered in Canterbury Cathedral on December 29, 1170. The next day he was dressed for burial in an extraordinary way. He had on a large brown mantle; under it was a white surplice, and below that was a lamb's-wool coat, then another woolen coat, and a third woolen coat below this; under this there was the black-cowled robe of the Benedictine order; and under this was a haircloth shirt covered with linen. As the body grew cold, the vermin that were living in the many layers of clothing started to crawl out, and it is told that "the vermin boiled over like water in a simmering cauldron, and the onlookers burst into alternate weeping and laughter."

The habit of shaving the head and wearing a wig probably came about in part to hold down the vermin population, but even the wigs could harbor head lice and their eggs, called nits. Samuel Pepys (1633-1703) wrote in his diary that he had to return to Westminster and complain to his barber that the wig he had newly purchased was so full of nits that he had to have it cleaned because they "vexed me cruelly." George Washington wrote: "Kill no vermin such as fleas and lice in the sight of others, and if you do see any carefully put your foot down on it; and if it is on the clothes of your companion put it off privately, and if it be on your own clothes, return thanks to him who put it off."

Even today lice are found in jails, mental institutions, schools, and facilities where cleanliness is less than optimal. Lice are more frequently encountered among the crowded poor in the city. For example, in England in 1939-1940, the incidence of lice in preschool and school children in large cities was 50%, but in rural areas the incidence was 5%. In the city of Riverside, California, the incidence of head lice today in some of the public schools is 30%. Although it has been claimed that people may be infested with thousands of lice, this appears to be an exaggeration; a more accurate number would be less than a dozen. In a recent survey carried out in Ethiopia among migrants, 36% had <10, while only 5% had >200 body lice; 39% had body lice, 65% head lice, 10% scabies mites, and 4% fleas.

The head louse and body louse are descendants of lice that were associated with our hairy forefathers and -mothers when we lived in caves. As we became more and more hairless, the lice became restricted to the hairier parts of the body—the pubic area and the head—and others lived in the clothing next to the skin. A person infested with dozens of body lice may remove his clothing and find not a single specimen on his body, but examination of the underwear or clothing will reveal them on the threads or in the seams.

The life cycle of the body louse begins with an egg laid in the folds of clothing (but nowhere else!). Since the body louse is susceptible to cold, the eggs are usually attached to the inner clothing close to the skin. The egg (Fig. 6.4) is held in place by a glue-like substance, and if the clothes are not removed, the egg will hatch in 6 to 9 days. The newly emerged louse moves onto the skin and feeds before returning to the clothing, where it remains until its next meal. Typically a louse will feed five times a day. The growing louse will shed its outer skin (exoskeleton) three times, at 3, 5, and 10 days after hatching; after the last shedding (called a molt) it is mature and can live for another 20 days. All stages feed on blood. An egg, attached to the hair, is called a nit (Fig. 6.4). A female louse can lay 10 eggs a day and may live 30 to 40 days. Theoretically, a pair of lice can produce 200 offspring during their 1-month life span.

Lice have been recognized as parasites of humans for thousands of years. Nits are about 1 mm in length, so a fine-tooth comb is necessary to remove them; combs were probably invented to remove nits (Fig. 6.6). Nits have been found associated with a 5,000-year-old Egyptian mummy, and preserved lice have been recorded among the remains of the inhabitants of the city of Pompeii (destroyed by the volcanic eruption of Mount Vesuvius in A.D. 79). The nit has become a part of our language: someone who deals in small details, and is petty, is called a nitpicker.

The female and male lice puncture the skin and feed on blood, and the red blood cells in the gut are digested rapidly and remain liquid. The louse gut, however, is susceptible to rupture, and the louse may turn red as its gut contents spill out into the rest of the body. This is most frequently seen when the louse is infected, because the rickettsias also destroy its gut cells. Feeding lice have the nasty habit of defecation, and louse feces are extremely dry and powdery.

Lice do not survive if starved, and they cannot live at high or very low temperatures, preferring temperatures between 29 and 32°C—approximately the temperature in the seams of clothing close to the body. If a host becomes too hot because of fever or exercise, infecting lice will leave, and since body lice die at 50°C, this temperature is critical when washing clothing, as soap and water on their own will not kill lice. Lice are also very susceptible to dehydration and require a humidity of 70 to 90%; death occurs at <40%. The only way the louse can maintain hydration is by feeding on blood, and lice must feed every day or else they die. Lice abandon the body of a dead person, and they tend to avoid temperatures >40°C.

Lice become infected with rickettsias after taking a blood meal from an infected human. Several days later the rickettsias have multiplied in

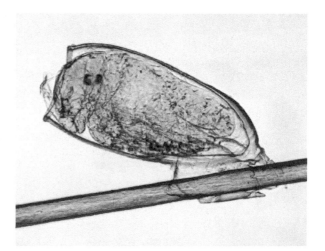

Figure 6.4 An unhatched nit containing a developing head louse. Courtesy CDC/Dr. Dennis Juranek, 1979.

the louse, and now the powdery feces contain many infective rickettsias. It is the feces of the louse that sets up the infection: the feeding is irritating to the host, and when the host scratches his or her skin, the feces carrying the rickettsias are rubbed into the wound. A typhus infection is thus the result of wound contamination, not inoculation through a louse bite. The destruction of the gut of the lice by the rickettsias results in its death 1 to 3 weeks after infection, but death does not occur before the louse is able to transmit the rickettsias.

Discrimination and Dissemination

Anne Frank wrote in her diary (*The Diary of Anne Frank*) how she and her family along with four others were driven into hiding during the Second World War. They remained hidden from the Nazis for 25 months by living in an annex of rooms above her father's business in Amsterdam. After being betrayed, all were arrested and deported to Nazi concentration camps. Nine months later Anne Frank, age 15, and her sister both died of typhus at Bergen-Belsen. Suffering from malnutrition and living in

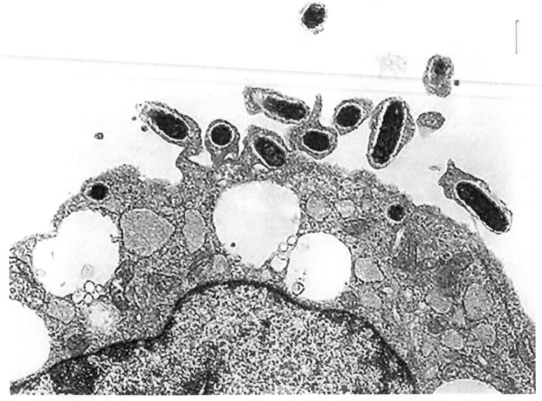

Figure 6.5 *Rickettsia prowazekii* as seen with transmission electron microscope, from the Pathogen Profile Dictionary, www.ppdictionary.com, with permission © 2010, *Journal of Undergraduate Biological Studies.*

the crowded and unsanitary conditions of the concentration camp, she was exposed to infected lice, contracted the disease, and succumbed to it. Millions of other Jews died in Nazi concentration camps, not from typhus but from being "deloused."

At the end of World War I and prior to the discovery and deployment of DDT (dichlorodiphenyltrichloroethane) as an insecticide (1939) and delousing agent (1943), it was a common practice to fumigate entire railroad cars and the holds of ships with steam or cyanide gas. In 1923, German chemists synthesized Zyklon B (a porous

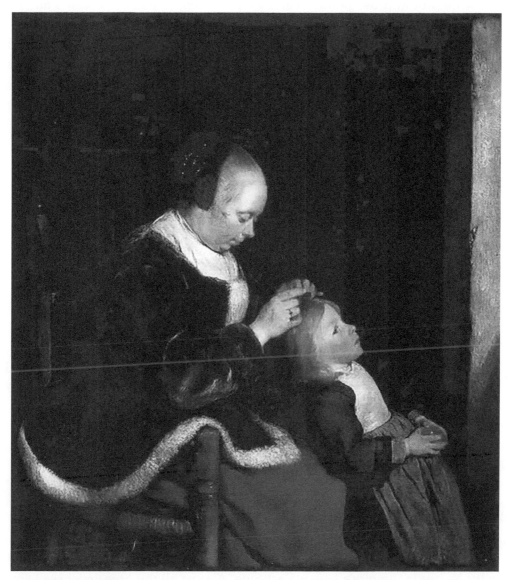

Figure 6.6 *The Louse Hunt* by Gerhard ter Bosch (1617-1681) Mauritshuis, The Hague, Courtesy Wikimedia Commons.

material soaked in hydrocyanic acid with a stabilizer), ostensibly as an improvement on these methods, since it did not require tanks of cyanide gas. Clothing, furs, and leather goods could be deloused without damage by steam heat. Zyklon B, however, was put to a more sinister purpose during World War II when the Germans disguised their extermination facilities at Bergen-Belsen, Buchenwald, Dachau, and Auschwitz as delousing stations with showers, barbers, and laundries to lull the filthy, over-crowded, and lice-ridden Jews, political prisoners, and other "undesirables" into gas chambers, which were then filled with Zyklon B. In this way, the delousing agent Zyklon B did not improve public health by controlling typhus; it became an instrument of death for the "Final Solution."

R. prowazekii can live in louse feces for up to 3 months, and Russian clothes cleaners have been reported to have acquired the infection by this route. Because body lice live in clothing, the prevalence of typhus is determined by weather, humidity, poverty, and lack of hygiene. Epidemic typhus occurs more frequently during the colder months, especially winter and early spring, since this is the time of the year when the inhabitants wear multiple layers of clothing, and in poor countries it is also a time when clothing is changed infrequently. Transmission is favored by sexual promiscuity, by crowding such as occurs in the trenches during war, and by conditions in prisons that are cramped and cold and where personal hygiene is limited.

Although nits may survive for up to 16 days if kept at low temperatures, leaving clothes unwashed and unworn for a week results in the death of the lice and their eggs. Historically, this may have been the basis for the weekly change of clothes on Sunday after the Saturday night bath. Eradication of the body louse can be achieved only by improving the general level of hygiene. The simplest method for control is delousing by frequent changes of clothing, or if this is not possible, then washing or dry cleaning can be used. Boiling is also effective. A more rapid approach is dusting of clothing with DDT (10%), malathion (1%), or permethrin (1%). Since lice visit the skin only to feed and otherwise spend their time in the clothing, it is unnecessary to delouse the individual. Neither bathing nor shaving is necessary for removal of adult lice, but the adults (not the nits) of head lice can be killed by an oral dose of ivermectin. Nits can be removed by combing (see Fig. 6.6) or using a chemical remedy such as Nix, Rid, or lindane.

Typhus and the "Immigrant Problem"

On January 30, 1892, the steamship S.S. *Massilia*, carrying 750 immigrants and a crew of 50, reached New York City. Shortly after the passengers disembarked, typhus fever broke out in the city's Lower East Side. Almost all the cases occurred among the newly arrived Russian Jewish immigrants. Public health authorities described its victims as foreigners with an odor that resembled the smell of rotting straw. To contain the further spread of typhus, the health authorities established a quarantine.

Avoiding and isolating the sick had been a human response to plagues since antiquity. When Hippocrates and Galen warned that it was "dangerous to associate with those afflicted," there was a clear appreciation of the fact that diseases could travel and that the path they would take followed human migrations; prior to the 20th century, however, physicians could offer little in the way of treatment or prevention of epidemic diseases. One approach at disease control—even before there was an appreciation of germs as agents of disease—was the quarantine of those discovered to be sick.

The word "quarantine," derived from the Italian *quaranta giorni*, or "40 days," has its origins in response to the Black Death (see p. 67). But why 40 days? The enforced detention of crew, passengers, and cargo when disembarking was not permitted by city ordinance for more than a month. The length of the quarantine may be based on the Hippocratic doctrine distinguishing acute and contagious diseases from those that are chronic, or it may refer to a number frequently mentioned in the Old Testament, while others ascribe it to the 40 days Jesus spent alone in the wilderness. The most likely explanation, however, is a practical one: during the Renaissance it was noted that after 40 days those stricken with the plague either died or recovered, and were then no longer contagious. The drive for establishing a quarantine tends to be greatest for those diseases for which the means of transmission is poorly understood. Because of this, any and all precautions against its spread seem reasonable. To wit: bubonic plague, SARS, AIDS, Ebola, and in 1892 the New York City typhus epidemic.

Quarantine may have had a medical rationale, but more often than not it has been used to isolate and stigmatize a particular social group. Dr. David Musto wrote that "quarantine is far more than … marking off or creation of a boundary to ward off a feared biological contaminant lest it penetrate a healthy population" and that it has "a deeper emotional and aggressive character." It blames, shames, and separates, but "when it hits hardest at the lowest social classes or other fringe groups, it provides that grain of sand on which the pearl of moralism can form." Typhus can be that grain of sand.

In September 1891 a group of Jews began their escape from the harsh conditions and anti-Semitism in the Russian province of Volhynia. The exiled Jews boarded a steamer in Odessa that took them across the Black Sea to Constantinople, Turkey, and there they remained as fugitives for 3 months. By Christmas, given expulsion orders, they traveled to Smyrna (Izmir), where some went by rail to Marseilles and there boarded the S.S. *Massilia* bound for New York. The voyage was interrupted by a stop in Naples, where 470 Italians, also immigrating to New York, boarded. Conditions on the ship—especially in steerage—were abominable. The S.S. *Massilia* was overcrowded and filthy. The food consisted of decaying herring, rotting potatoes, stale bread, rancid butter, and tea. Open troughs served as toilets that were only occasionally flushed with water or cleaned. Salt water was used for bathing, laundry, and the washing of plates and utensils. After 28 days at sea the immigrants—malnourished and debilitated—arrived in New York harbor. At a mandatory quarantine station off Staten Island the 800 passengers and crew were inspected by two physicians for evidence of typhus, cholera, yellow fever, plague, smallpox, and leprosy. The entire process took less than an hour. Because nothing "medically remarkable" was found, the ship and its passengers were admitted to the port of New York. A medical exam at the Ellis Island immigration station (the port of first landing for more than 75% of immigrants to the United States) was similarly perfunctory, and all were given a "clean bill of health." On February 1 the 268 Russian Jews from the S.S. *Massilia* were placed by the United Hebrew Charities in eight boardinghouses on New York City's Lower East Side. By February 11 there was an outbreak of typhus in this neighborhood that would spawn panic in New York City and beyond.

The social fabric of New York City in the 1890s hardly evokes the Gay Nineties celebrated in song and story. There were periods of economic depression, social upheavals coincided with the emergence of crowded cities bursting at their seams with factory workers, and there was labor unrest. Many Americans attributed these social evils to the newly arrived immigrants from Europe, especially those from Russia, Italy, and Austria-Hungary. Further, during this period there were pandemics of cholera and typhus raging in Europe and Asia, and with fast-moving steamships the wide Atlantic Ocean would no longer serve as an impenetrable barrier against the introduction of epidemic diseases by infected foreigners. In addition, the number of immigrants to the United States was accelerating: between 1881 and 1884 there were 3 million, and between 1885 and 1898 that number had doubled. These new immigrants, "wretched refuse," as they were called, were a threat not only to the nation's public health but to its economic well-being, since it was generally perceived that these foreigners would not only drive down wages but would also soon become public charges. Another objection to immigration from Eastern Europe came from the racist and anti-Semitic sentiments of "native" Americans.

As late as 1892 many New Yorkers were convinced that disease was a result of miasmas—malodorous vapors that could cause epidemics. It was "logical" (though unscientific) to propose that typhus could be spread from the foul-smelling, unkempt, and impoverished immigrants to the healthy and native-born Americans. The chief sanitary inspector of the New York City Health Department, Cyrus Edson, proposed a spontaneous generation theory for the miasma: "typhus cases could have arisen under the conditions such as existed on that ship; where the hatches have been battened down and the people were overcrowded and the conditions filthy, and they have breathed impure air for a certain length of time and had food which was not of a proper character to nourish them; these conditions would tend to breed typhus de novo." To stem this "typhus plague," Edson had to act swiftly. He ordered the sanitary policemen to find "every single Russian Jewish passenger of the *Massilia*." Rigid quarantine was established in the Lower East Side, but typhus continued to rage.

Edson then resorted to more-stringent measures. All healthy contacts were forcibly evacuated to Riverside Hospital on North Brother Island in the East River. (Ironically, this would be the same site where Mary Mallon—"Typhoid Mary"—would be held against her will from 1915 until her death in 1938; see p. 10). The overwhelming majority (about 1,150 people) sent to North Brother Island were healthy but were simply unfortunate enough to have resided in or near the boardinghouses where the sick *Massilia* passengers had lived. (The Italian passengers on the *Massilia* who had dispersed beyond New York City were much more difficult to track down, and those found were neither infected nor quarantined.) Adding insult to injury, the sanitary conditions on North Brother Island were far from optimal to contain an epidemic. Outhouse privies were rarely cleaned, and because there was limited access to soap and running water, many of the detainees waded into and bathed in the East River. Bed space was at such a premium that "patients" were crowded together in flimsy tents that provided little shelter against the elements.

Hearing of the typhus epidemic in New York City, many American newspapers published articles and cartoons depicting Russian Jews as vectors of the disease. The *Boston Medical and Surgical Journal* stated: "We open our doors to squalor and filth and misery—which means typhus." A *New York Times* editorial of February 13, 1892, ran: "Such immigrants are not wanted either in this city or any other part of the United States. They should be excluded. The doors should be shut against them." In late February these complaints led the port of New York to establish a policy that detained and quarantined all East European Jews coming through New York harbor regardless of their port of embarkation. Enforcement, however, was strictly along ethnic lines, since passengers from Scandinavia who traveled on the same ships as European Jews were allowed to land without delay.

Then, by April 1, 1892, the typhus epidemic seemed to be over. Among the *Massilia* passengers there had been 138 cases and 13 deaths, and among the 1,200 quarantined first on the Lower East Side and then on North Brother Island there had been 49 cases and 6 deaths, for an overall death rate of about 12%. No new cases occurred on the Lower East Side. Those who were quarantined and recovered or did not develop typhus after 3 weeks of observation were released. Quarantine was considered to be effective: typhus had been restricted to the Lower East Side and the disease spread no further. The *New York Times* credited Edson and the health department for "averting a pestilence," and the *New-York Tribune* claimed it was "the isolation of all Jewish suspects and their stringent quarantine that was responsible for the success." But quarantine's success had exacted a social cost: civil liberties were violated, there was overt discrimination and ethnic insensitivity, and the inadequate resources provided for medical care contributed to an excessive number of deaths.

There still remains a lingering question: why was it that only the Russian Jews of the S.S. *Massilia* were infected with typhus? It appears that the infection was not acquired in Odessa, since the Jews remained healthy for the 3 months they were in Turkey. Most likely the Jews contracted typhus in Constantinople, where the disease was endemic. Thus, the seeds of infection that were sown in Turkey blossomed in the crowded boardinghouses on New York's Lower East Side and set in motion a xenophobia that would restrict the entry of East European Jews and other undesirable immigrants. Active anti-immigration sentiments culminated in the passage of the Immigration Restriction Act of 1924 and similar legislation enacted through the 1930s. The message here is clear: epidemic disease can often inspire quarantines, and quarantines can be the tool for social scapegoating.

Typhus Today

Typhus occurs most frequently in the mountainous areas of Ethiopia, Burundi, and Rwanda in Africa; Peru in South America; and Nepal and Tibet in Asia, where the climate favors the presence of body lice. In the United States there is a reservoir for typhus in flying squirrels. Some claim that epidemic typhus occurred in the Americas sometime in the 16th century and that this coincided with the Spanish importing the body louse into the Americas. It may be that after human infections occurred, it passed into the flying squirrel population and then spread from Mexico, where it was described by the Aztecs.

Although typhus is frequently considered a disease of the past, it is not. In 1976, after a 12-year absence and following a civil war, typhus broke out in refugee camps in Burundi, and 100,000 were infected. Ethiopia reported 1,931 cases in 1983 and 4,076 cases in 1984. More than 45,000 cases of epidemic typhus occurred in Burundi in association with civil war during the 1990s. There were also sporadic cases in northern Africa and a smaller outbreak in Russia, and in 1998 there was a small outbreak in Peru.

When symptoms such as headache, fever, and rash occur in patients with body lice or persons living in cold, crowded, and unhygienic conditions, this should suggest typhus. Though traditional methods used in bacteriology cannot be used for growing the intracellular *R. prowazekii*, the rickettsias can be grown in embryonated eggs or tissue cultures. Once grown in chick embryos or tissue cultures, the rickettsias can be stained or they can be detected by molecular methods such as PCR.

Typhus Vaccine

At one time protection against typhus was achieved through vaccination. Rudolf Weigl (1883-1957) developed such a vaccine. Weigl obtained a doctorate in zoology and became acquainted with typhus during World War I. After the war he continued his studies of lice and typhus, and in 1921 he established a research institute in Lwow, Poland (now Lviv, Ukraine). There he discovered a method for injecting *R. prowazekii* intrarectally into lice—an essential element in the production of a vaccine. If the lice were to sustain the infection, they required human blood, and this was provided by "lice feeders"—that is, by healthy human volunteers who strapped as many as 40 cages containing lice onto their thighs and were admonished not to scratch. In this way the researchers fed 30,000 lice per month. Finally, the process required the removal of the blood-bloated intestine of each louse by a technician and the killing of the typhus. By 1931, Weigl had developed a vaccine by attenuating the *R. prowazekii* through passage in lice and animals. The preparation of the vaccine itself was not without risk, and several of Weigl's staff contracted the disease and died from it.

As World War II heated up across Europe, typhus was still the most dangerous and feared of diseases. France and the United States had produced typhus vaccines; Weigl's vaccine, however, was the only field-tested prophylactic. The Germans did not have access to the American vaccine, and Hitler and his underlings feared that typhus could imperil their wartime efforts as much as would Allied bullets. In the summer of 1941 the Germans seized Lwow, took over Weigl's laboratory, and forced him to develop large quantities of a typhus vaccine to protect the Wehrmacht. Weigl cooperated with the Nazis and provided them with a vaccine, but he also saw to it that 30,000 doses were smuggled into the Warsaw ghetto starting in November 1940. The Weigl Institute also became a haven for thousands of at-risk Poles—perhaps as many as 3,000 became lice feeders, lice injectors, and vaccine preparers.

Parallel to Weigl's typhus vaccine was one prepared by Ludwig Fleck (1896-1961), who earned a doctorate at Lwow University under the direction of Weigl. Fleck's specialty was immunology. In 1919 he joined Weigl's institute, and between 1921 and 1923 he developed a method to diagnose typhus using typhus antigens. Because Fleck was a Jew, however, and there was virulent anti-Semitism in Poland, this achievement did not earn him academic recognition. He worked in various governmental laboratories until 1935, when Polish anti-Semitic policies made it impossible for Jews to hold such governmental positions, so with his wife's dowry he opened a private laboratory. In August of 1942, Fleck, now in Lwow's Jewish ghetto, managed to create a typhus vaccine using the urine of typhus patients, and 6 months later he was sent to Auschwitz, where he continued to work on bacteriologic research. In December 1943 he was dispatched to the Buchenwald concentration camp to work on a typhus vaccine. At Buchenwald he conspired to produce two kinds of vaccine using rabbit lungs: large quantities of a useless vaccine were shipped to SS troops at the front, while smaller doses of an effective vaccine were used secretly to immunize concentration camp prisoners. The ruse was never discovered.

After the war Weigl was smeared as a collaborator by jealous rivals, and although he was allowed to teach, the government refused to appoint him to the Polish National Academy. At Weigl's 1957 memorial service a speaker said: "He transformed the louse, a symbol of dirt, misery and aversion, into a useful object of scientific research and a life-saving tool." With the end of the war Fleck assumed a microbiology professorship in Lublin, considered at the time a desert, both poor and culturally barren, where he remained until 1952. Despite his testifying at the Nuremberg trials, he too was accused of being a Nazi collaborator and he continued to face anti-Semitism. He moved to a bacteriology laboratory in Warsaw for 3 years, and in 1957 he emigrated to Israel, where he died not long afterwards.

Postscript: The Weigl vaccine was used in the 1920s and early 1930s, and later a phenol-killed vaccine developed in 1938 by Cox was used. Today, however, most patients are treated more efficiently with an antibiotic such as doxycycline, tetracycline, or chloramphenicol.

Coda

What lessons does typhus have to teach us now that we know its cause, its mode of transmission, how it can be controlled, and how it is treated? The words Hans Zinsser wrote 85 years ago still ring true: "Typhus is not dead. It will live on for centuries, and it will continue to break into the open whenever human stupidity and brutality give it a chance, as most likely they occasionally will. But its freedom of action is being restricted and more and more it will be confined, like other savage creatures, in the zoological gardens of controlled diseases."

doi: 10.1128/9781683670018.ch7

7
Malaria, Another Fever Plague

I wanted to sit up, but felt that I didn't have the strength to, that I was paralyzed. The first signal of an imminent attack is a feeling of anxiety, which comes on suddenly and for no clear reason. Something has happened to you, something bad. If you believe in spirits, you know what it is: someone has pronounced a curse, and an evil spirit has entered you, disabling you and rooting you to the ground. Hence the dullness, the weakness, the heaviness that comes over you. Everything is irritating. First and foremost, the light; you hate the light. And others are irritating—their loud voices, their revolting smell, their rough touch. But you don't have a lot of time for these repugnances and loathings. For the attack arrives quickly, sometimes quite abruptly, with few preliminaries. It is a sudden, violent onset of cold. A polar, arctic cold. Someone has taken you, naked, toasted in the hellish heat of the Sahel and the Sahara and has thrown you straight into the icy highlands of Greenland or Spitsbergen, amid the snows, winds, and blizzards. What a shock! You feel the cold in a split second, a terrifying, piercing, ghastly cold. You begin to tremble, to quake to thrash about. You immediately recognize, however, that this is not a trembling you are familiar with from earlier experiences—when you caught cold one winter in a frost; these tremors and convulsions tossing you around are of a kind that any moment now will tear you to shreds. Trying to save yourself, you begin to beg for help. What can bring relief? The only thing that really helps is if someone covers you. But not simply throws a blanket or quilt over you. The thing you are being covered with must crush you with its weight, squeeze you, flatten you. You dream of being pulverized. You desperately long for a steamroller to pass over you. A man right after a strong attack … is a human rag. He lies in a puddle of sweat, he is still feverish, and he can move neither hand nor foot. Everything hurts; he is dizzy and nauseous. He is exhausted, weak, and limp. Carried by someone else, he gives the impression of having no bones and muscles. And many days must pass before he can get up on his feet again.

Figure 7.1 (Left) A Thai mother attends her sick child suffering from cerebral malaria. Courtesy of Peter Charlesworth.

This is malaria as described by the Polish journalist Ryszard Kapuscinski in his personal story of Africa, *The Shadow of the Sun*. His experience is not unusual. Malaria is a fever plague, and it has been said this disease has killed more than half the people who have ever lived on this planet. It still kills (Fig. 7.1). Today, a person dies of malaria every 10 s—mostly children under the age of 5 living in Africa. The total number of cases of malaria is estimated to be 300 million to 500 million, with ~10% of these occurring outside Africa. Annually, millions of deaths are caused by malaria. Malaria infections are on the rise, and as the fever plague spreads, it will continue to affect us in the places we live, work, travel to, and fight in.

A Look Back

The antiquity of human malaria is reflected by the records in the Ebers papyrus (ca. 1570 B.C.), in clay tablets from the library of Ashurbanipal (2000 B.C.), and in the classic Chinese medical text the *Nei Ching* (2700 B.C.). These describe the typically enlarged spleen, periodic fevers, headache, chills, and fever. Malaria probably came to Europe from Africa via the Nile Valley or by closer contact between Europeans and the people of Asia Minor. The Greek physician Hippocrates (460-370 B.C.) discussed in his *Book of Epidemics* the two kinds of malaria: one with recurrent fevers every third day (benign tertian) and another with fevers on the fourth day (quartan). He also noted that those living near marshes had enlarged spleens. Although Hippocrates did not describe malignant tertian malaria in Greece, there is clear evidence of the presence of this malaria in the Roman Republic by 200 B.C. Indeed, the disease was so prevalent in the marshland of the Roman Campagna that the condition was called the Roman fever. Since it was believed that this fever recurred during the sickly summer season due to vapors emanating from the marshes, it was called by the Italian name *mal' aria*, literally, "bad air." Over the centuries malaria spread across Europe, reaching Spain and Russia by the 12th century; by the 14th century it was in England. Malaria was brought to the New World by European explorers, conquistadors, colonists, and African slaves. By the 1800s it was found worldwide.

On October 20, 1880, Charles Louis Alphonse Laveran (1845-1922), a physician in the French Foreign Legion in Algeria, examined under the microscope a drop of blood taken from a soldier suffering with malaria fever (Fig. 7.2). He found within the red blood cells transparent globules containing black-brown malaria pigment, and on occasion mobile filaments were seen emerging from clear spherical bodies (a process he called exflagellation). He also found that some malaria patients had blood cells shaped like crescents. In effect, Laveran had discovered an animal parasite with different developmental stages. Initially his work was received with skepticism, but by 1884 the Italian workers E. Marchiafava, A. Celli, and C. Tomassi-Crudeli, working in the malaria-infested Roman Campagna, confirmed his findings.

The significance of Laveran's observation of exflagellation went unappreciated until 1896-1897, when William MacCallum (1874-1944) and Eugene Opie (1873-1971), students at Johns Hopkins University, found that the blood of sparrows and crows infected with *Haemoproteus* (a bird parasite closely related to malaria) contained two kinds of crescent-shaped gametocytes (male and female sex cells), and that exflagellation reflected the release of microgametes from the male gametocyte. They also correctly interpreted their observations: the gametocytes in the blood, when ingested by a biting insect, release the gametes in the stomach, where fertilization takes place, producing a worm-like zygote, the ookinete.

There remained a mystery: how does "bad air" cause malaria? Ronald Ross, a surgeon-major in the Indian Medical Service, was probably the most

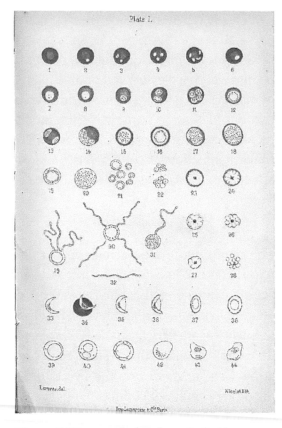

Figure 7.2 Laveran's drawing of what he saw under the light microscope when examining a drop of blood from a soldier with chills and fever.

unlikely person to solve the puzzle of how humans "catch" the disease. Ross was born on Friday, May 13, 1857, in the foothills of the Himalayas, where his father was an officer in the British army stationed in India, which at that time was a part of the British Empire. As a boy of 8 he was shipped by his parents to England to receive a proper British education. He was a dreamer, and, although he liked mathematics, he preferred wandering around the countryside, observing and collecting plants and animals. At Springhill Boarding School, which he attended from the age of 12, he began to write poetry, painted watercolors, and thought of becoming an artist. His father had other ideas and insisted that he study medicine in preparation for entry into the Indian Medical Service. So at age 17 (following his father's orders) young Ronald began his medical studies. He was not a good student, not because of laziness but because he had so many other interests and could not concentrate on medicine. He preferred composing music to learning anatomy, and he wrote epic dramas rather than writing prescriptions. Publishers rejected these "great works," and he had them printed at his own expense. He eventually did pass his medical examination (after failing the first time), worked as a ship's doctor, and then entered the Indian Medical Service. Although India was rife with disease—malaria, plague, and cholera—Ross busied himself writing mathematical equations, took long walks, wrote poetry, played the violin, and studied languages. Occasionally he used his microscope to look at the blood of soldiers ill with malaria, but he found nothing. He shouted to all who could hear: "Laveran is wrong. There is no germ of malaria."

In 1894, Ross returned to England on leave. By that time he had spent 13 years in India and had few scientific accomplishments: he had published a few papers on malaria in the *Indian Medical Gazette* and claimed that malaria was primarily an intestinal infection. On April 9 he visited with Patrick Manson (1844-1922), who, while working as a physician to the Customs Service in Taiwan (Formosa), had shown that mosquitoes carry filaria—the roundworms that cause the disease elephantiasis. Manson was an expert microscopist, and, taking a drop of blood from a sailor ill with malaria, he showed Ross Laveran's parasite peppered with the black-brown malaria pigment. One day, as the two were walking along Oxford Street, Manson said:

> Do you know Ross, I have formed the theory that mosquitoes carry malaria ... the mosquitoes suck the blood of people sick with malaria ... the blood has those crescents in it ... they get into the mosquito stomach, shoot out those whips ... the whips shake themselves free and get into the mosquito's carcass ... where they turn into a tough form like the spore of an anthrax bacillus ... the mosquitoes die ... they fall into water ... people drink a soup of dead mosquitoes and they become infected.

Though this was a romantic story and mostly a guess on the part of Manson, Ross took it as fact.

Manson became Ross's mentor and encouraged him to study mosquito transmission. Ross left England in March 1895 and reached Bombay a month later. In June 1895, encouraged by Manson's passionate plea, he captured various kinds of mosquitoes, although he hadn't a clue what kind they were. He set up an experiment using the water in which an infected mosquito had laid her eggs and in which her young had been swimming. The insects were allowed to die, and the water with its dead and decaying mosquitoes was then given to a volunteer to drink. The man came down with a fever, but after a few days no crescents could be found in his blood. The same experiment was repeated with two other men, who were paid for their services. Failure again. It appeared, then, that drinking water with mosquito-infected material did not produce the disease. Ross began to think that, perhaps, the mosquitoes had the disease but that they probably gave it to human beings by biting them and not by being eaten. He began to work with patients whose blood contained crescent-shaped malaria parasites, and with mosquitoes bred from larvae. The first task was to get the mosquitoes to bite the patients. It was like looking for a needle in a haystack. There are >2,500 different kinds of mosquitoes, and at the time there were no good means for identifying most of them. Initially, Ross worked mostly with the gray and striped-wing kinds. When these mosquitoes were dissected, although the whip-like extensions were found in their stomachs, there was no further development. This result was no more informative than what Laveran had seen in a drop of blood on a microscope slide nearly 20 years earlier.

Today, we understand (as Ross did not) why this was so: the gray mosquitoes are *Culex*, and those with striped wings are *Aedes*, and these do not carry human malaria. No, the one that he should have used was the brown, spotted-winged mosquito, *Anopheles*, but Ross did not recognize this for the entire year he dissected mosquitoes. Thousands of mosquito carcasses were all he had to show for his labors, and each mosquito dissection required hours of effort at the microscope.

Then Ross, now 40 years old and having spent 17 years in the Indian Medical Service, turned from the incompetent species of mosquito to the competent, brown, spotted-winged *Anopheles*. On August 16, 1897, his assistant brought him a bottle in which mosquitoes were being hatched from larvae. It contained "about a dozen big brown fellows, with fine tapered bodies hungrily trying to escape through the gauze covering of the flask which the angel of fate had given my humble retainer!" He wrote: "My mind was blank with the August heat; the screws of the microscope were rusted with sweat from my forehead and hands, while the last remaining eyepiece was cracked. I fed them on Husein Khan, a patient who had crescents in his blood. There had been some casualties among the mosquitoes, and only three of the *Anopheles* were left on the morning of August 20, 1897. One of these had died and swelled up with decay." At 7 a.m. Ross went to the hospital, examined patients, attended to correspondence, and dissected the dead mosquito, without result. Then—a significant reminder of one of those unknown quantities in the equation to be solved—he examined an *Anopheles*. No result again. He wrote in his notebook:

> At about 1 p.m., I determined to sacrifice the last mosquito. Was it worth bothering about the last one, I asked myself? And, I answered myself, better finish off the batch. A job worth doing at all is worth doing well. The dissection was excellent and I went carefully through the tissues, now so familiar to me, searching every micron with the same passion and care as one would have in searching some vast ruined palace for a little hidden treasure. Nothing. No, these new mosquitoes also were going to be a failure: there was something wrong with the theory. But the stomach tissues still remained to be examined—lying there, empty and flaccid, before me on the glass slide, a great white expanse of cells like a large courtyard of flagstones, each one of which must be scrutinized—half an hour's labor at least. I was tired and what was the use? I must have examined the stomachs of a thousand

mosquitoes by this time. But the angel of fate fortunately laid his hand on my head, and I had scarcely commenced the search again when I saw a clear and almost perfectly circular outline before me of about 12 microns in diameter. The outline was too sharp, the cell too small to be an ordinary stomach cell of a mosquito. I looked a little further. Here was another, and another exactly similar cell.

The afternoon was very hot and overcast; and I remember opening the diaphragm of the substage condenser of the microscope to admit more light and then changing the focus. In each of these, there was a cluster of small granules, black as jet, and exactly like the black pigment granules of the … crescents. I made little pen-and-ink drawings of the cells with black dots of malaria pigment in them [Fig. 7.3]). The next day, I wrote the following verses and sent these to my dear wife:

> This day relenting God
> Hath placed within my hand
> A wondrous thing; and God
> be praised. At his command,
> Seeking his secret deeds
> With tears and toiling breath,
> I find thy cunning seeds,
> O million-murdering death.
> I know this little thing
> A myriad men will save.
> O death, where is thy sting?
> Thy victory, O grave?

Here was the clue. I had shown that, four or five days after feeding on infected blood, the mosquito had wartlike oocysts on its stomach. But, did these keep on growing, and how did these mosquitoes become infective? I planned to answer these questions shortly but, before that work could begin, I reported my findings to the *British Medical Journal* in a paper entitled, "On some peculiar pigmented cells found in two mosquitoes fed on malarial blood." It appeared December 18, 1897.

Ross knew he could wrap up the unfinished work in a matter of a few weeks, but then was struck by a blow from the Indian Medical Service; he was ordered to proceed to Calcutta—immediately. As soon as he arrived in Calcutta, he set his hospital assistants the task of hunting for mosquito larvae and pupae. Soon he had a stock of the brown mosquitoes, and set about getting them to bite patients who were suffering from malaria. By flooding the ground outside the laboratory, he hoped to imitate rain puddles and see whether something could be learned about mosquito breeding. "If I am not on the pigmented cells again in a week or two," he wrote to Manson, "my language will be dreadful."

In Calcutta, Ross was given a laboratory. There were two Indian assistants who

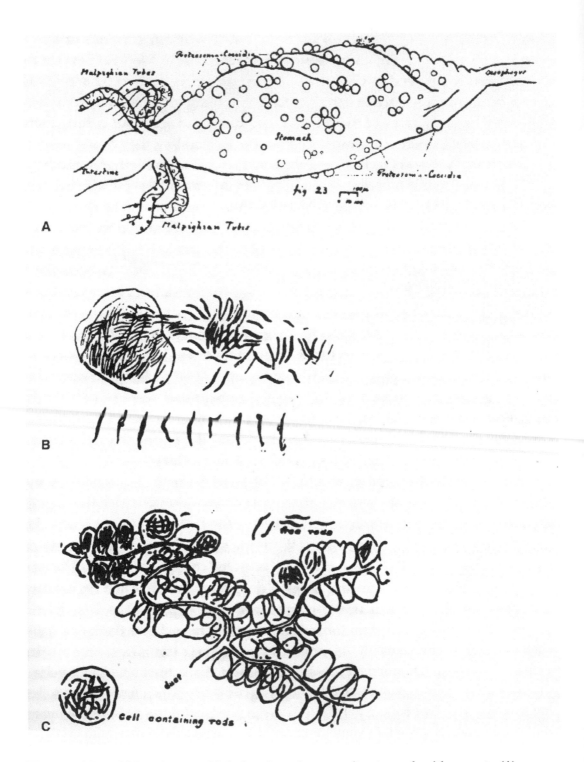

Figure 7.3 Ronald Ross' pen and ink drawing of a mosquito stomach with oocysts (A), an oocyst bursting to release sporozoites (B) and (C) a salivary gland with sporozoites (labeled by Ross as rods).

had already been working there when he arrived, but they were old men and not very intelligent, so he engaged two younger men, Purboona and Mahomed Bux. Both of these he paid out of his own monies. The Calcutta hospitals, however, did not have a large number of malaria cases. Ross accordingly turned to something that Manson had suggested earlier: the study of mosquitoes and malaria, as seen in birds. Pigeons, crows, larks, and sparrows were caught and placed in cages on two old hospital beds. Mosquito nets were put over the beds, and then at night infected mosquitoes were put under the nets. Before much time had passed, the crows and pigeons were found to harbor malaria parasites in their blood; Ross also found the pigmented cells on the stomachs of mosquitoes that had been fed on infected larks.

Ross became certain of the whole life history in the mosquito, except that he had not actually seen the zygotes, those spherical cells into which the flagella had penetrated, turning into oocysts. This was the last stage in the study. He found that the size of the pigmented cells on the stomach depended exactly on the length of time since the mosquitoes had been fed on infected blood. The parasites grew to their maximum size ~6 days after they had fed on infected blood. They left the stomach after this time, but what happened to them then?

One day, while studying some sparrows, he found that one was quite healthy, another contained a few of the malaria parasites, and the third had a large number of parasites in its blood. Each bird was put under a separate mosquito net and exposed to a group of mosquitoes from a batch that had been hatched out from grubs in the same bottle. Fifteen mosquitoes were fed on the healthy sparrow; on their stomachs not one parasite was found. Nineteen mosquitoes were fed on the second sparrow; every one of these contained some parasites, though, in some cases, not very many. Twenty insects were fed on the third, badly infected, sparrow; every one of these contained some parasites in their stomachs, and some contained huge numbers.

Ross wrote in his *Memoirs*:

> This delighted me! I asked the medical service for assistance and a leave, but was denied this. I wanted to provide the final story of the malaria parasite for this meeting; but, I knew that time was very short. I still did not have the full details of ... the change from the oocysts on the mosquito's stomach into the stages that could infect human beings and birds. Then, I found that some of the oocysts seemed to have stripes or ridges in them; this happened on the 7th or 8th day after the mosquito had been fed on infected blood. I spent hours every day peering into the microscope and wrote to Manson: "The constant strain on mind and eye at this temperature is making me thoroughly ill." I thought no doubt these oocysts with the stripes or rods burst—but then what happened to them? When they burst, did they produce the same stages that infected human blood?

Then, on July 4, 1898, Ross got something of value. Near a mosquito's head he found a large branch-looking gland. It led into the head of the mosquito. Ross said to himself, "It is a thousand to one that it is a salivary gland." Did this gland infect healthy creatures? Did it mean that if an infected mosquito fed off the blood of an uninfected human being or bird, then this gland would pour some of the parasites (called sporozoites [Fig. 7.3B]) … into the blood of the healthy creature?"

During July 21 and 22 of that year, Ross took some uninfected sparrows, allowed mosquitoes (which had been fed on malaria-infected sparrows) to bite them, and then, within a few days, was able to show that the healthy sparrows had become infected. This was the proof—this showed that malaria was not conveyed by dust or bad air. After all, men and birds don't go about eating dead mosquitoes. On July 25, now sure, he sent off a triumphant telegram to Sir Patrick Manson, reporting the complete solution; 3 days later Manson spoke at the British Medical Association meeting in Edinburgh, describing the long and painstaking piece of research that Ross had been carrying out for years. Ross's findings were communicated on July 28, 1898, and he left India for England in February 1899.

Ross's discovery of infectious stages in the mosquito salivary glands in a bird malaria appeared to be the critical element in understanding transmission of the disease in humans. Manson, however, rightly cautioned: "One can object that the fact determined for birds do not hold, necessarily, for humans." Ross and Manson wanted to grab the glory of discovery for themselves and for England, but they were not alone in such a quest. The German government dispatched a team of scientists under the leadership of Robert Koch to work in the Roman Campagna, an area notorious for endemic malaria. They isolated a bacillus from the air and the mud of the marshes and, rejecting the claims of Laveran, named the causative agent for malaria *Bacillus malariae*. When the bacillus could not be grown in the laboratory, however, Koch discarded it as the cause of the disease; undeterred, he continued to look further. Koch then visited the laboratory of Giovanni Battista Grassi (1854-1925) at the University of Rome and told him of his failure with the "germ" of malaria and mentioned Ross's communication. At that moment Grassi had what is today called an aha moment.

Where Ross was patient and perseverant and willing to carry out a seemingly endless series of trial-and-error experiments, Grassi was methodical and analytical. He was also able to distinguish the different kinds of mosquitoes. Grassi observed that "there was not a single place where there is malaria—where there aren't mosquitoes too, and either malaria is carried by one particular blood sucking mosquito out of the forty different kinds of mosquitoes in Italy—or it isn't carried by mosquitoes at all." He recognized that there were still two tasks left: to identify the mosquito that transmitted human malaria and then to demonstrate the mosquito cycle for human malaria. Working with Amico Bignami, Giovanni Bastianelli, Angelo Celli, and Antonio Dionisi, he went into the highly malarious Roman Campagna and the area surrounding it collecting mosquitoes and at the same time recording information on the incidence of malaria among the people. (Grassi was, in effect, carrying out an epidemiologic study.) It soon became apparent that most of the mosquitoes could be eliminated as carriers of the disease because they occurred where there was no malaria. But there was an exception. Where there were *zanzarone*, as the Italians called the large, brown, spotted-winged mosquitoes, there was always malaria. Grassi recognized that the *zanzarone* were *Anopheles* and stated: "It is the anopheles mosquito that carries malaria." Grassi and his team were able to infect clean *Anopheles* mosquitoes by having them feed on patients with crescents in their blood, and he was able to trace the development of the parasite from the mosquito stomach to the salivary glands. The life cycle in the human was, as Ross had correctly surmised, similar to that of the bird malaria with which he had worked. Grassi's work showed that the association of the disease with swampy, marshy areas of the world can be traced to their being ideal breeding sites for mosquitoes.

With this work Grassi demolished the theory of Koch and was able to prove that "it is not the mosquito's children, but only the mosquito who herself bites a malaria sufferer—it is only that mosquito who can give malaria to healthy people." Grassi wanted his beloved country Italy to receive recognition for his work, but it did not. Instead, it provoked a bitter and very nasty disagreement with Ross. Ross claimed that it was only after Grassi had read his work on the transmission of malaria using birds that he recognized that only in those areas where there was *Anopheles* was there human malaria, and that *Culex* was not involved, but Grassi did not publish this or the development of the parasite in these mosquitoes until late in 1898. Ross wrote in his *Memoirs*: "They … had this paper of mine before them when they wrote their note. Their statement was … a deliberate and intentional lie, told in order to discredit my work and so to obtain priority … Many of the items … are directly pirated from my … results … stolen straight from me …" For decades the arguments over priority and originality raged between Ross and Grassi. In truth, although Ross predicted that human malaria must also be transmitted in a similar way to bird malaria, it was Grassi and coworkers who actually conducted experiments on human malaria transmission and formulated and understood that there was a closed cycle between mosquito and human, and it was the Italians who were able to show the complete dependence on the *Anopheles* mosquito for the transmission of human malaria. In spite of this, Ross, not Grassi, received the Nobel Prize in 1902, and Laveran received the Prize in 1907.

Although malaria can be induced in a host by the introduction of sporozoites through the bite of an infectious female mosquito, the parasites do not immediately appear in the blood. This was surprising in view of the fact that in 1903 Fritz Schaudinn claimed to have seen sporozoites directly invade erythrocytes. Schaudinn was a dominant figure in parasitology, and so few doubted his word, but over time, since no one could repeat his observation, it was called into question. In 1948, H. E. Shortt, P. C. C. Garnham, and their colleagues in England inoculated rhesus monkeys with sporozoites that they had obtained from the salivary glands of mosquitoes infected with *Plasmodium cynomolgi* (a parasite similar to the benign tertian malaria, *Plasmodium vivax* of humans), and in a week parasites, called preerythrocytic stages, were found in the livers of the monkeys. Later, they demonstrated similar stages in biopsy material taken from the livers of human volunteers who had been infected by the bite of mosquitoes carrying *P. vivax*; and also, after infected mosquitoes had fed on other volunteers, this stage was found at the same site in malignant tertian malaria, *Plasmodium falciparum*. It was now clear that when an infected female anopheline mosquito feeds, it injects sporozoites that go first to the liver (where they live and multiply for several weeks).

The Disease Malaria

Although some 170 species of *Plasmodium* have been described, only five are specific for humans. The human malarias, *P. falciparum, P. vivax, P. ovale, P. malariae,* and *P. knowlesi*, are transmitted through the bite of an infected female anopheline mosquito when during blood feeding she injects sporozoites from her salivary glands. The number of sporozoites inoculated is usually <25. These travel via the bloodstream to the liver, where they enter liver cells. The entire process takes less than an hour. Within the liver cell the parasite multiplies asexually to produce 10,000 or more infective offspring. These do not return to their spawning ground, the liver, but instead invade erythrocytes. It is the asexual reproduction of parasites in red blood cells and their ultimate destruction with release of infectious offspring (merozoites) that is responsible for the pathogenesis of this disease. Merozoites released from erythrocytes can invade other red cells and continue the cycle of 10-fold parasite multiplication, with extensive red blood cell destruction. In some cases the merozoites enter red cells but do not divide. Instead, they differentiate into male or female gametocytes (the crescents that Laveran observed). When ingested by the female mosquito, the male gametocyte divides into eight flagellated microgametes, which escape from the enclosing red cell (exflagellation). These swim to the macrogamete, one fertilizes it, and the resultant motile zygote, the ookinete, moves either between or through the cells of the stomach wall. This encysted zygote, resembling a wart on the outside of the mosquito stomach, is an oocyst, and through asexual multiplication thread-like sporozoites are produced within it. The oocyst bursts, releasing its sporozoites into the body cavity of the mosquito, which quickly find their way to the salivary glands. When this female mosquito feeds again, the transmission cycle has been completed.

All the pathology of malaria is due to parasite multiplication in erythrocytes. The primary attack of malaria begins with headache, fever, anorexia, malaise, and myalgia. These symptoms are followed by paroxysms of chills, fever, and profuse sweating. There may be nausea, vomiting, and diarrhea. Such symptoms are not unusual for an infectious disease, and it is for this reason that malaria is frequently called "the Great Imitator." Then, depending on the species, the paroxysms tend to assume a characteristic periodicity. In *P. vivax*, *P. ovale*, and *P. falciparum* the periodicity is 48 h, and for *P. malariae* the periodicity is 72 h. The fever spike may reach up to 41°C and corresponds to the rupture of the red cell as merozoites are released from the infected cell. If the infection is not synchronous and there are several broods of parasites, the periodicity may occur at 24-h intervals. Anemia is the most immediate pathologic consequence of parasite multiplication and destruction of erythrocytes, and there can also be suppression of red cell production in the bone marrow. During the first few weeks of infection the spleen is swollen from the accumulation of parasitized red cells and the proliferation of white cells. At this time it is soft and easily ruptured. If the infection is treated, the spleen returns to normal size; in chronic infections, however, the spleen continues to enlarge, becoming hard and blackened due to the accumulation of malaria pigment. The long-term consequences of malaria infections are an enlarged spleen and liver.

Falciparum infections are more severe, and when untreated can result in a death rate of 25% in adults. Complications of malaria include kidney insufficiency, kidney failure, fluid-filled lungs, neurological disturbances, and severe anemia. In the pregnant female falciparum malaria may result in stillbirth, lower-than-normal birthweight, or abortion. Nonimmune individuals and children may develop cerebral malaria, a consequence of the mechanical blockage of microvessels in the brain as the result of the sequestration of infected red cells. If relapse occurs in falciparum malaria, it is the outcome of the increase in numbers of preexisting erythrocytic forms, previously too low to be detected microscopically; this type of relapse is termed recrudescence. Falciparum malaria accounts for half of all clinical malaria cases and is responsible for nearly all deaths.

P. vivax and *P. ovale* malarias also have the capacity to relapse; that is, parasites can reappear in the blood after a period when none were present. This type of relapse, called recurrence, results from the delayed liberation of merozoites from preerythrocytic stages in the liver, called hypnozoites. *P. vivax* results in severe and debilitating attacks but is rarely fatal. It accounts for ~45% of all clinical malaria cases.

The benign quartan malaria, *P. malariae*, may persist in the body for up to 4 decades without signs of pathology. *P. malariae* and *P. ovale* infections make up the balance of all clinical cases.

"Catching" Malaria

Malaria is by far the most important of the world's tropical parasitic diseases, but it can and does exist in temperate areas. At present 90 countries or territories in the world are considered malarious, almost half of them in Africa south of the Sahara. As Grassi correctly observed: "Mosquitoes without malaria … but never malaria without mosquitoes." Historically, malaria occurred anywhere that the vector and parasite could live, generally between latitudes 64°N and 32°S where the temperatures were between 16 and 33°C and the altitude was below 2,000 m. Mosquito transmission of malaria is dependent on a complex array of factors, including the incidence of infections in the human population, the suitability of the local anopheline population—density and breeding and biting habits—the availability of susceptible or non-immune hosts, climatic conditions, and the local geographic and hydrogeographic conditions that contribute to mosquito breeding sites. Malaria can also be transmitted without mosquitoes. Introduction of infected blood by transfusion or through contaminated needles are two of the less common ways of "catching" malaria. Malaria may also infect the fetus by the transplacental route.

Malaria Today

When the journalist Ryszard Kapuscinski described his experiences with malaria, he could not be certain whether he was suffering from some other disease such as a bad case of the flu. How could he be sure it was malaria? A drop of blood from the fingertip could serve as the principal specimen source for diagnosis. In expert hands a stained blood film examined by light microscopy (Fig. 7.4) can distinguish among the various species and can detect 5 to 10 parasites per µl. Several new techniques, however, have been developed: fluorescence microscopy, detection of parasite-specific nucleic acid sequences, and parasite antigen detection.

Once Grassi had flagged the *Anopheles* mosquito as the vector of human malaria, methods for controlling it were possible. Of the 450 species of *Anopheles*, only 50 are capable of transmitting the disease, and of these, only 30 are considered efficient vectors. In Africa the most efficient vector is *A. gambiae*, which can breed in small, temporary pools of water such as those formed by foot- or hoofprints or tire tracks. In other areas *A. stephensi*, which can breed in wells or cisterns, is the vector. In the 1900s larvicides in the form of oil and Paris green (bright green powdered copper acetoarsenite that is extremely poisonous and sometimes used as an insecticide or fungicide), as well as drainage, were introduced to limit mosquito-breeding sites in water. These steps had outstanding success in reducing transmission in some parts of the world. DDT (dichlorodiphenyltrichloroethane) was introduced later as a component of a World Health Organization (WHO) eradication campaign, though by the early 1960s it had became clear that eradication could not be accomplished due to the emergence of DDT-resistant mosquitoes and the negative ecological side effects of DDT. By 1969 the WHO formally abandoned its eradication campaign and recommended that countries employ control strategies. Today, attempts at control involve insecticide-impregnated bed nets and spraying with ecologically less disruptive, but more expensive, insecticides.

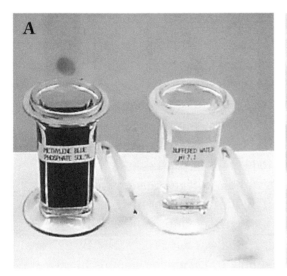
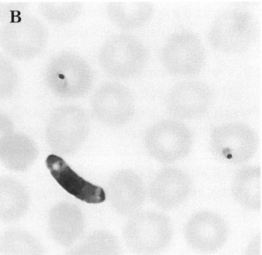
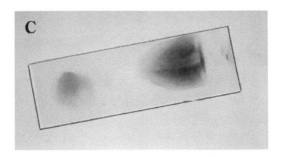

Figure 7.4 (A) Method of staining blood film (Courtesy CDC/Dr. Mae Melvin, 1977). (B) Stained blood smear from a patient with falciparum malaria showing a crescent (gametocyte) and 2 red cells infected with ring stages (Courtesy CDC/Steven Glenn, Laboratory Training & Consultation Division, 1979). (C) Thick and thin blood smear stained with Giemsa stain.

Treatment

Since all the pathology of malaria stems from parasites multiplying in the blood, most antimalarial drugs are directed to these rapidly dividing stages. In the 16th century a serendipitous discovery led to a treatment for malaria. The remedy, called by the Indians of Peru *quina-quina*, literally "bark of barks," was first introduced into Europe by the Jesuit missionaries who came to Peru 4 decades after Francisco Pizarro's conquest of the Incas. The Jesuits had easy access to the bark through their missions; arranged collection of the bark in Peru, Ecuador, and Bolivia; powdered it; and then sold it in Europe. Because the Jesuit fathers were the promoters and exporters of the remedy, it was called Jesuits' bark or Jesuits' powder. The Jesuits' powder, infused with white wine to disguise its bitter taste, was sprinkled with the juices of various herbs and flowers and given immediately after the malaria paroxysm. Rising demand for the new remedy and a desire to better understand the trees that produced the useful bark led to a series of botanical expeditions to the New World to find those trees with a predictable high-quality bark. Only the bark of one cinchona tree (of the 23 different kinds), *Calisaya officinalis*, was found to have any fever-reducing properties.

The active component of the cinchona bark was unknown until 1820, when two French apothecaries, Pierre Pelletier and Joseph Caventou, isolated the active alkaloid and named it quinine. By the 1840s export of the bark to Europe from the Andean republics amounted to millions of tons. In 1844, Bolivia passed laws that prohibited the collection and export of seeds and plants without a license, since 15% of the country's tax revenue came from bark exports. The idea was to protect their monopoly, discourage reckless stripping of the forests, and prevent smuggling. By 1850 the British had decided that a controlled supply of cinchona was necessary. The British army in India estimated that it needed an annual supply of 1.5 million pounds to prevent 2 million adults from dying of malaria in India, and they calculated that to rehabilitate the 25 million survivors of malaria would require 10 times as much. And then there was Africa, with a malaria fever rate of up to 60% in some regions.

The most successful collector of cinchona seeds was a short, barrel-chested Englishman, Charles Ledger (1818-1905). In 1865, Ledger sent 14 pounds of the high-quality seeds to his brother George in London, and George attempted to sell them to the British government. The British government was not interested, and so the remaining half was sold to the Dutch for about $20. Within 18 months the Dutch had 12,000 plants ready to set out, and 5 years later their analyses of bark showed the quinine content to be between 8 and 13%. To honor Ledger, this high-yielding cinchona species was named *Cinchona ledgeriana*. Experimenting with hybrids and grafting onto suitable rootstocks, the Dutch developed the world's best cinchona trees. The Dutch formed a cooperative, the Kina Bureau in Amsterdam, to control quinine production. Consequently, by 1918, 90% of the bark from Java, which represented 80% of the world's production, was sent to Amsterdam and distributed by the Kina Bureau. At the outbreak of World War II, Java had some 37,000 acres of cinchona trees, which produced >20 million pounds of bark a year. The Dutch quinine combine had created what amounted to the most effective crop monopoly of any kind.

On December 7, 1941, the Japanese attacked Pearl Harbor, and as a consequence the United States declared war. Because of the Japanese occupation of Java, the supply of quinine to the Allies was cut off. During World War II the Allies relied on an acridine compound, Atabrine, that had been synthesized in the 1930s by the German company I.G. Farben. Since the antimalaria activity of Atabrine was similar to that of quinine—i.e., it killed the parasites growing in red blood cells but did not affect the parasites in the liver—it could be used prophylactically to suppress the symptoms of malaria.

During World War II it was recognized that a victory by the United States would be tied to the development of a number of research and development programs: the Manhattan Project, which produced atomic bombs; the Radiation Laboratory, which developed radar; and a crash program for the development of new antimalarials. The latter, begun in 1942 under the auspices of the National Research Council, was a massive program coordinated by a loose network of panels, boards, and conferences, with the actual work done at universities, hospitals, and pharmaceutical industry laboratories as well as in U.S. Army and U.S. Navy facilities. At Johns Hopkins University canaries, chicks, and 60,000 ducks were infected with malaria parasites and used to screen and test 14,000 potential antimalarial compounds for activity.

On April 12, 1946, an article in the *New York Times* declared, "Cure for malaria revealed after 4-year, $7,000,000 research. The curtain of secrecy behind which the multimillion dollar Government anti-malaria program had been operating, in the most concentrated attack in history against this scourge, was completely lifted today for the first time, with the revelation of the most potent chemicals so far found." The next day an editorial in the same paper said: "When the scientific story of the war is written, we have here an epic that rivals that of the atomic bomb, the proximity fuse and radar." The drug receiving so much attention was chloroquine (marketed under the trade names Aralen and Nivaquine). In 1955 the WHO, confident in the power of chloroquine to kill malaria parasites and DDT to kill the malaria-carrying *Anopheles* mosquitoes, launched the Global Malaria Eradication Campaign. As a result, malaria was eliminated from Europe, Australia, and other developed countries, and large areas of tropical Asia and Latin America were freed from the risk of infection, but the campaign had excluded sub-Saharan Africa. The WHO estimated that the campaign saved 500 million human lives that otherwise would have been lost due to malaria, but the vast majority of the 500 million lives saved were not Africans. By the 1960s there were ominous reports of treatment failures with chloroquine. There were strains of falciparum malaria in South America and Southeast Asia that were no longer susceptible to the killing effects of chloroquine, and these were now spreading across the globe. As a result, the WHO eradication campaign was abandoned.

The Vietnam War began in 1959, and by 1962 it was already apparent to the U.S. military that the malaria situation in the country was serious. Chloroquine-resistant malaria was killing more troops than bullets. In 1963, Colonel William Tigertt, the director of the Walter Reed Army Institute for Research, set into motion the machinery for the U.S. Army Research Program on Malaria to identify and develop antimalarial agents effective against the emerging chloroquine-resistant strains of *P. falciparum*. The result was a Walter Reed compound, WR-142,490, named mefloquine, that was effective against chloroquine-resistant malaria. In 1979, Hoffmann-La Roche launched the drug by itself (as a monotherapy) under the trade name Lariam; Lariam was the drug of choice for travelers and visitors to areas where chloroquine-resistant malaria was present. As early as 1977, however, using a variety of rodent malarias, Wallace Peters (London School of Hygiene and Tropical Medicine) had suggested that emergence of resistance to mefloquine would be reduced if it were administered in combination with another antimalarial. Indeed, Peters wrote: "It is strongly recommended that mefloquine should be deployed only for the prevention or treatment of malaria in humans caused by chloroquine-resistant *P. falciparum* … and for large scale use, mefloquine should not be employed until a second antimalarial has been identified that will minimize the risk of parasites becoming resistant to this potentially valuable new compound." Indeed, in the early 1980s the first reports of resistance to mefloquine appeared. This prompted the WHO to issue a publication in 1984 expressing reservations concerning the widespread use of the drug as a monotherapy and suggesting that it be used in combination with another antimalarial; this, the WHO said, might preserve the effectiveness of mefloquine by delaying the emergence of resistance as well as potentiate the drug's activity. There were further complications with the prophylactic use of mefloquine: neuropsychiatric episodes, including insomnia, strange or vivid dreams, paranoia, dizziness or vertigo, depression, and attempts at suicide, presumably due to a blockade of ATP-sensitive potassium channels and connexins in the substantia nigra in the brain. (The overall risk varies with ethnicity and is higher in Caucasians and Africans than Asians, and involves differences in health and cultural and geographical background, and thus the actual reasons for the differential adverse reactions are not clear.) Due to concerns over the safety of mefloquine prophylaxis in Western countries, the packet insert was revised (July 2002) to declare that use of Lariam "is contraindicated in patients with known hypersensitivity to mefloquine or related compounds such as quinine and quinidine. Lariam should not be prescribed for prophylaxis in patients with active depression, generalized anxiety disorder, psychosis or other major psychiatric disorders or with a history of convulsions." For these reasons, as well as the development of mefloquine resistance in parts of Southeast Asia where malaria is endemic and the loss of efficacy, it is unlikely that mefloquine alone will ever regain its clinical efficacy in such areas.

During the Vietnam War (1959-1975), the Chinese government supported the Vietnamese Communists against the United States, which was becoming increasingly involved in the fighting. Malaria was rampant in Vietnam, causing casualties on both sides, and Ho Chi Minh asked Mao Zedong for new antimalarial drugs to assist the Vietnamese troops in their fight against the Western foe. On May 23, 1967, the China Institute of Sciences set up a top-secret military program called "Project 523" with the express purpose of discovering antimalarial drugs that could be used in the battlefield.

Professor Youyou Tu joined Project 523 in 1969, and she and her research group examined >2,000 Chinese herbals that might have antimalarial activity, and >200 recipes from Chinese traditional herbs and >380 extracts from these herbs were tested in a rodent malaria model. In 1971, Tu and her group identified *qinghaosu* (translated as "artemisinin"), meaning the active principle of Qinghao (the Chinese name for *Artemisia*), obtained from dried *Artemisia annua* by using low heat and ethyl ether extraction after those parts of the herbal extract that made it sour were discarded. In May 1972 the first official report stated that when the preparation was fed to mice (50 mg/kg daily for 3 days) infected with *Plasmodium berghei* malaria, it killed asexual-stage parasites and was 95 to 100% effective as a cure. Qinghaosu had no effect on liver-stage parasites. Tu and her colleagues bravely tested the active compound on themselves before knowing its chemical structure; 7 months later there was another report of the successful treatment in Beijing of 30 patients with malaria, >90% of whom were said to have recovered from infections with either *P. falciparum* or *P. vivax*. It was only then that a systematic characterization of the chemistry and pharmacology of the active compound, artemisinin, began.

Between 1975 and 1978 the Chinese performed successful clinical trials with artemisinin on >6,550 patients with malaria. New variants of artemisinin were developed in the country, such as the methyl ester artemether, soluble in oils and suitable for oral administration; and the hemisuccinate ester artesunate, which is more soluble than artemether and can be used for intravenous injection. Artemisinin was effective in chloroquine-resistant cases and produced a rapid recovery in 141 cases of cerebral malaria. There was no evidence of serious toxicity; there was, however, a high relapse rate. Artemisinin works quickly and is eliminated quickly. How does artemisinin act to kill the malaria parasite? It has been shown that in the parasite heme (in the form of hemozoin) catalyzes the activation of artemisinin to bind irreversibly to as many as 124 parasite proteins, especially to key metabolic enzymes. Resistance of *P. falciparum* to artemisinin has been shown both in laboratory experiments and in the field, and this resistance results from a mutation (*kelch13*) that prolongs the length of time the parasite spends as a ring stage. Because artemisinin activation is low at the ring stage, parasites with the *kelch13* mutation are able to overcome protein damage; thus, they are selected, since they have a higher capability to survive the drug treatment at the ring stage. On the other hand, the much higher level of drug activation at the later developmental stages (trophozoite and schizont) triggers extensive protein modifications that act like an exploding bomb, inhibiting multiple key biological processes and eventually resulting in the death of the parasite. It is thus less likely that the parasite will develop resistance at the later stages. There has been considerable concern that resistance in the field to the artemisinins would emerge, just as it has to almost every other class of antimalarials. Indeed, it has been possible in the laboratory to select strains of rodent malarias that are 5 to 10 times less susceptible to artemisinin, and by *in vitro* selection the same level of resistance has been accomplished with *P. falciparum*. After 2001 the use of artemisinins as a single drug for treatment was decreased significantly in order to prevent the emergence of resistance. In 2006 the WHO requested discontinuation of the manufacturing and marketing of all artemisinin monotherapies except for the treatment of severe malaria.

Investigators found that when artemisinin was combined with a partner drug that had not already engendered resistance, the new combinations (called artemisinin combination therapy, or ACT) were effective and well tolerated, although they were more expensive than the failed single-drug treatment. Further, the rapid clearance of artemisinin drugs when used to treat early cases of uncomplicated malaria may prevent progression to severe disease and reduce mortality. Because of their very rapid clearance from the blood, complete cure using artemisinin requires longer courses of therapy: given alone, a 7-day regimen is required to maximize cure rates. Adherence to a 7-day course of treatment, however, is frequently poor, so in the ACT the partner drug chosen is usually a more slowly eliminated drug. This allows a complete treatment course to be given in just 3 days. This being so, in 2001 the WHO endorsed ACT as the policy standard for all malaria infections in areas where *P. falciparum* is the predominant infecting species. And in 2005 the WHO recommended a switch to ACTs as a first-line malaria treatment.

In 1994, Novartis formed a collaborative agreement with the Chinese Academy of Military Sciences, Kunming Pharmaceutical Factory, and the China International Trust and Investment Corporation to develop artemisinin combinations further, and eventually it registered Coartem and Riamet. Coartem constitutes ~70% of all current clinically used ACTs. Coartem (Novartis), consisting of artemether/lumefantrine, in tablet form was approved by the FDA in April 2009. It is highly effective when given for 3 days (six doses) with a small amount of fat to ensure adequate absorption and thus efficacy. In an ACT the artemisinin component acts quickly over a 3-day course of treatment and provides antimalarial activity for two asexual parasite cycles, resulting in a staggering reduction of the billion blood parasites within an infected patient, while the parasites that remain are left for removal by the partner drug (a process variably assisted by an immune response). Therefore, the artemisinin component of the ACT reduces the probability that a mutant resistant to the partner drug will arise from the primary infection, and if effective, the partner should kill any artemisinin-resistant parasites that arise.

The major threat of malaria today is not an increasing range of endemicity, but rather a rise in the intensity of antimalarial drug resistance. Drug resistance in malaria has been defined as the ability of a parasite strain to survive and/or multiply despite the administration and absorption of a drug in doses equal to or higher than those usually recommended but within the limits of tolerance of the subject. Resistance is thus a characteristic of the particular parasite strain. In addition to resistance, there is another treatment constraint: cost. Quinine and chloroquine were cheap antimalarials; mefloquine and artemisinin are more expensive. (This is of particular concern in those countries where malaria is most prevalent and the economy is poor.) A third constraint for the drug treatment of malaria is the reluctance of the pharmaceutical industry to invest their capital in developing antimalarials that they believe will not yield profits.

Prevention of Malaria

Generally speaking, the measures for the prevention of malaria are to block infected mosquitoes from feeding on humans, eliminate the breeding sites of mosquitoes, and provide measures to kill mosquito larvae, as well as to reduce the life span of the blood-feeding adult. Prevention of contact with adult mosquitoes can be accomplished by using insect repellents, wearing protective clothing, using impregnated mosquito netting, and equipping houses with door and window screens. Breeding sites can be controlled by draining water, changing the salinity, flushing, altering water levels, and clearing vegetation. Destruction of adult mosquitoes can be accomplished by using sprays, and larvae can be destroyed by larvicides. Prevention can also be accomplished by education and treatment of the human population. Employing all these strategies has helped to eradicate malaria from many temperate parts of the world, but in the tropics and in developing countries, especially those with limited budgets for mounting public health campaigns where there is parasite and insecticide resistance, malaria is on the rise.

Genetic Resistance to Malaria: Sickle Cell Trait and Duffy Factor

In 1949, J. B. S. Haldane hypothesized that β-thalassemia, caused by mutations in the β-globin genes that result in a decrease or loss in hemoglobin production, may offer protection against malaria. Haldane's hypothesis is supported by the observation that the geographic distribution of the thalassemias overlaps that of endemic falciparum malaria, and that populations with particular thalassemias have less malaria (Fig. 7.5). It has been proposed that the protection afforded by thalassemia may result from a modification in surface receptors for *P. falciparum*.

In 1910, James Herrick, a physician in Chicago, examined a drop of blood from a medical student who came from the West Indies and who had pains in the joints and anemia. He placed a drop of the student's blood on a slide, covered it with a cover glass, and ringed this with Vaseline to prevent it from drying out and also to exclude oxygen; when he observed the blood under the microscope, the normally biconcave red cells had become sickle-shaped. It was shown later that this is due to the presence of sickle hemoglobin. Such patients with sickle cell anemia may show clogging of blood vessels, pneumonia, rheumatism, heart disease, painful episodes, inflammation of the hands and feet, and anemia.

One of the curiosities of malaria in Africa was that the native populations seemed to suffer less from the disease than the Europeans. Why? And if malaria is so lethal, why hasn't the human species been eliminated by this fever plague? Malaria as a disease can be controlled not only by drugs but also by the development of acquired immunity—which usually requires repeated exposure to infection. But there is also resistance to malaria that does not require exposure, infection, and recovery; this is natural or innate immunity and is a result of genetics.

A glance at the map of the world showing the distribution of malaria prior to the eradication campaigns of the 1940s and '50s shows that malaria is a tropical and semi-tropical disease (Fig. 7.6). In the 1950s, Anthony Allison observed that the distribution of the sickle hemoglobin gene overlapped the distribution of falciparum malaria in Africa (Fig. 7.5). Indeed in some parts of Africa the prevalence of the sickle cell gene was as high as 20% and in some areas could be 40%. But what is sickle cell trait and what is sickle hemoglobin?

When sickle cell disease was studied further, it was shown to be inherited and that persons who had the sickle cell gene produced a slightly different hemoglobin from those with normal hemoglobin. The Nobel laureate Linus Pauling called it a molecular disease. The hemoglobin molecule is a heme-containing protein that gives our blood its red color and is made up of four chains of amino acids: two α and two

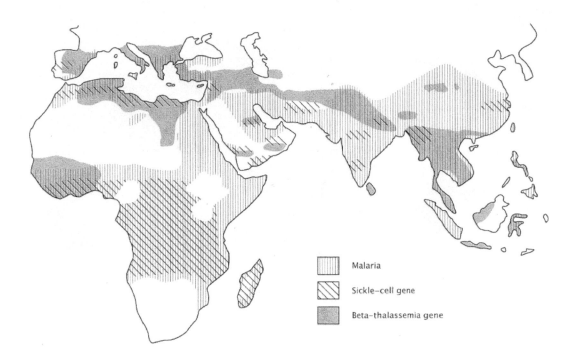

Figure 7.5 The world distribution of malaria (prior to the WHO eradication campaign of the 1950s) and that of sickle cell and beta-thalassemia genes. The geographic coincidence provided the suggestion that resistance to falciparum malaria might be of evolutionary significance and that the presence of malaria tended to maintain genes responsible for some deleterious blood diseases at higher frequencies in some populations.

β. The α chain has 141 amino acids and the β chain 146 amino acids. In people with sickle cell disease, their hemoglobin, called hemoglobin S, has only one amino acid in the β chain changed from its counterpart in normal hemoglobin, called hemoglobin A. Hemoglobin A contains the amino acid glutamic acid at the same position that hemoglobin S has the amino acid valine. Since the genetic code word for glutamic acid is CTC (cytidine thymidine cytidine) or CTT (cytidine thymidine thymidine) and valine is CAC (cytidine adenine cytidine) or CAT (cytidine adenine thymidine), it is clear that the change from hemoglobin A to S results from a single nucleic acid base change of T to A, a mutation in the DNA that leads to an alteration in 1 amino acid out of 287!

Now let us look at the inheritance of sickle cell hemoglobin. Humans have 23 pairs of chromosomes; 22 pairs are similar in size and shape in both males and females, and these are called autosomes. The remaining pair, called the sex chromosomes, X and Y, is similar in females but differs in size and shape in the male. The gene for hemoglobin is found on the autosomes; that is, it is not carried on the X or Y

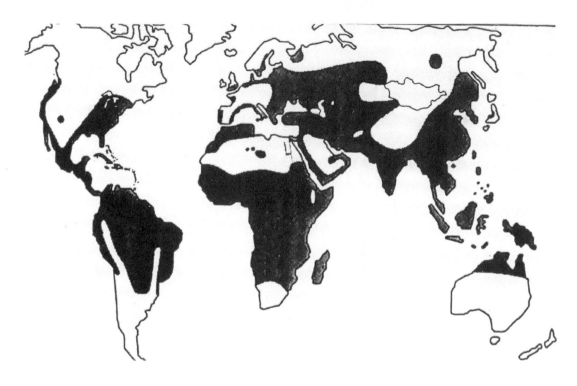

Figure 7.6 The worldwide distribution of malaria in the 1930s prior to the WHO eradication campaigns of the 1940s and 1950s. From I. W. Sherman, Malaria: Parasite Biology, Pathogenesis and Protection (Washington, DC: ASM Press, 1998)

chromosome and is transmitted without dominance—it is inherited in the same way that ABO blood groups are. An individual with a double dose of the S gene, called a homozygote, has sickle cell anemia; an individual with a double dose of the A gene, also called a homozygote, has normal hemoglobin A. An individual with one A and one S gene is a heterozygote, has sickle cell trait, and has equal amounts of hemoglobin S and A in the red blood cells.

The inheritance pattern when both parents have sickle cell trait is shown in Fig. 7.7.

The frequency of AS, or sickle cell trait, is 8% of Americans of African heritage, and for SS, or those with sickle cell anemia, the frequency is 1 out of 600 in the same population. In the case of the mating shown, the probability is that 50% of the offspring will have sickle cell trait, 25% will have sickle cell anemia, and 25% will be normal.

Sickle cell anemia, brought on by a double dose of the S gene, can be a debilitating disease, with blockage of blood vessels; severe pain in the extremities, chest, back, and abdomen; as well as anemia. Clearly such individuals are sick. Since that is the case, why wouldn't such a deleterious gene be eliminated from the population? The reason is that the heterozygote, the AS genotype, produces an individual with sickle cell trait who is protected against falciparum malaria. Because falciparum malaria generally kills children, this allows the most sensitive age group to survive, to develop acquired immunity, and to reproduce. This is an example of natural selection in action, that is, survival of those most fit to reproduce. In this case, natural immunity allows the child to survive and develop acquired immunity to falciparum malaria. Such a survivor can go on to reproduce.

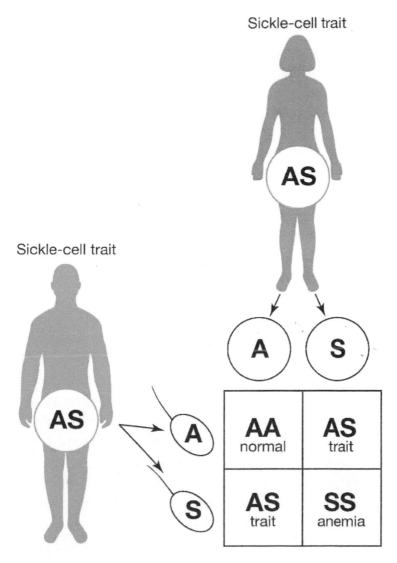

Figure 7.7 The inheritance of sickle cell hemoglobin. The mating of two individuals with the sickle cell trait (AS) produces three possible offspring: normal (AA), sickle cell trait (AS) and sickle cell anemia (SS).

This superior fitness of the AS individual (heterozygote) over both AA and SS individuals (homozygotes) is called hybrid vigor or heterosis. The S and A genes are both maintained in the population because the loss of the homozygotes due to death from either malaria or anemia is balanced by the gain that results from the enhanced survival and reproduction of the heterozygotes, a phenomenon known as balanced polymorphism.

Of course, without the selective pressure of malaria, the heterozygote has the same advantage as that of the normal homozygote, and under these conditions the frequency of the sickle hemoglobin gene would be expected to decline—and it does.

Today, it is possible to grow the most lethal of malaria parasites, *P. falciparum*, outside the human body. The "taming" of *P. falciparum* to a life in the laboratory took place in 1976 in the Rockefeller University laboratory of William Trager. In the 1970s, Trager began a systematic evaluation of the suitability of commercially available tissue culture media for growing the parasites, and one, RPMI 1640, was found to be superior to all others. He also found that an atmosphere containing 7% carbon dioxide, 5% oxygen, and the balance nitrogen was better than 7% carbon dioxide and 93% air. In 1975 a postdoctoral fellow, James Jensen, joined Trager, and together they set up stationary falciparum cultures in petri dishes and held these in a candle jar. (In the candle jar a white candle is placed in a desiccator equipped with a stopcock, the candle is lit, and the desiccator cover replaced; when the candle flame goes out, the stopcock is closed. The atmosphere in the candle jar is 1 to 3% carbon dioxide and 4 to 15% oxygen.) Maintained at 37°C, *P. falciparum* grows with a 48-h asexual cycle, as it does in the human body.

The Trager-Jensen method has been used to answer the question: how does sickle cell hemoglobin protect? When falciparum is grown in petri dishes and the O_2 levels are lowered to 3%, the sickle hemoglobin in SS cells forms solid rods, and those rigid "spears" kill the parasite; in AS cells the cells do not sickle but they lose their intracellular water, and this suppresses parasite growth; and in AA cells the parasite grows and divides normally. When a malaria parasite gets into an AS cell and that infected cell passes through the capillaries in the deep tissues where the oxygen concentration is low—very much like the drop of blood examined by Herrick—the infected cell may sickle prematurely or it may lose water. Phagocytes in organs like the spleen and liver recognize such infected cells and remove them. Consequently, the parasite is destroyed. In this way, when a person with sickle cell trait is infected with falciparum malaria, there is selective removal of the malaria-infected AS red cells. The protection—innate immunity—is especially important in the young, those <10 years of age, since they have not had sufficient time to build up immunity as a result of infection, and in these individuals such infections are also milder.

Let us now turn to another trait that protects against another kind of malaria—vivax malaria. The surfaces of our red cells bear certain distinguishing molecules; we call these antigens, and since the 1920s we have known that the success of a transfusion depends on the proper match of donor and recipient red cells. The AB and MN are antigens on the red cell surface. If one lacks the A or B antigen, one is of blood type O. And if one lacks rhesus antigen (called Rh factor), one is Rh negative. There is another antigen on the surface of our red cells, called Duffy factor. Persons with Duffy factor are Duffy positive, and those lacking it are Duffy negative. There is no health hazard associated with being Duffy negative. Where is Duffy negative most prevalent? The majority of West Africans, and many East Africans and Bedouins, are Duffy negative. When such individuals were studied retrospectively, it was observed that they were never infected with vivax malaria. Why not? The answer has to do with the manner by which a malaria parasite gains entry into the red cell. The invasive stage—the merozoite—must "dock" onto a surface receptor of the red cell before it can get into that cell. For *P. vivax* that receptor is Duffy factor. If the Duffy factor is missing—as it is in Duffy-negative people—the merozoite cannot attach or dock, and as a result it cannot invade the red cell. A red cell that cannot be invaded does not allow the parasite to reproduce itself, and as a consequence the individual does not suffer from malaria. Clearly, the incidence of Duffy-negative trait in Africa, and in those of African ancestry, suggests that the mutation that led to a loss of Duffy factor protected their ancestors from malaria and promoted their survival.

Persons who are Duffy negative are protected against vivax but not falciparum malaria. This is because there is a different receptor for the merozoites of *P. falciparum*, called glycophorin A. When this receptor is absent, the individual lacking it is resistant to falciparum but not to vivax malaria. Individuals who are genetically deficient in red cell glycophorin A are normal in all ways except that they cannot be infected by falciparum merozoites.

The Elusive Malaria Vaccine

What makes scientists believe that a vaccine against malaria is both practical and possible? First, there is abundant evidence from natural infections in humans that immunity is acquired, and, though incomplete and ineffective in preventing reinfection, that immunity does result in a reduction in mortality. Second, adults living in areas where malaria is endemic produce antibody and have low fatality rates. Third, immunization with radiation-attenuated sporozoites induces immunity and protects >90% of human recipients for >10 months. Finally, immunoglobulins, purified from adults who are immune, when passively transferred protect the recipient against disease.

A fully effective malaria vaccine has yet to be developed; when, however, it does become available, it will have to be safe and easy to administer. To be fully protective the vaccine will have to produce a strong immune response. In those parts of Africa where malaria is endemic and the value for R_0 is estimated to be 50 to 100, it would require 99% coverage of a lifelong vaccine given at 3 months of age to eliminate malaria. Making the vaccine situation even more difficult is that there is no good laboratory test to measure the level of immunity and monkeys suitable for testing a human vaccine are scarce.

Nevertheless, work goes on to develop a malaria vaccine. Some of the potential vaccines targeting different parasite stages are (i) liver-stage vaccines to reduce the chance of a person's becoming sick, which would be suitable for travelers and the military; (ii) blood-stage vaccines to reduce disease severity and the risk of death; and (iii) mosquito-stage vaccines to prevent the spread of malaria through the community. Most of the hundred or so vaccines being contemplated are not based on killed or attenuated stages, but are focused on subunit vaccines consisting of selected antigens. Such subunit vaccines may consist of synthetic peptides, recombinant proteins, or parasite DNA, either packaged in a virus or naked.

In late July of 2014, GlaxoSmithKline announced that after 30 years of effort and an investment of $350 million it had applied for regulatory approval for the world's first vaccine against malaria designed for children in Africa. This recombinant protein vaccine, called RTS,S, has an effectiveness of only about 30%, and so it is clearly not the final answer to eradicating malaria.

One of the more promising vaccines is not a subunit vaccine but one containing a whole sporozoite, named PfSPZ. PfSPZ is produced by a company named Sanaria, meaning "healthy air," a clever counterpoint to the Italian words *mal' aria* ("bad air"). Producing PfSPZ involves raising mosquitoes aseptically, supercharging them with far more parasites than nature does by membrane-feeding the mosquitoes on blood containing *in vitro*-grown gametocytes, allowing 2 weeks for the sporozoites to mature in the mosquito salivary glands, X-irradiating the infected mosquitoes, and finally dissecting out the sporozoites from salivary glands. These X-irradiated sporozoites are then placed in suspended animation in liquid nitrogen until needed for immunizing human volunteers. In 2012 the PfSPZ vaccine was administered four to six times intravenously to 40 adult human volunteers; 0 of 6 subjects receiving five doses over 20 weeks and 3 of 9 subjects receiving four doses developed malaria when bitten by infected mosquitoes. The results of the study were published in the September 20, 2013, issue of *Science* (and publicized in the August 8, 2013, issue of *Nature* News with the headline "Zapped malaria parasite raises vaccine hopes"). An August 12, 2013, *New York Times* article had a more cautious (and realistic) headline: "A Malaria Vaccine Works, With Limits." Although these results are encouraging, many challenges remain before this vaccine can be licensed for widespread use. For example, it required >600,000 sporozoites per subject to induce complete immunity, and even with this dosage, immunity persisted for only 1 year.

Coda

Today, malaria still ravages the continent of Africa. Its target is the indigenous peoples, and slowly and inexorably it is killing them. This is especially true in the rainy and low-lying agricultural areas, where transmission is high. In other areas the problem is famine and malnourishment. Malnourishment increases the susceptibility to malaria and a variety of other diseases, leads to lower productivity, and puts a further strain on the already fragile health care system. Sadly, in Africa the riders of the Apocalypse have as their principal target the children; one out of three die from malaria and a like number from AIDS. It is thus a certainty that in the 21st century Disease and Death will continue to ride together and influence the course of history.

doi: 10.1128/9781683670018.ch8

8
"King Cholera"

"It was 1901. ... The revolver was cold, like the grave. ... The bullets nestled in their chambers. Max von Pettenkofer caressed the metal with liver-spotted hands. The gun was heavy. He put the muzzle to his temple ... there was no audience. The shot echoed off the walls and down the hall. The last of the miasmatists slumped forward, blood pouring onto the desk."

What provoked Professor Max von Pettenkofer, the father of public health; head of the Hygienic Institute; favorite of Kings Ludwig and Maximillian II; and recipient of gold medals from the British Institute of Public Health, the German Chemical Society, and the city of Munich, to take his own life? It was *cholera morbus*, the "cholera of death," but it wasn't the disease itself but rather his failed theory of its cause. He was now without honor for his "insightful" theory for the causation of cholera—a theory he had staked his life upon to prove. Pettenkofer's theory postulated that there was a factor in the air, called x, but x could not cause disease by itself. What was required, he said, is factor y, which was to be found in the soil. By itself y too did not cause disease, but when x got into the soil and united with y, then it gave rise to z, a "miasma," which caused cholera.

Who were these fellow travelers of Pettenkofer, the miasmatists, who believed that disease arises from "bad vapors" and not contagious agents—animalcules and microbes? They were scientists, and their beliefs about disease held sway until 1900—until the microbe that caused cholera was identified. In 1835 cholera swept across Russia and into Munich, Pettenkofer's home city. The epidemic began in the slums, and the poor suffered the most, but it also killed enough of those with money that terror gripped the city, causing panic. Cholera was believed to travel in the air, breathed in by all. Physicians such as Max von Pettenkofer thought that its causes were filth, laziness, and neglected hygiene. He was certain it was due to an unhealthy environment. The solution to this disease, he reasoned, was not cure but prevention.

Figure 8.1 (Left) *Death's Dispensary*. From the Illustrated London News 1860.

In Munich, his city, the public water was so filthy that only the poor drank it, and they died from diarrhea and fever (Fig. 8.1). Those better off drank good München beer! Pettenkofer's theory was that stinking miasmas rose from the low-lying marshes and polluted streams to cause disease, and he devoted all his energies to improvements in the quality of the drinking water in the belief that the cholera epidemics, which physicians could do nothing to control, would be stopped. Cleaning up the water, as well as eating good food and breathing fresh air, would, he was convinced, restore health. In this approach to disease control he became the first scientific hygienist.

When cholera broke out in 1854, Pettenkofer sought to identify the environmental factors. He made a map of where the victims lived (and died) and found that those neighborhoods in low-lying marshlands had the highest incidence of disease. Clearly, he reasoned, it must be the soil. He proposed that clean air went into the damp soil, where it activated a chemical reaction that resulted in a poisonous vapor, and people who lived nearby came down with cholera. That same year John Snow, of England, wrote a small pamphlet, *On the Mode of Communication of Cholera*, in which he discussed a cholera outbreak in London's Soho neighborhood. Snow had made a map of the affected houses and recorded the number of deaths and the number of survivors in each building. Snow's conclusions on the cause of the London cholera epidemic, however, were contrary to those of Pettenkofer. Snow wrote: "I found that nearly all the deaths had taken place within a short distance of the Broad Street pump." Snow took a sample of water from the pump and, on examining it under the microscope, found that it contained "white flocculent particles." He was convinced that these were the source of the infection. On his suggestion the handle of the pump was removed, and the spread of cholera stopped. Pettenkofer dismissed Snow's findings, saying he had failed to consider the soil, the water table, and the amount of decaying material. He asked rhetorically: if the contagious agent was in the water, why didn't everyone get sick? (Indeed, he was correct: the number of people with cholera was actually declining before the handle of the Broad Street pump was removed.) Pettenkofer said in a derisive manner that Snow's explanation was too simple.

In 1857, Louis Pasteur, a French chemist, demonstrated that fermentation (in the production of beer and wine) required a living microbe (yeast) and that wine soured because of the action of certain bacteria. Pettenkofer dismissed these findings. Cholera, he said, had nothing to do with spoilage and fermentation. Pettenkofer, ever committed to his miasma theory, mounted a vigorous public health campaign and tried to bring clean water to Munich as the British had done in London. He proposed the building of an aqueduct that would carry fresh water from the foothills of the Alps to the faucets of the people of Munich. By 1865 the great aqueduct was finished. Now he wanted the wastewater removed by sewers, as it was in London. Appealing to German national pride, he called on his countrymen to demand clean water, and he was determined that Germany match or exceed the sanitary measures of London, where the death rate from cholera was lower than that of Munich. He was appointed head of the Hygienic Institute in 1879, and within its laboratories ventilation, poisons, nutrition, and the environment were studied, but not microbes.

Then, in 1882, a fellow German, Robert Koch, caused a stir. Koch had previously identified the microbe that causes tuberculosis (see p. 334). When Koch claimed that the soils of Bombay and Genoa, where cholera occurred, did not meet Pettenkofer's specifications, the two scientific titans were on a collision course. Cholera broke out again in India, and by 1883 it was in the Mediterranean region, brought there by Muslim pilgrims traveling from Mecca. Thousands were dying in Cairo, and Koch (not Pettenkofer) was dispatched to Egypt to isolate the microbe responsible. He did. Koch used his microscope to see bacteria (vibrios) swimming in the feces of 12 cholera patients and in the corpses of 10 people who had died from the disease.

Koch was celebrated as a hero when he returned to Berlin on May 2, 1884, and Pettenkofer raged. Pettenkofer insisted that the vibrios did not cause cholera, and he asked Koch to send him a culture prepared from a patient who had died from the disease. Koch complied with a flask, and Pettenkofer, after estimating the number of bacteria to be a billion, put the flask to his lips and drank it in front of an audience of his adoring students. Pettenkofer did not become ill. He wrote Koch: "Herr Doktor Pettenkofer presents his compliments to Herr Doktor Professor Koch and thanks him for the so-called cholera vibrios which he was kind enough to send. Herr Pettenkofer has now drunk the entire contents and is happy to inform Herr Doktor Professor Koch that he remains in good health." Koch did not reply. (Pettenkofer failed to mention that he did have diarrhea and that he had suffered a bout of cholera some years earlier. Perhaps the reason he did not did not die was his partial immunity by previous exposure to the disease.) This one demonstration, heroic and dramatic as it was, was insufficient to refute the germ theory of disease, i.e., that diseases are caused by tiny creatures—microbes—that invade the body, announced by Pasteur in 1862. The germ theory was supported by many years of accumulated evidence of other disease-causing microbes obtained in painstaking fashion by Pasteur and Koch and their associates. But now, in 1901, Pettenkofer was an old man with old ideas that stood against progress in understanding disease transmission. The scientific tide had turned, and Pettenkofer's theory of the cause of cholera was without support. Tragically, he felt there was little honor left, and so he took his own life. The sad story of Pettenkofer, the last of the miasmatists, illustrates how scientists—even very good ones—are sometimes unable to shake themselves free of old (and unsupported) theories and become blinded by their concept of Nature, a model they create in their own mind and with which they fall in love.

A Look Back

Considering the fact that epidemics have been such a scourge of mankind, it is sometimes difficult to understand why as late as the middle of the 19th century there was such reluctance on the part of scientists like Pettenkofer to believe in microbes as causative agents. In ancient times, outbreaks of disease were considered to be caused by divine intervention, to be a punishment for the wicked. Later, it was believed (as did Pettenkofer) that there were vapors or "plague miasmas" that caused disease. In Italy during the bubonic plague (1346-1353) this led to special physician's costumes—robes of linen cloth coated with a paste of wax so that the slippery surface would not allow the miasma or "bad vapors" to enter; the plague doctors wore a goggled mask with a beaklike nose that contained strong-smelling materials (perfumes and sweet-smelling herbs) to neutralize the disease spirits (see p. 75). People believed that aromatic materials could ward off disease, and so they burned tar and old shoes. One of the practical aspects that developed from miasmatist thinking was that disease might be avoided if one shunned the sick, or had a minimum of human contact with their "bad vapors" if the dead were disposed of quickly. Burning then became the preferred method of corpse disposal, and in some cases it actually worked to stop the spread of some diseases.

But why did people not make the connection between microbes and disease instead of implicating "bad vapors"? The reason is that for many centuries the microbial agents of disease could not be determined by our senses. We are all inclined to disbelieve that which we cannot detect by touch or with our eyes, ears, and nose. Humans, however, are primarily visual animals; we affirm this when we say that "seeing is believing." As early as 1500 the Italian Girolamo Fracastoro (1478-1553) wrote about contagion, claiming that the "seeds or germs" of disease could be spread by touching contaminated clothing or other inanimate objects, as well as by contact between uninfected and infected individuals. Few believed him, because the intellectual climate during the time he lived did not foster an understanding of the significance of his work, and the "seeds or germs" could not be seen. As a result, his theory of contagion was neglected for 3 centuries.

When did visible proof for the existence of microbes occur? First, for there to be proof there had to be a technological advance—the microscope. Once the microscope was invented, then microbe hunters could begin to find the agents of disease. Microbes were first seen in 1674 by Antonie van Leeuwenhoek (1632-1723), a Dutch linen merchant who had a hobby of grinding magnifying lenses. Using a crude microscope (with a magnification of 300×), Leeuwenhoek discovered bacteria and the first human parasite (*Giardia*). In a letter to the Royal Society of London on November 4, 1681, he wrote:

I weigh about 160 pounds ... and I ordinarily a-morning have a well-formed stool; but now and then hitherto I have had a looseness, at intervals of 2, 3, or 4 weeks, when I went to stool some 2, 3, or 4 times a day ... My excrement being so thin, I was at diverse times persuaded to examine it. ... I will only say that I have generally seen, in my excrement, many irregular particles of sundry size. All particles afore-mentioned lay in a clear and transparent medium wherein I have sometimes also seen animalcules a-moving very prettily. Their bodies were somewhat longer than broad, and their belly, which was flattened was furnished with sundry little paws, wherewith they made such a stir in the medium, and among the globules.

Leeuwenhoek had discovered the cause of his own diarrhea (though he didn't realize it). His diarrhea was caused by an infection by *Giardia*, and Leeuwenhoek's "animalcules" are today called protozoa. Presently, giardiasis (a disease caused by *Giardia*) is a cause of diarrhea, and there are millions of human infections each year. *Giardia* is endemic in the Rocky Mountains, in places such as Aspen and Snowmass. It is also a chronic disorder in St. Petersburg, Russia. An outbreak occurred in the American gymnastics team when it visited Leningrad in 1971, and another occurred in 1990 in workers at a Wisconsin insurance company because a food handler was infected.

Once it was possible to see that there were microscopic organisms—microbes such as bacteria and protozoa—that could cause a culture medium such as chicken soup or meat infusion to spoil and become rancid, the question became: where did these "germs" come from? Vapors, inanimate objects, or other bacteria? Up until the 1850s the theory of spontaneous generation, that is, life arising from the lifeless, was popular.

Francesco Redi (1626-1697) was an Italian physician. In 1668 he performed a simple but classic test that shook the foundations of the theory of spontaneous generation. Where others were content to observe nature and suggest imaginative explanations (such as those of Pettenkofer 200 years later), Redi was not content to observe natural phenomena as they occurred, but rather set out to test ideas and to arrange some of the components of nature so that analysis of phenomena could be made. In short, he did an experiment. Redi arranged three jars with decaying meat; one of the jars he covered with paper, one with gauze, and the other he left uncovered. Flies were attracted to the meat samples in the gauze- and paper-covered jars but could land only on the meat in the open jar; in this jar maggots developed, but not in any of the others. Decaying meat in itself, said Redi, did not give rise to maggots. It was necessary for flies to land on the meat and deposit their eggs, which gave rise to maggots (fly larvae). This simple refutation of the generation of life from substances such as rotting meat (lifeless matter) held sway, however, for only a short time.

After Leeuwenhoek discovered microbes in 1675, the proponents of spontaneous generation argued that, although one could not get flies from the nonliving world, one could get living microbes from nonliving broth, rainwater, hay infusions, and other such nutritive sources. The first attack on this micro-level doctrine of spontaneous generation came from Lazzaro Spallanzani (1729-1799), who showed in 1767 that if one boiled the meat broth in a flask, and then sealed the neck of the flask, the broth remained sterile. If, however, the neck of the flask was broken, in a short time the broth swarmed with microbes. Boiling the broth (sterilization) prevented the growth of microbes; contamination of it with unheated broth or another substance apparently provided the source of these tiny living creatures. Therefore, he concluded, microbes must arise from other microbes. The adherents of spontaneous generation, however, ever resourceful, stated that Spallanzani's closed flasks eliminated the "vital force" and that is why the broth remained sterile. The issue remained in doubt for more than a century, for it was well known that many microbes required oxygen, a so-called vital principle, and the absence of oxygen in a sealed flask could be expected to inhibit the spontaneous development of living bacteria or protozoans. For a short while it appeared as if the adherents of spontaneous generation would be able to refute the microbial origin of life.

In 1862, though, Louis Pasteur (1822-1895), the French chemist-microbiologist, dealt the spontaneous generation theory its deathblow by performing a simple and elegant experiment. Pasteur contended that microbes (yeast, bacteria, and protozoa) not only cause disease in animals and humans but are involved in decay and fermentation (such as occurs in beer and wine). Pasteur discovered the bacteria that fermented milk, made butter rancid, spoiled wine, and killed silkworms. And he devised a method for reducing spoilage by heating and then cooling—a technique called, after its discoverer, pasteurization. (Pasteurization is used to preserve milk, cheese, and beer without changing the flavor. This is done by heating to 161°F [72°C] for 15 s and then cooling; Pasteur's original method was to heat at 145°F [63°C] for a half hour and then to chill quickly to 50°F [10°C]. Milk must be kept cold after pasteurization because it does not kill all the bacteria but simply slows down much of their multiplication. To kill the bacteria would require boiling, but that would destroy the flavor of the milk.) Without microorganisms, Pasteur said, there should be no change in the meat broth. Pasteur placed some meat broth in a flask and boiled it until it was sterile; then he drew out the neck of the flask so that it was formed into the shape of an S. He did not seal the neck, and air could pass freely into the broth by moving through the twisting neck (Fig. 8.2). The long, curving, swan-like neck of the flask, however, trapped airborne microbes and prevented them from reaching the broth. In spite of the ease of access of the "vital force"—whatever it was—the flasks remained sterile. If, however, Pasteur tipped the flask so that some of the broth ran into the bend of the S and then back into the flask, the broth became contam-

Figure 8.2 Pasteur's swan neck flask. Courtesy Wellcome Library, London, CC-BY-4.0.

inated with microbes. Obviously, more than broth plus air was needed to produce life. His swan-necked flask experiment was convincing proof that a vital force did not exist and that microbes come from other microbes. A contemporary of Pasteur the Frenchman (and his archrival because of fierce nationalistic pride) was the German Robert Koch (1843-1910). He and his colleagues were able to identify the microbial agents responsible for the diseases anthrax, cholera, tuberculosis, diphtheria, and tetanus. He also defined a strategy, called Koch's postulates*, for unambiguously identifying the causative agent for a microbially induced disease. Together these great microbe hunters established the firm foundation upon which the germ theory of disease was built.

Thus, toward the end of the 19th century (a time when Pettenkofer was in his scientific prime), it was clear that life gives rise to life and that microbes can cause disease. "Science," one writer wrote, "is a study of errors slowly corrected." And so it was with germs and disease. It should be recognized, however, that at times even

---

*Koch's postulates for establishing the causal significance of "germs" can be stated:

1. The parasite occurs in every case of the disease.

2. The parasite does not occur in other diseases.

3. After isolation and repeated growth in pure culture, the parasite is able to produce the same disease when introduced into a healthy animal.

when scientific evidence is presented and may be persuasive to some, unless it can be implemented or accepted into the prevailing culture, there will be no error correction. This is how it was with cholera and Pettenkofer. The contagionists (exemplified by Pasteur and Koch) believed that diseases were spread from person to person by an infectious agent, and the miasmatists (exemplified by Pettenkofer) believed disease was caused by "bad vapors." Until the 1900s the view of disease causation espoused by the miasmatists (which was the theory of spontaneous generation in a new guise) was the prevailing one.

The Disease Cholera

The historian William McNeill described cholera: "The speed with which cholera killed was profoundly alarming, since perfectly healthy people could never feel safe from sudden death when the infection was anywhere near. In addition, the symptoms were particularly horrible: radical dehydration meant that a victim shrank into wizened caricature of his former self within a few hours while ruptured capillaries discolored the skin, turning it black and blue." This was what was called the blue stage of cholera. McNeill continued: "The effect was to make mortality uniquely visible: patterns of bodily decay were exacerbated, as in a time-lapse motion picture, to remind all who saw it of death's ugly horror and utter invincibility."

Cholera is a disease that causes severe diarrhea producing "rice water" stools (Fig. 8.3); there is also vomiting, convulsions, and muscle cramps, but little abdominal pain, and the diarrhea results in a loss of water and electrolytes but not of protein,

Figure 8.3 Rice water stools. Courtesy Wikimedia Commons/F1jmm, CC-BY-SA 3.0.

leading to shock if untreated. In some cases the individual may die within a day. The origin of the word "cholera" is generally ascribed to the Greek, meaning "gutter of a roof," probably because the symptoms of the disease suggested the heavy flow of water on roof gutters after a thunderstorm. Another possibility, however, is that it is from the Hebrew and Arabic *choleh rah*, meaning "bad illness."

Cholera is a water- or foodborne disease that does not manifest itself until the "germ" enters the human digestive tract. During ancient times cholera-like diseases were recorded in China, Europe, and Asia, but whether this was true cholera remains in doubt. The presence of cholera in India is clearly documented by the writings of early Portuguese settlers in the 16th century, and it is probable that it was there even earlier. Cholera was in Greece by 400 B.C. Hippocrates described the symptoms: "An Athens man was seized with cholera. He vomited and was purged … and neither vomiting nor purging could be stopped … his eyes were dark and hollow … he became cold." And Aretus of Cappadocia (A.D. 81-138?), writing in what is present-day Turkey, recorded similar symptoms, adding:

> What is first vomited is like water, but what passes by stool is dunglike fluid and of ill odor … there are spasms and drawing together of the muscles of the calves of the legs and of the arms. The fingers are twisted … there is cold refrigeration of the extremities … if he rejects everything by vomiting … and becomes cold and ash-colored, and the pulse approaches extinction … it is well under such circumstances (for physician) to make a graceful retreat.

In A.D.. 900, Rhazes, an Arab writer from Baghdad, said the disease was incurable. The English physician Thomas Sydenham (1624-1689) first used the term "cholera morbus" to distinguish the disease from "cholera," meaning "state of anger." Other accounts of outbreaks of severe diarrhea were recorded in France, Germany, Brazil, and England in the 17th century, but whether they were actually caused by cholera is unknown.

Pandemic Cholera

The history of cholera is usually described as the history of pandemics. Indeed, there is little to dispute the claim that the greatest epidemic disease of the 19th century was cholera. Prior to the first pandemic, cholera was associated with Hindu pilgrimages and holy festivals that drew great crowds to the river Ganges in India. Cholera has existed in India since "immemorial times," and its impact was so great that there was even a goddess of cholera, Oladevi, who was worshiped to protect the villages from the disease. For centuries the range of cholera was confined mostly to the Indian subcontinent (Fig. 8.4). In 1816 the first cholera epidemic began

8.4 Villagers from Pimpri fetch water from a polluted river which is not safe. With permission from the Watershed Organization Trust.

in India, when British ships and troops were already in the region. Their presence and troop movements in and out of Calcutta brought the disease to new places with naive populations. British soldiers fighting on the northern frontier of India carried cholera with them from their headquarters in Bengal; the troops then transmitted it to their Nepalese and Afghan enemies. Between 1816 and 1823 cholera moved eastward by ship to Sri Lanka, Burma, Thailand, Singapore, and Indonesia. It reached China by land in 1817 and again by sea from Burma and Bangkok, causing a major epidemic in mainland China in 1822-1824. Cholera arrived in Japan by ship in 1822. Muscat, in southern Arabia, encountered the disease in 1821 when British troops were sent there to suppress the slave trade. From there it moved south along the east coast of Africa following the slave traders. It entered the Persian Gulf, Mesopotamia (Iraq), Iran, Turkey, Syria, and Egypt and moved along the shores of the Caspian Sea to southern Russia. Then, mysteriously, its spread stopped.

In the next pandemic (1829-1851), either cholera emerged from its endemic base in India (Bengal) or it arose in Astrakhan (Russia) as a recrudescent infection that had persisted from the first pandemic. Cholera retraced its previous path, beginning in southern Russia, but this time it went farther, reaching Europe, where it became rampant. Russian troops fighting in Persia (1826-1828) and Turkey (1828-1829) and the Polish revolt (1830-1831) carried cholera to the Baltic region, and from there it spread by ship to England (1831) and Ireland (1832). Irish immigrants carried the disease to Canada, and then it moved southward to the United States (1832) and Mexico (1833). In North America, beginning in the 1830s (during the second pandemic) and onward, cholera outbreaks occurred with great regularity in port cities such as New Orleans, New York, and Philadelphia, and the death rates were so high it caused panics. In 1849 cholera was called "America's greatest scourge," but still no one understood its cause. In 1831 in Cairo, Egypt, 13% of the population died from cholera. In Europe, especially in Great Britain, the disease was known as "King Cholera" (Fig. 8.1). Between 1832 and 1833 more than 60,000 died in England of cholera. Because the number of deaths was so high and the fear among the people so great, victims of cholera were buried in separate graveyards.

Cholera also established itself in Mecca in 1831 at the time of the Muslim pilgrimages. And from here it spread, as it did in India, along the pilgrim routes from Mecca to Medina, and these followers of Mohammed then carried the disease back to their homeland. Until 1912, when cholera broke out in Mecca and Medina for the last time, epidemics of this dread disease were a common accompaniment of the Muslim pilgrimages, appearing no fewer than 40 times between 1831 and 1912, or every other year on average. On top of this, after midcentury the swifter movement of steamships and railroads became increasingly able to accelerate the global diffusion of cholera from any major world center.

During the third pandemic (1852-1859), cholera was found in Africa, the United States, and the Middle East, as well as in Europe and India. The fourth pandemic began in 1863, moved along its old path, and burned itself out 10 years later for unknown reasons. The fifth pandemic lasted 15 years (1881-1896) and was widespread in China, Japan, other countries in Asia, Egypt, Germany, and Russia. Hygienic measures helped to stop its spread in North America, but there were outbreaks in South America and East Africa. The sixth pandemic (1899-1923) missed the Western Hemisphere and most of Europe, but there were outbreaks in the Balkans, Hungary, Russia, and the Far East.

We are presently in the seventh pandemic, which began in 1961 in Indonesia; this spread in Asia and the Middle East and reached Africa in 1970. In the 1990s more than 200,000 people in Southeast Asia were sickened by cholera. In 1991, after an absence of almost 100 years, cholera reappeared in the Americas. The first confirmed cases occurred in Peru, and by year's end there were nearly 400,000 cases and 4,000 deaths worldwide. In 1992 there was a mild outbreak involving 76 people aboard a flight from South America to the United States, resulting from undercooked or raw fish and vegetables. In 1994 a cholera outbreak in Goma, Zaire, killed 50,000 of the half-million Rwandan refugees, mainly Hutus, who were escaping from the Tutsi rebels. This took only 21 days! In 2000 there was another epidemic in Africa, this time in an area of KwaZulu-Natal, South Africa; interventions such as the provision of chlorinated water curbed the spread of the disease and mortality was low (<1%) due to oral rehydration therapy. In 2001 there was a cholera outbreak in southern Africa (South Africa, Swaziland, Mozambique, Zambia, and Zimbabwe) coinciding with the start of the summer rains; thousands were infected and 63 died. This is the exception rather than the rule, however: mortality from cholera can be high, and between a quarter and half of all cases may end in death.

Cholera today takes advantage of breakdowns in sanitation and health infrastructure, often in the setting of natural and complex disasters. This was the case in Haiti after a devastating earthquake in 2010. By June of 2015, nearly 5 years after cholera's appearance in the country, Haiti was still grappling with a cholera epidemic. Cases spiked from 1,000 per month in the summer to 1,000 per week in the fall, according to Haiti's Ministry of Public Health and Population. There were nearly 8,400 cases in January and February alone and 82 deaths—twice as many cases and three and a half times as many deaths as over the same period the previous year. This wave of cases is like a recurring nightmare for the director of community care, whose job it is to prevent the disease's spread, which can be as simple a task as reminding patients to wash their hands with soap and water. But patient follow-through is the tricky part. They often ask him, "Where am I going to find soap and water?" That is the reality for many Haitians. According to the World Bank, ~40% of Haitians lack access to clean water, and only one in four have access to a sanitary toilet.

Haiti never knew cholera until 2010, when the United Nations flew in a group of peacekeepers from Nepal and set them up in a camp with poor plumbing. Contaminated sewage leaked into a tributary of the longest river in Haiti, the 200-mile Artibonite. Since the first person was diagnosed in October 2010, there have been more than 739,000 cases of cholera and 8,900 deaths, according to Haiti's Ministry of Public Health and Population. A sanitation systems plan, announced in 2012 and supported by an international community of donors, is still only 13% funded.

Death by Dehydration

In 1882 cholera broke out in India, and a year later it was in the Mediterranean region, brought there by Muslim pilgrims traveling from Mecca. Thousands were dying in Cairo. Robert Koch, the preeminent microbe hunter, was determined to find the "germ" of cholera, and in 1883, when cholera broke out in Egypt, Koch had the insight to suggest that the disease was caused by a microbe that produced a special poison and that this toxin caused the profuse, watery diarrhea. Obtaining fecal samples and examining these microscopically before and after staining, he discovered the comma-shaped bacillus (Fig. 8.5) and called the microbe *Vibrio cholerae*. He called it *Vibrio* because of its vibrating wiggles. Subsequently he was able to grow the vibrio in pure culture, thereby satisfying (in part) his postulates for identifying a disease-producing organism. The French had also sent a medical team from Pasteur's laboratory in Paris; they arrived in Egypt at about the same time as Koch and immediately began to take samples of blood from cholera victims and injected this into rats and guinea pigs. No disease resulted. When one of the members of the French team came down with cholera and died, it so unnerved his colleagues that the research was abandoned and they returned to Paris. German science was victorious over French science, and when Koch returned to Berlin on May 2, 1884, he was celebrated as a hero.

"Catching" Cholera

How might one get cholera? An infected individual excretes a trillion *V. cholerae* organisms each day, which can easily contaminate water and food. By eating vibrio-contaminated food and bathing in (and swallowing) contaminated water, one can develop cholera. Notwithstanding current advertising about the defects of a person

Figure 8.5 *Vibrio cholerae* stained and viewed with the light microscope. Courtesy CDC, 1979.

having high stomach acidity, it has been found that stomach gastric juice is lethal to *V. cholerae*. Therefore, it takes about a billion bacteria to infect a person with normal acidity, whereas in people with low stomach acidity only 100,000 would be needed to establish an infection.

A perplexing question has been: where do the bacteria go between epidemics? How could the disease arise spontaneously if people are not infected? The answer, according to microbiologist Rita Colwell, is that the bacteria go dormant and enter a spore-like state when conditions for their reproduction aren't favorable—if the water is too cold, for example, or is nutrient poor or has the wrong salinity. The shrunken vibrios cannot be grown in the laboratory, but they do survive in brackish water, where they are eaten by tiny crustaceans called copepods (Fig. 8.6). Inside the copepod intestine they cause no damage, but the copepods, using the world's ocean currents, may transport them to distant shores. Under the right conditions they come out of dormancy, multiply, and spread their misery. It may be that the seed for the next epidemic is a bloom of phytoplankton (triggered by global warming or El Niño) that leads to a rise in the cholera-causing bacteria in and around the Bay of Bengal or elsewhere.

Figure 8.6 The one-eyed copepod, *Cyclops* (blickwinkel/Alamy Stock Photo).

Controlling Cholera

In the early part of the 19th century, few people appreciated that cholera was spread from person to person via contaminated drinking water. Indeed, few (except perhaps for Pettenkofer and his students) appreciated the value of sanitation as a means of controlling disease. Quite remarkably, it was cholera that ultimately provoked the development of sanitary measures to control its spread. Health, as Pettenkofer, director of the Hygienic Institute, reminded the burghers of Munich and their Austrian kings, depends on an adequate supply of clean water. This, he said, could be provided by an aqueduct such as those built during the time of the Roman Empire. The architecturally magnificent aqueducts even to this day remain monuments to the grandeur that was Rome. The first aqueduct that brought clean water to Rome dates from 312 B.C. By A.D. 1 there were six aqueducts, and 100 years later the number had increased to ten. Together this system of aqueducts could supply the city with 250 million gallons daily and every Roman with 50 gallons—about the same amount Londoners and New Yorkers use today. The water brought by the aqueduct was used for more than drinking. It was also used for bathing. Baths were believed to have curative powers, and the Romans built baths throughout their empire. The baths of Caracalla (A.D. 200) could accommodate 1,500 bathers at a time, and a bath built during the reign of Diocletian (A.D. 245-313) had 3,000 rooms. The sanitation system of Ephesus, in present-day Turkey, one of the three largest cities in the Roman Empire, was known to Pettenkofer because its ruins had been excavated in the 19th century by British and Austrian archeologists. The seaport city of Ephesus was first settled in 1400 B.C. and then colonized by the Ionian Greeks (1000 B.C.). In a succession of battles the city fell to the Persians (546 B.C.) and then to the Romans (133 B.C.). By A.D. 27 Emperor Augustus declared Ephesus the capital of the Asian province. During this time there is evidence that the apostle John and Mary, the mother of Jesus, fled Jerusalem to escape persecution and found safe haven in Ephesus. Paul's confrontation with the Ephesians, who worshipped the fertility goddess Artemis, was one of the conflicts that led to his decapitation on the outskirts of Rome in A.D. 64. By A.D. 100 Ephesus had 250,000 inhabitants and streetlights! Water came to the Ephesians via four aqueducts, one with a source 25 miles away. The flow of water averaged a quart a second, and since the city possessed a reservoir, it never suffered from a lack of water. Ephesus had paved streets, elegant houses and shops, temples, a library, fountains, baths, and a public toilet; in the public toilet there were 40 seats with no separation between. The toilet walls were faced with marble, the floor was covered with mosaics, and a pool in the center received its water from spouting dolphin statues. Musicians played to accompany the patrons passing their effluvia.

In the 1800s the sanitary conditions in London, compared to the cities of the Roman Empire, were as different as chalk is to cheese. London's water supply was much worse than those in Rome and Ephesus more than 1,000 years earlier. Why? The explosive growth of the human population led to the crowding together of people in cities. But different as they were, these cities, ancient and modern, had a nexus: none of their citizens was aware that disease could be spread by the water that flowed along the aqueduct or in the river if that source were to become contaminated. Indeed, without such understanding the most magnificent baths and toilets were useless in preventing a waterborne disease—one of the greatest threats facing the inhabitants of Rome, Ephesus, and London.

Today, we tend to take clean water for granted. We obtain a drink of water simply by turning a faucet. When we want to bathe or shower or cook, we turn on the tap. To get rid of our bodily wastes, we flush the toilet. It wasn't always that way, however. Queen Elizabeth I, who was born September 7, 1533, and who reigned from 1558 until 1603, it is claimed, said, "I take a bath once a year whether I need it or not." In 1832 the local publicity in New York City was that its water was the best—not only could you use it for cooking, bathing, and drinking, but it was also a wonderful laxative! We marvel at the elegance of the Palace of Versailles built by Louis XIV, the Sun King (1638-1715), but when Versailles was built in 1661, it didn't have a single toilet. The guests at Versailles simply urinated below the staircase, and if they had to defecate, they went to the basement or relieved themselves in one of the outdoor privies. The cesspool, which was used to dispose of waste, was, once it was filled, covered up, and a new one dug nearby. There is a myth that the flush toilet was invented in England by a plumber named Thomas Crapper (1836-1910), because many of the porcelain toilets and manhole covers bore his name. In actual fact it was an Albert Giblin who received the patent for the flush toilet in 1819. (In England the flush toilet is frequently in a room with a "W.C." sign on its door; during World War II some Americans thought it stood for "Winston Churchill," but it is nothing more than an acronym for "water closet.") Much as the flush toilet appeared to provide a more sanitary means for disposing of human waste (and conserving water), it was a public health disaster since all of the effluents were dumped into the rivers, which were not only the exit for waste but the source of water for drinking and washing, and no one got the connection between infectious diseases, such as cholera, and sewage and drinking water.

Cholera, Sanitation, and Public Health

How did we get to this sorry state of hygiene? When human populations were much smaller and people lived in villages, water was drawn from the nearby streams and rivers or shallow wells were dug. Human excrement was disposed of in a privy or by spreading it over the field, transforming it into "night soil." The system worked reasonably well in most of Europe and North America until urbanization and industrialization came about in the 18th century. With the Industrial Revolution the human population soared, villages became towns, and towns became cities; as there were fewer places on the land for the night soil, its disposal became more difficult. The solution: dump it into the rivers and streams, where it would be carried away by the current. The swifter the flow of water in the stream and river, the quicker the pollutants were removed, but sometimes the rivers (such as the Thames) were tidal and the flow intermittent. More serious, the same rivers and streams that served for waste removal were also the source of water for drinking, washing, and bathing. The textile and steel mills of the expanding 18th- and 19th-century Industrial Revolution required more and more factory workers, creating an in-migration from the villages to the city. To accommodate these burgeoning numbers, houses were built; these urban dwellings were arranged in back-to-back rows separated by the narrowest of streets and alleys, thereby using the smallest amount of land for the largest number of people. Families, consisting at times of 10 or more people, occupied a single room, and the row houses had a common outside water supply and privy. Under such crowded conditions there was little privacy, cleanliness was difficult to achieve, and oftentimes clean water was unavailable (Fig. 8.7). Unlike the palace at Versailles, the privies were used by dozens of people, they were rarely cleaned out, and wastes overflowed into the surrounding back alleys and the shallow wells. Thus every evil of the industrial town originated in the agricultural village and resulted from the transfer not of people only but of their manner of living. This way of life did not become intolerably dangerous in the countryside because communities were small, cottages widely separated, and work performed in the open air. The disaster of epidemic illness became inevitable when the community was cramped for space, houses lay cheek by jowl and factory hands worked close together for long hours in an enclosed atmosphere. In this way, the urban residents in England and the Americas came to be deprived of the benefits of hygiene.

When cholera arrived in Liverpool, England, in 1832, death came quickly and horribly. Sewage and night soil were deposited in the streets and alleys, either to stagnate or to flow into the open ditches and then to the rivers. Drinking water, obtained from these same rivers, was untreated. Industrialization had created an overcrowded city with inadequate sanitation, and some people were forced to live below ground level in cellars. No one knew the cause of the disease. Physicians could do little for those afflicted save for dispensing opium and brandy, and their bloodletting and purgatives worsened the dehydration of those who were already sick. Three percent of the population of 165,000 was ill, and 1,523 died (a mortality of 31%). Those most affected were the cellar-dwelling poor, most of whom were Irish immigrants. Over a period of weeks the response to the outbreak was a series of riots called by the press "the Cholera Riots." The seeds of discontent and unrest among the people were the inability of those in the medical profession to render any real help. Indeed, there were so many more cholera victims dying in the hospitals than at home that people feared to enter the Liverpool hospital. There was also the suspicion that cholera victims who were removed to the hospital were likely to be killed by the doctors and then their bodies sold for use in anatomical dissection. Victims went into hiding to keep from being caught by the body-snatching doctors.

Over a 10-day period mob violence broke out. Most of the rioting was directed at the physicians, who were called murderers and "Bunkers," a term used for those who sold bodies for dissection. At times there were thousands in the street, mainly women and young boys. They threw stones at the hospital building, chased away the physicians, and prevented the removal of cholera patients from their homes to the hospital. Police were needed to restore order, and some rioters were arrested. Social unrest in response to cholera was not confined to England. Civil disturbances also occurred in Russia and elsewhere in Europe. In Hungary castles were attacked and nobles murdered by mobs, who believed the rich were responsible for the cholera deaths among the poor. The notoriously conservative Liverpool newspapers referred to the rioters as "misguided" and "ignorant" and said they represented the lower orders of society, particularly the Irish. Though the Cholera Riots of 1832 took place long ago and seem so remote from what we know today about this disease (and other infectious diseases), the events show how fragile the interface can be between physicians and the public.

The horror of cholera, never experienced before the 19th century in Europe, began to demand action from those in authority lest there be more riots. Max von Pettenkofer, the first sanitary scientist, said that sanitation, that is, the disposal of wastes and sewage in such a manner that water and food are incapable of spreading disease, is key to improving health. And forthwith he began to institute sanitary measures to restore, preserve, and improve the health of the public. Edwin Chadwick (1800-1890), Pettenkofer's equivalent in England, led the movement to improve sanitary conditions. In 1842 he produced a "Report into the Sanitary Conditions of the Labouring Population in Britain," demonstrating that the life expectancy in the cities was much lower than in the countryside. Chadwick and his boards of health produced "sanitary maps" showing the relationship of disease—especially cholera—to overcrowding, lack of drainage, and defective water supply. Chadwick was not entirely altruistic: he believed that a healthier population would be able to work harder and would cost less to support. In the 1840s piped water was in use, but the companies that supplied the water used bored elm trunks as conduits; since these split under the pressure required for efficiently moving water into the city, the water was supplied at very low pressure by ground-level standpipes. Sewage disposal was solved by the use of narrow-bore self-cleaning drains (instead of the brick channels with stone covers), but these drains required a constant flow of water to be cleansed. This led to the development and installation of flush toilets instead of privies. The outflow from the sewers was the river, however, and almost without exception this was also the source of the drinking water.

Over a 20-year period Edwin Chadwick, a lawyer-journalist and not a scientist, as Secretary to the Poor Law Commission, became the most powerful member of the General Board of Health, which had sweeping powers: house-to-house visits in search of infectious diseases, removal of the sick from overcrowded tenements, vaccination of smallpox contacts, investigations into the causes of disease outbreaks in schools and factories, inspection of sanitary improvements, and determination of the causes of unexpected deaths. Chadwick was arrogant, loudmouthed, and dictatorial in his approach, and his bigoted outlook was more characteristic of a totalitarian state than a democracy. He came to be hated for his Board of Health policies and the Poor Law Guardians, which had the sweeping powers to "clean stagnant pools and ditches, inspect lodging houses and prosecute any person failing to abate nuisances," and for his demands for vital statistics on deaths and numbers of cases of infectious diseases. Despite his and the board's unpopularity, however, these and other measures—the paving of roads, the cleaning the streets, the provision of clean water, and the carrying off of wastes—did contribute somewhat to improving the public health. He and other members of the Board of Health were dismissed, however, because of their inability to arrest the 1853-1854 cholera epidemic. And like so many of his time, Chadwick believed that cholera was caused by air pollution!

John Snow, the Father of Epidemiology

John Snow (1813-1858) was the son of a coal-yard laborer in York, England. As a boy he proved to be an exceptionally bright, methodical, and eager student, so his mother used a small inheritance to send him to a private school. Snow planned to become a physician, and at 14 he was apprenticed to Dr. William Hardcastle in Newcastle upon Tyne. Snow had an analytical mind that thrived on details that others often overlooked. During his early years as an apprentice, he filled notebooks with his thoughts and observations on scientific subjects. In the summer of 1831, when Snow was 18 and in his fourth year as an apprentice, a cholera epidemic struck London. By October the disease had spread north to Newcastle. Hardcastle had so many sick patients that he could not personally see them all, so he sent Snow to treat the many coal miners who had fallen sick. There was little Snow could do to help the stricken miners, because the usual treatments for disease—bleeding, laxatives, opium, peppermint, and brandy—were ineffective against cholera. Snow continued to treat cholera patients until February of 1832, when the epidemic ended as suddenly and mysteriously as it had begun. By that time it had left 50,000 people dead in Great Britain. During the next 16 years Snow earned a medical degree, moved to the Soho district of London, and became a practicing physician. He distinguished himself by making the first scientific studies of the effects of anesthetics. By testing the effects of precisely controlled doses of ether and chloroform on many species of animals, and on himself, as well as on his surgery patients, Snow made the use of those drugs safer and more effective.

In 1848, Snow turned his attention to cholera. In his day most physicians believed that cholera was caused by poisonous gases—miasmas—that were thought to arise from sewers, swamps, garbage pits, open graves, and other foul-smelling sites of organic decay. Snow felt that the miasma theory could not explain the spread of certain diseases, including cholera. In August 1849 there was, as he put it, "a most terrible outbreak of cholera which ever occurred in this kingdom." He searched through the south London municipal records and discovered that two private companies were supplying it with water. One, the Southwark and Vauxhall Water Company, was drawing water from an area along the River Thames that was known to be polluted by sewage, whereas the other company, the Lambeth Water Company, had recently moved its water intake facilities to a location above the sewer outlets. Snow decided to compare the mortality rates of consumers of the two sources of water. This was a chance, as Snow said, to conduct an experiment "on the grandest scale." Snow began by looking at two subdistricts of south London, Lambeth and Kennington, where there had been 44 cholera deaths prior to August 12. Determining which customers were served by which water company proved difficult. Most tenants did not know, and their landlords often lived elsewhere. Snow called at each house, traced landlords to their homes, and determined that 38 of the 44 deaths had occurred in houses supplied by the Southwark and Vauxhall Water Company—the company whose water came from a polluted source. Snow decided to expand his survey, but he needed help to canvass all the homes where cholera deaths had occurred. He engaged Dr. Joseph J. Whiting to visit half of the homes, while he visited the other half. When the two men tallied their figures, they learned that in the 4-week period between July 8 and August 5, 286 of the 334 victims had used Southwark and Vauxhall water, whereas just 14 of the victims had used Lambeth water. Snow took large statistical samples from other districts and discovered that deaths related to the two companies stood at a ratio of 71:5. Excited by his results, Snow believed that he had obtained "very strong evidence of the powerful influence which the drinking water containing the sewage of a town exerts on the spread of cholera when that disease is present." His critics, however, were not impressed by the results of the survey. They continued to believe that cholera was caused by miasmas, and they asserted that the enormous quantity of water in the River Thames would sufficiently dilute any poison to render it harmless.

The self-effacing Snow tried to agree with his opponents as much as possible. He downplayed his beliefs, asking his colleagues only to consider that cholera was spread by a type of poison that could somehow remain potent after dilution. He wrote, "The poison consists probably of organized particles, extremely small no doubt, but not incapable of indefinite division, so long as they keep their properties." Snow's critics pointed out that he had no evidence of the actual presence of the cholera "poison" in the water. In fact, Snow had accumulated statistical evidence that, by today's standards, would be considered worth acting upon, but because he was the first person to make use of a survey of the incidence and distribution of an epidemic in an effort to determine its cause, his novel epidemiologic findings were seen as unsound. He needed more-convincing evidence.

In late August of 1853, cholera broke out suddenly and devastatingly in a neighborhood just a 5-min walk from Snow's home in Soho. There were upwards of 500 fatal attacks of cholera in 10 days. The mortality in this limited area equaled even that caused in England by the plague, and it was more sudden, as greater numbers of cases terminated even in a few hours. The mortality would have even been greater had it not been for the flight of the population. Persons in furnished lodgings left first; then other lodgers went away, leaving their furniture to be sent for when they could meet with a place to put it. Many houses were closed altogether, owing to the death of the proprieters, and in a number of instances the tradesmen who remained had sent away their families, so that in less than 6 days from the beginning of the outbreak three-quarters of the inhabitants had fled the most afflicted streets. Snow wrote:

> As soon as I became acquainted with the situation and the extent of this cholera outbreak I suspected some contamination of the water of the much-frequented street pump in Broad Street. But on examining the water I found so little impurity in it of an organic nature that I hesitated to come to a conclusion. Further inquiry showed me that there was no other circumstance or agents common to the circumscribed locality in which this sudden increase of cholera occurred and not extending beyond it except the water of the above mentioned pump. I found moreover that the water varied during the next two days in the amount of organic impurity visible to the naked eye. On close inspection in the form of white, flocculent particles; and I concluded that, at the commencement of the outbreak, it might possibly have been still more impure. I requested permission, therefore, to take a list, at the general Register Office, of the deaths from cholera, registered during the week ending September 2nd in the subdistricts of Golden Square, Berwick Street and St Ann's, Soho, which was kindly granted. Eighty-nine deaths were registered during the week in these three subdistricts. On proceeding to the spot, I found that nearly all the deaths had taken place within a short distance of the Broad Street pump. There were only ten deaths in houses sitiated decidedly nearer to another street pump. In five of these cases the families of the deceased persons informed me that they always went to the pump in Broad Street because they preferred the water to that of the pump that was nearer; in three other cases the deceased were children who

went to school near the Broad Street pump. Two of them were known to drink the water and the parent of the third thinks it probable that it did too. The other two deaths beyond the district which this pump supplies represent only the amount of mortality from cholera that was occurring before the eruption took place. With regard to the deaths occurring in the locality belonging to the pump there were sixty-one instances in which I was informed that the deceased persons used to drink the pump water from Broad Street either constantly or occasionally. The result of the inquiry then was that there had been no particular outbreak or increase in cholera in this part of London except among persons who were in the habit of drinking the water of the above mentioned pump well. I had an interview with the Board of Guardians of St James parish on the evening of Thursday 7th September and represented the above circumstances to them. I said to them: Remove the handle on the Broad Street pump. The handle was removed on the following day.

Snow went on to write:

There are certain circumstances bearing on the subject of this outbreak of cholera which require to be mentioned. The workhouse in Poland Street is more than three-fourths surrounded by houses in which the deaths from cholera occurred yet out of 535 inmates only five died of cholera. The workhouse has a pump well on the premises and the inmates never sent to Broad Street for water. If the mortality in the workhouse had been equal to that in the streets immediately surrounding it I estimate one hundred persons would have died. There is a brewery in Broad Street near to the pump and on perceiving that no brewer's men were registered as having died of cholera I called on Mr. Huggins the proprieter. He informed me that there were seventy workmen employed in the brewery and that none of them suffered from cholera, at least in a severe form. The men are allowed a certain quantity of malt liquor and Mr. Huggins believes they do not drink water at all and he is certain that the workmen never obtained water from the Broad Street pump.

Snow then produced a map of the area and plotted on it the deaths, where they occurred, and the location of the Broad Street pump as well as the surrounding pumps to which the public had access at the time.

Snow wrote:

The water at the pump at Marlborough Street was so impure that many people avoided using it, and those who did die near this pump in September had water from the Broad Street pump. Further, the mortality appears to have fallen pretty much among all classes in proportion to their numbers. When the pump well was opened there was no hole or crevice in the brickwork of the well however a sewer passes within a few yards of the well. Mr Eley of 37 Broad Street informed me that he had long noticed that the water became offensive, both to smell and taste after it had been kept about two days. Whether the impurities of the water were derived from the sewers, the drains or the cesspools of which there are a number in the neighborhood I cannot tell. As there had been deaths from cholera just before the outbreak not far from this pump well the evacuations from the patients might of course be among the impurities finding their way into the water, and judging by the considerations previously detailed we must conclude that such was the case.

Snow published his findings in the second edition of *On the Mode of Communication of Cholera* (1855).

John Snow wrote that the measures that are required for the prevention of cholera, and all diseases that are communicated in the same way as cholera, are of a very simple kind.

During an epidemic these are the measures that should be adopted:

1. The strictest cleanliness should be observed. By those about the sick. There should be a hand-basin, water and towel in every room where there is a cholera patient and care should be taken that they wash their hands frequently used by the nurse and other attendants, more particularly before touching any food.
2. The soiled bed linen and body linens of the patient should be immersed in water as soon as they are removed until such time as they can be washed lest the evacuations become dry and be wafted about as a fine dust. Articles of bedding and clothing which cannot be washed should be exposed for some time to a temperature of 212°F or upwards.
3. Care should be taken that water employed for drinking and preparing food is not contaminated with the contents of cesspools, house drains or sewers or in the event that water free from suspicion cannot be obtained it should be boiled and if possible also filtered.
4. When cholera prevails very much in the neighborhood all provisions which are brought into the house should be washed with clean water and exposed to a temperature of 212°F or at least should undergo these processes and purified either by water or fire.
5. When a case of cholera appears among persons living in a crowded room the healthy should be removed to another apartment where it is practicable leaving only those who are useful to wait on the sick.
6. The communicability of cholera ought not to be disguised from people under the knowledge of it would cause a panic or occasion the sick to be deserted.

Before an epidemic the following measures to be taken are:

7. To effect good and perfect drainage
8. To provide an ample supply of water quite free from contamination with the contents of sewers, cesspools and house drains

9. To provide model lodging-houses for the vagrant class and sufficient house room for the poor generally

10. To inculcate habits of personal and domestic cleanliness among people everywhere

11. Some attention should be undoubtedly directed to persons, and especially ships, arriving from infected places in order to segregate the sick from the healthy. In general supervision would generally not be of long duration.

And what was the result of these sensible admonitions during the time of Snow? Neglect largely, because the prevailing notion was that cholera was a disease caused by miasmas that seeped from the soil.

John Snow, age 45, died suddenly on the morning of June 9, 1858, while working on his manuscript "Chloroform and Other Anesthetics." The health of John Snow was never the best. In 1844 he suffered from tuberculosis of the lungs, and in 1845 he had an acute attack of renal disease. He died prematurely, presumably due to excessive and inappropriate self-experimentation with anesthetics (e.g., ether, chloroform, ethyl nitrate, benzene, bromoform, and ethyl bromide).

The legacy of John Snow is that through patient and thorough analysis he had carried out the first epidemiological study and was able to cast aside that which is accidental and coincidental in a disease outbreak. He discovered that cholera was a waterborne disease and was able to suggest the means for control using a map, a register of persons dying from cholera, and a pencil and paper. And he did this 35 years before the cause of cholera was specifically identified by Koch.

Cholera Today

The diagnosis of cholera remains much as it was in the time of Koch. The examination of stools in the light microscope, though, usually does not show distinctive features. To see the vibrios requires special methods such as dark-field or phase-contrast microscopy. When viewed with the dark-field microscope, the vibrios move about using their whiplike flagellum and look like shooting stars. But microscopy and culture from stools or a rectal swab are not necessary in order to treat cholera. Most often clinical diagnosis is based on the history of the time of appearance of acute symptoms such as "rice water" diarrhea. Then treatment is begun.

In Hanover, Germany, in 1831 health authorities recommended that to treat the violent diarrhea (see Fig. 8.3) a variety of nostrums and quack remedies be used: vinegar, camphor, wine, horseradish, mint, mustard plaster, leeches, bloodletting, laudanum, calomel, steam baths, and hot baths. None were effective, and most patients, treated or not, died. Since cholera was believed to be due to a miasma, the houses where cholera victims lived (and died) were fumigated using a smoke pot. This too did not curb the spread of disease. In that same year, however, a 22-year-old British physician, William O'Shaughnessy, boldly proposed that by the application of chemistry cholera might be cured. After analyzing the blood of a patient with the "blue stage" of cholera, he suggested that therapy be undertaken "to restore its deficient saline matters … by absorption, by inbibition, or by injection aqueous fluid into the veins." Dr. Thomas Latta carried out the first practical application of intravenous injections of a salt solution on May 15, 1832, in patients suffering with cholera. All improved. He cautioned: "although by injection of water and salts … we may restore the efficient fluids of the body and bring back the blood to its normal state … we must still remember that the unknown remote cause, and other agents are still in operation, and must be remedied before a perfect cure can be performed." Though the cause of cholera was still unknown (and would be for 50 more years), this practical treatment saved lives (8 out of 25), and it was regarded as "the working of a miraculous and supernatural agent." When this pandemic subsided, however, the therapy fell into disuse. Clearly, it was a simple and effective therapy that was far ahead of its time. Latta died in 1833; O'Shaughnessy joined the East India Company, went to India, and never again concerned himself with cholera. He died in 1899, and his obituary mentions nothing of his pioneering work on rehydration therapy for cholera.

One hundred sixty years later, research showed that *V. cholerae* organisms secrete a mucinase that digests mucus so that they can attach to the intestinal cell; once attached to the intestinal epithelial cell, the bacteria release cholera toxin, which consists of two parts, an A subunit and a B subunit. It is the B subunit that allows attachment to the epithelial cell and permits the entry of the A segment, which results in the stimulation of cyclic AMP, which in turn leads to secretion of sodium and chloride as well as inhibiting sodium and chloride absorption. Fluid, bicarbonate, and potassium are lost, and with the osmotic pull of the sodium chloride it leads to shock from fluid loss. The actual cause of death in cholera is dehydration.

Although the toxin acts on the absorptive cells of the intestine to shut down one major route of sodium transport, which allows sodium into the cells together with chloride, the toxin leaves unaffected the transport system that brings sodium and glucose simultaneously into the cells from the lumen of the intestine. This transport system works only when both sodium and glucose are present, and it remains active during diarrhea. Armed with this, it was possible (in 1940) for physicians to make the appropriate solution with salts and glucose to rehydrate the patient suffering from severe cholera-induced diarrhea. Today, oral rehydration treatment (ORT) involves oral or intravenous administration of a solution containing glucose, sodium chloride, potassium, and lactate. The cost is only $5 per quart. Another type of ORT, called food-based ORT, substitutes starches and proteins for the glucose. Cereal grains and beans are the source of starches and proteins, and they work effectively to reduce both diarrhea and mortality.

Cholera and Nursing

Florence Nightingale (1820-1910) was the founder of modern nursing as we know it today. At the time she announced she wanted to become a nurse, the nursing profession was equivalent to the "oldest profession"—which was prostitution. In fact, most nurses of that time were women of loose morals, and they drank excessively. They slept on the wards with their patients, and the surgeons had sex available on demand. The hospitals themselves were filthy, and there was virtually no sanitation. Without any real means for alleviating pain, most patients died in anguish on the crowded, stench-filled, and dirty wards.

Florence Nightingale began her nursing training (such as it was) at the Institution for the Care of Sick Gentlewomen in London in 1853. When the Crimean War broke out, she was able to put her nursing skills to the test. The Crimean War raged from October 1853 until February 1856. The direct causes of the war were Russia's demands to exercise protection of Orthodox subjects in Turkey and the privileges of Russian and Roman Catholic monks in the holy places of Palestine. The Turks resisted, however, and were able to take a firm stand against the Russians largely because the Russian threat to take the Dardanelles in Turkey also threatened British sea routes in the Mediterranean Sea. Consequently, the British backed up the Turks. When the Russians moved into Romania, the British fleet was sent to Constantinople. The Turks then declared war against the Russians, but the Turkish fleet was quickly defeated by Russia in the Black Sea. To protect Turkish shipping, British and French troops entered this region of the Black Sea. With the British, French, and Turkish troops allied against Russia, and hoping to prevent Austria from entering the war, the Russians abandoned Romania. The British landed in the Crimea in 1854, beginning a yearlong siege of the Russian fortress at Sevastopol. One of the major battles, at Balaklava, is immortalized in Alfred, Lord Tennyson's poem "The Charge of the Light Brigade":

> Theirs not to make reply,
> Theirs not to reason why,
> Theirs but to do and die.
> Into the Valley of Death
> Rode the six hundred.

Though the poem honors the soldier's responsibility to his duty and willingness to charge to an inevitable death knowing that someone above him has blundered, the British cavalry brigade that attempted to maintain a hold on Balaklava failed because the Russians sent in a large reserve force from Sevastopol. Less than 30% of the British troops survived, partly due to rivalry between the two highest-ranking British officers.

How did Florence Nightingale get involved in nursing? In 1837 she believed she heard the word of God informing her of her mission. She was not exactly clear about what that mission was, but in 1850 she attended the German School of Nursing, and by 1853 she had risen through the ranks and became the superintendent of the London Institution for the Care of Sick Gentlewomen. She assisted with the cholera epidemic in London in 1854 and she instituted changes that were revolutionary: a bell whereby the patient could call for help, bulk buying of supplies, and an improved diet. When the Crimean War broke out, she was appointed Superintendent of Nursing for Army Military Hospitals. The Crimea was rat and flea infested. Most of the casualties could be attributed to gunshot wounds and disease—mostly cholera. At one point >70% of the British troops were ill from disease and unfit for duty. Clean clothing was largely unavailable, and the hospital bedding consisted of straw mattresses. In addition, the hospital water allowance was limited to a pint per day per

A COURT FOR KING CHOLERA.

Figure 8.7 "A Court for King Cholera" from Punch in 1852. Courtesy of the Wellcome Library, London, CC-BY 4.0.

patient. Florence Nightingale attributed the defects in hospital care to four things: (i) the confinement of a large number of sick under the same roof, (ii) a deficiency of space, (iii) deficiencies in ventilation, and (iv) deficient lighting. She set about to correct these defects. She recruited 38 nurses, asked for 200 scrubbing brushes, and washed the patients' clothes outside the hospital building. She organized the hospital kitchens, cleaned and painted the wards, disposed of the wastes in a more sanitary way, and spent time visiting the patients in the evening—hence she became known as "the Lady with the Lamp." In 2 weeks the death rate had dropped by 80%. Pettenkofer would have been proud of her accomplishments.

After the war Nightingale returned to England, and in 1860 she established the Nursing School for the Training of Nurses, the forerunner of all nursing training. The wartime experiences of Florence Nightingale convinced her that the health administration of the British army was in need of reform. She had studied mathematics and statistics as a young woman, and she put these to good use in assembling data on how administrative inadequacies affected patients' health. Her analysis of mortality for men showed that even in peacetime, mortality was higher in the military than among civilian males of similar age. This higher mortality she ascribed to contagious disease, and she emphasized that this could be prevented by improved sanitary conditions. She published a 1,000-page report, *Notes on Matters Affecting the Health, Efficiency and Hospital Administration of the British Army* using graphs to illustrate statistical relationships. Although it is unlikely that she would have considered herself an epidemiologist, she was one of the first to apply statistics to health data and to use this to promote health care reform in the interest of preventing disease and death.

Cholera and the "Immigrant Problem"

Grosse-Ile, lying 30 miles east of Quebec in the middle of the St. Lawrence River, is today a picturesque island whose background of majestic peaks is a tranquil national park; its significance, however, lies in its being Canada's most visible link with Ireland's Great Famine.

The quarantine station at Grosse-Ile was established after the Napoleonic Wars had ended in 1815 and when growing numbers left the British Isles, especially the Irish, to make new lives for themselves in North America. The second cholera epidemic (1829-1837) had taken 5 years to travel from India to Moscow and on to the British Isles, where in late 1831 it developed into a severe public health problem. Reports to the colonial government in Canada that people from the Old World with the dreaded disease were about to arrive via the St. Lawrence River prompted the Assembly of Lower Canada (as Quebec was then called) to pass a resolution on February 23, 1832, that made Grosse-Ile a place of detention for cholera, the most feared epidemic disease of the 19th century. In early June of 1832 the number of immigrants arriving in Canada had doubled, and with each ship's arrival so did the risk of a cholera epidemic beginning in North America. The Irish immigrants were profitable for the ships' owners, who had lost the African slave trade and now were able to fill the cargo bays on the outward-bound journey with other "wretched humans" in exchange for carrying Canadian timber to England. Because the numbers of arriving passengers was so great and the space on Grosse-Ile so limited, many ships passed the island and entered Quebec harbor directly. By mid-June more than 25,000 people had landed, and by summer's end the numbers exceeded 50,000. Soon cholera was in Quebec City itself and had been spread beyond its borders by those newly arrived immigrants, who, believing they were healthy, made their way to Montreal and the cities of Ontario.

Grosse-Ile lacked a proper jetty, so some of the debilitated passengers who had endured a miserable passage of more than 60 days were forced to wade ashore; many drowned and the sandy bottom became their grave. For those who reached the island, death might have been a better fate. The writer Susannah Moodie, who emigrated to Canada from England in 1832, described her own prejudices:

> I looked up and down the glorious river; never had I beheld so many striking objects blended into one mighty whole! Nature had lavished all her noblest features in producing that enchanted scene. The rocky isle in front, with its farmhouses at the eastern point, and its high bluff at the western extremity, crowned with the telegraph … the middle space occupied by tents and sheds for cholera patients and its wooded shores dotted over with motley groups added to the picturesque scene. … Never shall I forget the extraordinary spectacle that met my sight the moment we passed the low range bushes which formed a screen in front of the river. A crowd of many hundred Irish emigrants had been landed during the present and former day and all … men, women and children, who were not confined to the sheds (which resembled cattle pens) … were employed in washing clothes or spreading them out on rocks and bushes to dry. The people appeared perfectly destitute of shame or a sense of common decency. Many were almost naked, still more partially clothed. We turned in disgust from the revolting scene. … Could we have shut out the profane sounds which came to us on every breeze, how deeply we should have enjoyed an hour amid the tranquil beauties of the … lovely spot.

That same summer Catharine Traill passed the island and described it in a letter home:

> There are several vessels lying at anchor close to the shore; one bears the melancholy symbol of disease, the yellow flag; she is a passenger ship and has the

smallpox and measles among her crew. When any infectious complaint appears on board, the yellow flag is hoisted, and the invalids conveyed to the cholera hospital, a wooden building ... surrounded with palisades and a guard of soldiers. There is also a temporary fort at some distance from the hospital, containing a garrison of soldiers, who are there to enforce the quarantine rules. The rules are ... very defective and in some respects quite absurd. ... When the passengers and crew of a vessel do not exceed a certain number they are not allowed to land, under penalty to both the captain and to the offender; but if ... they should exceed the stated number, ill or well, passengers and crew must turn out and go ashore, taking with them their bedding and clothes, which are spread out on the shore, to be washed, aired and fumigated, giving the healthy every chance of taking the infection from the invalids near them.

The quarantine of the healthy and sick together might be accepted as an example of the lack of appreciation of the means of transmission of cholera (since the *Vibrio* responsible would not be discovered for another 50 years); its imposition, however, was not the result of scientific ignorance at Grosse-Ile but rather an application of the practice of class distinction and scapegoating of the "wretched refuse" of Ireland. Indeed, cholera was felt by the English to be a vulgar and demeaning disease—a disease of the poor Irish—and this strengthened its repulsiveness and justified the quarantine. Quarantine shamed and blamed and contributed to the increased mortality on the island, estimated that summer at 2,000 deaths. By 1833 the number of sick immigrants arriving in Canada had dwindled, and in the intervening years up until 1846 all remained quiet at the Grosse-Ile quarantine station.

The tragedy at Grosse-Ile in 1847 stemmed from the famine in Ireland that began in 1845 and coincided with the arrival in Europe of the third cholera pandemic. Between 1845 and 1849, the time of the Great Irish Famine, the population of Ireland (~8 million) declined by more than 2 million. Half died of starvation, disease, and malnutrition, while the other half emigrated. The United States was the traditional route for Irish immigrants, but in 1847 the U.S. enforced an increase in the cost of passage, and ships that were overloaded were to be confiscated. This opened up new routes to the United States from Canada as ship owners sought a cheaper option for their passengers. There was also an economic incentive for the English landlords to support the emigration of the Irish tenant farmers. It was reasoned that those tenants who remained on the land and were unable to pay their rents would become paupers. The cost of a landlord's maintaining a pauper in the poorhouse was 71 pounds sterling per year, whereas the cost of passage was 31 pounds sterling. The saving to the landlord in the first year was 40 pounds sterling, and after that the entire cost of supporting a pauper was saved.

In 1847, fleeing the terrible failure of the potato crop and the rampant cholera in famine-stricken Ireland were hundreds of thousands of Irish who sailed for Quebec. Weakened by malnutrition and starvation, they were crowded aboard unsanitary sailing ships unfit for transporting human beings. During the voyage of these "pest ships," those that died, along with their possessions, were hastily wrapped in canvas and thrown overboard as if they were dead birds or garbage.

On May 15 the curtain was raised on the drama of death that would play out on Grosse-Ile. The first immigrant ship, Syria, arrived with 241 passengers. Fever and dysentery had broken out a few days after the ship had left the port of Liverpool (March 24). Nine passengers had died en route, and of those who landed, 84 were immediately hospitalized; another 24 would be admitted shortly thereafter. The first death in quarantine was 4-year-old Ellen Keane, and within a week 16 more were dead; before the end of May 70 Irish had died. During that summer 398 ships were inspected at Grosse-Ile. Although ships usually took 45 days to cross the Atlantic Ocean, 26 of those that set sail in 1847 had taken over 60 days to reach Grosse-Ile. More than 5,000 people died en route, and 5,424 were buried in a mass grave on the island. Other gravesites on the island replicated the "lazy bed" in which potatoes were grown in Ireland. Four physicians at Grosse-Ile were aided by a crew of eight, working from dawn until dark every day digging trenches and burying the dead three deep. By August dirt had to be imported to the rocky island to bury more bodies. In spite of this, rats were coming off the ships to feed on the cadavers. All told, the number of deaths on the island probably exceeded 9,000, and many thousands more died elsewhere in colonial Canada during that "Summer of Sorrow."

The quarantine operation at Grosse-Ile was characterized by haste, improvisation, trial and error, and physicians who lacked any understanding of what caused cholera, how it was transmitted, or how it might be treated. The island's facilities were ill equipped to accommodate and safely handle the large number of immigrants, especially those who were already ill. These problems were compounded by the effects of crowding and the unsanitary conditions aboard the ship during the long ocean crossing. No quarantine could meet this challenge and contain the spread of cholera. Many historians, though, feel that there were other factors that led to the deaths of those who were quarantined on Grosse-Ile: they were Irish, they were Catholic, and they were the subjects of an English government that had considered them to be less than human for close to 1,000 years. In memory of that tragic summer a Celtic cross was placed on a steep hill at Grosse-Ile in 1909. Its inscription reads in part: "Children of the Gael died in the thousands on this island having fled from the laws of foreign tyrants and an artificial famine. … God's blessing on them."

It has been claimed that pollution, especially sewage, has claimed more lives than smallpox and bubonic plague. "King Cholera," the most feared of all the sewage-related diseases, provoked global horror and terror. Yet it also changed the way disease was looked at, how it was spread, and the manner by which human health could be preserved. In those countries where an adequate sewage and water supply system has been established, cholera has been eradicated. Cholera, however, remains endemic in areas of Asia and Africa and in countries of South and Central America. Today, quarantining and treatment of the sick, as well as preventing those who are infected from contaminating drinking water and food, accomplish the control of cholera. Provision of potable water almost always reduces epidemic spread. Additionally, there are now two oral cholera vaccines (OCVs), Sanchol and Dukoral. Dukoral is the older and more expensive of the two and is adminstered in a glass of buffer solution to preserve the B subunit of the vaccine; Sanchol, licensed in India in 2009, has no B subunit, needs no buffer, is less costly, and consists of 1.5 ml of liquid. When Sanchol was used in Haiti, coverage was 75% and efficacy was close to 97%. In other published studies, the use of either OCV provided short-term protection yielding 77% efficacy for two doses, whereas with one dose effectiveness was about 45%. It has been predicted by epidemiologists that with 50% coverage by OCV, transmission could be avoided in areas where the disease is endemic. Although OCV protects vaccinated individuals, it also lowers the risk for those who are not vaccinated if coverage is sufficiently high (a phenomenon termed herd immunity). The problem that exists today is having enough vaccine stockpiled to treat an outbreak.

Coda

Cholera is a disease that resulted in the institution of sanitary reforms and the rise of public health, but it has also been used to scapegoat those suspected of being the carriers of disease. At times, the public health organizations designed to protect the people at large became the "health police," and those who were already marginalized were labeled as bearers of disease and a threat to all of society. When some of the immigrants who arrived in developed countries became ill with cholera, their plight was sometimes extrapolated to all members of their ethnic or cultural minority, and the entire group was stigmatized. Naturally, the burden and blame fell, as it always does, on those without resources or political power—on the immigrants and the urban poor. But the cholera pandemics also had a positive effect: the fear and horror of cholera promoted the establishment of a public health system within many countries and inspired the formation of international bodies to monitor and control the global spread of all infectious diseases.

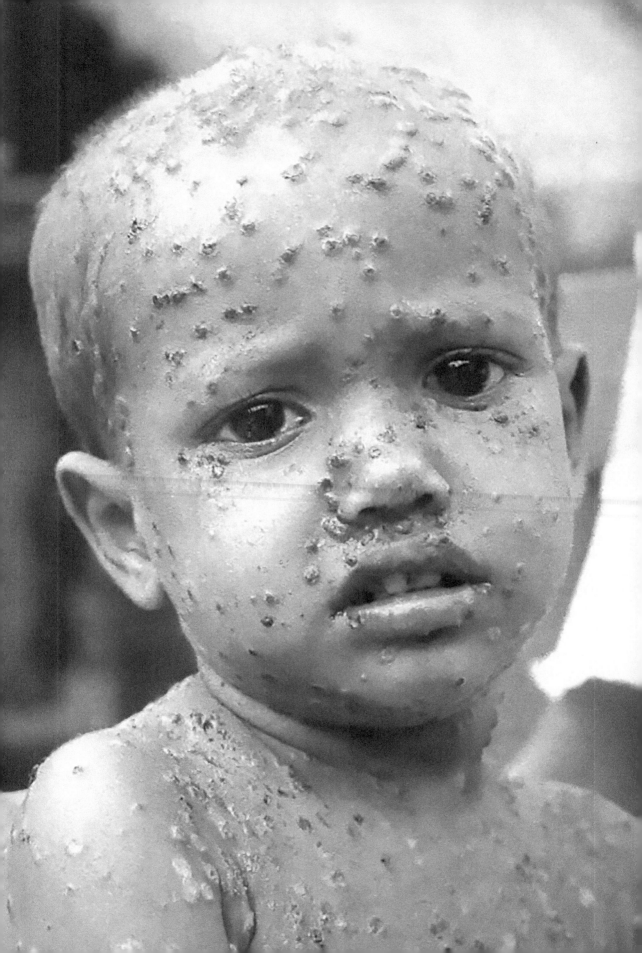

doi: 10.1128/9781683670018.ch9

9
Smallpox, The Spotted Plague

In 1521 the subjects of the Aztec Empire numbered in the millions. Incredibly, Hernan Cortes, with fewer than 600 troops, was able to topple it. The Aztecs were militaristic and wealthy, having subjugated other indigenous Indian tribes and then extracted tribute from them. Cortes and his Spanish conquistadors set out to explore and claim Mexico for their king, Charles V. They landed in the Yucatan on the eastern coast of Mexico. With their armor plate, swords, horses, rifles, cannons, and attack dogs, they appeared to the Aztecs as a formidable fighting force. They moved toward the Aztec capital Tenochtitlán (now Mexico City) without much incident, but at Tenochtitlán there were encounters that showed the limitations of guns, steel, and horseflesh. The outnumbered Spaniards lost one-third of their troops, and Cortes and his army were forced to retreat. Cortes expected a final and crushing offensive by the Aztecs, one that would result in their complete defeat. The attack never came. On August 21 the Spaniards stormed the city, only to find that a greater force had ensured their victory. Bernal Diaz, witness to the scene, wrote:

> I solemnly swear that all the houses and stockades in the lake were full of heads and corpses. It was the same in the streets and courts. ... We could not walk without treading on the bodies and heads of the dead Indians. I have read about the destruction of Jerusalem, but I do not think the mortality was greater there than here in Mexico. ... Indeed the stench was so bad that no one could endure it ... and even Cortes was ill from the odors which assailed his nostrils.

Cortes's ally was a very efficient killer, the disease smallpox (Fig. 9.1). The disease had preceded Cortes's arrival in Mexico. In 1520 an expedition arrived from Spanish Cuba. Led by Panfilo de Narvaez, it was under orders to seize control from Cortes. Among the crew was a smallpox-infected slave. From this

Figure 9.1 (Left) Child with smallpox. Courtesy CDC, Dr. Stan Foster, 1975.

initial infection smallpox spread from village to village throughout the Yucatan. So great was the die-off that there were not enough people to farm the land or protect the cities. Famine and havoc resulted. The paralyzing effect of the epidemic explains why the Aztecs did not pursue the demoralized Spaniards. It allowed the latter to rest and regroup, and in time permitted them to gather Indian allies. Smallpox was a devastatingly selective disease— only the Aztecs died, while the Spaniards were left unharmed—and it so demoralized the survivors that the Amerindians had no doubt that the Spaniards were the beneficiaries of divine favor. It was this perceived superior power of the God the Spaniards worshipped that led the Aztecs and other Amerindians to accept Christianity. Without further resistance they submitted to Spanish control. In the time of Cortes and his conquistadors 3 million Amerindians, an estimated one-third of the total population of Mexico, were killed by smallpox.

Twelve years later (1532), another Spanish conquistador, Francisco Pizarro, led a group of 168 soldiers and captured Atahualpa, the leader of the Inca Empire. Pizarro held Atahualpa prisoner for 8 months, demanded a ransom of gold from his subjects for a promise of freedom, and, after the gold was delivered, reneged on the promise and had Atahualpa executed. Again, smallpox had allowed Pizarro's success in South America. Smallpox had arrived in Peru by land in 1526, killing much of the Inca population. When Pizarro landed on the coast of Peru in 1531, the stage for the Inca conquest had already been set; the smallpox epidemic that had preceded the conquistadors had weakened the Inca Empire, there was civil war, and Atahualpa's army was vulnerable and in disarray. Victory for Pizarro's conquistadors had been ensured by disease.

Smallpox was an Old World disease to which the New World Amerindians had never been exposed, and against which they had no immunity. Smallpox spread in advance of the Spaniards' entry into Mexico and Peru. In its wake followed death, devastation, and demoralization. Guns and steel were not the critical armaments the Spanish conquistadors used in destroying the Inca and Aztec Empires. Disease moved ahead of them, and, in effect, smallpox cleared the way.

Smallpox is no longer with us, but over the centuries it killed hundreds of millions of people. In the 20th century alone it killed 300 million people—three times the number of deaths from all 20th-century wars. Smallpox has been involved with war, with exploration, and with migration. As a result, it changed the course of human history. Smallpox was indiscriminate, with no respect for social class, occupation, or age; it killed or disfigured princes and paupers, kings and queens, children and adults, farmers and city dwellers, generals and their enemies, rich and poor.

A Look Back

Smallpox was at one time one of the most devastating of all human diseases, yet no one really knows when smallpox began to infect humans. It is suspected that humans acquired the infectious agent from one of the pox-like diseases of domesticated animals in the earliest concentrated agricultural settlements of Asia or Africa, when humans began to maintain herds of livestock, sometime after 10,000 B.C. The best evidence of smallpox in humans is found in three Egyptian mummies, dating from 1570 to 1085 B.C., one of which is Pharaoh Ramses V, who died as a young man in 1155 B.C. The mummified face, neck, and shoulders of the pharaoh bear the telltale scars of the disease: pock-marks. (The word "pox" comes from the Anglo-Saxon word *pocca*, meaning "pocket" or "pouch," and describes the smallpox fingerprint—crater-like scars [Fig. 9.2].)

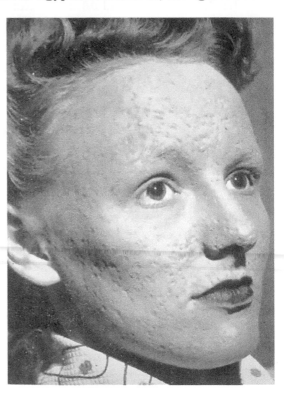

Figure 9.2 Smallpox 6 months after infection with residual facial scarring (pock marks) and loss of eyebrows and eyelashes. From: C. W. Dixon Smallpox, London: Churchill 1962.

From its origins in the dense agricultural valleys of the great rivers in Africa and India, smallpox spread from the west to China, first appearing about 200 B.C. Trade caravans assisted in the spread of smallpox, but at the time of the birth of Christ it was probably not established in Europe because the populations were too small and greatly dispersed. Smallpox was apparently known in Greece and Rome, but it does not seem to have been a major health threat until about A.D. 100, when there was a catastrophic epidemic called the Antonine plague (see p. 61). The epidemic started in Mesopotamia, and the returning soldiers brought it home to Italy, where it raged for 15 years and brought about 2,000 deaths daily in Rome. Emperor Marcus Aurelius died of smallpox in A.D. 180. Smallpox, along with malaria, may have contributed to the decline of the Roman Empire.

There are records of smallpox in the Korean peninsula from A.D. 583. It had reached Japan by A.D. 585. In the western part of Eurasia the major spread of smallpox occurred in the 8th and 9th centuries during the Islamic expansion across North

Africa and into Spain and Portugal. As the disease moved into central Asia, the Huns were infected in either Persia or India. In the 5th century, when the Huns descended into Europe, they may have carried smallpox with them, but from other records it appears that by this time the disease had already established itself in Italy and France. In A.D. 622 a Christian priest, Ahrun, living in Alexandria 30 years before the Arab conquest of Egypt, wrote a clear description of the disease, differentiating it from measles, as did the Persian physician Al-Razi, living in Baghdad in A.D. 910. By A.D. 1000 smallpox was probably endemic in the more densely populated parts of Eurasia, from Spain to Japan, as well as the African countries bordering on the Mediterranean Sea. The spread of smallpox was assisted by the caravans that crossed the Sahara to the more densely populated kingdoms of West Africa, and the disease was repeatedly introduced into the port cities of East Africa by Arab traders and slavers. The movement of people to and from Asia Minor during the Crusades in the 12th and 13th centuries helped reintroduce smallpox to Europe; in 1300, England and Germany suffered epidemics. By the 15th century smallpox was in Scandinavia, and by the 16th century it was established in all of Europe except for Russia. As Europe's population increased and people were crowded into the cities, smallpox epidemics appeared with increasing intensity and frequency.

Smallpox was a serious disease in England and Europe in the 16th century. As British and European explorers and colonists moved into the newly discovered continents of the Americas, Australia, and Africa, smallpox went with them. Besides the crucial role it played in the Spanish conquests in the New World, smallpox also contributed to the settlement of North America by the French and English. The English settlers who arrived in the American colonies in 1617 set off an epidemic among the Indians, thereby clearing a place for the settlers who came from Plymouth in 1620. Later, the English used smallpox as a weapon of germ warfare. In the war of 1763 between England and France for control of North America, under orders of Sir Jeffery Amherst (commander-in-chief of North America), the British troops were asked: "Could it not be contrived to send the smallpox among those disaffected tribes of Indians? We must on this occasion, use every stratagem in our power to reduce them." The ranking British officer for the Pennsylvania frontier, Colonel Henry Bouquet, wrote back: "I will try to inoculate the Indians with some blankets that may fall into their hands and take care not to get the disease myself." Blankets were deliberately contaminated with scabby material from the smallpox pustules and delivered to the Indians, allowing the spread of the disease among these highly susceptible individuals. This led to extensive numbers of deaths and ensured their defeat. The westward movement of smallpox among the Indian tribes far outpaced that of European and British settlements.

In West Africa smallpox traveled with the caravans that moved from North Africa to the Guinea Coast. In 1490 it was spread by the Portuguese into the more southerly regions of West Africa. Smallpox was first introduced into South Africa in 1713 by a ship docking in Cape Town that carried contaminated bed linen from a ship returning from India. Smallpox was again introduced into Cape Town by a ship from Sri Lanka (1755) and later by one from Denmark (1767). The first outbreak of smallpox in the Americas was among African slaves on the island of Hispaniola (present-day Haiti and the Dominican Republic) in 1518. Here and elsewhere the Amerindian population was so decimated that by the time of Pizarro's conquest it is estimated that 200,000 had died. With so much of the Amerindian labor force lost to disease, there was an increased need for replacements to do the backbreaking work in the mines and plantations of the West Indies, the Dominican Republic, and Cuba. This need for human beasts of burden stimulated, in part, the slave trade from West Africa to the Americas. Slaves also brought smallpox to the Portuguese colony of Brazil. Smallpox was first recorded in Sydney, Australia, in 1789, a year after the British had established a penal colony there. It soon decimated many of the aboriginal tribes. As in the Americas, this destruction of the native population by smallpox paved the way for easy expansion by the British.

The Disease of Smallpox

The cause of smallpox is a virus, one of the largest of the viruses, large enough that with proper illumination it can actually be seen with a light microscope. Much of its detailed structure can be visualized, however, only by using an electron microscope. The outer surface (capsid) of the smallpox virus resembles the facets of a diamond, and its inner dumbbell-shaped core contains the genetic material, double-stranded DNA (Fig. 9.3). The virus has about 200 genes, 35 of which are believed to be involved in virulence. Other poxviruses that can infect humans are monkeypox, cowpox, milker's node, tanapox, and chickenpox. Most of these cause mild diseases in humans, with the exception of the zoonotic monkeypox, which is virtually indistinguishable from smallpox.

Most commonly the smallpox virus enters the body through droplet infection by inhalation. It can also be acquired, however, by direct contact or through contaminated fomites (inanimate objects) such as clothing, bedding, blankets, and dust. The infectious material from the pustules can remain infectious for months. In the mucous membranes of the mouth and nose the virus multiplies. During the first week of infection there is no sign of illness, though the virus can still be spread by coughs or by nasal mucus. The virus moves on to the lymph nodes and then to the internal organs via the bloodstream. Here the virus multiplies again. Then the virus

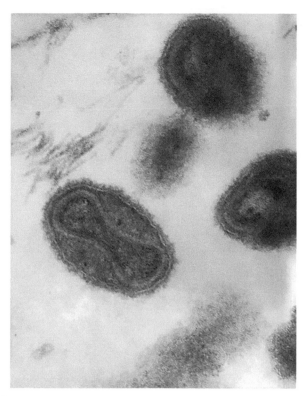

Figure 9.3 The smallpox virus as seen with the transmission electron microscope. The dumb-bell shaped structure is the core containing the genetic material. Courtesy of CDC/ Steven Glenn, Laboratory Training & Consultation Division, 1979.

reenters the bloodstream. Around the 9th day the first symptoms appear: headache, fever, chills, nausea, muscle aches, and sometimes convulsions. A few days later a characteristic rash appears (Fig. 9.4).

Richard Preston, in *The Demon in the Freezer*, described the onset of smallpox:

The red areas spread into blotches across (the) face and arms, and within hours the blotches broke out into seas of tiny pimples. They were sharp feeling, not itchy, and by nightfall they covered (the) face, arms, hands and feet. Pimples were rising out of the soles of (the) feet and on the palms of the hands, too. During the night, the pimples developed tiny, blistered heads, and the heads continued to grow larger ... rising all over the body, at the same speed, like a field of barley sprouting after the rain. They hurt dreadfully, and they were enlarging into boils. They had a waxy, hard look and they seemed unripe ... fever soared abruptly and began to rage. The rubbing of pajamas on (the) skin felt like a roasting fire. By dawn, the body had become a mass of knob-like blisters. They were everywhere, all over ... but clustered most thickly on the face and extremities. ... The inside of the mouth and ear canals and sinuses had pustulated ... it felt as if the skin was pulling off the body, that it would

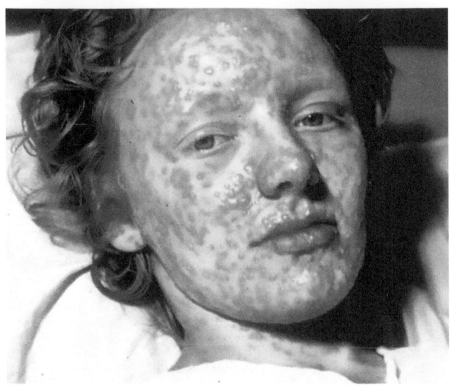

Figure 9.4 Individual with smallpox pustules. From: C. W. Dixon. Smallpox, London: Churchill, 1962.

split and rupture. The blisters were hard and dry, and they didn't leak. They were like ball bearings embedded in the skin, with a soft velvety feel on the surface. Each pustule had a dimple in the center. They were pressurized with an opalescent pus.

The pustules began to touch one another, and finally they merged into confluent sheets that covered the body, like a cobblestone street. The skin was torn away ... across ... the body, and the pustules on the face combined into a bubbled mass filled with fluid until the skin of the face essentially detached from its underlayers and became a bag surrounding the tissues of the head ... tongue, gums, and hard palate were studded with pustules ... the mouth dry ... The virus had stripped the skin off the body, both inside and out, and the pain ... seemed almost beyond the capacity of human nature to endure.

The individual is infectious a day before the rash appears and until all the scabs have fallen off. Many die a few days or a week after the rash appears. Not infrequently there are complications from secondary infections. The infection results in a destruction of the sebaceous (oil) glands of the skin, leaving permanent crater-like scars in the skin—pockmarks.

By the beginning of the 18th century nearly 10% of the world's population had been killed, crippled, or disfigured by smallpox. The 18th century in Europe has been called the "age of powder and patches" because pockmarks were so common. The "beauty patch" (a bit of colored material) seen in so many portraits was designed to hide skin scars. In this way the pockmark set a fashion.

There are two pathologic varieties of the smallpox virus: *Variola major* and *Variola minor*. They can be distinguished from one another by differences in their genes. *V. major*, the deadlier form, frequently killed up to a quarter of its victims, although in naive populations such as the Amerindians the fatality rate could exceed one-half. *V. minor*, a milder and kinder pathogen, had a fatality rate of about 2% and was more common in Europe until the 17th century, when it mutated to the more lethal form, perhaps as a result of a reintroduction from the Spanish colonies in the Americas. In the 17th century smallpox was Europe's most common and devastating disease, killing an estimated 400,000 each year. It caused one-third of all cases of blindness. England suffered most severely in the second half of the 17th century, and between 1650 and 1699 there were 20 deaths a week. In 1660 in London alone more than 57,000 died of smallpox out of a population of 500,000.

"Catching" Smallpox

Smallpox is a contagious disease of civilization that is spread from person to person, and there are no animal reservoirs; in other words, it is not a zoonotic disease. It can exist in a community only so long as there are susceptible humans. It has been estimated that a minimum population on the order of 100,000 persons is needed to ensure that there are enough susceptible persons born annually to sustain the chain of infection indefinitely. The R_0 value (i.e., the number of secondary infections that result from a single primary infection [see p 12]) has been calculated to be between 5 and 7 for epidemic smallpox. Smallpox spreads more rapidly in temperate climates during the winter and during the dry season in the tropics.

Smallpox caused large epidemics when it first arrived in a virgin community where everyone was susceptible, but afterwards most individuals who were susceptible either recovered and were immune, or died; then the disease died down. If the infection was reintroduced later, that is, after a new crop of susceptible individuals had been born or migrated into the area, then another epidemic would occur. After a period of time, however, the disease would reappear so frequently that only newborns would be susceptible to infection, older individuals being immune from previous exposure. In this situation smallpox became a disease of childhood similar to measles, chickenpox, mumps, whooping cough, and diphtheria. Sometimes smallpox did not disappear altogether, especially in larger communities, and remained at a low level. The infection would break out as an epidemic every 5 to 15 years, however, when enough susceptible individuals had accumulated. Smallpox affected both sexes and all ages. Mortality was highest among infants, and smallpox can induce abortions in all trimesters of pregnancy.

Variolation

Early treatments for smallpox involved prayer and quack remedies. In ancient Africa and Asia there were smallpox gods and goddesses that could be enlisted for protection. In France one could appeal to St. Nicaise, the Bishop of Rheims, who recovered from the disease only to be killed by the Huns in 452. Because smallpox produces a red rash, a folklore remedy of "like cures like" arose in medieval times. In 1314 the Englishman John of Gaddesden followed this recipe and recommended that smallpox victims could be helped by the color red, and so those who were infected were dressed in red. In 1834, Niels Finsen, the Dane who pioneered light therapy, gave "like cures like" a pseudoscientific status, prescribing that those afflicted with smallpox be screened from UV light and exposed only to red light. Other charlatans suggested that a cure would come about by eating red foods and imbibing red drinks. These remedies, which of course did no good, persisted in some quarters until the 1930s! Today, prevention of smallpox is often associated with Edward Jenner, but even before Jenner's scheme of vaccination, techniques were developed to induce a mild smallpox infection. The educated classes called this practice of folk medicine "inoculation," from the Latin *inoculare*, meaning "to graft," since it was induced by cutting. It was also called "variolation," since the scholarly name for smallpox was variola, from the Latin *varus*, meaning "pimple." The method of protection by variolation may have been inspired by the folklore belief of "like cures like," since in some versions of this remedy secretions from an infected individual would be inoculated into the individual to be protected from that same infection.

There were numerous techniques of variolation. The Chinese avoided contact with the smallpox-infected individual, preferring that the person either inhale a powder from the dried scabs shed by a recovering patient or be given powdered scabs placed on cotton swabs and inserted into the nostrils. In the Near East and Africa material from the smallpox pustule was rubbed into a cut or a scratch in the skin. The first scholarly account of variolation (or immunization against smallpox) was written in 1675 by Thomas Bartholin, physician to Charles V, king of Denmark and Norway. Forty years later, this work came to the notice of the Turkish physician to Charles XII, Emmanuel Timoni. He in turn communicated a description of their method of variolation to the Royal Society of London: material was taken from a ripe pustule and rubbed into a scratch or an incision in the arm or leg of the person being inoculated.

Mary Pierpont was a highborn English beauty who eloped with Edward Wortley

Montagu, a grandson of the first Earl of Sandwich. After Edward became a member of Parliament, the couple soon became favorites at the court of King George I. In London in 1715, at age 26, Lady Mary Montagu contracted smallpox. She recovered, but smallpox-induced facial scars and the loss of her eyelashes marred her beauty forever. Her 22-year-old brother was less fortunate, however, and died from smallpox. The following year Lady Montagu's husband was appointed ambassador to Turkey, and she accompanied him. There she learned of the Turkish practice of variolation. She was so impressed by what she saw that she had her son inoculated by Timoni. Upon returning to England in 1718, she propagandized the practice, and she had her daughter inoculated against smallpox at age 4. She convinced Caroline, the Princess of Wales (later queen of England during the reign of George II), that her children should be variolated, but before that could happen, the royal family required a demonstration of its efficacy. In 1721 the "Royal Experiment" began with six prisoners who had been condemned to death by hanging and were promised their freedom if they subjected themselves to variolation. They were inoculated, suffered no ill effects, and upon recovery they were released as promised. Despite this, Princess Caroline wanted more proof and insisted that all the orphan population of St. James Parish be inoculated. In the end only six children were variolated, but with great success. Princes Caroline, now convinced, had both her children, Amelia and Caroline, variolated. This too was successful. Lady Montagu, though, met with sustained and not entirely inappropriate opposition to the practice of variolation. The clergy denounced it as an act against God's will, and physicians cautioned that the practice could be dangerous to those inoculated as well as others with whom they came in contact. The danger of contagion was clearly demonstrated when the physician who had previously inoculated Lady Montagu's children variolated a young girl who then proceeded to infect six of the servants in the house; although all seven recovered, it did cause a local epidemic. Although inoculation was embraced by the upper classes in England and Lady Montagu's physician was called upon to variolate Prince Frederick of Hanover, it never caught on with the general population there or elsewhere. This had little to do with some of its demonstrated successes but rather with the fear of contagion, the risks involved in inoculation with smallpox, and the lack of evidence that the protection would be long-lasting. It was found later that variolation did provide lifelong protection, though the danger of death remained at about 2%.

Variolation came into practice quite independently in the English colonies in America. Cotton Mather, a Boston clergyman and scholar, was a member of the Royal Society of London. Although he read the account of Timoni, he wrote to the Society that he had already learned of the process of variolation from one of his African slaves, Onesimus. In April 1721 there was smallpox outbreak in Boston. Half the inhabitants fell ill, and mortality was 15%. During this time Mather tried to encourage the physicians in the city to carry out variolation of the population. The response from the physicians was negative save for one, Zabdiel Boylston. On June 26, Boylston variolated his 6-year-old son, one of his slaves, and the slave's 3-year-old son without complications. By 1722, Mather, in collaboration with Boylston, had inoculated 242 Bostonians and had found that it protected them. His data showed a mortality rate of 2.5% in those variolated, compared to the 15 to 20% death rate during epidemics. Boylston's principal opponents, however, were not the clergy but his fellow physicians.

In the crowded cities of England during the 18th century most adults had acquired some measure of immunity to smallpox through exposure during childhood. In the sparsely settled villages in the American colonies, however, there was little exposure and hence little immunity. With the outbreak of war in 1775, the most dangerous enemy the colonists had to fight was not the British but smallpox. Following the Battle of Bunker Hill, in June 1775, General Howe's forces occupied Boston, and Washington's troops were deployed on the surrounding hills. Howe was unable to attack because smallpox was so rampant among the Bostonians and to a lesser degree among his troops (perhaps because of their practice of variolation). Washington, however, did not attack for fear that his troops would be decimated by smallpox. So he first ordered his troops to be variolated. When the British evacuated the city on March 17, 1776, Washington sent 1,000 of his men who were immune to smallpox to take possession of the city.

Earlier (in 1775) the American colonists feared that the British forces from Quebec would attack New York. The Americans sent 2,000 colonial troops under General Benedict Arnold to attack the poorly fortified garrison at Quebec. Although General Arnold's army was superior and better equipped, the colonials halted their advance on Quebec when they were suddenly stricken with smallpox. Over half of Arnold's troops developed the disease, and mortality was very high. Few of the others were well enough to fight. In contrast, the much smaller British forces were well variolated and fit, so they could hold out until reinforcements arrived. The colonials abandoned their siege of the garrison and retreated to Ticonderoga and Crown Point on Lake Champlain, where they continued to die at a high rate. The Americans never made another serious attempt to intrude into Canada. Ten years later, having learned from this experience, Washington ordered that his entire army be variolated, but by that time it was too late—Canada had been preserved for the British Empire.

Vaccination

Today, smallpox has been eradicated, thanks in part to the development of the first vaccine by Edward Jenner (1749-1823), an English physician. Jenner did not know what caused smallpox or how the immune system worked, but nevertheless he was able to devise a practical and effective method for immune protection against attack by the lethal virus *V. major*. Essentially Jenner took advantage of a local folktale and turned it into a reliable and practical device against the ravages of smallpox. The farmers of Gloucestershire believed that if a person contracted cowpox, they were assured of immunity to smallpox. In milk cows cowpox shows up as blisters on the udders that clear up quickly and produce no serious illness (Fig. 9.5). Farm workers who come into contact with cowpox-infected cows develop a mild reaction, with the eruption of a few blisters on the hands or lower arm. Once exposed to cowpox, neither cows nor humans develop any further symptoms. It was noted that hardly a milkmaid or a farmer who had contracted cowpox showed any of the disfiguring scars of smallpox, and most milkmaids were reputed to have fair and almost perfect complexions, as suggested in the nursery rhyme:

> Where are you going, my pretty maid?
> I'm going a-milking, sir, she said.
> May I go with you, my pretty maid?

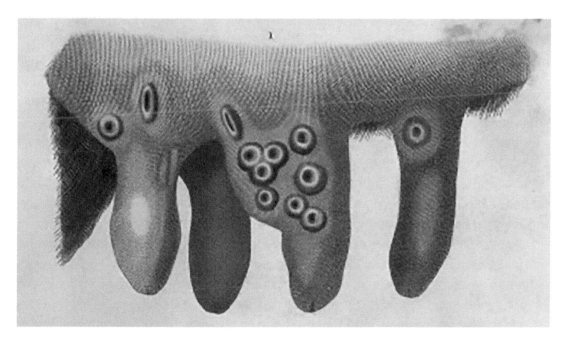

Figure 9.5 Cowpox on cow udder. Courtesy Wikipedia.

You're kindly welcome, sir, she said.
What is your father, my pretty maid?
My father's a farmer, sir, she said.
What is your fortune, my pretty maid?
My face is my fortune, sir, she said.

In 1774 a cattle breeder named Benjamin Jesty contracted cowpox from his herd, thereby immunizing himself. He then deliberately inoculated his wife and two children with cowpox. They remained immune even 15 years later when they were deliberately exposed to smallpox. Jenner, however, and not Jesty, is given credit for vaccination because he carried out experiments in a systematic manner over a period of 25 years to test the farmer's tale. Jenner wrote:

It is necessary to observe, that the utmost care was taken to ascertain, with most scrupulous precision, that no one whose case is here adduced had gone through smallpox previous to these attempts to produce the disease. Had these experiments been conducted in a large city, or in a populous neighborhood, some doubts might have been entertained; but here where the population is thin, and where such an event as a person's having had the smallpox is always faithfully recorded, no risk of inaccuracy in this particular case can arise.

As with the earlier "Royal Experiment," Jenner felt free to carry out human experimentation without any reservations, although it should be noted that all his patients wanted the vaccination for themselves or their children. On May 14, 1796, he took a small drop of fluid from a pustule on the wrist of Sarah Nelms, a milkmaid who had an active case of cowpox. Jenner smeared the material from the cowpox pustule onto the unbroken skin of a small boy, James Phipps (Fig. 9.6). Six weeks later Jenner tested his "vaccine" (from the Latin *vacca*, meaning "cow") and its ability to protect against smallpox by deliberately inoculating the boy with material from a smallpox pustule. The boy showed no reaction—he was immune to smallpox. In the years that followed, "poor Phipps" (as Jenner called him) was tested for immunity to smallpox about a dozen times, but he never contracted the disease. Jenner wrote up his findings and submitted them to the Royal Society, but to his great disappointment, his manuscript was rejected. It has been assumed that the reason for this was that he was simply a country doctor and not a part of the scientific establishment. In 1798, after a few more years of testing, he published a 70-page pamphlet, *An Inquiry into the Causes and Effects of the Variola Vaccinae*, in which he reported that the inoculation with cowpox produced a mild form of smallpox that would protect against severe smallpox, as did variolation. He correctly observed that the disease produced by vaccination would be so mild that the infected individual would not be a source of infection to others—a discovery of immense significance.

Figure 9.6 Edward Jenner (1749-1823) performing the first vaccination against smallpox in 1796. The woman wrapping her wrist is Sarah Nelms. A 1879 oil on canvas painting by Gaston Melingue (1840-1914)/Academic Nacionale de Medicine, Paris, France/Archives Charmet/Bridgeman Images.

The reactions to Jenner's pamphlet were slow, and many physicians rejected his ideas. Cartoons appeared in the popular press showing children being vaccinated and growing horns (Fig. 9.7). But within several years some highly respected physicians began to use what they called the "Jennerian technique" with great success. By the turn of the century the advantages of vaccination over variolation were clear, and Jenner became famous. Parliament awarded him a prize of 10,000 pounds in 1802 and another 20,000 pounds in 1807. Napoleon had a medal struck in his honor, and he received honors from governments around the world.

By the end of 1801 the methods of Jenner were coming into worldwide use, and it became more and more difficult to supply sufficient cowpox lymph or to ensure its potency when shipped over long distances. To meet this need, Great Britain instituted an Animal Vaccination Establishment in which calves were deliberately infected with cowpox and lymph was collected. Initially this lymph was of variable quality, but

Figure 9.7 Cartoon of Edward Jenner. Courtesy of the Wellcome Library, London, CC-BY 4.0.

when it was found that addition of glycerin prolonged preservation, this became the standard method of production. The first "glycerinated calf's lymph" was sent out in 1895. With general acceptance of Jenner's findings, the hide of the cow called Blossom, which Jenner used as the source of cowpox in his experiments, was enclosed in a glass case and placed on the wall of the library in St. George's Hospital in London, where it remains to this day.

Why does vaccination work? As mentioned earlier, there are two species of smallpox virus in humans: *V. major*, which can kill, and the less virulent *V. minor*. Variolation or inoculation with *V. minor* protects against the lethal kind of virus. Vaccination, however, involves cowpox, and this disease is caused by a different virus, named by Jenner *Variola vaccinae*. The exact relationship between the virus used in vaccines today and Jenner's cowpox is obscure. Indeed, in 1939 it was shown that the viruses present in the vaccines used then were distinct from contemporary cowpox as well as smallpox. Samples of Jenner's original vaccine are not available, and so it cannot be determined what he actually used, but current thinking is that the available strains of vaccine are derived from neither cowpox nor smallpox, but from a virus that became extinct in its natural host, the horse.

V. vaccinae and *V. major* are 95% identical and differ from each other by no more than a dozen genes. In order for there to be protection, there must be cross-reactivity between the two such that antibodies are produced in the vaccinated human that can neutralize the *V. major* virus, should the need arise. Fortunately, this does happen. Recovery from a smallpox attack gives lifelong immunity, but vaccination ordinarily does not. The advantages of vaccination over variolation are 2-fold. First, the recipient of a vaccination is not infectious to others, and second, death occurs very rarely, whereas with variolation not only is the individual infectious, but the death rate is 1 to 3%.

Some consider smallpox a disease of historical interest because it was certified as eradicated by the World Health Assembly on May 8, 1980. This feat was accomplished 184 years after Jenner introduced vaccination. The program began in 1967. By 1970 smallpox had been eliminated from 20 countries in western and central Africa, by 1971 from Brazil, and by 1975 from all of Asia, and in 1976 from Ethiopia. The last natural case was reported from Somalia in 1977. Eradication of smallpox was possible for three reasons: (i) there were no animal reservoirs, (ii) the methods of preserving the vaccine proved to be effective, and (iii) the vaccine was easily administered. Smallpox is the first and only naturally occurring human disease to be eradicated by human intervention.

With eradication there arose another question. What should be done with smallpox virus stocks? Should they be destroyed or transferred to a few reference laboratories? In 1979 the World Health Organization recommended transfer. Transfer, however, provides for hazards to the persons working with the virus, and if stocks were obtained by terrorist groups or rogue governments, they could be used as an agent of biological warfare.

A nightmare biohazard scenario is that smallpox is released into a major city. The virus would spread rapidly through the entire unprotected population. The death rate in those without any immunity would be 30%. Anyone born after 1970, when most countries stopped vaccinating against smallpox, would be vulnerable. Imagine the worldwide catastrophe of millions of people dying from smallpox as a result of a bioterrorist attack. In the United States alone the 120 million people born after the end of routine vaccination would be highly susceptible, and there could be 40 million deaths! What we have learned about biological weapons is this: they have a high destructive potential, they can be unpredictable in their consequences, they are cheap and easy to prepare with a minimum of scientific knowledge, they may be difficult to detect, and they would certainly have a devastating psychological impact. So perhaps destruction of the stocks of smallpox would be better—or would it? Destroying the stocks would result in a loss of scientific knowledge of smallpox pathogenesis and would prevent the design of new antiviral agents to prevent large outbreaks of smallpox should they ever recur. This is the dilemma. As yet it remains unresolved.

Vaccination and Its Social Context

In 1799 news of Jenner's vaccine reached the American colonies. Benjamin Waterhouse (1754-1846), a physician and professor at the newly established Harvard Medical School, was one of the first to send for the vaccine, and he popularized its use in Boston. His first trial was in 1800, when he vaccinated his son; then he sent a pamphlet and list of successes to Vice President Thomas Jefferson. Jefferson enthusiastically endorsed the "Jennerian technique" and had his own family and his slaves vaccinated. Jefferson expected the public to follow his lead, but they did not. Indeed, it wasn't until the term of James Madison, the fourth president of the United States, who was familiar with the positions of Waterhouse and Jefferson, that a law encouraging vaccination was signed. But Waterhouse's success with Jefferson turned out to be grist for the mill of his adversaries. Waterhouse, a practicing Quaker, was a pacifist and opposed the Revolutionary War. Born in Rhode Island, he was considered an outsider to the conservative Boston community, and he further incurred the wrath of the Boston elite (who were supporters of Federalism) when he allied himself with the populist democracy espoused by Jefferson. His political enemies, consisting of a coalition of physicians from Harvard, the clergy, and others from the Boston area, arranged for his dismissal from the Harvard Medical School. In 1820 the vaccine law was repealed, and in 1822, Dr. James Smith, the federal agent for the distribution of cowpox, was dismissed from office. Politics was favored over good public health policy, and the result was that by 1840 epidemics and deaths from smallpox once again increased in the United States.

Some European states did make vaccination compulsory (Bavaria, for example, in 1807), and by 1869 it was required in Germany that all citizens and especially soldiers be vaccinated. This proved to be of great benefit to Germany during the Franco-Prussian War (1870-1871), when the number of cases of smallpox in the German army was slightly more than 8,000, with fewer than 300 deaths, whereas among the French, who did not practice vaccination, there were 28,000 cases and 2,300 deaths. Compulsory vaccination never became a federal policy in the United States, however, and communities were free to vaccinate or not. With the expansion of elementary and secondary education after the Civil War and eventually the compulsory attendance of children, though, the most effective way of enforcing vaccination was for public schools to require a vaccination scar on the arm before any child could attend.

This should have solved the problem, but it did not. Up until the 1900s there were Anti-Vaccination Societies, whose members believed the practice of vaccination to be dangerous, ineffective, and a violation of their civil liberties. The strength of the movement was clearly demonstrated by the "Milwaukee Riots" in 1890. The city of Milwaukee was home to an increasing number of German and Polish immigrants, who worked in the factories and lived in the poorer sections. During a smallpox epidemic in 1894 the public health authorities pursued a strict policy of enforcement, in some cases actually removing children suspected of being infected from their homes and placing them in the city's isolation hospital. The immigrant community believed not only that their civil rights were being violated but that they were being discriminated against because they were poor immigrants. Some believed they were being poisoned, and others held to the belief that home care was better than hospitalization. They asked why, if the city's smallpox hospital was so safe, it had been built in the part of the city where the poorest people lived. When the health department ambulance came to collect a child suspected of having the disease, the people threw stones and scalding water at the ambulance horses, and they threatened the guards with baseball bats, potato mashers, clubs, bed slats, and butcher knives. In response, the ambulances withdrew. By the end of the epidemic the health commissioner had been impeached and the health department lost its police powers, which it never regained.

The vaccination controversy did not end in Milwaukee. In May 1901 there was an outbreak of smallpox in Boston. The fatality rate was 17%; there were about 1,500 cases in a population of about 550,000. By the fall the Boston Board of Health took steps to control the epidemic: a free and voluntary vaccination program was started. By December 400,000 people had been vaccinated, but outbreaks of smallpox continued to be reported. The Board of Health established "virus squads" with the following orders: all inhabitants of the city must be vaccinated or revaccinated, there would be a house-to-house program of vaccination, and vaccine should be administered to all. Persons who refused vaccination were fined $5 or sentenced to 15 days in jail. As is frequently the case, the disenfranchised of society were blamed for spreading the disease. In this instance it was the homeless. Virus squads were dispatched to cheap rooming houses to forcibly restrain and vaccinate the occupants. The opponents of vaccination (Anti-Compulsory Vaccination League) swung into action, complaining that vaccination was a violation of civil liberties; however, other groups favored vaccination and warned of dire consequences were the practice to be abandoned. By January 1902 legislation was proposed to repeal the state's compulsory vaccination law. This led to a landmark legal case on the constitutionality of compulsory vaccination (Johnson v. Massachusetts). In 1905 the U.S. Supreme Court voted 7 to 2 in favor of the state, ruling that although the state could not pass laws requiring vaccination in order to protect an individual, it could do so to protect the public in the case of a dangerous communicable disease. The epidemic ended in March 1903.

The Boston epidemic of 1901-1903 illustrates the benefits of educating the public about the benefits of disease prevention and the value of having a public debate on the pros and cons of public health policies. It also shows how well-meaning boards of health can institute policies that disregard civil liberties and ethical concerns.

Coda

Smallpox was endemic worldwide for more than 3,000 years. The fatality rate was particularly high in naive populations. The disease affected the course of British, European, and American history. It contributed to the exploitation of the Amerindians, helped to destroy their native culture, promoted their conversion to Christianity, accelerated the slave trade to the New World, affected the outcome of battles during the Revolutionary War, and paved the way for European and British colonial settlements. Smallpox also provided the incentive for the development of protective measures (variolation and vaccination), affected the cultural responses to disease, and contributed to the establishment of more-humane public health policies. The eradication of smallpox has demonstrated that there is the possibility through immunization for the elimination of other infectious diseases that continue to plague humankind.

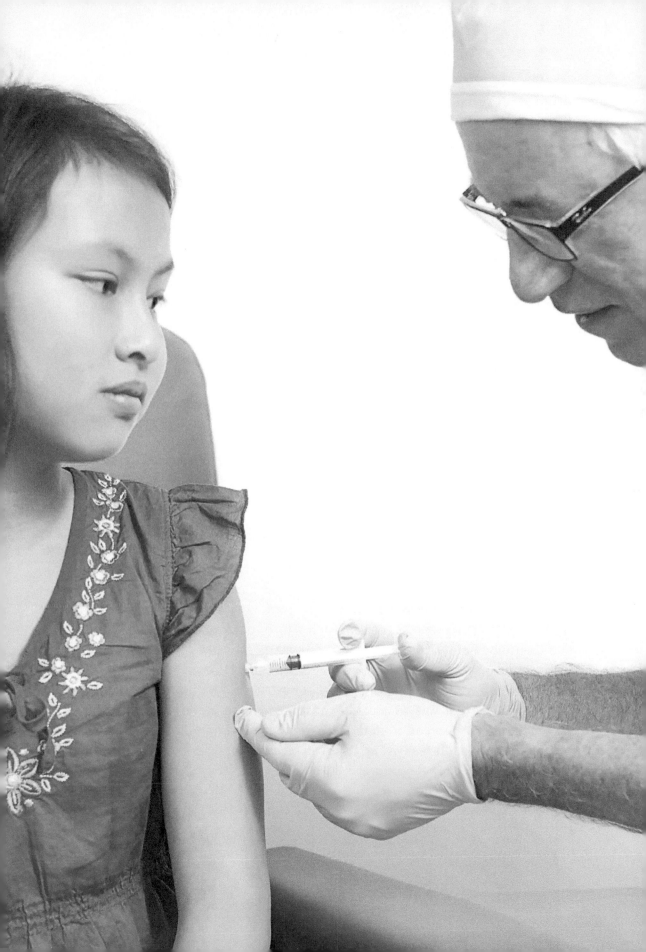

doi: 10.1128/9781683670018.ch10

10
Preventing Plagues: Immunization

The year was 430 b.c., and Athens was being devastated by a plague. The Greek historian Thucydides wrote:

> The infection first began, it is said, in parts of Ethiopia above Egypt and thence descended into Egypt and Libya and into most of the king's country (i.e., Persia). Suddenly falling upon Athens, it first attacked the population in Piraeus ... and afterwards appeared in the upper city, when the deaths became more frequent. Fear of gods or laws of men there was none to restrain them. As for the first, they judged it to be just the same whether they worshipped them or not, as they saw all alike perishing. As for the latter, no one expected to live to be brought to trial for his offenses ... even the most staid and respectable citizens devoted themselves to nothing but gluttony, drunkenness and licentiousness.

The plague of Athens, described by Thucydides, raged for 4 years, during which time a quarter of the population died. Many historians regard this plague as a "turning point" in the history of Western civilization. Thucydides' description, however, does not allow for a clear identification of the specific disease that caused this plague in Athens. Some have argued that it must have been typhus because it broke out during wartime; its clinical course lasted 7 to 10 days; and there was a rash, fever, and delirium. Hans Zinsser, an expert on typhus, was not persuaded that its cause was typhus since Thuycidides makes no mention of lice or a cloudy mental state, and furthermore, typhus usually does not have such a high fatality rate. Others argue that the disease was smallpox because Thucydides describes skin pustules and hints that the rash spread from the face to the trunk; however, he does not mention pockmarks among the survivors. Its cause remains a mystery. But Thuycidides, who was stricken with this plague and recovered, stated: "Yet still the ones who felt most pity for the sick and dying were those who had the plague themselves and recovered from it. They knew what it was like and at the same time felt themselves to be safe,

Figure 10.1 (Left) Immunization against measles. Courtesy CDC/Amanda Mills, 2011.

for no one caught the disease twice, or if he did, the second attack was never fatal." There is no mystery here: those who survived this plague were immune.

What does it mean when we say someone is immune? The dictionary definition of "immune" states that it comes from the Latin *immunis* and is derived from the time in ancient Rome when citizens were given freedom from some onerous duty owed to the empire. But in medicine, to say that someone is "immune" means that he or she is protected from a disease.

The Immune System

A key to understanding the protective mechanisms of the immune system is an appreciation of the roles of the cells of the system (Fig. 10.2). Blood consists of a salty protein-containing fluid, the plasma, in which float red blood cells and white blood cells. (When blood plasma clots, the remaining fluid, which is deficient in clotting molecules, is called serum.) The red cells carry oxygen and carbon dioxide and are not involved in immunity. The white cells, on the other hand, are a critical part of the immune system. White cells come in two varieties: the granule-containing granulocytes (called, whether they stain with eosin [an acidic red dye] or a basic blue dye or both, eosinophils, basophils, and neutrophils, respectively), and the non-granule-containing cells called monocytes and lymphocytes (Fig. 10.3). The "plumbing" of our bloodstream does not exist as an entirely closed system of "pipes." It leaks fluid but not cells. This is because blood, headed for the tissues and organs, is pumped by the heart into the arteries. The outgoing arteries branch into smaller and smaller vessels the farther and farther they are from the heart, and finally they ramify into the smallest of vessels, the capillaries. Fluid, consisting essentially of plasma with much less protein and without blood cells, seeps out of the capillaries into the tissue spaces bathing the cells. Cells are microscopic manufacturing plants, and to accomplish their energy-demanding tasks they contain an internal furnace that burns fuel (glucose) and dumps ashes (carbon dioxide and other waste products). The cell's bathing solution (called interstitial fluid) serves as a waterway for carrying fuel to and removing ashes from the cells. But what happens to all the fluid that has leaked out into the tissues? How does it return to the bloodstream? The fluid cannot return in the direction from which it came because the pumping of the heart creates an opposing hydrostatic pressure. The only means by which the fluid can return to the bloodstream is by an auxiliary drainage system that collects the seepage and returns it. (If the drainage is faulty or limb movement is limited, then fluid accumulates in the tissue spaces, especially in the lower limbs by gravity, and the result is swollen feet!) The drainage system consists of permeable "pipes" (lymph vessels) of the lymphatic system (Fig. 10.2). Just as the capillaries are widely distributed throughout all the tis-

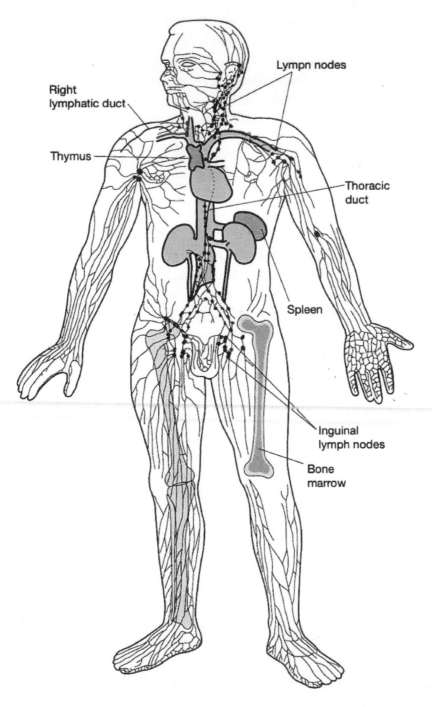

Figure 10.2 The immune system consists of tissues containing lymphocytes and lymphatic vessels that transport the cells from the lymphatic tissues and the antibodies secreted by lymphocytes returning them to the bloodstream via the lymphatic and the thoracic duct in the neck region. Lymphocytes are manufactured in bone marrow and multiply in the thymus, the spleen and the lymph nodes.

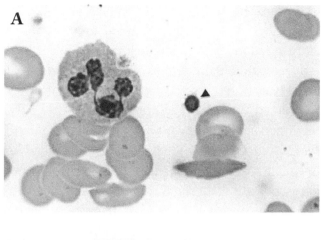

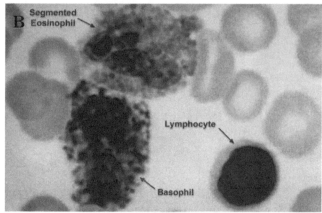

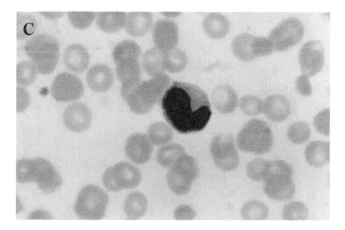

Figure 10.3 (A) Neutrophil with blood platelet (arrow) Courtesy CDC Public Health Image Library #6645/Dr. F. Gilbert, 1972; (B) Eosinophil, basophil and lymphocyte Courtesy CDC Public Health Image Library #15133/Dr. Candler Ballard, 1974; (C) Monocyte. Courtesy CDC Public Health Image Library #12084/Dr. Mae Melvin, 1965.

sues of the body, so too are the lymph vessels. Lymph vessels pick up the fluid from the tissue spaces (and such movements are aided by muscular movements of the limbs) and return it to the blood circulation upstream from the heart by a "connector pipe" (the thoracic duct located in the region of the collarbone), and in this way there is a complete circulatory system.

Mammals have coopted the lymphatic system for another purpose: immunologic defense. This has been accomplished by placing clusters of immune tissues at "checkpoints" or "surveillance stations" along the lymphatic routes, and here the fluids (now called lymph in the lymph vessels) are examined. The surveillance stations are called lymph nodes. Lymph nodes are scattered throughout the body but tend to be clustered in the region of the groin, armpits, and chest. In these locations the lymph nodes act as filters. The filter is principally composed of two kinds of white cells: macrophages and lymphocytes. Our body contains 10^{10} to 10^{12} lymphocytes, equivalent in mass to the brain or liver! These also circulate in the bloodstream and are also found in the spleen, a kind of lymph node. Now that we have an idea of the plumbing and surveillance properties of the immune system, let us consider how it works to protect us against disease.

There are several levels of protective human immune mechanisms. The first involves the physical and physiological barriers of the body: the close-fitting epithelial cells of the skin, secretions such as tears containing the antibacterial lysozyme, mucus, sebum (the oil associated with hair follicles), and the acidic gastric juice of the stomach and urine. These tend to act as frontline defenses. But if the body's physical defenses are breached by an invading pathogen, there is a second level of immune responses to protect: white blood cells capable of ingesting and digesting foreign substances. We call these "foot soldiers" of the immune system phagocytes, or "eating cells," because they feed on, digest, and destroy foreign particles, especially bacteria, and they also clean up the victims—the dead and damaged cells. The remnants of living and dying bacteria, the healthy and wounded phagocytes and the fluid that surrounds them, is the material called pus. Pus is characteristic of the site where a battle is taking place: blood vessels dilate, the capillaries become more permeable, fluid leaks and produces swelling (edema), there is an influx of phagocytes (neutrophils and macrophages), histamine is released, and the battle area becomes heated up and reddened. (This sequence of cellular events, called the innate inflammatory response, is mobilized quickly in order to ward off an invader and until the somewhat slower adaptive response can come into play.)

Inflammation and innate immunity

The inflammatory response is localized, swift, and characterized by redness, puffiness, tenderness, and warmth. It prevents invading bacteria and viruses from spreading through the body and is critical to our well-being until acquired immunity can develop. How does inflammation work? There are two kinds of cells involved: macrophages and dendritic cells, each of which has 10 different kinds of receptors (or docking stations) on its outer surface. (Dendritic cells have a maze of long extensions, and although they resemble nerve cells, they are not nerve cells but are more closely related to macrophages of the immune system.) The receptors on these cells are able to recognize ~20 different pathogen-associated molecular patterns (PAMPs). Some PAMP molecules are the lipopolysaccharide in the cell wall of Gram-positive bacteria; the viral coat protein; the glycolipid in the cell wall of mycobacteria; and flagellin, a protein that makes up the bacterial flagellum. Once a PAMP is fitted into the docking station of the dendritic cell or macrophage (that is, bound to the receptor), chemical messengers called lymphokines are released. These lymphokines make the blood vessels leaky; there is redness and swelling; and neutrophils and lymphocytes (of the T cell kind, which will be considered shortly) move out of the blood vessels and are attracted to the area where they begin to work. This painful, swollen, warm, and reddened site may become filled with pus (when it is called a boil) and consists of an accumulation of live or dead infectious bacteria and live or dead neutrophils, the debris of battle. Neutrophils are the foot soldiers of the immune system that are activated immediately, and they fight and die on the inflammatory battlefield. Pus can be milky and thick in a staphylococcal infection, watery and blood-stained in a streptococcal infection, and foul-smelling when there are anaerobic bacteria present. In the good old days a poultice, a word dating back to the 16th century, which consisted of a soft paste spread on muslin or some other cloth, was heated and placed on the boil or the inflamed area. The goal of using a poultice was to increase the blood flow and to enhance the movement of white cells (neutrophils and lymphocytes) into the area and to produce pus.

Acquired immunity

Let us now turn our attention to another level of defense, the kind of immunity that develops as a consequence of exposure to a foreign intruder—acquired immunity. Imagine that the infected area on your hand is filled with bacteria called streptococci. The inflammatory reaction has been set in motion, but all the bacteria are not destroyed, and some enter the lymph vessels that drain the inflamed site. These are carried to the regional lymph node, which enlarges and gets sore. (Recall the buboes of bubonic plague.) Sometimes the pathway is marked by a reddish streak of inflammation from the site where the pathogenic bacteria entered a break in the skin and have been carried to the regional lymph node. The bacteria contain foreign materials—"not self" antigens. Some of these will now trigger the production of antibodies.

How are antibodies made? Antibodies are made by lymphocytes called B lympho-cytes or B cells. Their discovery came about in an interesting way. It all began with a study of immunity in chickens. In the 1950s, Bruce Glick was a graduate student who sought to discover the function of a lymphoid organ called the bursa of Fabricius, lo-cated at the lower end of a chicken's digestive tract. (The bursa, Italian for "purse," is named after the 16th-century anatomist Hieronymus Fabricius, who first described it.) Glick surgically removed the bursa and found no difference in the survival or health between chickens with a bursa and those lacking it. He left the project tempo-rarily. Then another graduate student in Glick's department needed some chickens to demonstrate the production of antibodies. He used Glick's chickens, both those with a bursa and those without one. After the chickens were injected with antigen, he discovered that the chickens lacking a bursa made no antibody. Subsequently it was found that the cells in the bursa (essentially a lymph node) that were involved in the production of antibody were lymphocytes, and in honor of their location they were named B cells after the first letter of "bursa." (In mammals, which lack a bursa, the B cells are made in the bone marrow, which fortuitously also begins with a B.) The work was submitted for publication to *Science* magazine but rejected because it was "uninteresting." It was published later, however, in the journal *Poultry Science*, where it went unnoticed for several years. Glick and his colleagues then made another inter-esting observation regarding the chickens that lacked a bursa: although they could not make antibody, they could reject skin grafts and were able to overcome a viral infection. This suggested that although the bursa played a role in antibody-based im-munity, there must be additional mechanisms that could protect against infection or foreign materials. As we shall soon find out, this additional immune response would be discovered to involve another kind of lymphocyte, called the T lymphocyte.

How do B cells make antibody? On the surface of the B lymphocyte is an anti-body molecule that serves as a receptor for the foreign antigen. Once bound to the B cell surface, the B cell (called a plasmablast) is triggered to divide asexually, and the population of identical offspring from such division is a clone. The members of the clone are called plasma cells. Division to produce an adequate number of plasma cells may take several days, since the division rate is about 10 h per division. Plasma cells are cellular factories that can manufacture large quantities of antigen-specific antibody, which they secrete into the blood plasma. A single plasma cell can secrete >2,000 molecules of antibody per s, or about 10^6 molecules each h, and it is possible for different plasma cells to synthesize 10^{15} different antibodies, each having a differ-ent specificity for the antigen that triggered its division in the first place. The specific antibody produced can protect in several ways: it can neutralize the antigen, or it can aggregate (clump) bacteria and free virus in the blood, or it may lyse (disintegrate) virus-infected cells or bacteria, or it may block virus entry into cells, or it may make the bacteria more attractive so the phagocytes readily eat the microbe. Because the antibody is in the liquid fraction (plasma or serum) of the blood, this is called humoral immunity after the four humors (black and yellow bile, phlegm, and blood) that the ancient Greeks and Romans believed were able to control our behavior. We still use these terms when we describe someone who is sluggish as "phlegmatic" or we say we are in a bad humor or in a good humor.

Not all the B cells differentiate into plasma cells; some remain quiescent and persist as memory cells. When there is a reencounter with the same antigen, the memory cell undergoes rapid division, plasma cells are produced, and antibody is synthesized and released. This rapid response is due to memory of a past antigenic experience and is the basis of immunization with booster shots.

Clonal selection theory

Now you might ask: how does the body learn to distinguish its own cells and its own antigens from those that are foreign or "nonself"? In the 1950s, Macfarlane Burnet, an Australian immunologist, proposed a theory to explain how the immune system is able to distinguish "us" from "them." He called it the clonal selection theory. According to Burnet's theory, each embryo contains progenitor lymphocytes, and each is capable of interacting with a different antigen, including those antigens of its own body—"self-antigens." In the embryo, however, interaction of a self-antigen with a progenitor lymphocyte specific to it results in the death of that particular progenitor cell. As a result, by the time of birth the immune system is purged of the lymphocytes that can react with self-antigens. In short, the lymphocytes that react with "self" have been selected against. The remaining lymphocytes may react to an assault by a foreign, or "not self," antigen because they are able to recognize only "not self" antigens. When an encounter with a nonself-antigen does take place, the binding of antigen to the antibody molecules on the surface (now called a receptor) of the B lymphocyte causes the B cell to multiply; a clone of plasma cells is produced; and these synthesize and secrete antibody specific for that particular antigen. Thus, only the antigen-selected cells multiply to make an antibody, and the levels of that specific antibody increase in the blood as a result of the expansion of the plasma cell population. Antibody levels usually rise over a 2- to 4-week period following exposure to antigen and may last for quite a long time. As noted above, secondary exposure to antigen (a booster) produces an even more rapid response because now there are more plasma cells making that particular antibody. Prolonged exposure to antigen is no longer required. Furthermore, the binding ability of an antibody improves with time because the antigen "selects" for the replication of those cells that carry the receptor with the better match.

In effect, the immune system "learns." In 1960, Burnet received the Nobel Prize for his work on the clonal selection theory, which was fully substantiated by further experimental work over the course of the next decade and beyond.

Autoimmunity

The immune system has as its prime function surveillance; that is, it is constantly "looking about" for foreign substances, whether these be cancer cells or bacteria or viruses or toxins. The system is not perfect, however, and things can go wrong, especially when it reacts with a foreign antigen that resembles a self-antigen. This is a case of mistaken identity in an antigen and can result in what are called autoimmune diseases, literally "immunity against oneself." Rheumatic fever is an example of an autoimmune disorder. Rheumatic fever, a disease almost unknown today thanks to the effectiveness of the antibiotic penicillin, starts as an inflammation in the throat, pharyngitis. It may, however, result in heart valve destruction with permanent scarring, so that the individual can suffer from a heart murmur for life. The cause of the pharyngitis is a group A β-hemolytic streptococcus. One strain of streptococcus causes a benign infection of the skin called pyoderma or impetigo, but others—especially those heavily encapsulated by hyaluronic acid and M protein—are invaders of the throat. It is not, however, the direct action of the streptococcus that results in heart damage. Instead, the M protein antigens are swallowed in large amounts during pharyngitis, and this triggers an immune response, producing antibodies that cross-react with heart valves as well as autoimmune cell-mediated cytotoxic cell reactions (see below). These all act to destroy the heart valves, resulting in rheumatic fever disease. Other autoimmune diseases are insulin-dependent diabetes, rheumatoid arthritis, multiple sclerosis, lupus erythematosus, myasthenia gravis, and Addison's and Graves' diseases.

Diphtheria

Diphtheria is a horrible disease. On January 12, 2010, a 7.0 magnitude earthquake hit Haiti. It was estimated that 300,000 died. By February more than 880,000 people were living in displaced persons camps. It was on the morning of May 3, 2010, when a 15-year-old boy named Oriel showed up at the clinic with a scratchy throat and fever. Noting a gray hue at the back of his throat, the doctors made a startling diagnosis: diphtheria. Few foreign doctors would have ever seen this childhood disease: it has all but been eradicated in the developed world, but not in Haiti, where it has remained endemic.

The boy's symptoms had been going on for 6 days. He needed to get to a hospital, fast. At any time the infection could close his throat, suffocating him to death. They went to the General Hospital in Port-au-Prince. But the hospital administrator was reluctant to introduce a highly contagious infection into a damaged facility housing patients with tuberculosis and AIDS. Those hesitant to take the patient were overruled. Others argued that so long as Oriel was isolated, the risk would be manageable. The standard treatment would be to administer a dose of diphtheria antiserum (see below), which in Haiti was housed only at a warehouse operated by the health ministry and World Health Organization (WHO). Doctors at the hospital said that Oriel had to be admitted before the warehouse would release the antiserum. After a confusing couple of hours, Oriel finally arrived at the General Hospital's green gates—reportedly just after 5 P.M. The warehouse had just closed.

Even in the United States the anti-diphtheria serum takes a while to procure. The Centers for Disease Control and Prevention (CDC), which usually quarantines it at major airports, promises to deliver it only "within hours." In Haiti, despite infinitely more difficult circumstances, they were able to reopen the warehouse and get the serum. But Oriel's story had a heartbreaking end. As his days-old condition deteriorated, the doctors put him on a respirator. At some point overnight, perhaps during a shift change, Oriel was left alone. When doctors returned, the breathing tube was dangling and Oriel was in a coma. The boy had probably awoken, panicked, and ripped the life-sustaining tube out. The doctors revived him, but the lack of oxygen to his brain likely dealt an irreversible blow. Two days later, he was dead from diphtheria.

The incubation period with diphtheria is ~7 days, followed by fever, malaise, a sore throat, hoarseness, wheezing, a blood-tinged nasal discharge, nausea, vomiting, inflamed tonsils, and headache. The infected child is quite pale and has a sticky gray-brown membrane (made of white cells, red cells, dead epithelial cells, and clots) that adheres to the palate, throat, windpipe, and voice box. This membrane impairs the child's breathing. If disturbed, it can bleed and break off and block the airways of the upper respiratory tract. Within 2 weeks this upper respiratory infection can spread to the heart, leading to inflammation and failure of the heart, circulatory collapse, and paralysis. During a diphtheria epidemic up to one-half of untreated children die, but more commonly the death rate is 20%.

In 1812, Francisco de Goya (1746-1828) painted *El Garrotillo* (*The Strangler*), which was the common name for diphtheria in the late 18th and early 19th century. In Goya's time the treatment of diphtheria consisted of efforts to remove the adherent pseudomembrane obstructing the upper airway, as is shown in the painting, where a bearded man violently opens the boy's mouth to proceed with a horrific operation (cauterization) on the boy's pharynx (Fig. 10.4).

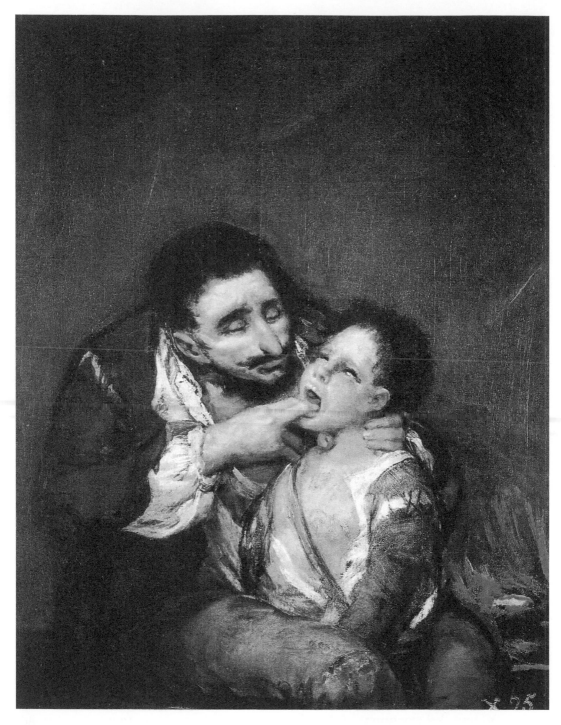

Figure 10.4 Detail from the 1812 painting "El Garratillo" (*The Strangler*), attributed to Francisco de Goya (1746-1828). The bearded man is attempting to remove the adherent membrane in the throat that is blocking the breathing of the child. Courtesy Wikimedia Commons.

Emil Behring and serum therapy

In 1883 two German scientists, Edwin Klebs and Friedrich Löffler, working in Robert Koch's laboratory in Berlin, discovered the microbe behind diphtheria in throat swabs taken from an infected child; they were able to grow the bacterium in pure culture and named it *Corynebacterium diphtheriae*. Then, in 1888, Émile Roux and Alexandre Yersin of the Pasteur Institute in Paris found that the broth in which the *Corynebacterium* bacteria were growing contained a substance that produced all the symptoms of diphtheria when injected into a rabbit, and at high enough doses it killed the rabbit. This killing agent was a poisonous toxin. In humans it is this toxin, a neurotoxin, that leads to heart muscle degeneration; neuritis; paralysis; and hemorrhages in the kidneys, adrenal glands and liver. The toxin is produced because the diphtheria bacterium itself is infected with a virus. In order for the toxin to be synthesized, iron molecules must be present in its surroundings—provided by the body tissues or in the culture medium.

Emil Behring, born in 1854 in West Prussia, studied medicine in the medical academy in Berlin and after graduation worked in several military commands until 1867, when he was posted to the Institute of Pharmacology in Bonn for further training. It was there that he became interested in bacteriology research. In 1889 he joined Koch's laboratory, where he studied the effects of iodoform, as well as immunity to anthrax in rats. He also infected guinea pigs with *C. diphtheriae,* and when he treated them subsequently with iodoform was able to allow a few animals to survive. He then treated the survivors with a dose of diphtheria that would have killed normal animals, but none developed pathology. Behring asked, "Were they immune due to therapy, due to an externally applied chemical or intrinsic activity?"

Shibasaburo Kitasato (1853-1931) was already at the Institute of Hygiene when Behring arrived. He had come to Berlin in 1886 by order of the Japanese government to receive training as a bacteriologist. Kitasato's main interests were cholera, typhus, and above all tetanus. Working together, Kitasato and Behring found that if small amounts of tetanus toxin were injected into rabbits, the rabbit serum would contain a substance that would protect another animal from a subsequent lethal dose of toxin. They called the serum that was able to neutralize the toxin "immune serum." Behring and Kitasato found that the same situation also occurred with diphtheria toxin, and a week after the publication of their work Behring alone presented the results in the *German Medical Journal* paper "Investigations on the Manifestation of Diphtheria Immunity in Animals." He posited "the possibility of cure, even of acutely progressing disease." In effect, by toxin inoculation the rabbit had been protected against a disease. This process was called immunization, although today in Jenner's honor any immunization is also called a vaccination. Further, the immune protection seen by Behring and Kitasato could be transferred from one animal to another by injection of this immune serum. In other words, it was possible to have passive immunity; that is, immunity could be borrowed from another animal that had been immunized with the active foreign substance when its serum was transferred (by injection) to a nonimmunized animal. (Passive immunization, or "immunity on loan," can be lifesaving in the case of toxins such as bee or spider or snake venoms or tetanus, since it is possible to counteract the deadly effects of the poison by injection of serum containing the appropriate antitoxin.) The implications of the research on immune serums for treating human disease were quite obvious to Behring and others in the Berlin laboratory. There were, however, two main obstacles for therapy. One was theoretical and the other practical. Elie Metchnikoff in 1884 had proposed that immunity was dependent not on serum but on phagocytes (white cells), and this was not consistent with Behring's thesis. There was a vehement dispute between the two. The problem would be solved (in part) with the production of a strong protective serum. In this Behring was assisted by the commitment of a company (Hoechst) and the acquaintance of Paul Ehrlich.

Paul Ehrlich

Paul Ehrlich (1854-1915) was the same age as Behring and had come to Koch's laboratory in 1880. Ehrlich had shown that feeding small quantities of toxins to animals and then increasing them rendered the animals immune to an otherwise lethal dose, and that they tolerated a dose up to a 1,000-fold higher than one that would kill an untreated animal. The association of Ehrlich and Behring would eventually result in a falling out. Ehrlich would say that Behring cheated him:

> I believe it was in the autumn of 1901 that trial experiments with Behring's serum was tried in the childrens ward of a hospital and no positive results were obtained. I was asked to make experiments with my serum, which immediately gave good results. It was under these circumstances that Behring asked me to collaborate. When we began working together he showed me a bottle containing five quarts of diphtheria toxin. He believed this would be sufficient for 50 years of immunization work. It would have hardly been enough for one horse. On the other hand, my collaborators and I were convinced of the necessity of using progressively increasing doses of a very active toxin in order to obtain the desired level of immunity. It was difficult for him to accept my view and to set him on the right track, but then after six months of experimenting he finally succeeded in producing active sera. So it was I who, at this critical time, straightened out the confusion and put Behring on the right track which enabled him to achieve success. And when he finally arrived at this point and was covered with honor by the Nobel committee in 1901 "for his

work on serum therapy and especially for use against diphtheria and which opened up a new path in the field of medical science and gave the physician a powerful weapon with which to combat disease and death," his first deed was to rob me of the rewards of many years of work. Behring did not mention me or his Japanese collaborator's name even once in his entire Nobel lecture.

Without waiting for the identification of the active ingredient in immune serum, the German government began to support the construction of factories to produce immune sera. Initially, Behring estimated that a diphtheria-stricken child would require two 50-ml injections of immune serum; after Ehrlich's contribution, however, the requirement was <20 ml per child. The first child was sucessfully treated with diphtheria antitoxin in December 1891. This stirred more-ambitious treatment at the Institute. In March and April 1893, of 11 children treated, 9 survived. In contrast, >65% of the children admitted with diphtheria to the Institute in 1891-1892 died. Examination of 5,000 reports of children from 1883 to 1967 revealed that before the use of serum therapy more than half of all the children's deaths were from diphtheria. In the years preceding World War II more than 55,000 cases of diphtheria were still recorded per year in England and Wales, of which 3,000 were fatal.

Toxin, antitoxin, and toxoid

C. diphtheriae colonizes the pharynx, where it releases a powerful toxin into the bloodstream. The exotoxin is composed of two subunits encoded by a virus (called a bacteriophage). The B subunit allows the bacteria to bind to the cells of the pharynx, and this allows the A subunit to enter. Once inside the cell, the A subunit inhibits protein synthesis by blocking the elongation factor that is involved in the translation of mRNA into protein. Unable to make proteins, the host cell dies.

Initially, it was believed that all bacterial infections could be treated with antitoxin therapy, but unfortunately, only the bacteria that cause diphtheria and tetanus and few others produce and secrete toxins, so the therapeutic application of immune serum against most bacterial infections remains limited. An antitoxin is one of many kinds of active materials found in immune serum after a foreign substance is injected into humans or rabbits or horses or chickens or mice or guinea pigs or monkeys or rats; the general term for the active material is "antibody." Simply put, antitoxin is one kind of antibody. The potency or strength of an antibody is called its titer and is reflected by the degree of dilution that must be made to have a specific amount of antigen that will bind to an equal amount of antibody. (The higher the titer, the greater the potency of the serum, or put another way, a high-titer immune serum can be diluted to a greater degree to achieve the same neutralizing effect as would a weaker immune serum.) Ehrlich's principle of titer was not only of theoretical interest; it enabled antitoxins (and other therapeutic biologics such as insulin and other hormones) to be provided in standardized amounts and strengths. Ehrlich also found that although the diphtheria toxin lost its poisoning capacity with storage, it still retained its ability to induce immune serum. He called this altered toxin (usually produced now by treatment with formalin) "toxoid." Toxoids made from diphtheria, pertussis, and tetanus toxins are used as the DPT vaccine. The Schick test for susceptibility to diphtheria also involves toxoid: a small amount of diphtheria toxoid is injected just under the skin surface; failure to react (by an absence of a red swelling—inflammation—at the site of injection) indicates a lack of protective immunity. Individuals who are immunosuppressed, such as AIDS patients or those on anticancer chemotherapy or receiving high doses of steroids, cannot mount an immune response and give false-negatives with such tests.

Diphtheria today

During 1980 to 2001, a total of 53 cases of probable or confirmed diphtheria were reported to the CDC. In October 2003 the Pennsylvania Department of Health and CDC were notified of a suspected case of diphtheria in a previously healthy Pennsylvania man, aged 63 years, who reported that he had never been vaccinated against diphtheria. He and seven other men from New York, Pennsylvania, and West Virginia had returned from a weeklong trip to rural Haiti, where they helped build a church. One day before leaving Haiti the patient had a sore throat. Two days after his return to Pennsylvania he visited a local emergency department complaining of a persistent sore throat and difficulty swallowing. He received oral amoxicillin and clavulanate potassium. On the fourth day of illness the patient returned to the hospital with chills, sweating, restlessness, difficulty swallowing and breathing, nausea, and vomiting. On examination, he had rasping breath and a swollen neck. Radiographs of the neck and chest showed prevertebral soft-tissue swelling, enlargement of the epiglottis, and opacity of the left lung base. The initial diagnosis was acute epiglottitis with airway obstruction and impending respiratory failure. The patient was admitted to the intensive care unit; a laryngoscopy was performed that revealed a yellow exudate on the tonsils, posterior pharynx, and soft palate, and sloughing of the anterior pharyngeal folds. During the next 4 days the patient was treated with antibiotics (azithromycin, ceftriaxone, and nafcillin) and steroids, but he became hypotensive and febrile (100.9°F, or 38.3°C). On the eighth day of illness the patient was transferred to a specialized medical care facility. A chest radiograph showed infiltrates in the right and left lung bases. During tracheostomy his physicians observed a white exudate consistent with *C. diphtheriae* infection. The pseudomembrane covered the epiglottis, the postcricoid region, and the glottic inlet. The patient continued to receive multiple antibiotics, including penicillin, vancomycin, and gentamicin; diphtheria antitoxin was administered on the 9th day of illness. Two days later a sample of the pseudomembrane sent to the CDC and tested by PCR was negative by culture but positive for *C. diphtheriae tox* genes. After 17 days of illness the patient had cardiac complications and died.

Investigations of the patient's close contacts were conducted in New York, Pennsylvania, and West Virginia. Close contacts were defined as persons who had been exposed to the patient's respiratory secretions or who lived in the same household as the patient. These persons included his wife, health care providers, Haiti traveling companions, and two other persons with whom he shared accommodations on the second day of his illness. Specimens were obtained for isolation of *C. diphtheriae* and PCR testing; all culture and PCR results were negative. Close contacts were administered antibiotic prophylaxis and offered vaccine if they had not been immunized in the past 5 years.

Diphtheria, whooping cough, and the DTaP vaccine

Efforts to develop an inactivated whole-cell whooping cough vaccine (see below) began soon after 1906, when Jules Bordet and Octave Gengou published their method of culturing the bacterium *Bordetella pertussis*. In 1942, Grace Eldering and Pearl Kendrick, working in the State of Michigan Health Department, combined their improved killed whole-cell pertussis vaccine with diphtheria and tetanus toxoids to generate the first diphtheria, tetanus, and pertussis (DTP) combination vaccine. To minimize the frequent side effects caused by the pertussis component, Yuji Sato in Japan developed in 1981 an acellular pertussis vaccine (aP) consisting of a purified filamentous hemagglutinin that is surface localized as well as secreted by *B. pertussis*; the hemagglutinin mediates the adherence to epithelial cells and is essential for bacterial colonization of the trachea. First introduced in Japan, the vaccine was approved in the United States in 1992 for use in the combination DTaP vaccine. The acellular pertussis vaccine has a rate of adverse events similar to that of a TD vaccine (a tetanus-diphtheria vaccine containing no pertussis vaccine). The DTaP vaccine is given in several doses, with the first recommended at 2 months and the last between the ages of 4 and 6 years.

Since universal vaccination began in the 1940s, diphtheria has been uncommon in the United States. The vaccination coverage rate among children aged 19 to 35 months who received more than three doses of diphtheria toxoid-containing vaccine was ~95%. Testing of serum samples indicates that the percentage of U.S. residents with protective levels (≥0.1 IU/ml) of diphtheria antibodies decreases progressively with age, from 91% at ages 6 to 11 years to ~30% at ages 60 to 69 years.

In the 1970s and 1980s controversy erupted over whether the killed whole-cell pertussis vaccine caused permanent brain injury. Extensive studies, however, showed no connection of any type between the DPT vaccine and permanent brain injury. The alleged vaccine-induced brain damage proved to be due to an unrelated condition, infantile epilepsy. But before the refutation, a few well-publicized anecdotal reports of permanent disability were blamed on the DPT vaccine and gave rise in the 1970s to anti-DPT movements. In the United States low profit margins and an increase in vaccine-related lawsuits led many pharmaceutical companies to stop producing the DPT vaccine by the early 1980s. In 1982 the television documentary "DPT: Vaccine Roulette" depicted the lives of children whose severe disabilities were incorrectly blamed on the DPT vaccine. The ensuing negative publicity led to many lawsuits against vaccine manufacturers. By 1985 vaccine manufacturers had difficulty obtaining liability insurance. The price of the DPT vaccine skyrocketed, leading providers to curtail purchases and limiting availability. Only one manufacturer remained in the United States by the end of 1985. To correct the situation, Congress in 1986 passed the National Childhood Vaccine Injury Act (NCVIA), which established a federal no-fault system to compensate victims of injury caused by mandated vaccines. The majority of claims that have been filed through the NCVIA have been related to injuries allegedly caused by the whole-cell DPT, not by the DaPT vaccine.

In the past decade there were <5 cases of diphtheria in the United States reported to the CDC. In countries where there is a lower uptake of booster vaccines, however, such as in India, there remain thousands of cases each year. In 2014 there were 6,094 cases of diphtheria reported in India, 1,079 in Nepal, and 35 in Bangladesh.

Why Do Antibodies Work?

Paul Ehrlich formulated a theory of how antigen and antibody interact with one another. It was called the "lock and key" theory. He visualized it in the following way: the tumblers of the lock were the antigen and the teeth of the key were the antibody. One specific key would work to open only one particular lock. Another way of looking at the interaction of antigen with antibody is the way a tailor-made glove is fitted to the hand. One such glove will fit only one specifically shaped hand. Similarly, only when the fit is perfect can the antigen combine with the antibody. Yet despite the fact that antigens are large molecules, only a small surface region is needed to determine the binding of an antibody to it. These antigenic determinants are called epitopes, and they can be thought of as a small patch on the surface of the much larger molecule of antigen. Earlier, when vaccination was discussed (see p 208), it was said, without any explanation, that protection occurred because the antibodies produced against *Variola vaccinae* were able to neutralize the *Variola major* virus. Simply stated, it is because the epitopes of the cowpox antigens are so similar to those of smallpox that when the body produces antibodies to cowpox antigens these are also able to fit exactly to the smallpox epitopes (i.e., the antibodies are said to be cross-reactive) and *V. major* is prevented from multiplying.

Attenuation and Immunization

In 1875, Louis Pasteur attempted to induce immunity to cholera. He was able to grow the cholera bacterium that causes death in chickens and to reproduce the disease when healthy chickens were injected with it. The story is told that he placed his cholera cultures on a shelf in his laboratory exposed to the air and went on summer vacation; when he returned to his laboratory, he injected chickens with an old culture. The chickens became ill, but to Pasteur's surprise, they recovered. Pasteur then grew a fresh culture of the bacterium with the intention of injecting another batch of chickens, but the story goes that he was low on chickens and used those he had previously injected. Again he was surprised to find that the chickens did not die from cholera; they were protected from the disease. The prepared mind of Pasteur recognized that aging and exposure to air (and possibly the heat of summer) had attenuated or weakened the virulence of the bacterium and that such a strain might be used to protect against disease. In honor of Jenner's work a century earlier, Pasteur called his attenuated cholera strain a vaccine.

After Pasteur's success in creating standardized and reproducible vaccines at will, the next major step came in 1886 from the United States. Theobald Smith and Edmund Salmon published a report on their development of a heat-killed cholera vaccine that immunized and protected pigeons. Two years later, these two investigators claimed priority for having prepared the first killed vaccine. Though their work appeared in print 16 months before a publication by Charles Chamberland and Roux (working in the laboratory of Louis Pasteur) and with identical results, the fame and prestige of Pasteur was so widespread that the claim of developing the first killed vaccine by Smith and Salmon was lost in the aura accorded those who worked in Pasteur's institute in Paris.

Influenza

The 1918 outbreak of influenza remains one of the world's greatest public health disasters. Some have called it the 20th century's weapon of mass destruction. It killed more people than the Nazis and far more than did the two atomic bombs dropped on Japan. Before it faded away, it had affected 500 million people worldwide and 20 million to 40 million had perished. This epidemic killed more people in a single year than the Black Death did in 100 years. In 24 weeks influenza destroyed the lives of more people than AIDS did in 24 years. As with AIDS, it killed those in the prime of their life: young men and women in their 20s and 30s. During its 2-year course more people died from influenza ("flu") than from any other disease in recorded history.

The 1918-1920 flu epidemic brought more than human deaths: civilian populations were thrown into panic; public health measures were ineffectual or misleading; there was a government-inspired campaign of disinformation; and people began to lose faith in the medical profession. Because the flu pandemic killed more than twice the number who died on the battlefields of World War I, it hastened the armistice that ended the Great War in Europe.

Where did this global killer come from? Epidemiologic evidence suggests that the outbreak was due to a novel form of the influenza virus that arose among the 60,000 soldiers billeted in the army camps of Kansas. The barracks and tents were overflowing with men, and the lack of adequate heating and warm clothing forced the recruits to huddle together around small stoves. Under these conditions they shared both the breathable air and the virus that it carried. By mid-1918 flu-infected soldiers were carrying the disease by rail to army and navy centers on the East Coast and in the South. Then the flu moved inland to the Midwest and onward to the Pacific states. In its transit across America, cases of influenza began to appear among the civilian population.

People in cities such as Philadelphia, New York, Boston, and New Orleans began to ask: What should I do? How long will this plague last? To minimize panic, the health authorities and newspapers claimed it was "la grippe" and said there was little cause for alarm. This was an outright lie. As the numbers of civilian cases kept increasing, however, and when there were hundreds of thousands sick and hundreds of deaths each day, it was clear that this flu season was nothing to be sneezed at and that the peak of the epidemic had not been reached. In Philadelphia, undertakers had no place to put the bodies and there was a scarcity of coffins. Gravediggers were either too sick or too frightened to bury influenza victims. Entire families were stricken, with almost no one to care for them. There were no vaccines or drugs. None of the folk remedies were effective in stemming the spread of the virus, and the only effective treatment was good nursing. In cities, as the numbers of sick continued to soar, public gatherings were forbidden and gauze masks had to be worn as a public health measure. The law in San Francisco was: if you do not wear a mask, you will fined or jailed.

Conditions in the United States were exacerbated as the country prepared to enter the war in Europe. President Woodrow Wilson's aggressive campaign to wage total war led indirectly to the nation becoming "a tinder box for epidemic disease." A massive army was mobilized, and millions of workers crowded into the factory towns and cities, where they breathed the same air and ate and drank using common utensils. With an airborne disease such as the flu, this was a prescription for disaster. The war effort also consumed the supply of practicing physicians as well as nurses, and so medical and nursing care for the civilian population deteriorated. "All this added kindling to the tinderbox."

The epidemic spread globally, moving outward in ever-enlarging waves. The hundreds of thousands U.S. soldiers who disembarked in Brest, France, carried the virus to Europe and the British Isles. Flu then moved to Africa via Sierra Leone, where the British had a major coaling center. The dockworkers who refueled the ships contracted the infection, and they spread the highly contagious virus to other parts of Africa when they returned to their homes. A ship arrived in Samoa from New Zealand, and within 3 months >21% of the population had died. Similar figures for deaths occurred in Tahiti and Fiji. In a few short years the flu was worldwide and death followed in its wake.

During the flu epidemic there was a rhyme to which little girls jumped rope:

I had a little bird,
And its name was Enza.
I opened the window,
And in flew Enza.

This delightful singsong rhyme did not describe Enza's symptoms, some of which you've probably experienced yourself: fever, chills, sore throat, lack of energy, muscular pain, headaches, nasal congestion, and a lack of appetite. It can escalate quickly to produce bronchitis and secondary infections including pneumonia, and can lead to heart failure and, in some cases, death.

The influenza virus (Fig. 10.5) is highly infectious (it has an R_0 value of 10) and is spread from person to person by droplet infection through coughing and sneezing. Each droplet can contain between 50,000 and 500,000 virus particles. The virus, though, does not begin its "island hopping" in humans. Instead, aquatic wild birds such as ducks and other waterfowl maintain the flu viruses that cause human disease. Because these aquatic birds, which carry the viruses in their intestines, do not get sick and can migrate thousands of miles, the virus can be carried across the face of the earth before humans enter the picture. The flu virus found in these waterfowl, however, is usually unable to replicate itself well in humans, and therefore if it is to become a human pathogen it must first move to intermediate hosts—usually domestic fowl (chickens, geese, or ducks) or pigs—that drink water contaminated with the virus-containing feces of the wild waterfowl. The domestic fowl tend to be dead-end hosts, because some sicken and die, but pigs live long enough to serve as "virus mixing vessels"—and here the flu genes of bird, pig, and human can be mixed because pig cells have receptors for both bird and human viruses. It is by this coming together of virus genes (reassortment) that new strains of the flu are produced.

Let us consider the eight genes of the flu virus as if they were a small deck of playing cards. If the eight different cards are shuffled and two cards are removed from the pack at the same time, how many different combinations are there? The answer: 2^8, or 256. In the same way, if two different viruses infect the same cell and genes are exchanged randomly, there can be 256 different virus offspring. This viral gene mixing tends to occur where birds, pigs, and humans live in close proximity—

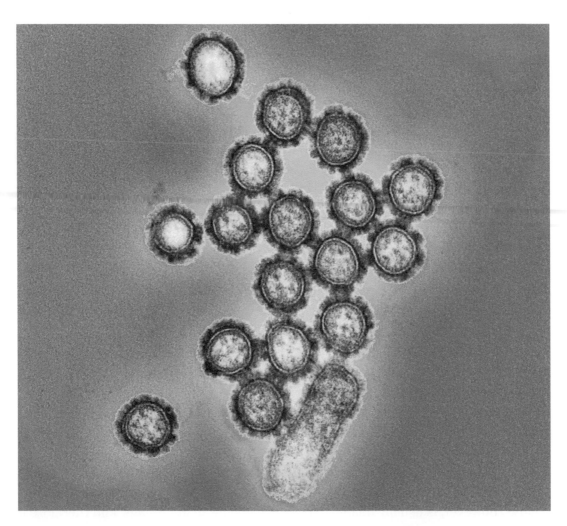

Figure 10.5 Digitally colored transmission electron micrograph (TEM) of H1N1 influenza virus particle. Surface proteins are shown in black. Courtesy CDC Public Health Image Library #18156/ NIAID, 2013.

predominantly in China (where the 2002-2003 SARS outbreak also began). The 1997 Hong Kong "bird flu" was a flu virus that became virulent by acquiring genes from geese, quail, and teal, because these birds were housed together in Hong Kong poultry markets where mixing could occur quite easily. This flu strain killed thousands of chickens before humans were infected. Eighteen people did acquire the infection from contact with the feces of infected chickens, not from other people. Fortunately, the spread to other humans was halted before person-to-person (droplet) transmission could result in a full-blown flu epidemic, because health authorities enforced the slaughter of more than a million fowl in Hong Kong's markets. Had this not occurred, one-third of the human population may have sickened and died.

What genes produce the novel and virulent flu strains? Two genes for surface proteins are principally involved, both critical to virus entry into cells where the mixing of genes takes place. One is called hemagglutinin (H), and the other is neuraminidase (N). The job of the surface spikes of H is to act like grappling hooks to anchor the influenza virus to host cell receptors called sialic acid. After binding, the virus can enter the host cell, replicate its RNA, and produce new viruses. The emerging viruses are coated with sialic acid, the substance that enabled them to attach to the host cell in the first place. If the sialic acid were allowed to remain on the virus and on the host cell, then these new virus particles with H on their surface would be clumped together and trapped much like flies sticking on flypaper. The N, which resembles lollipops on the virus surface, allows the newly formed viruses to dissolve the sialic acid "glue"; this separates the viruses from the host cell and allows them to plow through the mucus between the cells in the airways along the respiratory tract and to move from cell to cell. The entire process—from anchoring to release— takes about 10 h, and in that time 100,000 to 1 million viruses can be produced.

There are 15 different H's and 9 different N's, all of which are found in bird flu viruses. A letter and a number designate each flu strain. For example, H1N1 is the strain that caused the 1918 pandemic, and H5N1 is the 1997 Hong Kong flu strain. Flu epidemics occur when either H or N undergoes a genetic change due to a mutation in one of the virus genes. If a virus never changed its surface antigens, as is the case with measles and mumps, then the body could react with an immune response—antibodies and cell-mediated—to the foreign antigens (see p. 223) during an infection or by a vaccine, and there would be long-lasting protection. Indeed, if a person encountered the same virus, the immune system, having been primed, could swiftly eliminate that virus and prevent infection. In the case of flu, however, there may be no immune response because the virus changes the N and H molecules—sites where the antibodies would ordinarily bind to neutralize the virus. This mutational change of the flu virus (called antigenic drift) ensures escape from immune surveillance and allows the virus to circumvent the body's ability to defend itself. Such slight changes in antigens leads to repeated outbreaks during interpandemic years. Every 10 to 30 years a more radical and dramatic change in the virus antigens (called antigenic shift) may occur, and with this change an influenza virus emerges approximately every quarter of a century that the human immune system has never before encountered. When this happens, a pandemic can occur. There have been flu pandemics in 1957 (Asian flu) and 1968 (Hong Kong flu), and scares in 1976 (swine flu) and 1977 (Russian flu). In 2004 the H5N1 strain resulted in 27 cases and 20 deaths in Vietnam and 16 cases and 11 deaths in Thailand. Still to be explained is why the 1918 outbreak of influenza was so pathogenic. It may have had something to do with an exaggerated innate inflammatory immune response with release of lymphokines such as tumor necrosis factor; this can lead to a toxic shock-like syndrome including fever, chills, vomiting, and headache, and ultimately can result in death. Alternatively, the H1N1 virus that led to the 1918-1920 pandemic could have had a very different kind of H, one more closely related to that of a swine flu, not one from birds, and for this there was little in the way of immune recognition.

Globally, influenza remains an important contagious disease, with 20% of all children and 5% of adults who are infected in any one flu season developing symptoms. The death rate can be as high as 5%. Each fall we are encouraged to get a "flu shot" so that when the flu season arrives in winter we will be protected. But flu shots can protect only against targeted or known strains—those whose antigenic type has been determined by scientists at the WHO—not from unexpected or unidentified types. The flu vaccines most commonly used today do protect against a known type, and they do not produce disease because after the virus has been grown in chicken embryos they are purified and inactivated. Although some vaccines consist of only a portion of the virus, such as H or N antigens, and these serve to activate the immune response, weakened live-virus vaccines may give a stronger and longer-lasting protection. But for these live-virus vaccines it is critical that a return to virulence does not occur. Although these live-virus vaccines may be administered as a nasal spray, avoiding the pain of inoculation, they may also generate disease symptoms.

Flu outbreaks will continue to plague humankind so long as there is "viral mixing," and in those at high risk (i.e., those over 50 years of age, the very young, the chronically ill, and the immunosuppressed) it may be impossible to prevent infection under any circumstances. In addition, those infected with the flu are at greater risk for pneumococcal pneumonia (a bacterial disease), which annually kills thousands of elderly people in the United States. Flu combined with pneumonia can result in skyrocketing mortality.

For a disease such as the flu, with its high R_0 value, quarantining those who show symptoms is not enough to bring the average number of new infections caused by each case to below 1—the level necessary for an epidemic to go into decline. Therefore, other measures, such as treatment and immunization, have to be employed. Flu treatments involve antiviral drugs (zanamivir and oseltamivir) that block the synthesis of the neuraminidase. When such drugs are administered early enough, virus cannot be released and infection is aborted. In the case of other antiflu drugs (amantidine and rimantidine), the viruses quickly acquire resistance, and so these are less effective in preventing infection or in reducing the severity and duration of symptoms.

Another pandemic of influenza is inevitable. Modern means of transportation—especially jet airplanes—ensure that the virus can be spread by an infected traveler across the globe in a matter of hours or within a day. Surveillance may alert us to an epidemic's possibility, and drugs may reduce the severity of illness, but neither can guarantee when or where or how lethal the next pandemic will be. What is predictable is that it will impact our lives: hospital facilities will be overwhelmed because medical personnel will also become sick; vaccine production will be slower because many of the personnel in pharmaceutical companies will be too ill to work; and reserves of vaccines and drugs will soon be depleted, leaving most people vulnerable to infection. There will be social and economic disruptions. How can we prepare for a "future shock" such as that of 1918-1920 pandemic? Stockpiling anti-infective drugs, promoting the development of new drugs, and increasing the methods for surveillance may all help blunt the effects, but these measures cannot eliminate them.

Measles

Measles spreads from person to person through sneezing and coughing; the virus particles are hardy and can survive for several hours on doorknobs, handrails, elevator buttons, and even in air. For the first 10 to 14 days after infection there are no signs or symptoms. A mild to moderate fever, often accompanied by a persistent cough, runny nose, inflamed eyes (conjunctivitis), and sore throat, follows. This relatively mild illness may last 2 or 3 days. Over the next few days the rash spreads down the arms and trunk, then over the thighs, lower legs, and feet (Fig. 10.6b). At the same time fever rises sharply, often as high as 104 to 105.8°F (40 to 41°C). The rash gradually recedes, and usually lifelong immunity follows recovery. Complications occur in about 30% of cases and may include diarrhea, blindness, inflammation of the brain, and pneumonia. Between 1912 and 1916 there were 5,300 measles deaths per year in the United States. Yet all that changed in 1968 with the introduction of the measles vaccine; in the United States measles was declared eliminated in 2000.

And although a measles vaccine has been available worldwide for decades, according to the WHO about 400 people a day died of measles in 2013.

The measles virus (Fig. 10.6a) is unlike the bacteria that cause diphtheria and tetanus in that it cannot be grown in anything but living cells. To make the measles vaccine first required growing the virus in the laboratory in tissue culture—glass vessels seeded with living cells—a procedure first developed in 1928 by a Canadian

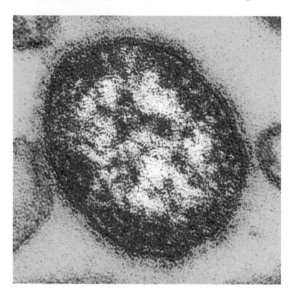

Figure 10.6a Measles virus as seen with the transmission electron microscope. Courtesy CDC Public Health Image Library #10707/C. Goldsmith/W. Bellini.

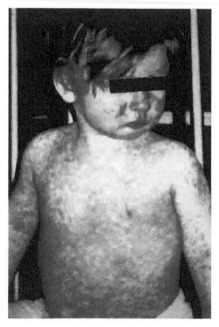

Figure 10.6b Child with 4 day rash of measles. Courtesy CDC Public Health Image Library #132/NIP/Barbara Rice.

husband-and-wife team, Hugh and Mary Maitland, working in Manchester, England. Using minced kidney, serum, glucose, and mineral salts, they were able to grow smallpox virus and other viruses. This was a great breakthrough, since until their study all work with viruses had to be conducted using laboratory animals (mice or monkeys). In 1931, Ernest Goodpasture and Alice Woodruff developed another technique for growing viruses: chick embryos. This too has been used ever since for the manufacture of some vaccines.

In January 1954, John Enders at the Boston Children's Hospital assigned a young physician, Thomas Peebles, the task of capturing the measles virus and then propagating it in tissue culture for future use as a vaccine. When it was learned that there was an outbreak of measles at the all-boys Fay School, Peebles went to the school and convinced the principal to allow him to collect blood samples. On February 8, 1954, Peebles collected blood from a 13-year-old student, David Edmonston, who was clearly suffering with measles—he had nausea, a fever, and a telltale red rash. Edmonston's blood was added to a culture of human kidney cells, and within a few days it was clear that the measles virus was growing and killing the cells. Following the work of Pasteur on attenuation, Peebles hoped that by forcing the virus to grow in a hodgepodge of different kinds of cells the measles virus would become sufficiently weakened that it could serve as a potential vaccine. Over many months the measles virus was passed serially: 24 times in human kidney cells, 28 times in cultures of cells from the placenta, and 6 times in minced chick embryos. Believing that the virus had been sufficiently attenuated, in 1958 the researchers tested the putative measles vaccine in 11 mentally handicapped children at the Fernald School, where measles occurred annually with a high morbidity and mortality. In the trial all the children developed antibodies to the virus, but some experienced fever and had a mild rash. Clearly, the measles virus had not been sufficiently attenuated. In 1958-1959 the chick embryo cell-cultured measles virus was injected into monkeys; they developed antibodies but no illness, and no virus was detectable in the blood. This attenuated virus was then administered to immune adults with no ill effects. In February 1960 this vaccine was tested at the Willowbrook State School in mentally handicapped children. Twenty-three children received the vaccine and 23 other children received nothing. Six weeks later there was an outbreak of measles at the Willowbrook School, with hundreds of children becoming infected; none of those vaccinated, however, came down with measles. Many but not all of the unvaccinated also became infected. Again there was a high rate of side effects, including fever and rash. It was suggested that the toxic effects of the vaccine might be ameliorated by the addition of a small amount of gamma globulin along with the vaccine. This was tested at a woman's prison in New Jersey, Clinton Farms for Women, where there was also a nursery full of their babies. With parental approval, six of the infants were given the vaccine in one arm and the gamma globulin in the other. None of the babies had high fevers, and only one had a mild rash. Subsequent studies were carried out with susceptible children, after parental permission, chosen on the basis of an absence of antibodies to the measles virus. The success of these studies led to further trials among home-dwelling children in five U.S. cities. The Edmonston measles virus was further attenuated by Maurice Hilleman at the Merck Research Laboratories by passage 40 more times in chick embryos growing at lower temperatures; this attenuated strain, called Moraten (for more attenuated Enders), which did not require gamma globulin, was licensed in 1968 and has been used throughout the world. Before the development of the measles vaccine, 7 million to 8 million children around the world died from measles every year. In the United States between 1968 and 2006, hundreds of millions of doses were given and the number of measles cases decreased from 4 million to <50. It is estimated by the WHO that between 2000 and 2003 the measles vaccine had prevented 15.6 million deaths worldwide.

You might wonder why the measles and other candidate vaccines would be tested first in mentally handicapped children. The reason is that up until the 1960s this was a common practice for determining the efficacy of a candidate vaccine. Critics suggest that this was the practice because these children were considered to be expendable subjects and were more vulnerable, being without protected rights. The vaccine researchers, on the other hand, justified the use of the mentally handicapped children because they were confined to institutions where hygiene was poor, care was negligent, crowding was the norm, and consequently they were at greater risk of catching and dying from infectious diseases than other children. One vaccinologist put it this way: "They were tested not because they were more expendable; they were tested because they were more vulnerable." Today, because of ethical considerations, experiments using mentally handicapped children (as well as prisoners or others who are unable to give voluntary informed consent) are no longer included in vaccine or drug studies.

Mumps

In the 1960s the virus causing mumps infected a million people annually, mostly children, in the United States. The mumps virus (Fig. 10.7a) attacks the salivary glands in the front of the ears, causing swollen cheeks (Fig. 10.7b) so that the children look like chipmunks. Sometimes the virus infects the brain and the spinal cord, causing meningitis, paralysis, and deafness. In the male it might also infect the testes and result in sterility; in the pregnant female it can result in birth defects and fetal death. If the virus attacks the pancreas, it can cause diabetes. The spherical mumps virus contains a single strand of RNA that codes for nine proteins. The HN (hemaggluti-

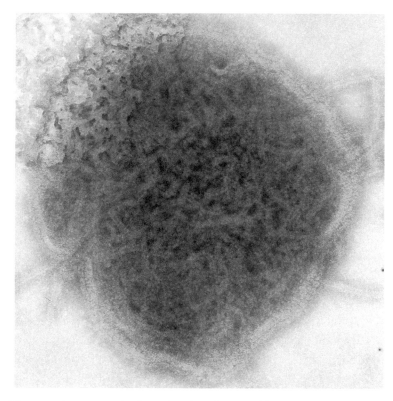

Figure 10.7a Mumps virus negatively stained and viewed by transmission electron microscope. Courtesy CDC Public Health Image Library #8578/F.A. Murphy, 1973.

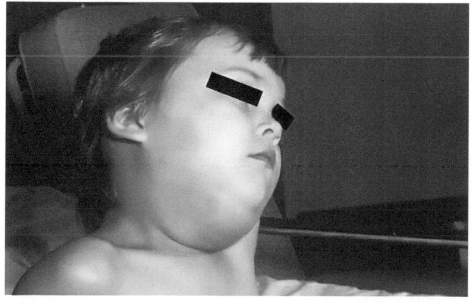

Figure 10.7b Child with mumps. Courtesy CDC Public Health Image Library #130, 1976.

nin-neuraminidase) proteins are essential for viral attachment to the host cell and release from the cell, and the F fusion protein is essential for viral penetration into the host cell. The NP nucleocapsid protein encloses the RNA (ribonucleoprotein; RNP) in a capsid, and there is a matrix protein, M. The nonstructural NS1 and NS2 proteins have no known function, and the V protein is involved in producing the pathogenic effects.

On March 23, 1963, Maurice Hilleman's 5-year-old daughter Jeryl Lynn came down with mumps. Hilleman, then the director of virus and cell biology at the Merck Research Laboratories, stroked the back of her throat with a swab and placed it in a vial of nutrient broth, which was then used to inoculate an egg containing a chick embryo. The virus grew, and after he had passed it several times in embryos, Hilleman placed it into a flask containing chick embryo brain cells. This procedure was repeated over and over again to attenuate the mumps virus. To test the protective capacity of the attenuated virus, Hilleman (who had a Ph.D. degree, not an M.D.) contacted the pediatrician Robert Weibel and Joseph Stokes, chairman of pediatrics at Children's Hospital in Philadelphia. In June 1965, Weibel tested the candidate vaccine in 16 mentally handicapped children at the Trendler School in Bristol, Pennsylvania, where Weibel's own son with Down syndrome was a resident. The vaccine was found to be safe and to give rise to antibodies against the virus. Encouraged by the findings, Weibel carried out additional tests in August with 60 severely handicapped children. The results were the same—antibodies to the mumps virus and no illness. To prove that the vaccine provided full protection, Stokes and Weibel recruited parents of kindergartners to participate in a more extensive clinical trial. Four hundred children were enrolled in the study; 200 received the vaccine (the attenuated Jeryl Lynn strain of the mumps virus) and 200 received no vaccine. Sixty-three came down with mumps, but only two of these came from the vaccinated group. On March 30, 1967—4 years after Hilleman's daughter came down with mumps—the vaccine was licensed. Since then hundreds of million doses have been distributed in the United States, and today mumps is a rare childhood disease.

Rubella

A German physician originally described rubella, or German measles, at the end of the 18th century. For years the skin rash of rubella was confused with measles and scarlet fever, but in 1881 a consensus was finally reached that rubella was a specific viral disease. Rubella begins as an upper respiratory infection, with virus multiplication in the nasopharynx and then in the lymph nodes. After 14 to 21 days there is a fine rash that begins on the face and spreads out over the entire body. After the rash subsides, there may be complications such as arthritis, encephalitis, and platelet depression. The most serious complication may occur when the infection is acquired during the first trimester of pregnancy, when a woman has a better than a two in three chance of giving birth to an infant that is deaf, blind, or mentally handicapped. A famous case was that of the movie actress Gene Tierney. When she gave birth to a severely deformed child 8 months after being kissed during a USO tour by a fan with German measles, the infant was placed in an institution and Tierney had to be hospitalized with severe depression.

In 1961, during a rubella outbreak at Fort Dix in New Jersey, two Walter Reed Army Institute physicians took throat washings from some hospitalized recruits and placed these into cultures of African green monkey kidney (AGMK) cells. The virus that grew was named the Parkman strain after one of the physicians. More than 77 passages in AGMK cells were needed to attenuate the virus so that it could be tested in human subjects. The Walter Reed researchers gave this high-passage Parkman strain (named HPV-77) to Maurice Hilleman at Merck, who then adapted it to duck embryo cultures. Before Hilleman could get the attenuated rubella strain into production as a vaccine, however, Mary Lasker, the widow of the advertising millionaire Albert Lasker and a philanthropist involved in health funding, paid him a visit. Lasker noted that Parkman, then at the National Institutes of Health, was already working on a rubella vaccine, and she warned Hilleman: "You should get together and make one vaccine or else you'll have trouble getting yours licensed." Hillman agreed and modified the Parkman HPV-77 vaccine, and by 1969 the rubella vaccine was ready for marketing in the United States.

A competitor to Hilleman was Stanley Plotkin. Plotkin had grown up in the Bronx; went to the Bronx High School of Science, a highly competitive school for gifted students; and was enamored by reading Paul de Kruif's *Microbe Hunters*. He attended New York University (NYU) and the Downstate Medical Center in Brooklyn, both on full scholarships. Plotkin then worked at the CDC and then at Philadelphia's Wistar Institute. In 1963, during an outbreak of rubella, there were thousands of damaged fetuses from women who were anxious about their pregnancies and were having therapeutic abortions. In 1964, Plotkin obtained an aborted fetus from an infected mother. Because this was the 27th aborted fetus that Plotkin had received, and because he was able to isolate rubella from the third organ he tested—the kidney—Plotkin called his virus Rubella Abortus 27/3 (RA27/3). Plotkin attenuated RA27/3 by growing it for 25 consecutive passages at low temperature in fetal cells named WI-38 (for Wistar Institute-38). It was immunogenic when given by the intranasal route. Clinical trials were carried out between 1967 and 1969, and it was licensed in the United Kingdom in 1970. Because the eminent American virologist Albert Sabin claimed there might be some unknown agents lurking in WI-38, acceptance of Plotkin's rubella vaccine in the United States was nil until 1978, when Plotkin received a phone call from Hilleman suggesting that the latter's vaccine (HPV-77) be replaced with Plotkin's RA27/3. This was agreed to; Merck carried out large-scale clinical trials, and the vaccine was licensed in the United States in 1979. Today, this is the only rubella vaccine used throughout the world.

The impact of the rubella vaccine can be seen in a comparison of the number of cases. In 1964-1965 there were 12,500,000 cases of rubella in the United States, accompanied by 20,000 cases of congenital rubella syndrome (CRS). Between 1970 and 1979 there were 1,064 cases of CRS—or only 106 cases per year—a number that declined further between 1980 and 1985, when there were only 20 CRS cases yearly. Since that time rubella has been eliminated from the United States; however, worldwide 100,000 babies are born with CRS every year.

Whooping Cough

An epidemic of whooping cough was first described in the 16th century by the Parisian Guillaume de Baillou, who wrote: "The lung is so irritated by every attempt to expel that which is causing the trouble it neither admits the air nor again easily expels it. The patient is seen to swell up and as if strangled holds his breath tightly in the middle of his throat ... For they are without the troublesome coughing for the space of four or five hours at a time, then this paroxysm of coughing returns, now so severe that blood is expelled with force through the nose and through the mouth." By 1678 its name "pertussis," from the Latin meaning "intensive cough," was in common use in England. In 1906, Jules Bordet, the director of the Pasteur Institute in Brussels, and his brother-in-law Octave Gengou isolated from a child with whooping cough the bacterium that causes the disease and named it *Bordetella pertussis*. Pertussis is a highly contagious disease. Once a person is infected, it takes about 7 to 10 days for signs and symptoms to appear, though it can sometimes take longer. At first it resembles a common cold—patients have a runny nose, nasal congestion, watery eyes, and fever—but after a week or two thick mucus accumulates inside their airways, causing uncontrollable fits of coughing. Following a fit of coughing, a high-pitched whoop sound or gasp may occur as the person breathes in. The coughing may last for 10 or more weeks; hence the phrase "hundred-day cough." A person may cough so hard that he or she vomits, breaks ribs, or becomes extremely fatigued. The disease spreads easily through coughs and sneezes, and people are infectious to others from the start of symptoms until about 3 weeks into the coughing fits. It is estimated that 16 million people worldwide are infected per year. Most cases occur in developing countries, and people of all ages may be affected. In 2013 whooping cough resulted in 61,000 deaths. Pertussis is fatal in an estimated 1.6% of hospitalized U.S. infants under 1 year of age.

Efforts to develop an inactivated whole-cell whooping cough vaccine began soon after Bordet and Gengou published their method of culturing *B. pertussis*. In 1925 the director of the Danish State Serum Institute, Thorvald Madsen, used a killed whole-cell vaccine to control outbreaks in the Faroe Islands in the North Sea. In 1942, Grace Eldering and Pearl Kendrick, working in the State of Michigan Health Department, combined their improved killed whole-cell pertussis vaccine with diphtheria and tetanus toxoids to generate the first DTP combination vaccine. In 1981, Yuji Sato in Japan developed an acellular pertussis vaccine (aP). The vaccine was approved in the United States in 1992 for use in the combination DTaP vaccine. The aP vaccine has a rate of adverse events similar to that of a TD vaccine (a tetanus-diphtheria vaccine containing no pertussis vaccine).

Before the introduction of pertussis vaccines, an average of 178,171 cases of whooping cough were reported annually in the United States, with peaks reported every 2 to 5 years; >93% of reported cases occurred in children under 10 years of age. After vaccinations were introduced in the 1940s, the annual incidence fell dramatically to less than 1,000 by 1976. Incidence rates have increased, however, since 1980. In 2012 rates in the United States reached a high of 41,880 people—the highest it has been since 1955, when the numbers reached 62,786.

In 2010, 10 infants in California died from whooping cough, and health authorities declared there to be an epidemic encompassing 9,120 children. Demographic analysis identified a significant overlap in communities with a cluster of nonmedical child immunization exemptions and cases. The number of exemptions varied widely among communities. In some schools more than three-fourths of parents filed for vaccination exemptions. Vaccine refusal based on nonmedical reasons and personal belief exacerbated this outbreak of whooping cough, as it did with the Disneyland measles outbreak (see p. 15).

Although vaccination is the preferred method for preventing whooping cough, antibiotics may be used to treat whooping cough in those who have been exposed and are at risk of severe disease. Antibiotics (erythromycin, azithromycin, or trimethoprim-sulfamethoxazole) are useful if they are started within 3 weeks of the initial symptoms, but otherwise they have little effect in most people. In children younger than a year of age and in those who are pregnant, antibiotics are recommended within 6 weeks of the onset of symptoms.

Chickenpox

Chickenpox is caused not by a poxvirus (such as vaccinia) but by a herpes virus, *Varicella-zoster* (Fig. 10.8a). Chickenpox was not recognized as a separate illness until the mid-18th century. Indeed, for many years it was considered a milder form of

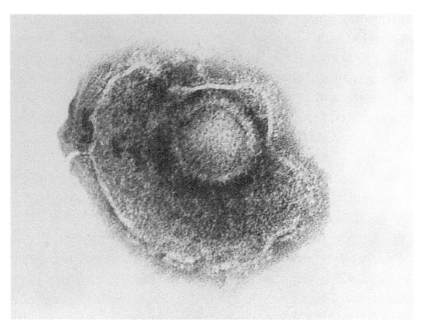

Figure 10.8a Transmission electron micrograph of chicken pox virus. Courtesy CDC Public Health Image Library #1878/Erskine Palmer, B. G. Partin, 1982.

smallpox. Prior to the development of the vaccine for chickenpox, the virus infected 4 million people each year in the United States and 100 million people worldwide.

The virus, varicella, is shed from the pustules in the skin of an infected individual and is spread from person to person by the airborne route. Infection occurs first in the tonsils and then in lymphocytes, and then the virus moves to the skin. Over a period of 2 weeks there is fever and malaise and finally the appearance of the rash (Fig. 10.8b) that lasts about a week. Recovery usually results in immunity to reinfection. Although chickenpox is considered a mild disease, complications can be severe and can include encephalitis, hepatitis, pneumonia, and bacterial skin infections from *Staphylococcus* and *Streptococcus*. About three-quarters of those infected develop latent infections that persist for a lifetime; if the virus is reactivated (usually in people over 50 years of age), it spreads from the site of latency (the sensory ganglia) down the nerve to the skin, resulting in the painful and itching skin rash called shingles.

In 1951, Thomas Weller's 5-year-old son Peter came down with chickenpox. Weller broke open the blisters, collected the pus, and was able to grow the virus in human fibroblast cultures. Although he was the first to grow varicella in tissue culture, Weller found it difficult to propagate the virus serially, and so he was unable to produce a vaccine. In 1974, Michiaki Takahashi in the microbiology department at Osaka University isolated and attenuated the virus from a 3-year-old boy named Oka. Like Weller, Takahashi took fluid from the blister and passed it 11 times in human embryonic lung fibroblasts at low temperature, followed by 12 passages in guinea pig embryo cells and then 10 passages in WI-38 and MRC-5 human cells. Because there is no animal model for chickenpox, Takahashi and colleagues took a risky approach in administering the candidate vaccine to healthy children; luckily, it was found to be immunogenic and safe. It was licensed for use in Japan and Korea in 1988, and by 1993 1.8 million children had been vaccinated. Maurice Hilleman at Merck also tried to isolate and attenuate the virus but was unsuccessful, and so he developed the Oka strain as a vaccine for the U.S. market in 1995. Today's live varicella vaccine is the Oka strain and is produced by Merck and GlaxoSmithKline. Before the chickenpox vaccine became avail-

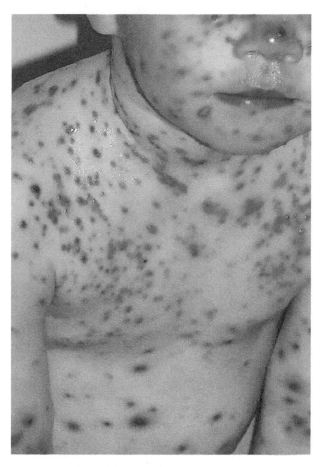

Figure 10.8b Child with chicken pox. Courtesy Wikipedia Commons, 1995.

able, chickenpox caused about 10,000 hospitalizations and a hundred deaths annually in the United States. Since 1995 one dose of the vaccine has been recommended as part of the standard immunization schedule for children in the United State ages 1 to 12 years. Because of immunization, chickenpox has become much less common as an illness in children and there have been few fatalities. Indeed, by 1999, when 60% of toddlers received the vaccine, only 1 child in 10 was getting chickenpox. For adults and children over 13 years of age, two routine doses are prescribed. The varicella vaccine was combined with MMR in 2006; reports of febrile seizures in 2008, however, led to the combination MMRV being not recommended for use.

The attenuated varicella vaccine (Zostavax) also protects against shingles.

And it should be noted that infections with chickenpox, but not shingles, can be treated with acyclovir.

Polio

During the first half of the 20th century no illness inspired more dread and panic than did polio. Summer was an especially bad time for children, when it was known as the "polio season." Children were among the most susceptible to paralytic poliomyelitis (also known as infantile paralysis). Many victims were left paralyzed for life. When exposed to the poliovirus in the first months of life, infants usually showed mild symptoms because they were protected from paralysis by maternal antibodies still present in their bodies; as hygienic conditions improved, however, and fewer newborns were exposed to the virus, paralytic poliomyelitis began to appear in older children and adults who did not have any immunity. Perhaps the most famous victim of polio was President Franklin Delano Roosevelt, who at the age of 39 contracted the disease that left him crippled for life (Fig. 10.9a).

Figure 10.9a President Franklin Roosevelt confined to a wheelchair due to contracting polio in 1921 at age 39. By FDR Presidential Library & Museum photograph by Margaret Suckley - 73-113 61, CC BY 2.0.

The poliovirus (Fig. 10.9b) enters the body through the mouth and initially multiplies in the intestine. During the early stages of infection the symptoms are fever, fatigue, headache, vomiting, stiffness in the neck, and pain in the limbs. Most infected patients recover, though in a minority of patients the virus attacks the nervous system. One in 200 infections leads to irreversible paralysis (usually in the legs). Among those paralyzed, 5 to 10% die when their breathing muscles become immobilized. Although the names most associated with a vaccine for polio are Jonas Salk and Albert Sabin, their work would not have been possible without the studies of John Enders.

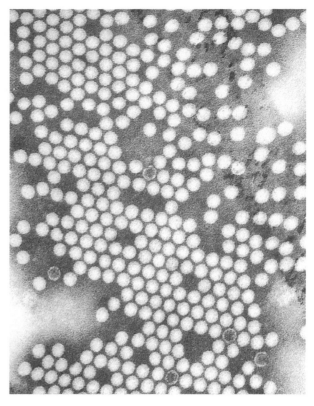

Figure 10.9b Polio virus as seen with the transmission electron microscope. Courtesy CDC Public Health Image Library #1875/Dr. Fred Murphy, Sylvia Whitfield, 1975.

John Enders

Enders (1897-1985), the son of a banker, was born on February 10, 1897, in West Hartford, Connecticut; he was educated at St. Paul's School in Concord, New Hampshire. Finishing school in 1915, he went to Yale University, but in 1917 left his studies there to become, in 1918, a pilot in the U.S. Air Force. After World War I he returned to Yale and was awarded a B.A. in 1920. He then went into business in real estate in Hartford, but, becoming dissatisfied with this, entered Harvard University. For 4 years he studied English literature and Germanic and Celtic languages with the idea of becoming a teacher of English. He did not decide on a life in microbiology until he shared a room in a boardinghouse with several Harvard University medical students, one of whom was working with the legendary bacteriologist/immunologist Professor Hans Zinsser. Enders was captivated by the stories the medical students told of their laboratory work with Zinsser, and when Enders attended some of Zinsser's lectures and read his writings, he abandoned his graduate studies in literature. In 1929, a year before he earned his doctorate, Enders became Zinsser's teaching assistant. At the time, Zinsser was trying to develop a vaccine against the bacteria-like rickettsiae that cause typhus, Rocky Mountain spotted fever, and scrub typhus. Rickettsiae are similar to viruses in that they are smaller than bacteria and cannot be grown in culture as bacteria can; instead, rickettsiae require a living cell for their growth and reproduction. In Zinsser's laboratory Enders learned the methods for cultivating large quantities of the rickettsiae in minced chick embryo tissues. After Enders spent 15 years with Zinsser in the department of bacteriology and immunology at the Harvard Medical School studying mumps, and where he prepared a vaccine against feline leukopenia virus, he established his own laboratory at the Children's Hospital in Boston. There two former medical school roommates, Weller and Fred Robbins, joined Enders. Prior to World War II, Enders and Weller were able to grow the vaccinia virus in chick embryo tissues in flask cultures or in roller tubes. In 1948, Weller succeeded in growing mumps virus in minced amniotic membranes obtained from aborted fetuses. With support from the National Foundation for Infantile Paralysis, Enders suggested that Weller and Robbins inoculate some roller tube cultures with poliovirus that was stored in the freezer. Weller set up tissue cultures containing fibroblasts from human foreskin and embryonic tissues from stillborn or premature babies who had died shortly before birth in the Boston Lying-In Hospital. Some of these were seeded with throat washings from his son Peter, suffering with the chickenpox (*Varicella*) that was one of Weller's particular interests. The remaining roller tubes were inoculated with a mouse-adapted strain of type 2 polio. With Enders's encouragement, Weller maintained them for long periods of time and changed the media weekly. The chickenpox virus failed to grow well; when, however, the culture fluid from the polio culture was injected into the brains of mice, much to their surprise they found that the mice became paralyzed, indicating that the poliovirus was multiplying in the roller tube cultures. This was an historic first, since until then no one had been able to grow poliovirus in human non-nervous tissue, in part because no one had tried and in part because the prevailing dogma held that poliomyelitis was exclusively a disease of the central nervous system and infected only neural tissues. They repeated their work with cultures of nervous tissue, as well as with skin, muscle, kidney, and intestine. These seminal observations were published in January 1949 in *Science* magazine: "Cultivation of the Lansing Strain of Poliomyelitis Virus in Cultures of Various Human Embryonic Tissues." Later, they were able to grow poliovirus types 1 and 3 in similar culture. Because the polioviruses kill the cells in which they grow and multiply, and this was obvious microscopically, they had a convenient assay for virus multiplication and did not need to inoculate monkeys (for types 1 and 3) or mice (for type 2 poliovirus). It was also possible using this *in vitro* assay to assess the presence or absence of virus-specific antibodies obtained from monkeys that had been infected with poliovirus.

In 1954 the Nobel Prize for Physiology or Medicine was awarded for this work to Enders, Weller, and Robbins. As was characteristic of Enders, who could have easily been the sole recipient, the prize was shared with his younger collaborators, who he said were full participants. Enders was a scientist with a green thumb and a great heart. He was unwavering in his honesty and exceptional in his generosity, sharing reagents, cultures, viruses, and know-how with all who requested them. Within 4 years of the Enders, Weller, and Robbins publication, Jonas Salk was able to report success for a polio vaccine containing a formalin-killed preparation of the three types of poliovirus.

Jonas Salk

Salk (1914-1995) was born in New York City. His parents were Russian Jewish immigrants who, although they themselves lacked formal education, were determined to see their children succeed and encouraged them to study hard. When he was 13, Salk entered Townshend Harris High School, a public school for intellectually gifted students that was "a launching pad for the talented sons of immigrant parents who lacked the money—and pedigree—to attend a top private school." In high school "he was known as a perfectionist … who read everything he could lay his hands on," according to one of his fellow students. Students had to cram a 4-year curriculum into just 3 years. As a result, most dropped out or flunked out, despite the school's motto, "Study, study, study." Of the students who graduated, however, most would have the grades to enroll in the City College of New York (CCNY), noted for being a highly competitive college. At age 15 Salk entered CCNY intending to study law, but at his mother's urging he put aside aspirations of becoming a lawyer and instead concentrated on classes necessary for admission to medical school. Salk managed to squeeze through the quota limiting Jewish admissions and entered NYU Medical School in 1934. According to Salk, "My intention was to go to medical school, and then become a medical scientist. I did not intend to practice medicine, although in medical school, and in my internship, I did all the things that were necessary to qualify me in that regard." During medical school Salk was invited to spend a year researching influenza. After completing medical school and his internship, Salk returned to the study of influenza viruses. He then joined his mentor, Dr. Thomas Francis, as a research fellow at the University of Michigan. There, at the behest of the U.S. Army, he worked to develop an influenza vaccine. By 1947, Salk decided to find an institution where he could direct his own laboratory. After three institutions turned him down, he received an offer from the University of Pittsburgh, with a promise that he would run his own lab. He accepted, and in the fall of that year he left Michigan and relocated to Pennsylvania. The promise, though, was not quite what he expected. After

Salk arrived at Pittsburgh, "he discovered that he had been relegated to cramped, unequipped quarters in the basement of the old Municipal Hospital." Salk, a driven, obstinate, and self-assured individual, began securing grants from the Mellon family and over time was able to build a working virology laboratory, where he continued his research on flu vaccines. It was in Pittsburgh that Salk began to put together the techniques that would lead to his polio vaccine. Salk used the Enders group's technique to grow poliovirus in monkey kidney cells. Then he purified the virus and inactivated it with formaldehyde but kept it intact enough to trigger the necessary immune response. Salk's research caught the attention of Basil O'Connor, president of the National Foundation for Infantile Paralysis (now known as the March of Dimes Birth Defects Foundation) (Fig. 10.9c) and at the time President Franklin D. Roos-

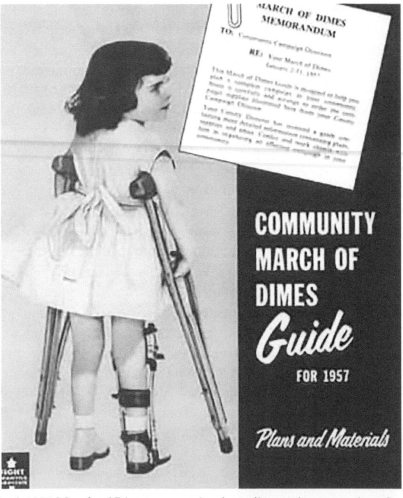

Figure 10.9c A 1957 March of Dimes poster for the polio vaccine campaign. Courtesy Wikipedia Commons.

evelt's lawyer. Salk's killed injectable polio vaccine (IPV) was tested first in monkeys and then in patients at the D.T. Watson Home for Crippled Children (now the Watson Institute) who already had polio. The tests were successful. The vaccine was given next to volunteers who had not had polio, including Salk, his laboratory staff, his wife, and their children. The volunteers developed anti-polio antibodies and none had adverse reactions to the vaccine. Finally, in 1954, national testing began on 1 million children, ages 6 to 9, who became known as the Polio Pioneers: half received the IPV and half received a placebo. One-third of the children, who lived in areas where vaccine was not available, were observed to evaluate the background level of polio in this age group. On April 12, 1955, the results were announced: the IPV was safe and effective.

After Salk made his successful vaccine, five pharmaceutical companies stepped forward to make it, and each of those companies was permitted to sell the vaccine to the public. One company, Cutter Laboratories in Berkeley, California, made it badly. As a result of Cutter's not filtering out the cells in which the poliovirus was growing, some virus escaped the killing effects of the formaldehyde treatment. As a consequence, more than 100,000 children were inadvertently injected with live, dangerous polio. Worse still, those children injected with the Cutter-produced vaccine were able to infect 200,000 people, resulting in 70,000 mild cases of polio; 200 people were severely and permanently paralyzed, and 10 died. It was the first and only man-made polio epidemic and one of the worst biological disasters in American history. Federal regulators quickly identified the problem with Cutter's vaccine and established better standards for vaccine manufacture and safety. Cutter Laboratories never made another polio vaccine.

Within 2 years of the Salk vaccine's release, more than 100 million Americans had been immunized, as well as millions more around the globe. In the 2 years before the vaccine was widely available, the average number of polio cases in the United States was more than 45,000. By 1962 that number had dropped to 910. Salk never patented the vaccine, nor did he earn any money from his discovery, preferring to see it distributed as widely as possible.

Following the announcement that Salk's IPV worked, Americans named hospitals after him; schools, streets, and babies were also named after Salk. Universities offered him honorary degrees, and countries issued proclamations in his honor. He was on the radio, on the cover of *Time* magazine, and made TV appearances. The vaccine's success made Salk an international hero, and he spent the late 1950s refining the vaccine and establishing the scientific principles behind it. By 1960, however, Salk was ready to move on. Salk's dream was to create an independent research center where a community of scholars interested in different aspects of biology—the study of life—could come together to follow their curiosity. In 1960 the dream was realized with the establishment of the Salk Institute for Biological Studies in La Jolla, California. At the Institute, Salk tried to prepare killed vaccines against HIV and cancer without success. He died of heart failure in 1995.

Albert Sabin

Albert Sabin (1906-1993) developed the "other" polio vaccine. Sabin was born in Bialystok, Russia (now a part of Poland), on August 26, 1906. In 1921, Albert and his family immigrated to the United States to avoid the pogroms and the rabid anti-Semitism prevalent in Imperial Russia. Because the Sabin family was poor, Albert would not have been able to acquire a degree in higher education were it not for Albert's uncle, a dentist, who agreed to finance his college education, provided he study dentistry. In 1923, Sabin entered NYU as a pre-dental student, but he switched to medicine after reading Sinclair Lewis's medical novel *Arrowsmith* and Paul de Kruif's *Microbe Hunters*. Thus began a lifelong interest in virology and public health. In 1928, upon graduation from NYU with a B.S. degree, he enrolled in the NYU School of Medicine and completed an M.D. in 1931—the year of a great polio epidemic in New York City and the worst since 1916. He then spent 2 years as an intern at Bellevue Hospital; was a National Research Council fellow at the Lister Institute in London, where he studied virology; and in 1935 was on the staff at the Rockefeller Institute, where he worked in the virus laboratory of Peter Olitsky. Shortly before Sabin's arrival at Rockefeller, Olitsky was warned that hiring him would be a mistake, since Sabin had a difficult personality. Olitsky was unmoved and regarded Sabin as a genius and, having worked successfully with geniuses before, was anxious to have him as an associate. At the time, Olitsky's laboratory was concerned with immunity to viruses, particularly poliomyelitis. At Rockefeller, Sabin worked tirelessly, refusing to take even Sundays or holidays off. He worked with infinite patience, had most careful technique, engaged in precise planning, and carried out detailed and elaborate recording of observations; he was especially incisive in the analysis of a problem and conducted skillful tests with rigid controls. Sabin's interest in active immunization was nurtured by another scientist at Rockefeller, Max Theiler, whose work was focused on developing a live-virus vaccine for yellow fever. In 1938 the first field trials of Theiler's live-attenuated yellow fever vaccine were being tested in Brazil, and their success had a profound impact on Sabin. (Theiler was awarded the Nobel Prize in 1951 for the yellow fever vaccine.)

In 1939, Sabin established his own research laboratory at the University of Cincinnati Children's Hospital. World War II interrupted Sabin's work on polio vaccines. He served in the U.S. Army Medical Corps and developed experimental vaccines against dengue fever and Japanese encephalitis virus. At war's end he returned to Cincinnati and to polio research.

Sabin had a longstanding interest in polio, and his first published article, in 1931, dealt with the purification of the poliovirus. This was an important step in that it provided for the possibility that the infection came about via the oral route, which was contrary to the prevailing belief that the disease was primarily neurological. In 1936, Sabin, in collaboration with Olitsky, was able to grow the poliovirus in human embryo brain tissue.

One January day in 1948, Hilary Koprowski, then at the Lederle Laboratories in Pearl River, New York, macerated brain material in an ordinary kitchen blender. He poured the result—thick, cold, gray, and greasy—into a beaker, lifted it to his lips, and drank. It tasted, he later said, like cod liver oil. He suffered no ill effects. Koprowski had set out to attenuate the polio virus through adaptation to mouse brain. Starting with a type 2 mouse strain, he achieved attenuation of neurovirulence in monkeys. After ingesting the orally administered vaccine, he arranged to vaccinate 20 mentally handicapped children in collaboration with the physician in charge of the institution in which they resided. The justification for this experimentation on humans unable to give their own consent was the fear that poliovirus might enter the institution, a common occurrence at the time. Although this first trial showed safety and immunogenicity of the strain, the presentation of the results at a later scientific meeting was greeted with shock. Two years later Koprowski received a call from Letchworth Village, a home for mentally disabled children in Rockland County, New York. Fearing an outbreak of polio, the administrators of the home asked him to vaccinate its children. In February 1950, in the first human trial of a live polio vaccine, Koprowski vaccinated 20 children there. At the time, approval from the federal government was required to market drugs but not to test them. Seventeen of the children developed antibodies against polio. (The other three turned out to have antibodies already.) None of the children experienced complications. Unfortunately, the father of a child receiving the vaccine developed paralysis and died. Upon autopsy, virus was recovered from the man's brain tissue; the vaccine was withdrawn. Although Koprowski was the first to produce a live oral polio vaccine (OPV), and his strains were tested extensively in the former Belgian Congo, his native Poland, and elsewhere, they were never approved for use in the United States because they were regarded as too virulent.

With the production of attenuated poliovirus strains, it became apparent to Sabin that oral exposure might be a promising direction for development of a vaccine. He was convinced that an oral vaccine would be more easily administered and better tolerated as a public health measure. Further, he felt a live oral vaccine would mimic the natural infection to produce an asymptomatic infection in the gut that would stimulate the immune response and lead to the production of a systemic immune response. By 1954, Sabin had obtained three mutant strains of the poliovirus that appeared to stimulate antibody production without paralysis. Sabin entered into an agreement with Pfizer to produce the OPV, and the pharmaceutical company began to perfect its production techniques in its U.K. facilities. From 1957 to 1959 Sabin successfully tested the OPV on human subjects—himself, his family, research associates, and hundreds of prisoners from the nearby Chillicothe Penitentiary. Since during this time the Salk vaccine was being used in the United States, Sabin was unable to get support for a large-scale field trial for his OPV. But because polio was widespread in the Soviet Union, Sabin was able to convince the Soviet Health Ministry to conduct trials with his OPV. It was used in Russia, Estonia, Poland, Lithuania, Hungary, and East Germany, and by 1959 more than 15 million people there, mainly children, had been given the Sabin live oral vaccine.

On the morning of April 24, 1960, more than 20,000 children in the greater Cincinnati area lined up to receive the Sabin OPV in its first public distribution in the United States. An additional 180,000 children in the surrounding area received the vaccine during the next several weeks in what became known as "Sabin Oral Sundays." In 1960, Sabin published a landmark article, "Live, Orally Given Poliovirus Vaccine," in the *Journal of the American Medical Association* (*JAMA*). In the article he described the results of these large studies with children under the age of 11. Due to the self-limiting nature of the poliovirus, a second dose was needed to achieve full protection. In 1961 the trivalent OPV (administered as drops or on a sugar cube) was licensed in the United States, and by 1963 it was the polio vaccine of choice. Between 1962 and 1964 more than 100 million Americans of all ages received the Sabin OPV. By the 1980s, after large public health field trials were conducted and found to be successful, the WHO adopted a goal of polio eradication. In 1994 the WHO declared that naturally occurring polio had been eradicated from the Western Hemisphere. By 1995 80% of children worldwide had received the requisite three doses of the vaccine in the first year of life, and it is estimated that it prevented half a million cases of polio annually. During his lifetime Sabin staunchly defended his OPV, refusing to believe that it could cause paralysis. Despite this belief, the risk, though slight, does exist. As a consequence, in 1999 a federal advisory panel recommended that the United States return to the Salk IPV because of its lower risk in causing disease. On the basis of a decade of additional evidence this recommendation was affirmed in 2009.

Sabin was very severe and demanding of himself and those with whom he worked. He monitored everything and knew everything that happened in the laboratory during that day, and the personnel would hear about it the next morning if things weren't right. He had rigorous standards. Although he never saw his research as a race to the finish line, he once said his mission was to kill the killed (Salk) vaccine because he believed that a live oral vaccine was superior. Although there was rivalry between Salk and Sabin, in truth Sabin simply did not have a very high opinion of Salk as a scientist. He was just another guy. Sabin was a stubborn but eloquent speaker, and it was often difficult to defeat him in scientific arguments. Of the Salk vaccine he once declared it was "pure kitchen chemistry."

Sabin published more than 350 scientific papers. Although he never received the Nobel Prize, he received numerous other awards including more than 40 honorary degrees, the U.S. National Medal of Science, the Presidential Medal of Freedom, the Medal of Liberty, the Order of Friendship among Peoples (Russia), the Lasker Clinical Research Award, and election to the U.S. National Academy of Sciences (1951). In 1970 he became the president of the Weizmann Institute. He retired from full-time work in 1986 at the age of 80, although he continued publishing right up until his death from congestive heart failure in 1993. He is buried in Arlington National Cemetery.

Aftermath

Of all the pioneering polio researchers, only Enders, Weller, and Robbins were awarded the Nobel Prize. Salk, Koprowski, and Sabin were never selected for the honor. The reasons for this remain controversial. In the case of Koprowski, it may be because his vaccine was not a universal success. Examination of the Nobel archives reveals that Sven Gard, professor of virology at the Karolinska Institute, convinced the Nobel Committee to name Enders and his colleagues recipients of the 1954 prize, because, as he wrote, "the discovery had had a revolutionary effect on the discipline of virology." Salk was nominated for the prize in 1955 and in 1956. The first time, it was decided to wait for the results of the clinical trial of Salk's killed polio vaccine, which was in progress. In 1956, Gard wrote an eight-page analysis of Salk's work, in which he concluded, "Salk has not in the development of his methods introduced anything that is principally new, but only exploited discoveries made by others." He wrote: "Salk's publications on the poliomyelitis vaccine cannot be considered as Prize worthy." Few of the scientific societies honored Salk, and he received little recognition from his peers. Some in the scientific community considered his posing with movie stars and giving television interviews to be behavior unbecoming of a prominent researcher. Although Salk was the public's darling, he remained a pariah in the scientific community. He received the Lasker Award but was never elected to the U.S. National Academy of Sciences, possibly because he was blackballed by Sabin, who sniped, "He never had an original idea in his life." Salk's rejection by the Academy may also have been fueled by jealousy of his success in the public arena. In effect, Salk had broken two of the commandments of scientific research. Thou shalt give credit to others. Thou shalt not discuss one's work in newspapers and magazines.

Cell-Mediated Acquired Immunity

Thus far we have considered the humoral (antibody), or the B-cell, component of the immune system involved in protective immunity. Now let us look at the other part. This alternative immune pathway involves another set of lymphocytes, the T-cell variety, and is concerned with nonhumoral or cellular immunity. In the 1960s it was discovered that the bone marrow and thymus are the "master organs" of our (and other mammals') immune system. The bone marrow is the site where all the different cell types of the blood and the immune system are made, and it is here that the blood stem cells also occur. Stem cells are primitive, unspecialized cells that have no function other than to divide and to make other cells. When a stem cell divides to produce two daughter cells, though, the offspring are unalike: one daughter cell is an identical replica of the stem cell, while the other can grow and become a very specialized cell. In this way stem cells are both self-renewing and give rise to a variety of specialized cell types, including B and T lymphocytes. A T lymphocyte arises in the bone marrow but migrates via the bloodstream from the bone marrow to a gland in the neck called the thymus. (The thymus was so named because in the 2nd century Galen, in an imaginative mood, thought the shape of the gland resembled the leaves of the thyme plant.)

The thymus is made of lymphoid tissue. In children the gland, located just above the heart, is rather large. As we age, it tends to atrophy and shrink. The thymus is the place where the T lymphocytes from the bone marrow establish residence for some time and where they mature (or, as it is said, become educated) and multiply. It is because of their maturation in the thymus that they are called "thymus-educated" or T cells. While in the thymus some T cells are selected to become either helper T cells (also called CD4 cells) or killer T cells (also called CD8 cells). The killer cells are also referred to as cytotoxic T lymphocytes, or CTLs. Some of the progeny of these T cells leave the thymus, and by way of the bloodstream they reach and settle down in other lymphoid organs—the spleen, the tonsils, and the variously distributed lymph nodes—or they circulate in the blood. The CTLs patrol the body in search of altered cells (cancer), foreign cells (grafts of tissues), or cells compromised by an internal pathogen such as a virus or a rickettsia. Because they are capable of attacking and destroying any cell in the body that appears to be "not self," they are aptly called "killers." But in the attack the killer direct its action not against the virus itself but rather against the cell (no longer recognized as self) in which the virus is hiding. Killing may occur in one of two ways: CTLs have the molecule CD8 on their surface; when a foreign cell is encountered, they attach to it and punch in a "security code" that then triggers a self-destruct mechanism in the nonself cell. In effect, the CTL does not have to shoot the invader with a "silver arrow"; it simply tells the pathogen-infected cell, "Die." The other mechanism is for the CTL to punch a hole, using a pore-forming molecule, and then to inject it with a lethal cocktail of destructive enzymes. The CTL response is strong within 5 days after exposure to foreign antigen, peaks between 7 and 10 days, and then declines over time. (Antibody formation, when it does occur, usually follows the CTL response.)

What roles do T cells play in protective immunity? The first line of defense in cell-mediated immunity is the macrophage, or "big eater," described by Metchnikoff more than a century ago. Macrophages, the "foot soldiers" of the immune system, are also the cellular "vacuum cleaners" or "filter feeders" of the immune system. They ingest, digest, and then regurgitate the foreign pathogens (virus or bacterium) or even the dead and dying pathogen-containing cells they have eaten. In this way, they are able to display the broken antigenic bits on their outer surface in association with another molecule called the major histocompatibility complex class II (MHC-II). This combination of MHC-II and foreign antigen is information that is offered to a helper T lymphocyte. The T helper cell has the CD4 receptor (this receptor is not antibody), and once the CD4 binds to the combination, and thus receives information, it is triggered to divide, forming a clone. The T cell, however, instead of secreting antibody, secretes chemical messengers called lymphokines. The chemical messages then activate other T lymphocytes or macrophages, or they can activate the B cells, which are also MHC-II, so that they begin the process of antibody synthesis and secretion. In addition, as with B cells, there are memory T cells left after an encounter with a foreign antigen, so that the response to a second encounter is swift and specific. The CD4 helper T cell is the keystone that maintains the connection between the two branches of the immune system: cell-mediated and humoral. The humoral branch produces antibodies that prevent foreign invaders from infecting cells, while the members of the cell-mediated branch search out and destroy infected or foreign cells. The central role of the CD4 helper T cell in protective immunity is most clearly seen in AIDS: the destruction of the helper T cell by HIV results in a collapse of the immune system, and now the uncoordinated and defenseless body can be assaulted by opportunistic pathogens and cancer, the syndrome of AIDS.

CTLs are critical in organ transplantation. If the graft is "not self," then the CTLs kill it and the graft is rejected. Graft rejection does not involve antibody. Other examples of cell-mediated immunity are seen in delayed-type hypersensitivity such as reactivity to tuberculin, an inflammatory response that takes 24 to 72 h. Contact dermatitis, as after a person touches poison ivy or poison oak, is also cell mediated, but because this reaction occurs in a matter of hours, it is called immediate-type hypersensitivity. Immediate-type hypersensitivity reactions are hay fever, asthma, hives, eczema, food allergies, sensitivity to sulfonamide and penicillin and bee venom, and transfusion reactions; all are cell mediated and not due to antibody.

Why does a CTL not destroy our own cells or the helper T cells? Because all the cells in our body (except for red blood cells) express MHC-I. This allows infected cells to signal their plight to the CTL, which is also MHC-I. Once this information is received by the CTL, the infected cell is killed. But because the helper T cell is MHC-II, either it is not recognized by the CTL or the information received tells the CTL "self" and killing does not occur. Not all MHC molecules are alike, however. The many different kinds of MHC-I molecules on the surface of our nucleated cells are the basis for matching tissues (called tissue typing) to be used in grafting. When a proper match is made, compatibility is ensured and the graft will not be rejected.

Coda

It is difficult to underestimate the contribution immunization has made to our well-being. It has been claimed that were it not for childhood vaccinations against chickenpox, diphtheria, measles, mumps, pertussis, polio, and rubella, childhood death rates would probably hover in the range of 20 to 50%. Indeed, in those countries where vaccination is not practiced, the death rates of infants and young children remain at that level.

doi: 10.1128/9781683670018.ch11

11
The Plague Protectors:
Antiseptics and Antibiotics

As he watched his beloved wife Virginia dying from tuberculosis, Edgar Allan Poe penned the following lines:

> "The Red Death" had long devastated the country. No pestilence had ever been so fatal, or so hideous. Blood was its Avatar and its seal—the redness and the horror of blood. There were sharp pains, and sudden dizziness, and then profuse bleeding at the pores, with dissolution. The scarlet stains upon the body and especially upon the face of the victim, were the pest ban which shut him out from aid and from the sympathy of his fellow men. And the whole seizure, progress and termination of the disease, were the incidents of half an hour.

For Poe's wife and millions of others, the catastrophic effects of the disease were clear, but there was little that could be done to protect or treat or cure them (Fig. 11.1).

Although some of the prehistoric diseases that afflicted our ancestors are recorded in their bones, little or no evidence exists to document the treatments used for the illnesses they suffered. Studies of present-day primitive cultures, however, make it clear that the medical treatments would have been inseparable from religion and magic. At the core was the priest or healer, who was given the task of restoring the ill to health by driving out evil spirits, luring back the good ones, or begging forgiveness from an offended god. On occasion there was a direct attack on sickness: sucking, bleeding, cupping, fumigating, steam baths, drugs to deaden the senses or to reduce fever, and surgery. Most of these treatments were based on trial and error, and few had a rational or pharmacologic basis— for the most part, the benefits to the afflicted were psychological. Indeed, the greatest ally these "medicine men" had was the body's ability to cure itself.

11.1 (Left) *The Doctor* (Detail) by Sir L. Fildes (1843-1927). Commissioned by Henry Tate for his new National Gallery of British Art. The painting shows the artist's son attended by Doctor Murray, who though he showed care, could do little to cure the dying child. Courtesy of the Wellcome Library of Medicine, London, CC-BY 4.0.

Secular medicine, which developed in parallel to the religious approach, is usually dated from the time of the Greeks in the 5th century B.C. and most often is personified by the physician Hippocrates of Cos. It should be recognized that Western medicine did not spring forth fully formed from the forehead of Zeus as did Athena, but was dependent on knowledge accumulated over hundreds and even thousands of years. Let us consider the healing arts and the science of medicine that have been, and continue to be, used to protect humans from plagues, and thereby to understand how attitudes toward disease influenced prevention, treatment, and cure.

Barbers, Bloodletting, and Antisepsis

Archeological records show that cleanliness was important to the ancients. Their motivation, however, was not the prevention of disease but rather their belief that cleanliness would appease the gods and prevent them from sending down plagues. The word "hygiene," meaning "healthy," comes from the name of the Greek goddess Hygeia, who was the daughter of the god of medicine Asclepios. Asclepios is depicted in paintings and sculpture (Fig. 11.2) as an older and bearded wise man holding a staff, around which is wrapped a serpent (signifying death), suggesting that he was able to ward off illness. His daughter Hygeia is represented as a young and beautiful woman (another daughter, Panacea, represented treatment). The most famous of the healing temples dedicated to Asclepios were those in Epidauros and Pergamon; these temples were not hospitals but were a mixture of a religious shrine and a health spa. The teachings of healthy living, first by the Greeks and later by the Romans, produced aqueducts to bring clean water, bathhouses, toilets, and sewage systems. Since effective remedies for disease were almost nonexistent, however, the "cure" provided in the temples of Asclepios and the bathhouses was faith, not medicine.

The Greek physician Hippocrates, the father of modern medicine, believed there was a rational basis for disease. His theory on the mechanism of disease was based on an imbalance in the body of the four humors (fluids)—blood, phlegm, yellow bile, and black bile. Key to his theory was that the body contained varying amounts of blood (warm and moist), phlegm (cold and moist), yellow bile (warm and dry), and black bile (cold and dry). Illness was due to poor diet, bad air, and injuries that disturbed the balance. Cure, he believed, could be effected by restoring the balance. At the time, it was accepted as fact that blood was made from food in the liver and that overeating led to an overproduction of blood. This excess of blood resulted in a hot and wet condition (fever, sweating, coughs, headaches, rheumatism, and heart disease). In order to cure such a patient, purging the humor causing the condition was required. In the case of blood overproduction, bloodletting would, it was claimed, eliminate the overbalance of blood and result in cure. Hippocrates said: What cannot be cured with medicines is cured by the knife; what the knife cannot cure is cured with the searing iron; and whatever this cannot cure must be considered incurable. The grisly art of bloodletting (based on Hippocrates' theory) flourished for 1,000 years, during Late Antiquity (476-1453). Its practice would last well into the 19th century. Indeed, a bimonthly bloodletting was, if one could afford it, a preventive medicine. George Washington died from a throat infection in 1799 after being drained of 9 pints of blood in a 24-h period!

During the first millennium of the Christian era, few people were able to read or write. The majority of people were treated by the literate clergy, especially the monks (who became physicians), women who were acquainted with medicinal plants, and lay surgeons who treated wounds. The physicians enlisted barbers (from the Latin *barba*, meaning "beard") to act as assistants in bloodletting. In 1163, when the Council of Tours ruled by papal decree that it was sacrilege for the clergy to draw blood from humans, the barbers became the only individuals capable of performing bloodletting. The practice was so successful that barber-surgeons began to thrive throughout Europe. Soon the royals as well as commoners went to the barbers to

Figure 11.2 Asclepios, the God of Medicine. Wrapped around his staff is a serpent signifying death and suggesting that Asclepios was able to ward off illness and prevent death. The tiny figure standing to his right foot is Telesphorous, the child god of convalescence. From a ceremonial ivory diptych carved in the late Roman style.

be shaved, to have their hair cut, and to have their illnesses cured or prevented through bloodletting. Barbers also used their razors for lancing boils and were able to set fractures and amputate limbs. Later, the barbers expanded on their reputations as surgeons and began practicing dentistry.

Bloodletting involved slicing small cuts in the flesh of the arm through which blood was allowed to drain into a bowl. To encourage blood flow, the barber-surgeon commonly gave his patient a pole to hold on to and squeeze. Bleeding would continue until the patient fainted. The barber-surgeons also used a strip of cloth as a tourniquet, which was later applied as a bandage when the operation was complete. Afterwards, the bloody bandage was rinsed and hung on the pole outside, flapping in the breeze. As it dried, it twisted around the pole in the spiral pattern that we now associate with the sign of a barbershop, the barber pole.

In the middle of the 13th century the barbers of Paris, also known as the Brotherhoods of St. Cosmos and St. Damian, established the first school for instruction in the practice of surgery; later this was expanded and served as a model for other surgical schools in the Middle Ages. The oldest barber organization in England (the Worshipful Company of Barbers) was established in London in 1308. For the most part, barber-surgeons were unskilled, uneducated, and lacked an appreciation not only of contagious diseases but how they might be prevented; as a consequence, in their hands postoperative infections tended to become more and more common. Up until 1416 barbers had a free hand in the practice of surgery, but complaints of barber-surgeon incompetence and the sicknesses their cutting induced led the mayor of London in that year to forbid them from taking care of anyone "in danger of death or maiming unless within three days after being called in they presented the patient to one of the masters of the Barber-Surgeon guild." To facilitate improvement in their surgical skills, Henry VIII in 1540 allowed a merger of the barbers and surgeons and decreed that barbers were to receive the bodies of four criminals yearly for the purposes of dissection. The merger, however, was an uneasy one, and in 1745 a bill was passed to separate the barbers and surgeons. The Company of Surgeons was formed with its own guildhall close to the Old Bailey (the law courts) and Newgate Prison (where the bodies of criminals to be used in anatomical dissection could be obtained), and in 1800 the Company of Surgeons was granted a royal charter to become the Royal College of Surgeons of England. Barbers, on the other hand, began to decline in stature as men of medicine, and by the end of the 1800s they had lost all rights to perform surgery and were restricted to cutting hair, shaving beards, and fashioning wigs.

Although attention to cleanliness in surgery can be traced as far back as 600 B.C. to the Hindu surgeon Sursuta, as well as the writings of Hippocrates, who advised that surgeons thoroughly cleanse their hands before operating and that boiled water be used to irrigate a wound, these recommendations were later largely forgotten. Indeed, in Rome, Galen (A.D. 129-210) proposed a "laudable pus theory," claiming that pus was an essential part of the healing process. As a result, wounds were encouraged to become contaminated and filled with pus. It may seem inconceivable to us today to think that microbe-induced rotting and pus-filled flesh would be accepted as a part of wound recovery, but because Galen was regarded as the ultimate medical authority, sepsis (literally "putrefaction" or "decay") was considered for more than 1,000 years as an indicator of a process that would eventually lead back to health.

The disease puerperal fever (also called childbed fever) was known since the time of Hippocrates, who described it thus: "The disease usually commenced with a diarrhea; the uterus became hard, dry and painful … pain in the head and sometimes a cough. On opening the bodies, curdled milk was found on the surface of the intestines, a milky … fluid in the stomach, the intestines and the uterus when carefully examined, appear to have been inflamed." In 1772 in Paris, Vienna, and other European cities, outbreaks of puerperal fever were common, and one-fifth of all delivered mothers died from it. Alexander Gordon (1752-1799), a Scottish surgeon, was the first to suggest its cause, writing:

> I will not venture positively to assert that puerperal fever and erysipelas [a streptococcal skin infection causing redness and swelling and also known as St. Anthony's fire] are precisely the same in specific nature (but) that they are concomitant epidemics I have unquestionable proof. They began at the same time and afterwards kept pace together; they arrived at their acme together and they both ceased at the same time. After delivery the infectious matter is readily and copiously admitted by numerous … orifices, which are open to imbibe it, by separation of the placenta from the uterus.

Gordon established that in 77 cases of the disease, 28 patients had died. The cause, he claimed, was not a noxious constituent in the atmosphere but was transmitted from one patient to another by doctors or midwives as carriers, as they themselves were not affected by the disease. He argued that "the cause of puerperal fever … was contagion of infection altogether unconnected with … a miasma." He suggested, "Nurses and physicians who have attended patients affected with puerperal fever ought to carefully wash themselves and to get their apparel properly fumigated before it be put on again."

Despite these observations and remedies, the standards of care in hospitals remained poor. One surgeon wrote in 1801: "In hospital … the patient sinks almost inevitably under the formation of pus of a compound fracture … and some surgeons continue to believe that the formation of pus is one of the essential processes of healing—a legacy from the days of Galen." By 1840, 3,000 mothers in Britain were dying every year from puerperal fever, and yet doctors remained unconvinced that they carried the disease from one patient to another. Oliver Wendell Holmes (1809-1894),

Figure 11.3 Thomas Eakins, American - Portrait of Dr. Samuel D. Gross (*The Gross Clinic*) - Google Art Project (Courtesy Wikipedia) (Public Domain).

the father of the great American jurist, was not a surgeon but a professor of anatomy and physiology at Harvard University. He collected anecdotal evidence on puerperal fever, describing two cases. "A practitioner opened the body of a woman who had died of puerperal fever, and continued to wear the same clothes. A lady whom he delivered a few days afterwards was attacked with and died of a similar disease; two more of his lying in patients, in rapid succession, met with the same fate; struck by the thought that he might have carried the contagion in his clothes he instantly changed them and met with no more cases of the kind." In the second case, "Dr. Campbell … assisted in the post-mortem examination of a patient who died with puerperal fever. He carried the pelvic viscera in his pocket to the classroom. The same evening he attended a woman in labor without previously changing his clothes; this patient died. The next morning he delivered a woman with forceps; she also died … and within a few weeks, three shared the same fate in succession." (The surgeon's clothing and the operating theater are shown [Fig. 11.3] in Thomas Eakins's painting *The Gross Clinic*, from 1875.) In 1843, Holmes summarized his findings in an essay, "The Contagiousness of Puerperal Fever," in which he clearly stated the case for communication from one person to another, directly or indirectly. Holmes became so engrossed by the disease that he attended several women who developed puerperal fever; he died of a blood infection he had contracted from one of them.

Despite Holmes's persuasive contagionist view, his work attracted little attention save for that of the obstetrician Ignaz Semmelweis. Semmelweis (1818-1865) was a Hungarian acting as an assistant in the obstetrics department of the General Hospital in Vienna, Austria. The department had two clinics—the First Clinic was attended by medical students and the Second Clinic was attended by midwives. The overall death rate of pregnant women who had been admitted to the hospital was between 5 and 10%. Semmelweis found that for the period 1841-1846 the death rate for the two clinics was quite different:

	First Clinic			Second Clinic	
Year	Births	Death rate (%)		Births	Death rate (%)
1841	3,036	7.7		2,442	3.5
1842	3,287	15.8		2,659	7.5
1843	3,060	8.9		2,739	5.9
1844	3,157	8.2		2,956	2.3
1845	3,492	6.8		3,241	2.0
1846	4,010	11.4		3,754	2.7
Total	**20,000**	**9.92**		**17,791**	**3.35**

In 1846 a colleague of Semmelweis, Jakob Kolletschka, professor of forensic medicine, age 43, died of septicemia from a scalpel wound he received during a post-mortem exam. The knife had slipped, the skin of his finger was pierced, and infection set in. He developed a severe inflammation in his hand, followed by fever, delirium, a rapid pulse, difficulty in breathing, a stiffness in the neck, peritonitis, and multiple abscesses over his body. Semmelweis described his colleague's death: "Professor Kolletschka … frequently participated with his pupils in the performance of medic-legal autopsies; during such an exercise, he was stuck in the finger by a student with a knife … used during the post-mortem … (he) became ill with lymphangitis and phlebitis … died … of … pleuritis, pericarditis, peritonitis, meningitis." Semmelweis noted that the cause of Kolletschka's death was similar to that of women who had died from puerperal fever. He was reminded of how the medical students did their autopsies and then, with only a quick wipe of a towel, attended the women in delivery. He reasoned that an agent—a hand-borne cadaveric particle—was transferred from the dead by the particles clinging to the knife in the case of Kolletschka or on the hands of those physicians who examined pregnant women. He set up an experiment on his observations: "In order to destroy the cadaveric particles adhering to the hand … I began to use Chlorina liquida with which I and every student were obliged to wash our hands before making an examination. After some time I abandoned Chlorina liquida because of its high price and changed to the considerably cheaper chlorinated lime." The death rate in the First Clinic was 12/100 in April, but 2 months later, after Semmelweis had instituted the hand washing with chlorine, the death rate had dropped to 2/100. Semmelweis concluded that decaying animal-organic matter was conveyed to the individual from external sources. "These are the cases represented as epidemics of childbed fever; these are the cases that can be prevented. The source of decaying-organic matter can be a corpse of any age, either sex … whether the corpse is a pregnant woman or not. Decaying matter is carried by examining fingers, operating hands, instruments, bed linen, sponges, basins and attendants. In other words, anything that is contaminated … and then comes in contact with the genitals of the patients." Semmelweis wrote: "Puerperal fever is caused by conveyance to the pregnant woman of putrid particles derived from living organisms through the agency of contaminating fingers. Consequently must I make my confession that God only knows the number of women whom I have consigned prematurely to the grave." Semmelweis did not identify the cause of childbed fever (that task was left to Pasteur, who in 1879 identified it as a streptococcus), but he did develop a method for preventing it—asepsis, literally "without putrefaction or decay." Semmelweis encountered strong opposition from the hospital officials in Vienna, and in 1850 left for the University of Pest, Hungary. As professor of obstetrics at the hospital, he enforced aseptic practices, and the death rate from puerperal fever dropped to 0.85%. His findings and publications, however, were resisted by hospital and medical authorities in Hungary and elsewhere. Semmelweis was a manic-depressive, and on August 1, 1865, he was admitted to a lunatic asylum, where a minor cut on his right hand, contracted during a gynecologic operation, was noted. He died of septicemia from the cut.

Semmelweis is almost unknown today, and much of the credit for aseptic technique is given to Joseph Lister (1827-1912), whose name is familiar to us on the bottle of mouthwash—Listerine. Lister actually used another method to prevent infection—antisepsis—developed 18 years after Semmelweis's chlorine hand-washing experiment and 35 years before the establishment of the germ theory of disease. Earlier, John Pringle, a British army surgeon, coined the term "antisepsis" to describe substances such as mercuric chloride, which stopped the rotting of wounds. Although Lister lived at a time when it was already recognized that microbes could be responsible for infectious diseases, he was unaware of the work of Pasteur on fermentation and putrefaction, and he developed his techniques of antisepsis independent of these ideas on infection. In 1861, Lister was made head of the new surgical ward in Glasgow's Royal Infirmary and set out to prevent putrefaction in a wound by chemically destroying the offending microbes. Lister, however, erroneously believed that such germs were carried in the air, and so he decided to attack these before they entered the body through the open wound. In his search for antiseptics he found that carbolic acid (creosote or phenol) was quite effective. On August 12, 1865, Lister treated an 11-year-old boy, James Greenlees, whose leg had been crushed by a cart, by covering the wound with a bandage soaked in linseed oil and carbolic acid and enclosing this with tinfoil to prevent evaporation. The wound did not become infected, the leg fracture healed perfectly, and the boy left the clinic 6 weeks later. After success in treating the wounds of several other patients, Lister published his findings in the British medical journal *The Lancet*, explaining that germs were the cause of infection and that infection and pus formation were neither beneficial nor a normal part of the healing process. Lister remained concerned, though, that exposure of the wound to "septic air" while the dressing was changed could result in infection, and so he used a hand pump to spray a 2.5% solution of carbolic acid during bandage changes. Lister was an expert practical surgeon and insisted on hand washing, cleansing of the wound with carbolic acid, and spraying carbolic acid during surgery. In effect, Lister used both antisepsis (killing the infective agents already in the wound) and asepsis (preventing bacteria from getting into the wound).

Lister's techniques were put to the test during the Franco-Prussian War (1870-1871) when Prussian military surgeons used his carbolic acid treatment for battle wounds. Their success in reducing infection was far greater than that of the French, who completely rejected Lister's methods, and Lister was hailed as a hero in Germany. Lister later developed a steam-powered spray that was able to fill the entire operating theater with carbolic acid; because British surgeons hated being drenched with carbolic acid, as well as the smell, the method was soon abandoned. By 1880, Lawson Tait had replaced Lister's antiseptic method by a thorough cleansing of the wound area with soapy water; instead of soaking his hands in carbolic acid, Tait washed his hands with soap and hot water. During surgery Tait wore a clean coat; used clean towels and instruments; and, instead of treating sponges with carbolic acid to mop up the blood and pus, used washing soda. For suturing he used silk sterilized by boiling instead of Lister's carbolic acid-treated catgut. He also replaced Lister's carbolic acid-impregnated bandage with clean, dry dressings. In 1886 steam sterilization of instruments was introduced, which allowed surgery to move toward a germ-free environment from the very beginning of an operation. By 1918 few "old guard Listerians" still believed in antiseptic surgery. Today, aseptic surgery—that is, germ exclusion rather than germ destruction—is practiced: the operating theater is disinfected, as are the patient and the surgical team, and the surgeons employ sterile instruments, gowns, masks, and gloves.

Surgery, Disease, and Anesthesia

During the Middle Ages and until the 19th century cleanliness was a priority neither of medicine nor of people in general. Communal wells provided water to the streets but there were few facilities for removing sewage. Wastes polluted the wells. The streets were dumping grounds for garbage and human wastes and domestic animals including pigs roamed the streets. Uneaten food from the table was thrown on the floor to be devoured by the dog and cat or to rot ... and draw swarms of flies from the stable. The cesspool in the rear of the house would have spoiled ... the appetite ... if the sight of the dining room had not.

At this time (and even before), surgery was an unsafe practice, since control of infection (by asepsis and antisepsis) and anesthesia were unknown. Surgery was regarded as a manual art and was looked upon by the public with fear and hatred because of the suffering and almost certain death from infections (Fig. 11.4). In general, surgery was restricted to surface cutting, and opening the body cavity was avoided because it usually meant sure disaster. There were two kinds of surgeons: "true" surgeons, who carried out major operations such as the removal of tumors, the repair of fistulas, operations on the face, and amputations; and the barber-surgeons, who would assist surgeons by carrying out bloodletting, extraction of teeth, resetting of bone fractures, and removal of skin ulcers. Gunshot wounds and injured limbs, frequently gangrenous, were amputated and then treated with cautery or boiling oil. The settings for early practitioners of the "cutting art" were the kitchen table, the battlefield, or below deck on a ship, not the hospital or operating theater. Prior to the 1840s hospitals were not institutions for the care of the sick as they are today, but existed primarily to provide a refuge for the poor, orphans, and cholera victims. Hospitals—dirty, poorly ventilated, and overcrowded—were perpetrators of diseases and were more like prisons than places where health might be restored.

The methods for minimizing pain during surgery were few—opium, nightshade, strapping, compression of nerve trunks, distracting noises, and hypnotism. Among the surgeons themselves the notion existed that anesthesia was a threat to their profession; as somebody said, "It was unworthy to attempt to try by artificial sleep to transform the body into an insensitive cadaver before beginning to use the knife." Some even claimed that the pain endured during surgery was critical to recovery. The best surgeons were speed artists. Indeed, one of the most famous was the London surgeon Robert Liston (1794-1847). Liston was famous for his nerves of steel, his

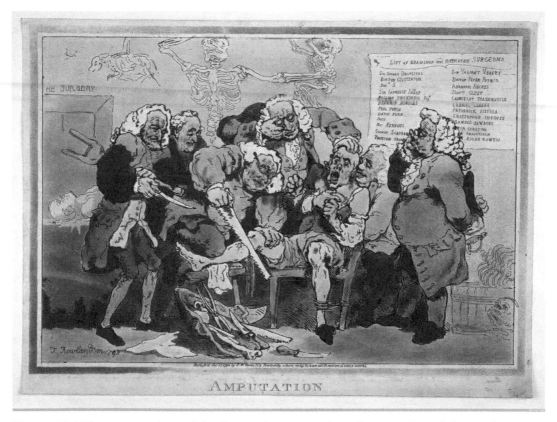

Figure 11.4 Five surgeons participating in the amputation of a man's leg while another oversees them. Coloured Aquatint 1793 By: Thomas Rowlandson. Wellcome image L034242. Courtesy of Wellcome Library, London, CC-BY 4.0.

satirical tongue, and an uncertain temper. He was renowned for his speed and dexterity in the performance of an operation. It was said that that when he performed an amputation the gleam of his knife was followed so instantaneously by the sound of the saw as to make the two actions appear almost simultaneous. Liston removed a 45-lb scrotal tumor in 4 min, but in his enthusiasm cut off the patient's testicles as well. He amputated a leg in two and a half minutes, but the patient died shortly thereafter from gangrene (as was often the case in those days); in addition, he amputated the fingers of his young assistant, who also died later from gangrene, and slashed the coattails of a distinguished surgical spectator, who was so terrified that the knife had pierced his vital organs that he dropped dead in fright. So ended the only operation in history with 300% mortality! In the early 1840s, however, attitudes toward pain began to change, though this new perspective came from dentists, not surgeons. Why were dentists so inclined to use agents that were able to produce insensibility to pain? Some have suggested that unlike surgeons, who would have contact with their patients only once (sometimes with fatal consequences), dentists would have to minimize pain in order to keep their patients coming back.

Laughing gas

The first dental anesthetic to be used was nitrous oxide (N_2O), or "laughing gas." In 1798, Sir Humphrey Davy performed research on nitrous oxide, a gas that had been identified by Joseph Priestley in 1772. He found that severe pain from dental inflammation would be relieved by breathing the gas and correctly observed "that it is capable of destroying pain, (and) it may probably be used with advantage during surgical operations." Yet very few European surgeons used it. They believed that the pain endured during an operation was beneficial to the patient. In the United States, however, nitrous oxide was used to amuse, and demonstrations accompanied lectures for the purpose of entertainment. The American dentist Horace Wells attended one of these lecture-demonstrations. He saw the possibilities, borrowed some of the gas, and had a fellow dentist extract one of Wells's own teeth painlessly during inhalation. There followed dozens of extractions using "painless dentistry," and with the help of another dentist, William T. G. Morton, Wells arranged to demonstrate the anesthetic properties at Harvard University. One of the Harvard students volunteered to have his tooth removed, but because the bag of laughing gas was removed too early, he screamed from excruciating pain. Wells was shattered by the experience. He became the laughingstock of the medical community and left the practice of dentistry to engage in a variety of odd pursuits, including the selling of canaries and fake paintings and etchings.

In 1847, Wells began self-experimenting with the newer anesthetic chloroform. The experiments he conducted eventually led to his becoming addicted to chloroform, and he began to suffer from bouts of depression and hallucinations. During one of his delusional fits he threw sulfuric acid on two prostitutes on Broadway and was promptly arrested and imprisoned in the Tombs in New York City. While in prison he managed to obtain a bottle of chloroform and a razor. On January 24, 1848, under the influence of the chloroform, he cut the femoral artery in his left thigh and bled to death. He was only 33. A few days after his death the Paris Medical Society recognized Wells's contribution to general anesthesia and to painless surgery, and today in a park in Paris there stands a statue of Wells that commemorates his discovery. In 1875, Wells's hometown (Bushnell Park) erected a statue in his honor. At its base are a walking stick, a book, a scroll, and a bag (for nitrous oxide). The practice of nitrous oxide anesthesia, however, was abandoned for 10 years until until J. H. Smith, another dentist, anesthetized several hundred patients and removed 2,000 teeth painlessly. By 1886 nitrous oxide was in common use in dentistry in the United States, and it is still used today.

Ether

Ether (named by August Sigmund Frobenius in 1730) is a distilled mixture of sulfuric acid and alcohol. In 1605, Paracelsus described its properties: "If it is taken even by chickens … they fall asleep from it … but awaken later without harm. It quiets all suffering without any harm and relieves all pain and quenches all fevers." In 1818, Michael Faraday (1791-1867), who was working in Sir Humphrey Davy's laboratory, reported the effects of inhaling ether. He found that after inhalation of ether vapors the lethargic state was followed by a great depression of spirits and a lowered pulse— conditions that might threaten one's life. In England, John Snow (who later achieved fame for his theory of cholera transmission) was among the first to witness the use of ether for tooth extraction and was captivated by the gas's power to produce "quietude" in the patient. But unlike others, Snow did not rush to treat patients. Instead, he began chemical and physiological experiments to establish the parameters of the new technique and developed an inhaler that took into account the relation between temperature and the "strength" of the vapor. In just 6 months Snow described the different degrees of anesthesia that marked ether's sequential suspension of consciousness and volition. Snow mastered the intricacies of ether, whereas other physicians struggled. Ether's irritant qualities made it difficult to breathe, and ill-designed inhalers that gave too little of the vapor stimulated rather than subdued the patient.

In the United States ether inhalation became a part of social entertainment, and there were "ether parties." On March 30, 1842, Crawford Long (1815-1844), a physician who had used ether for amusement when he was a student at the University of

Pennsylvania, successfully employed it for the painless surgical removal of a tumor of the neck after the patient, a student, inhaled the ether. A second patient, a child who required amputation of two severely burned fingers, was given ether, but when the child awoke prematurely and was in great pain, he had to be tied down to complete the operation. Under popular pressure, Long was forced to abandon his use of ether.

In the late 1840s American dentists, especially William T. G. Morton, who had previously employed ether for tooth extractions, promoted its use in surgery. The first public demonstration of its successful use took place in the surgical amphitheater (now named the Ether Dome) at the Massachusetts General Hospital on October 16, 1846 (Fig. 11.5). A large vascular tumor of the neck was removed painlessly by the surgeon John Collins Warren, as Morton administered ether using a special but simple device: a sponge soaked in ether inside a glass vessel with a spout that

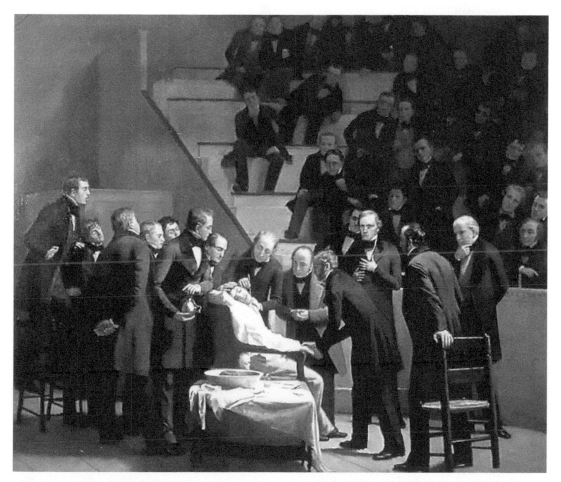

Figure 11.5 The first operation under ether by Robert C. Hinckley. Courtesy of the Boston Medical Library in the Francis A. Countway Library of Medicine.

was placed in the patient's mouth. That day is now called Ether Day. Morton concealed the chemical nature of the ether by adding fragrances and called his drug Letheon, from the Greek mythology in which a drink from the river Lethe could expunge all painful memories. He applied for and received a U.S. patent for Letheon, but the patent was later denied. He also applied for a monetary reward ($100,000) for his discovery from the federal government, and this too was denied. Charles Jackson, a Boston physician who had advised Morton to use ether, also claimed to have played a large part in its discovery, and they pressed for credit all the way to Congress, but on August 28, 1852, the U.S. Senate voted 28-17 to deny Morton's claim of priority in the discovery of the anesthetic properties of ether, and he was forced to reveal that Letheon was nothing other than ether. Morton was crushed, and in 1853 he was expelled from the American Medical Society and disowned by fellow dentists for his unseemly and greedy conduct. He then moved to a cottage near Boston that he named Etherton and died on July 15, 1868.

Jackson claimed that as early as 1845 he had used ether to banish pain and for the extraction of teeth. But since so few in the profession rallied to support his claim, he turned to an old-fashioned anesthetic, alcohol. In 1873 he was involuntarily committed to McLean Asylum, a mental institution, where he remained until his death on August 28, 1880. For Jackson's contributions to the discovery of ether anesthesia Massachusetts General Hospital underwrote all of the expenses during his stay at McLean.

Wells, Long, Jackson, and Morton, but especially Morton, deserve some measure of credit for the discovery of etherization. Despite Morton's greed, he probably should receive most of the credit for his public demonstration of general anesthesia. It was through all their efforts that dentists and surgeons soon began to catch on to the benefits of ether anesthesia, and on December 21, 1847, Liston (the surgical speed artist) performed the first painless major operation (a leg amputation) using ether.

Chloroform

Chloroform is a highly volatile substance with an agreeable taste and smell. It was discovered simultaneously in 1831 by the chemists Simon Guthrie in New York, Eugène Soubeiran in France, and Justus von Liebig in Germany. None of the chemists expected the substance, given the name "chloroform" by Jean-Baptiste Dumas, to be of any practical use until James Simpson (1811-1870) tried inhaling its fumes. Simpson, an Edinburgh obstetrician, had experimented with chloroform at a "chloroform frolic party," attempting to produce the same effects as nitrous oxide, but when all the guests fell asleep, he realized the potential of the substance as an anesthetic. Upon recovering consciousness, Simpson said, "This is stronger than ether." On November 4, 1847, Simpson used chloroform to relieve the pain of labor during childbirth. Other successes in both Britain and the United States soon followed Simpson's public announcement in 1848. The Scottish Church, however, was vehemently opposed to this use of anesthesia to relieve pain during childbirth, because of the biblical verse "in sorrow thou shalt bring forth children." Simpson responded in kind with: "And the Lord caused a deep sleep to fall upon Adam; and he slept; and He took one of his ribs, and closed up the flesh instead thereof."

The suffragettes, however, knew that the discrimination against painless childbirth came from the male-dominated medical establishment and the clergy, and they promoted the use of chloroform and ether. In 1850, Simpson used ether to deliver Queen Victoria's seventh child, Arthur. The London physician John Snow used chloroform during the birth of Queen Victoria's eighth child, Leopold, in 1853, and Queen Victoria described it as "the blessed chloroform." And when chloroform was used at the birth of her ninth child, Beatrice, in 1857, Victoria said, "We are having a child and we are having chloroform." Snow anesthetized 77 obstetric patients with chloroform, and in 1858 he wrote a treatise, "Chloroform and Other Anesthetics," in which he wrote: "The most important discovery that has been made in the practice of medicine since the introduction of vaccination is undoubtedly the power of making persons perfectly insensible to the most painful surgical operations, by the inhalation of ether, chloroform and other agents of the same kind." He also described 50 deaths during chloroform administration and recommended means for prevention. Simpson was knighted for his development of chloroform anesthesia, but not Snow (who had died in 1858). Because of its quick action and ease of administration, chloroform revolutionized battlefield surgery. During the U.S. Civil War the production of ether and chloroform by Squibb and Company (established in 1858) enabled it to reap substantial profits.

Anesthetics were medicine's original wonder drugs. Surgeons used ether and chloroform 40 years before Louis Pasteur made them realize that washing hands and boiling instruments might prevent disease and death. Semmelweis developed a practical method for asepsis before the germ theory was established. Sanitation, antisepsis, asepsis, and anesthesia prevented disease from needlessly destroying human life, but perhaps the most important point of these advances in health and medicine was their influence on the social climate and the acceptance and recognition of scientific truths.

Disease, Dyes, and Drugs

Today, we take for granted that when we become sick there is, with rare exception, a drug to cure us. How did it happen that the drug salvarsan was developed for the treatment of syphilis, or AZT was designed to cripple HIV, and what provided the foundation for the discovery of antibiotics such as penicillin and streptomycin? It all began with the German dye industry and the microbe hunter Paul Ehrlich, who said: "My dear colleagues, for seven years of misfortune I had one moment of good luck!"

In the first 3 decades of the 19th century studies of the chemistry of dyes began. This was the result of the large-scale commercial manufacture of illuminating gas (ethylene) for lighting purposes. In England illuminating gas was obtained by distillation from coal; it was not the gas itself that the English were after, however, but rather the tar. Tar was needed by the British navy as a result of the loss of their American colonies, which had supplied it. Unwilling to be an economic captive to the colonies, the British attempted to become self-sufficient and to make the tar from coal. For several decades illuminating gas, the first by-product of coal distillation, was of little value; after 1812, however, the importance of the gas outweighed that of tar. As a consequence of the stepped-up production of illuminating gas, the tar would have accumulated in great quantities had not the chemists discovered a new use for it. By boiling and distilling the tar it was possible to obtain light oils and heavier oils, called creosote and pitch. The creosote was used for preserving wood, and the pitch was sold for asphalt. A sample of the light oil was sent to Germany for analysis by a brilliant young German chemist, August von Hofmann. He found that it contained benzene and aniline. Hofmann made a variety of dyes by starting with aniline. By 1850, due to the efforts of Hofmann and his students, Germany had become the center of the European dye industry, and aniline dyes in all shades of the rainbow could be had. It was therefore natural that another German, Paul Ehrlich, would become interested in dyes and their uses.

Ehrlich (whom we met ealier; see p. 229) was born in 1854 in the city of Strehlen, Silesia, now a part of Poland. He studied at the University of Strasbourg, where his tutor, Wilhelm von Waldeyer, indulged his experimentation with various kinds of dyes, and where "little Ehrlich," as Waldeyer called him, discovered a new type of cell—the mast cell. (Later, it would be found that mast cells secrete the chemical histamine, making nearby capillaries leaky and in some cases provoking an allergic response. Drugs designed to blunt this effect are antihistamines.) Ehrlich graduated as a doctor of medicine at age 24. His thesis topic was the value and significance of staining tissues with aniline dyes. That certain dyes stained only certain tissues and not others suggested to Ehrlich that there was chemical specificity of binding. This notion became the major theme in his scientific life and led to a search for "magic bullets"—drugs (derived from dyes) that would specifically strike and kill parasites.

Ehrlich wrote that

> curative substances—a priori—must directly destroy the microbes provoking the disease; not by an "action from distance," but only when the chemical compound

is fixed by the parasites. The parasites can only be killed if the chemical has a specific affinity for them and binds to them. This is a very difficult task because it is necessary to find chemical compounds, which have a strong destructive effect upon the parasites, but which do not at all, or only to a minimum extent, attack or damage the organs of the body. There must be a planned chemical synthesis: proceeding from a chemical substance with a recognizable activity, making derivatives from it, and then trying each one of these to discover the degree of its activity and effectiveness. This we call chemotherapy.

Ehrlich's first chemotherapeutic experiments were carried out in 1904 with mice infected with the trypanosomes that cause African sleeping sickness. He was able by injections of a red dye he called trypan red to cure these mice of their infection. This aroused some interest, but because the drug was inactive in human sleeping sickness, he turned his attention to arsenicals. He began with the compound atoxyl, an arsenic compound that was supposedly a curative for sleeping sickness, but he found the drug was useless: it destroyed the optic nerve, so that when patients were treated with the drug, they not only were not cured of sleeping sickness but became blind.

Ehrlich was unable to produce a single drug that was of use in humans, but in 1905 all that changed. That year Fritz Schaudinn and Erich Hoffmann described the germ of syphilis (*Treponema pallidum*), and shortly thereafter Sahachiro Hata, working at the Kitasato Institute for Infectious Disease in Tokyo, discovered how to reproduce the disease in rabbits. It was intuition rather than logic that led Ehrlich to believe that arsenicals would kill *Treponema*. (His reasoning was that since trypanosomes and treponemes are both active swimmers, they must have a very high rate of metabolism and that an arsenical would kill the parasite by crippling its energy-generating ability.) And now, thanks to Hata's rabbit model for human syphilis, Ehrlich's drugs could be tested for their chemotherapeutic effectiveness. In the spring of 1909, Shibasaburo Kitasato of Tokyo sent his pupil, Hata, to Germany to work with Ehrlich, and a year later Hata successfully treated syphilis in rabbits with a dioxy-diamino-arseno-benzene compound, named 606. The synthesis of 606, reported in 1912, was patented, and the drug was manufactured by Hoechst and sold by the company under the name salvarsan. The newspapers took up the announcement of a cure for syphilis in humans, and overnight Ehrlich became a world celebrity.

Salvarsan was a yellow, acidic powder, packaged in 0.6-g ampoules. One difficulty with using it was the problem of dissolving this powder and neutralizing it under aseptic conditions prior to injection in order to avoid oxidation that could alter its toxicity. Ehrlich issued precise directions for its preparation in sterile water for its neutralization, and he advised that intravenous injection be carried out without delay. These directions were often not adhered to, and the resulting deaths attracted much unfavorable publicity. Further, it was dangerous to make salvarsan because the ether vapors used in its preparation could cause fires and explosion. There were problems with side effects of salvarsan—such as gastritis, headaches, chills, fever, nausea, vomiting, and skin eruptions—and, at times, the syphilis had progressed so far that it was not effective as a cure. Moreover, salvarsan was not selective when given as a single dose, and treatment had to be spread out over several months. This meant that fewer than one-fourth of the patients ever completed the course of treatment. Despite these problems, salvarsan and its successor (neoarsphenamine, or 914, a more soluble derivative of 606) remained the best available treatment for 40 years. Although Ehrlich was awarded the 1908 Nobel Prize (with von Behring) for work on immunity, he regarded his greatest contribution to be the development of a "magic bullet," salvarsan. At the Nobel ceremony he modestly said, "My dear colleagues, for seven years of misfortune I had one moment of good luck!"

How did humans come to use dyes in the first place? Humans, even from prehistoric times, have been fascinated with color. For example, the discovery of red ochre (a pigment from iron ore) in ancient burial sites and in the cave paintings at Lascaux in France reveals that color was used for aesthetic purposes more than 15,000 years ago. The art of dyeing cloth goes back to China of about 3000 B.C., and at about this time the principal dyes were obtained from plant roots and other plant parts, as well as from insects or shellfish. But despite their use for thousands of years, only a dozen natural dyes have found a practical use. Why, one might ask, are humans so fixated on color? It lies in our biology. Perhaps if we did not have color vision, that is, if we lived in a black-and-white world as do dogs, cats, horses, cows, and pigs (most mammals are in fact color-blind), then colored objects would have little demand; but primates—that is, apes, monkeys, humans—and birds see the colors of the rainbow, and so from the earliest times we have done what Ted Turner did to film—colorizing.

Natural dyes come in the colors red, yellow, blue, purple, and black. By mixing some of these, such as blue and yellow, one can get green. The red dyes have two principal sources: madder and cochineal. Madder is obtained from the roots of the herb madder; the dye is also called alizarin because "alizari" means "root." Coats and trousers dyed using madder have been used by the armies of Britain and France for several centuries. Indeed, the British were called redcoats after their red jackets. It may be that the color red was selected for military clothing to minimize the visual effect of a wounded soldier seeing his own blood. The red dye cochineal comes from the female scale insect *Coccus cacti* that lives on the prickly pear cactus. It takes 200,000 insects to produce 2 lb of dye. Yellow dyes, though the largest group of natural dyes, are distinctly inferior to the reds, blues, and blacks because they have weak coloring properties and they bleach out or fade easily. The most important yellow dye in the Middle Ages was weld, which comes from the seeds, stems, and leaves of *Reseda luteola*, a plant called dyer's rocket. Other yellow dyes are saffron from the anthers of the plant *Crocus sativus*, safflower petals, and the inner bark of the black oak (*Quercus velutina*). This latter dye bleaches or fades more easily than weld. Blue colors come from the leaves of the indigo plant, *Indigofera tinctoria*, also called woad. This plant probably originated in India and is used for coloring blue jeans. Weld in conjunction with woad gives Lincoln green, a color made famous by Robin Hood and his Merry Men. Related to the blue dyes are the purple dyes. The most famous is that obtained from the sea snail *Murex*, called Tyrian purple because it was widely distributed by the Phoenicians out of the city of Tyre (now in Lebanon). It is said that the Romans paid $300 for 2 lb of wool dyed with Tyrian purple, an amount so extravagant that only kings and priests could afford it. Thus, the phrase "born to the purple" signifies people of wealth or position. The only black dye of any importance comes from the heartwood of the tree *Haematoxylon campechiancum*, found in Central America; it gives a deep color when combined with metallic salts. Obtaining dyes from natural sources was slow, inefficient, wasteful, and very labor-intensive. The natural dyes were usually impure, the proportions of mixtures so variable that reproducibility was hard to come by, and few natural dyes are fast (fade-resistant).

After 1700 a complex series of social and economic events took place in England and Germany, many of them a result of the development of the technology for using steam-powered engines. Coal was used for powering these engines, as well as for smelting iron ore into iron and steel. Coal came into increased demand because in Britain woodlands had been depleted and supplies of charcoal were scarce. At this time, the coke oven was developed for heating coal to high temperature in the absence of air to produce a dense, smokeless, high-temperature-yielding fuel. By the 18th century the textile industry, once dependent on human power for weaving, began to use machines. In Britain the rising wool and cotton-textile industries characterized the Industrial Revolution. All this brought a sudden increase in interest on the part of the chemical industry, since textiles required bleaching, and natural processes and substances such as sunlight, rain, and urine were too slow. This need for coloring textiles was responsible for the development of the synthetic dyestuffs industry. Indeed, the synthesis of aniline dyes paralleled the early advances of organic chemistry, and it is doubtful that one would have succeeded without the other.

The founding father of the synthetic dye industry was the Englishman William Perkin (1838-1907), who at age 18 produced the purple dye mauveine. The story actually begins in Germany, however, with the appointment of Justus Liebig (1803-1873) as professor of chemistry at Giessen in 1824, the same year that coal tar (a viscous by-product of coke production) was discovered. This advance received little notice until a brilliant pupil of Liebig's, August Hofmann (1818-1892), was appointed director of the Royal College of Chemistry in London in 1845. Hofmann's main research interest was in aniline, a yellowish to brownish oily liquid present in coal tar.

Perkin entered the Royal College in 1854 at age 16, and in 1856, Hofmann gave him the task of making quinine from aniline by oxidation with potassium dichromate in sulfuric acid. If Perkin and Hofmann had known the chemical structure of quinine, they certainly would have abandoned this route. Perkin oxidized aniline with dichromate and got a black precipitate. He boiled the black sludge with ethanol and it gave a striking purple solution, which he called aniline purple or mauveine. This he recognized as a valuable dye, and he was successful in dyeing silk. From this discovery the synthetic dyestuffs industry began. Perkin left school at 18, patented his discovery, built a factory, and retired in 1873, at age 35, a rich man. He devoted his "retirement" to basic science studies at Glasgow University.

The magazine *Punch* wrote this of Perkin's discovery:

> There's hardly a thing a man can name
> Of beauty or use in life's small game,
> But you can extract in alembro or jar,
> From the physical basis of black coal tar.
> Oil and ointment, and wax and wine,
> And the lovely colors called aniline:
> You can make anything from salve to a star
> (If you only know how) from black coal tar.

Ironically, the Royal College of Chemistry lost its support because it could not show a contribution to agriculture, and so in 1865, Hofmann returned to Berlin, where he held a chair in chemistry. As a consequence, the center of the dyestuffs industry moved from Britain to Germany. By World War I about 90% of all dyestuffs were manufactured in Germany.

Prontosil

In the early morning hours of October 26, 1939, Gerhard Domagk was roused from slumber by the incessant ringing of the telephone. Picking up the phone, he heard the caller in a Swedish-accented voice say that he had been awarded the Nobel Prize for Physiology or Medicine for "the discovery of the antibacterial effects of Prontosil." Domagk was stunned and elated. His euphoria, however, was soon tempered by the fact that the Führer, Adolf Hitler, had forbidden any German to accept a Nobel Prize or deliver the lecture in Stockholm, Sweden. Domagk was forced to sign a letter addressed to the appropriate Nobel committee that said it was against his nation's law to accept the award, and that the award also represented an attempt to provoke and disobey his Führer. At the end of World War II, Domagk was reinvited to travel to Stockholm to accept, on behalf of Karolinska Institute and the king, the Nobel Prize, consisting of a gold medal and diploma, and to deliver his speech. The money, 140,000 Swedish crowns—a rich award worth several years' salary for most scientists—he would never see, since in accordance with the rules of the Nobel Foundation unclaimed prize money reverted back to the main prize fund. In the introductory speech Nanna Svartz of the Karolinska Institute, in 1947, said:

During the past 15-20 years a great deal of work has been carried out by various drug manufacturers with a view to producing less toxic but at the same time therapeutically effective … preparations. Professor Gerhard Domagk … planned and directed the investigations involving experimental animals. The discovery of Prontosil opened up undreamed of prospects for the treatment of infectious diseases. What Paul Ehrlich dreamed of, and also made a reality using Salvarsan in an exceptional case has now, through your work become a widely recognized fact. We can now justifiably believe that in the future infectious diseases will be eradicated by means of chemical compounds.

As with Paul Ehrlich's discovery of "magic bullets" such as methylene blue and trypan red, the development of Prontosil began with the synthesis of a group of red-colored dyes that contain nitrogen as the azo group –N=N-. Initially, the azo dyes were of interest to the chemists at I.G. Farben for their ability to contribute to fastness for dyeing wool, not for their antimicrobial activity! Although azo dyes similar to Prontosil had been synthesized almost a decade earlier (1919) by Michael Heidelberger and Walter Jacobs at the Rockefeller Institute in New York City, they were discarded after they showed poor antimicrobial activity in the test tube. Indeed, it was fortuitous that in 1933 to 1935 Domagk, in trying to understand the lack of correlation between test tube and animal antibacterial tests, resorted to examining drugs such as Prontosil in an animal model.

In 1921, Domagk, age 26, graduated from medical school. His first position was at the hospital in Kiel, where it was possible to observe infectious diseases firsthand, especially those caused by bacteria. The hospital was not well equipped, however, and there was poor access to microscopes; the equipment for chemistry was limited and laboratory facilities were nil. At the time, physicians knew what bacteria were and that they might cause disease; there was little understanding, however, as to how they actually produced the disease state. Moreover, they hadn't a clue as to how to stop the disease-producing bacteria, save for carbolic acid antisepsis. Unfortunately, carbolic acid (phenol) was indiscriminate in action, killing not only the bacteria but normal tissues as well. The harsh antiseptics could be used only externally, and once an infection had been established, little could be done to control it. What was needed were substances for use within the body after an infection had established itself, a kind of internal antisepsis. Eager to learn the latest techniques and findings, in 1923 Domagk moved to the Institute of Pathology at the University of Greifswald, and then in 1925 he moved to what appeared to be a better post at the University of Münster. Two years later he found himself dissatisfied with his ability to carry out research in a university setting. Then fortune smiled on him.

In 1927, Domagk met with Carl Duisberg, who told him that the pharmaceutical company Bayer was expanding its drug research and was looking for someone with experience in both medicine and animal testing to screen their chemicals for medical effects in animals. Domagk was, according to Duisberg, the perfect candidate, and not only would he be given a large laboratory, he would be appointed as director of the Institute of Experimental Pathology. At Bayer, Domagk's laboratories were carved out of the space occupied by the head of the tropical disease unit, Wilhelm Roehl. Roehl, a young physician, was a former assistant to Paul Ehrlich at Speyer House who had joined Bayer at the end of 1909. Roehl showed Domagk how he ran the animal tests and how his results with the compounds he had screened were correlated. Indeed, Roehl already had some success in developing plasmochin (plasmoquin), using his malaria-infected canary screening method and the compounds provided to him by the Bayer chemists. Things were beginning to look up for Domagk, who was to take over the work initiated by the chemist Robert Schnitzer at Hoechst (already a part of the I.G. Farben cartel) in his search for a drug against a generalized bacterial infection, particularly the hemolytic *Streptococcus pyogenes*, which caused meningitis, rheumatic fever, middle-ear infections (otitis media), tonsillitis, scarlet fever, and nephritis. During World War I this streptococcus was also responsible for the many deaths after wounding, and oftentimes hemolytic streptococcal infections were a consequence of burns and scalding. In addition, in concert with the influenza virus, it caused the pneumonia responsible for many deaths during the worldwide epidemic in 1918-1919. The appearance of streptococci in the blood of a patient, septicemia, was often a prelude to death. Indeed, in 1929, while Roehl was traveling in Egypt, he noticed a boil on his neck while shaving. It turned out to be infected with *S. pyogenes*. Roehl developed septicemia, for which there was no cure, and in a few days, at age 48, he was dead.

In the 1920s, Rebecca Lancefield, at the Rockefeller Institute, discovered that there were many types of hemolytic streptococci, though not every one was a danger to human health. With this knowledge Domagk decided to select a particularly virulent strain of hemolytic streptococcus, one isolated from a patient who had died from septicemia. After inoculation of mice, this strain reliably killed 100% within 4 days. Domagk reasoned that only an exceptional drug would enable the infected mice to survive his "drop-dead" test. Over time Domagk would test the thousands of compounds synthesized by the chemists Joseph Klarer (hired by Bayer at the same time as Domagk) and Fritz Mietsch (who had come to the company to work with Roehl). In their syntheses, Klarer and Mietsch followed the trail blazed by Ehrlich, concentrating their efforts on dyes that specifically bound to and killed bacteria. Domagk examined several of the acridines synthesized by Klarer and Mietsch for activity against streptococcal infections in mice. Optimum activity was found with an acridine with an amino group on the carbon opposite the nitrogen atom rather than to the adjacent ring, as was the case with pamaquin. This 9-aminoacridine, when given orally or by injection to streptococcal-infected mice, was able to control an acute infection, though this particular nitroacridine was not potent enough to be used clinically. Another analog, however, was found to be more effective as an antibacterial; it was marketed as Entozon (or Nitroakridin 3582) and found use in the treatment of bovine mastitis and uterine infections.

Domagk tested two other classes of compounds that had been reported clinically to have antibacterial properties: gold compounds and azo dyes. In 1926, Hoechst's gold compound, gold sodium thiosulfate (Sanochrysine), had been shown to be effective in treating bovine tuberculosis, and it protected mice against strep. Domagk also found it to be effective in the drop-dead test; it could not, however, be used for treating humans, because when the dosage necessary to cure patients of their strep infections was used, kidney damage resulted. Domagk then tried an azo dye. The synthesis of azo dyes was simple, chemical variations were easy to make, and they were less toxic. Domagk tried the azo dye Pyridium, and although it inhibited bacterial growth, it turned the urine a red color, which would be unacceptable to patients. In September 1931 an azo compound made by Klarer, called chrysodine, which incorporated a chlorine atom, worked to cure mice even when given by mouth, but it wasn't powerful enough for clinical trials. A year later Klarer began attaching sulfur atoms, and he produced a 4-amino-benzene-sulfonamide.

This compound, number 695 in the Klarer azo series, completely protected mice when given by injection as well as by mouth. It protected mice at every dose level; much to Domagk's surprise, however, it did not kill the streptococci in the test tube, only in the living animal. Further, it was specific, acting on only streptococci and not other bacteria. Klarer kept making other modifications. Domagk found that one, number 730, a dark-red azo dye, almost insoluble in water, was tolerated by mice at very high doses and brought about no side effects. It was like all the others in the series of azo dyes in that it was selective for streptococci and did not kill tuberculosis-causing bacteria or the pneumonia-causing pneumococci or staphylococci. On December 27, 1932, Domagk concluded that this compound, named Streptozon, was the best for treating streptococcal infections. And as was the practice at I.G. Farben, a patent was filed under the names of the chemists, Klarer and Mietsch, and not the pathologist Domagk. (And as was also the practice, only the chemists received royalties.) While Streptozon was shown to be clinically effective, there was a drawback: the brick-red powder could be made into tablets and ingested but was impossible to dissolve. A liquid formulation was needed, and in 1934 Mietsch produced a beautiful port-red fluid that could be injected. It was just as active as the powder. Streptozon solubile, as it was called, was patented and renamed Prontosil because it worked quickly.

Prontosil was given the chemical name sulfonamidochrysoidine, although this was almost never used. Prontosil treatment saved many from the death sentence imposed by *S. pyogenes*, including Domagk's own 6-year-old daughter, Hildegarde. The little girl had fallen down the stairs at home, and the needle she was carrying had punctured and broken in her hand. The needle was removed surgically, though the wound became infected and she developed septicemia. To avoid amputation of the arm and possible death, Domagk gave his daughter Prontosil by mouth and rectally. She recovered.

In April 1935, when the Pasteur Institute in Paris requested a sample of Prontosil, and this was denied, Ernest Forneau (1872-1949), a leading French chemist and head of the Laboratory of Therapeutic Chemistry at the Institute, instructed his staff to decipher what they could of the German patent application and then synthesize the compound. (Under French law, Prontosil could be patented only as a dye and not as a medicine.) Using a method described in the literature in 1908 by Paul Gelmo of Vienna, who had prepared it for his doctoral thesis, the French found a way to duplicate Prontosil's synthesis and began to manufacture it. It was marketed as Rubriazol because of its red color. There remained, however, a puzzle. Why did Prontosil not follow Paul Ehrlich's axiom *Corpora non agunt nisi fixata* ("A drug will not work unless it is bound"), which guided much of chemotherapy? Domagk wrote: "It is remarkable that in vitro it shows no noticeable effect against *Streptococcus* or *Staphylococcus*. It exerts a true chemotherapeutic effect only in the living animal."

Forneau's group at the Pasteur Institute continued to study the drug and found something unexpected: mice infected with streptococci but treated with an non-azo compound, i.e., pure sulfanilamide, survived. Simple sulfanilamide was colorless and unpatentable; it was, however, as effective as the red wonder drugs Prontosil and Rubriazol. The French finding explained why attaching sulfa to many types of azo dyes resulted in activity against strep, whereas dyes without sulfa were less active. The mystery of Prontosil's activity in the live animal and not in the test tube was solved. In order to become active, Prontosil had to be split to release the 4-amino-benzene-sulfonamide, and this was accomplished by enzymes in the body. In the test tube, where there were no enzymes to split the Prontosil, no sulfa could be released. Thus, Prontosil had to be "bioactivated." In short, the azo dye stained, whereas the sulfa moiety cured.

Drug firms were able to produce chemical variants of sulfa that could work much better, were less toxic, and had a wider range of activity against many different kinds of bacteria. May and Baker in England produced a version with an attached aminopyridine, called M&B 693 or sulfapyridine. Sulfapyridine was also shown to be effective against streptococcal, meningococcal, staphylococcal, and gonococcal infections. It had unprecedented efficacy in mice with lobar pneumonia, and in human trials it reduced the mortality from 1 in 4 to 1 in 25. Among others, it saved the life of Great Britain's prime minister Winston Churchill. On December 11, 1943, as Churchill was flying to Dwight Eisenhower's villa in Tunis, following an exhausting trip during which he had conferred with Roosevelt and Stalin at Yalta to firm up plans for the invasion of Italy, he fell ill. By the time he arrived in Tunis, his throat was sore and his temperature was 101°F. An X ray revealed a shadow on his lung, and when his lungs became more congested, they suggested lobar pneumonia. He suffered two bouts of atrial fibrillation and an episode of cardiac failure. As he hovered near death, Churchill was given M&B 693, and it worked. By Christmas he was back in action, planning the invasion of France, and within 2 weeks he was back home. Most were convinced that he owed his recovery to the new medicine, although he joked that in using M&B he was referring to his physicians, Moran and Bedford. Churchill said, "There is no doubt that pneumonia is a very different illness from what it was before this marvelous drug was discovered." He might have also said it was sulfa's finest hour!

Today, thousands of sulfa drugs have appeared on the market. The number of patents issued for sulfa drugs was 1,300 in 1940 and increased to 6,500 in 1951. In 1943, in the United States alone, 4,500 tons of sulfonamides were produced and millions of patients treated. Ironically, by the time Domagk received the Nobel Prize, many drug makers seemed to be less interested in sulfa drugs, and they began to turn their attention to the more powerful antibiotics such as penicillin and streptomycin. Sadly, Domagk died in the spring of 1964 from a bacterial infection that did not respond to either sulfa drugs or antibiotics. The microbes that killed him were able to repel the chemically designed "magic bullets."

What remained unanswered by Domagk and Forneau was why sulfa drugs worked. In the 1940s, Donald Woods and Paul Fildes, working in London, found that sulfa was less a magic bullet than a clever imposter. They started from the observation that sulfa never worked as well in the presence of pus or dead tissue. Woods went looking for the mysterious anti-sulfa substance that interfered with its action and found it as well in yeast extract. He characterized the anti-sulfa substance as a small molecule, approximately the size of sulfanilamide. In fact, the mystery substance looked like sulfa's molecular twin, *para*-aminobenzoic acid (PABA).

PABA is a chemical involved in the nutrition of certain kinds of bacteria, although today it is more familiarly known as a sunscreen ingredient. Some bacteria can make PABA, while others (such as streptococci) cannot. For those microbes unable to synthesize PABA from scratch it is an essential nutrient—a vitamin—and if their environment does not provide it, they starve to death. What Woods and Fildes showed was that sulfa worked because it was a molecular mimic of PABA, and when the sulfa was around, the bacteria would try to metabolize it instead of PABA. The sulfa could not be utilized, however, and once it bound to a specific bacterial enzyme (dihydropteroate synthase), that enzyme could no longer function, and then there could be no synthesis of DNA. Absent DNA synthesis, the bacteria cannot reproduce and the infection is stopped in its tracks.

Penicillin

The next era of "magic bullets" was the development of antibiotics. It was March 14, 1942, and Anne Miller, a 33-year-old mother, was dying from a streptococcal infection. For weeks she had a fever that ranged from 103 to 104°F. She could not eat and was losing weight; she went in and of coma. Blood transfusions and sulfa drugs all failed to produce a cure. In desperation, her physician contacted John Fulton, who was at Yale University, who in turn knew the chairman of the Committee on Chemotherapy in Washington, D.C., which at the time controlled all the important medicines during World War II. After the call the chairman authorized the pharmaceutical company Merck to release 5.5 g—a teaspoonful—of penicillin—and half the entire amount available in the United States. Not knowing the proper dosage, her physicians gave it to her in several small doses during the day, and within 24 h the deadly streptococcal infection had been cleared from her blood. She convalesced for a month, received more Merck-manufactured penicillin, went home, and lived a full and productive life until 1999, when she died at age 90. Anne Miller was the first patient in the United States pulled from death's door by the "miracle drug" penicillin. Where did penicillin come from, and who discovered it?

The discovery of penicillin is usually attributed to Alexander Fleming, who was born in Scotland in 1881. Fleming was an unimposing, short man with a shock of white hair and a flattened nose, but with bright blue eyes. In 1901 he won a tuition scholarship to St. Mary's Hospital medical school in London; he qualified as a physician in 1908 and a year later joined the inoculation department at St. Mary's, where he remained for the next 49 years. In 1909, Fleming described a method for testing for syphilis, using only a drop or two of blood, and a year after Ehrlich developed salvarsan (see p. 282) Fleming was giving shots of salvarsan to his private patients.

During World War I, as a lieutenant in the British army, Fleming experienced and was concerned with the manner in which open wounds were treated using cloth bandages soaked in the antiseptic carbolic acid. Indeed, of the 10 million soldiers killed in World War I, half died not from mustard gas or bullets or bombs but from bacterial infections. Fleming recognized that although the antiseptic was effective in killing the bacteria in the pus-laden surface wounds, the bacteria remained deeper within the wound and could cause tissue destruction and death. Surely, he reasoned, there must be some treatment that could prevent an infection from becoming established rather than requiring it to be overcome only when it was already present. The search for a material to prevent infection now became the focus of Fleming's life's work.

In November 1921, Fleming was at St. Mary's and suffering with a head cold; he decided to put the mucus from his runny nose onto a petri dish that had been seeded with bacteria. Later, when Fleming noticed that the colonies of bacteria were dissolving, he tried to isolate the bacteriolytic substance, which he believed to be an enzyme. Between 1922 and 1932 Fleming published eight papers on the substance he called lysozyme (which he found also occurred in egg white, turnips, human tears, and bronchial mucus). Fleming had limited skills as a writer, and his papers on lysozyme were short and had few details. Further, he was a dreadful speaker. Indeed, when he presented his work on lysozyme to the Medical Research Club, those in the audience responded with silence. Most in attendance felt his discovery to be irrelevant, especially so when Fleming noted that lysozyme had no effect on the bacteria that caused serious illness. As a consequence, the papers Fleming published on lysozyme were soon forgotten.

In 1928, when Fleming was playing with microbes, he made a serendipitous discovery. At the end of July, before he left on vacation, he piled two or three dozen petri dishes seeded with *Staphyloccocus* at the end of his workbench. Sometime after he returned in August or September he looked at the dishes his research assistant had not discarded and found that a blue-green mold had contaminated one of the dishes. But the odd thing was that within the sea of staphylococci there were no bacteria surrounding the mold. Fleming mused that something in the contaminating mold was making the bacteria lyse. It reminded him of lysozyme, and, as the baseball player Yogi Berra would have said, "It was déjà vu all over again." Two months later, when Fleming recorded his observation in his laboratory notebook and was in need of a photograph to prove the original observation, he grew the mold for 5 days at room temperature and then seeded the dish with staphylococci; only then did he put the petri dish in the incubator. In this case there was an obvious zone of inhibition of bacterial growth. Why did Fleming make such a change in the sequence of events? Because when he tried to reproduce the original finding by putting the mold directly on the lawn of bacteria and keeping the dish at room temperature, there was no inhibition. If he did the reverse, that is, if he grew the staphylococci first and then added the mold, there was no zone of inhibition and nothing noteworthy to photograph. (Later, Fleming and other researchers would find that staphylococcal growth would be inhibited only if the mold was grown first; this is because the mold grows best at room temperature, whereas the bacteria need body temperature to achieve their best growth.)

Fleming's story of stray spore of a green mold landing on a dish of staphylococci and producing an inhibitory substance, which he called penicillin, after the mold's name, *Penicillium*, was not published until 1944, and it appears to be more fiction than fact. In 1968, Ronald Hare, who worked in the inoculation department at St. Mary's during the same time as did Fleming, dismissed the notion of a stray spore coming in through the window of Fleming's 10-foot-by-12-foot laboratory, since the window was seldom if ever opened and no good bacteriologist would ever work near an open window for fear of contaminating the cultures. A far more likely source, according to Hare, was that the mold spores came up through the stairwell from the floor below, where an Irish scientist, C. J. LaTouche, was working with molds. Further, Hare found that LaTouche had in his collection the very same *Penicillium* mold that contaminated Fleming's staphylococci plate. Repeated attempts by Fleming and others to repeat his original finding failed repeatedly. Hare suspects that this is because from July 28 to August 7 (the time Fleming was away) the temperature in London rose above 68°F only twice. Because *Penicillium* grows best at room temperature, whereas the bacteria need body temperature (98°F) for good growth, if Fleming had not incubated the dish with the staphylococci they would not have grown, but the dish with the mold would have had a chance to establish itself before the weather turned warmer, and only then would the bacteria have grown. Is Hare's speculation correct? We shall never know, since Fleming's paper describing the action of the mold on the staphylococci is short on several important details, such as the species of staphylococci, how long the dishes had been on the workbench, the medium of the culture, and whether and when he incubated the petri dish. Some suggest that Fleming was actually looking for lysozyme activity (not penicillin) when he cultured the mold. When he tested the mold against bacteria known to be sensitive to lysozyme, as well as against *Staphylococcus* and other pathogenic bacteria, and found that the mold worked on pathogens on which lysozyme had no effect, Fleming had found something totally unexpected—a mold that produced an antibacterial substance.

Fleming was fond of saying, "When you have acquired knowledge and experience it is very pleasant … to be able to find something nobody had thought of." Fleming continued to work with *Penicillium*. He grew the mold, made large batches of broth, and collected the "mold juice"; to his disappointment he found that the antibacterial properties of the juice were quickly lost, meaning that penicillin was unstable. He found that besides *Staphylococcus*, penicillin was effective against *Streptococcus*, as well as against the bacteria that cause gonorrhoea, meningitis, and diphtheria. Penicillin was nontoxic when injected into healthy rabbits. But for some strange reason Fleming never tested the mold juice on an animal infected with *Staphylococcus* or *Streptococcus*.

On Feb 13, 1929, Fleming delivered a paper, "A Medium for the Isolation of Pfeiffer's Bacillus" (presumably about penicillin's inhibiting some kinds of bacteria) to the Medical Research Club. The evening was a replay of his 1921 talk on lysozyme in that there was little reaction by the audience. In May he submitted a paper, "On the Antibacterial Action of Cultures of Penicillin with Special Reference to Their Use in the Isolation of *B. influenzae*," to the *British Journal of Experimental Pathology*, and it was published a month later.

It is likely that Fleming would have remained an obscure bacteriologist and that penicillin's main application would be in the isolation of penicillin-insensitive bacteria in mixed cultures, instead of a useful therapeutic, were it not for the subsequent work at the Dunn School of Pathology at Oxford University by Howard Florey, Ernst Chain, and Norman Heatley. Howard Florey was born in Adelaide, Australia, on September 24, 1898; entered Adelaide Medical School in 1916; graduated first in his class in 1921; and then with a Rhodes scholarship left Australia for Oxford University (where he shared classes with another Rhodes scholar, John Fulton. At Oxford he was taken under the wing of Professor of Physiology Charles Sherrington (who would in 1932 share the Nobel Prize in Physiology or Medicine for his work on the nervous system). Thanks to Sherrington's efforts, Florey was awarded a prestigious studentship at Cambridge University, where he became interested in mucus secretions. After one year at Cambridge he traveled to the United States with a Rockefeller Foundation fellowship; worked in several different laboratories; returned to Cambridge; and became lecturer in special pathology, teaching physiology and conducting research in pathology. Florey was short, with an arresting presence: a lean face, piercing eyes, and hair neatly parted in the middle. In 1935 he was appointed professor of pathology at the Dunn School of Pathology at Oxford University, where he was determined to revitalize its teaching and research. Florey was outspoken and had the roughness of a no-nonsense Australian. Within a week at Oxford, Florey recruited the German-born Ernst Chain, a Jewish immigrant with a Ph.D. in biochemistry, earned in 1930. Chain has been described a better-kempt version of Albert Einstein, with brushed-back curly black hair and a moustache. He was flamboyant in appearance and in temperament. Chain studied for a second Ph.D. at Cambridge University with the biochemist Sir Frederick G. Hopkins, a 1929 Nobel laureate, working on snake venoms and the manner by which they cause central nervous system paralysis. Also joining Chain was Norman Heatley, who had a doctorate in biochemistry; his thesis was called "The Application of Microchemical Methods to Biological Problems." Chain and Heatley were opposites in temperament, personality, and scientific approach. Heatley wrote meticulous notes, whereas Chain took few notes. Heatley was reluctant to seek credit, whereas Chain was quick to claim his due. Yet despite these differences, the two began to work together on cancerous tumors; they found no differences in the metabolism between normal skin and malignant cells. Even before they completed the studies on cancerous cells, Florey, with an abiding interest in mucus, asked Chain to study lysozyme because of his previous work on snake venom enzymes. Chain, in collaboration with an American Rhodes scholar, Leslie Epstein, showed that lysozyme worked by digesting complex carbohydrates, or polysaccharides. Florey, recognizing Chain's interest in the action of lysozyme, told him that since 1929 bacterial antagonism had been an interest of his. Stimulated by Florey's interest, Chain took as his approach to the problem of bacterial antagonism reading all the literature on the subject. Chain collected 200 references on growth inhibitors caused by the action of bacteria and molds; according to the literature, however, nothing was known about the chemical nature of the inhibitory substances. Then one day in 1938, by sheer luck, Chain came across Fleming's paper on penicillin. Florey was already familiar with Fleming's work on penicillin, since he had been the *British Journal of Experimental Pathology* editor of Fleming's paper. Chain thought that perhaps penicillin was also an enzyme and was deserving of further study. Chain tells the story that he took some *Penicillium* mold back to his lab and did some preliminary experiments and then brought them to the attention of Florey. Florey disputes this and claims that the aim of the research was to study bacterial antagonism, and that they started with penicillin because he already knew it was potent against staphylococci.

With the lysozyme study coming to an end and 3 days after Britain's entry into the war on September 6, 1939, Florey asked for monetary support from the MRC (Medical Research Council) and the Rockefeller Foundation for studies of antibacterial substances, specifically penicillin, suggesting that there might be something of therapeutic value for humans. The MRC provided 25 pounds sterling and the possibility of 100 pounds sterling later, and the Rockefeller Foundation granted $5,000 (about 1,200 pounds sterling) per annum for 5 years. Now the penicillin project became a team effort.

Heatley's first task was to find a more productive means for growing the *Penicillium* and then to devise a quick way to assay for the antibacterial activity of penicillin. The assay he settled on was to make little wells in the agar in a petri dish and then to seed the dish with bacteria. He put whatever fluid to be tested into the wells, and then he incubated the dish so that the bacteria could grow. The distance the material diffused out of the well to inhibit the growth of the bacteria would be a measure of potency. Heatley then devised a better medium for the growth of the mold and was able to get greater production of penicillin. Chain found penicillin to be most stable from pH 5 to pH 8—close to neutrality. By mid-March of 1940, Heatley, through prodigious effort, was able to provide Chain with 100 mg of penicillin. From the spring of 1940 until early 1941 Heatley was able to improve on the extraction process by fashioning out of glass tubing an automated countercurrent apparatus. Once the filtrate was collected, it was acidified and then extracted with ether and then distilled. With the countercurrent apparatus, 12 liters of "mold juice" could be processed in an hour; the final product was freeze-dried to give a very nice brown powder with undiminished activity. The powder could be diluted a millionfold, and although quite impure, it was still able to inhibit bacterial growth.

Edward Abraham, another biochemist who was employed to step up the production of penicillin, used the newly developed technique of alumina column chromatography to remove the chemical impurities. When the filtrate was poured through the alumina column, four colored bands were seen: the top, brownish-orange band had little penicillin activity; the second, pale yellow band was 80% penicillin and free of contaminants that could cause fever; the third, orange layer had some or all of the fever-causing activity; and the bottom, brownish layer was full of impurities and had no penicillin activity. The yellow band was removed, washed in a neutral pH solution, and when diluted was a pale yellow color with a faint smell and a bitter taste. Eventually this product was used to produce crystals for X-ray crystallography. In 1945, Chain and Abraham, in collaboration with the X-ray crystallographer Dorothy Hodgkin, also at Oxford University, firmly established the structure of penicillin: a four-membered, highly labile β-lactam ring fused to a thiazolidine ring.

In the meantime, animal tests in mice showed that the penicillin was nontoxic. When a mouse was injected intravenously with the compound, the penicillin passed through the kidneys into the urine; a drop of urine using Heatley's well diffusion assay showed that the penicillin still had antibacterial activity. Chain said, "We knew that we had a substance that was non-toxic and not destroyed in the body and therefore was certain to act against bacteria in vivo." Penicillin now looked like a drug.

On May 25, 1940, there was enough penicillin, still <1% pure, to determine whether it could protect mice from an otherwise lethal bacterial infection. Eight mice were infected with virulent streptococci, and an hour later four of these were given the crude penicillin. All four untreated mice died within 16 h; all treated mice were alive and well the next day. Florey's remark, "It looks promising," was a typically laconic assessment of one of the most important experiments in medical history. The results of a large series of *in vitro* experiments with bacteria that cause gonorrhea, boils, diphtheria, anthrax, lobar pneumonia, tetanus, gas gangrene, and meningitis were published in the journal *Lancet* on August 24, 1940, in a two-page article, "Penicillin as a Chemotherapeutic Agent."

Florey tried to persuade British drug firms to produce enough penicillin to treat human infections; they were, however, already hard-pressed with other wartime needs. So he turned his own department into a small factory. Here again Heatley's technical ingenuity came into play. He was responsible for the design of ceramic bed-pans in which the *Penicillium* was grown with the help of "penicillin girls" specially employed for the purpose of preparing medium, adding spores, and decanting media.

By January 1941 there was enough penicillin for a limited trial on a patient. The first case was a 43-year-old retired policeman, Albert Alexander, who was near death's door from an overwhelming streptococcal and staphylococcal infection that resulted from a scratch while he was pruning a rose bush. He was given 200 mg of penicillin, followed by three doses of 100 mg every 3 h, and made a dramatic improvement over 24 h; after eight injections his temperature was reduced, and pus stopped oozing from the sores, but after 10 days the entire supply of penicillin (~4.4 g, or 200,000 units) was exhausted, and he relapsed and died. Subsequently, six other similar hopeless cases were treated by intravenous injection of penicillin; in all cases the results were conclusive: penicillin was able to overcome infections that ordinarily would be fatal. The results were published in the August 1941 issue of *The Lancet*.

In the summer of 1941, Heatley and Florey, with the support of the Rockefeller Foundation, traveled to the United States to see if they could interest U.S. pharmaceutical companies in producing penicillin on a large scale. Through the efforts and contacts of John Fulton at Yale University, they were put in contact with the Northern Regional Research Laboratory of the Department of Agriculture in Peoria, Illinois. By adding lactose instead of sucrose, they were able to induce the mold cultures to produce more penicillin; and then by using corn-steep liquor, a product of the corn wet-milling process, they were able to increase the yields further. It was recognized that the Oxford method (using Heatley's ceramic bedpans) of growing the mold on the surface of the nutrient medium was inefficient and that growth in a submerged culture would be superior. (In submerged cultures the mold is grown in large tanks and is constantly agitated and aerated.) Florey's isolate of *Penicillium* did not grow well in submerged culture, and so a more productive strain was sought. It came from a moldy (*Penicillium chrysogenum*) cantaloupe found in a Peoria market. In the United States the resultant purified penicillin product was named penicillin G.

To promote the development of penicillin, the U.S. government encouraged drug companies to collaborate without fear of potential antitrust violations. By February 1942, Merck and E.R. Squibb and Sons had signed an agreement to share research and production information. They also signed a joint ownership of their inventions to include other firms that made definite contributions, and in September, Charles Pfizer and Company joined the group. Later, Abbott Laboratories and Winthrop Chemical Company joined them. Eventually there were 21 other pharmaceutical companies. In the first 5 months of 1943, 400 million units of penicillin were produced in the United States—enough to treat about 180 severe cases; in the following 7 months, 20.5 billion units were produced. By the end of 1943 the production of penicillin was the second-highest priority at the War Department. Only the development of the atomic bomb was considered more important. In March 1944, Pfizer opened the first commercial large-scale production plant in Brooklyn, New York. By the end of 1945, Pfizer was making more than half of all the penicillin in the world. Between 1943 and 1945 the price per million units of penicillin—enough to treat one average case—dropped from $200 to $6. By 1949 the annual production in the United States was 133,229 billion units, and the price had dropped to less than 10 cents per unit.

Aftermath

In 1945 the Nobel Prize in Physiology or Medicine was awarded to Alexander Fleming, Ernst Chain, and Howard Florey for "the discovery of penicillin and its curative effect in various infectious diseases." The Nobel committee might have added that penicillin was the ideal nontoxic therapeutic in that it interfered with a specific chemical reaction necessary for the synthesis of the bacterial cell wall.

The committee awarding the Nobel Prize did not recognize the contributions made by Norman Heatley to the development of penicillin as a clinical drug. Heatley continued to work at the Dunn School, and in the 1950s he worked with Edward Abraham on cephalosporin, but as he confessed years later, no work ever matched the results with penicillin. He retired in 1976, and received no honorary degrees or awards save for a 1978 appointment to the Order of the British Empire. He died in Oxford at age 92.

The relationship of Ernst Chain with Howard Florey continued to deteriorate; in 1947, Chain went to Italy to give several lectures on penicillin, and when in 1948 he was invited to direct a research center at the Istituto Superiore di Sanità in Rome, he accepted; he did not return to the Dunn School for 30 years. In 1949 he was elected a Fellow of the Royal Society, and he received numerous honorary degrees and memberships in learned societies. In 1961 he was made professor of biochemistry at Imperial College, London; he was knighted in 1969, became a consultant to pharmaceutical companies, and was a valuable supporter of the Weizmann Institute in Israel. In 1971 he built a house in County Mayo, Ireland, where died in 1979 at age 73.

After the Nobel ceremony Florey returned to his laboratory. He was awarded honorary degrees from more than a dozen universities and honorary memberships in learned scientific societies. In 1944 he was made a Knight Bachelor, and in 1960 he was honored by the Royal Society of Medicine. Florey followed in the footsteps of his mentor Charles Sherrington when he was elected president of the Royal Society for 5 years, and in 1966 he was made provost of Oxford's Queen's College. He was made a life peer, to become Baron Florey, and was appointed a member of the Order of Merit, one of Britain's greatest honors. Even as his health declined, Florey continued to work until the evening of February 21, 1968, when at age 69 he died from a heart attack. In 1982 a commemorative plaque in his honor was put in the floor of the nave in Westminster Abbey, near the one for Charles Darwin.

Florey never wrote his memoirs. If he had, he might well have claimed that the greatest beneficiaries of penicillin were not the patients who were cured of their bacterial infections, but in actual fact Alexander Fleming. Indeed, without Florey's work, Fleming would have gone down as a somewhat eccentric microbiologist. After the Nobel ceremony Fleming's popularity soared, and over the next 10 years he averaged more than one tribute a month and was awarded more than a score of honorary degrees. He was made Lord Rector of Edinburgh University; received more than 50 medals, prizes, and decorations; and was an honorary member of more than 90 professional societies. Fleming retired in 1948, was an emeritus professor until 1954, and died suddenly from a heart attack on March 11, 1955.

Why did Fleming, who was an undistinguished metropolitan laboratory scientist until age 61, and whom Ronald Hare called a "third rate scientist," become such an international celebrity? It was, of course, because his name became synonymous with penicillin and the antibiotic revolution. The Fleming Myth, as it has been called, was promoted by the publicity machine at St. Mary's and by the efforts of Lord Beaverbrook, the press baron, who was a generous benefactor and patron of St. Mary's Hospital and who considered Fleming a genius; he felt it his duty to inform the world of Fleming's achievements so that Britain could bask in reflected glory. Charles Wilson, later Lord Moran, who was president of the Royal College of Physicians as well as Churchill's wartime physician, was also a supporter of Fleming's contribution. Edward Mellanby, secretary of the MRC and one of the most powerful men in Britain, was also a promoter of Fleming. Mellanby, a virulent anti-Semite, considered Chain a money-grubbing Jew and unfit for public acclaim. And because Florey was a gruff, rough colonial, he too, in Mellanby's eyes, did not qualify. Fleming, on the other hand, was the perfect hero, a modest, quiet Scotsman. For his part, Fleming did not make unreasonable claims about what he had discovered, but neither did he go out of his way to contradict some of the grossly exaggerated claims made by St. Mary's in the popular press.

Florey was a no-nonsense scientist who received far less public acclaim than Fleming and had little regard for Fleming or self-promotion. Henry Harris, who worked with Florey at Oxford University and succeeded him as professor of pathology, said: "Florey was a practical scientist, not a great one. He was not a seer, or a conceptualist, he was no Darwin or Pasteur or Einstein or Ehrlich and he never thought of himself in that league … he had one supreme virtue: he knew what had to be done and he got it done." Where Fleming gave up quickly, Florey persisted and came up with solutions. Florey, unlike the loner Fleming, was a team player and a great leader who, despite his temperament and habit of sometimes throwing off casual remarks without considering the seriousness with which others took them, inspired the members of his team to follow his example.

Antimicrobial Resistance

Perhaps the earliest notion that tolerance to a drug could result from prolonged use was that of the Turkish King Mithridates (119-63 B.C.), who "in order to protect himself from the effects of poison had the habit of taking such substances prophylactically with the result that, when ultimately defeated in battle by Pompey and in danger of capture by Rome, was unable to take his own life by poison and was obliged to perish by the sword." Although it was known since ancient times that drugs gradually lost their effects by repeated usage, the recognition of the problem of antimicrobial resistance (AMR) began with Paul Ehrlich's use of "magic bullets." Working with mice infected with trypanosomes (which caused the cattle disease nagana), he gave the mice a red aniline dye (trypan red) that was curative; i.e., the parasites disappeared from the blood. Shortly thereafter, however, they reappeared. Further treatment of infected mice showed the dye to be ineffective, with mice dying rapidly. The dye-treated strain of trypanosomes, when inoculated into a batch of normal mice, produced infections that even in the absence of drug were tolerant to the killing power of the dye, suggesting that a genetic change in the parasite had resulted in their renewed strength.

In the flush of the discovery of the antimicrobial properties of sulfa drugs and penicillin, some hoped that resistance could be circumvented. If one antibiotic did not work to cure the infectious disease, one could try another, and if that failed, another one, and so on *ad infinitum*. For a while that approach seemed to be the solution. But no more. From the 1960s through the early 1980s the pharmaceutical industry introduced many new antibiotics to solve the resistance problem, but after that the pipeline began to dry up and fewer and fewer new drugs were introduced. Today, the pharmaceutical industry no longer considers antibiotic development a wise economic investment. Because antibiotics are used for a relatively short time and are curative, they are not as profitable as drugs that treat chronic diseases such as diabetes, asthma, and psychiatric disorders. Furthermore, antibiotics tend to be priced low, and so the return on investment by the pharmaceutical company is less. This makes shareholders unhappy.

Here are some sobering facts. Each year in the United States at least 2 million people acquire a serious bacterial infection that is resistant to one of the antibiotics designed to treat that infection. At least 23,000 people die each year as a result of antibiotic-resistant infections. Many more die from conditions related to these antibiotic-resistant infections. The total cost of AMR to the U.S. economy has been estimated to be as high as $20 billion in direct health costs, and the estimated loss of productivity is close to $40 billion. AMR is not simply a personal problem; it is a problem of populations of people, and it is not confined to the United States. AMR is a worldwide problem, crossing international boundaries with remarkable speed. It poses a catastrophic threat to each and every person and affects all the countries of the world.

What underlies the appearance of AMR? Just as there is resiliency in our species to adapt to new environmental challenges, there is also a genetic resiliency in other species that enables them to survive the onslaught of a drug. Oftentimes resiliency lies in favorable mutations that permit an organism to survive an environmental threat. This capacity is then passed on to its offspring, i.e., survival of the fittest. The presence of the drug acts as a selective agent—a sieve, if you will—that culls those that are susceptible and allows only the resistant ones to pass through to the next generation. AMR is the result of natural selection: that is, those with a particular genetic makeup that are the most able to survive and reproduce pass on their genes to future generations, and in so doing increase in frequency over time. Antibiotic-resistant pathogens are not more virulent than susceptible ones; the same numbers of resistant and susceptible bacterial cells are required to produce disease, but the resistant forms are harder to destroy.

Sulfonamide-resistant *S. pyogenes* (the cause of strep throat, impetigo, rheumatic fever, and toxic shock syndrome) emerged in military hospitals in the 1930s. The successor to sulfonamide, penicillin, was successful in controlling bacterial infections in soldiers during World War II, but as early as 1945 Fleming warned that "the public will demand (penicillin) … and then will begin an era … of abuses." Indeed, there is a direct relationship between antibiotic consumption and the emergence and dissemination of resistant bacterial strains. Multiple-drug resistance was first detected in the 1950s and 1960s in *Escherichia coli* and *Salmonella*. When this occurred, it cost lives in developing countries; in the developed world, however, it was regarded as a curiosity and of little concern. But in the 1970s the attitude changed when *Haemophilus influenzae* (causing pneumonia and meningitis) and *Neisseria gonorrhoeae* (causing gonorrhea) emerged as resistant to ampicillin. And with *Haemophilus* there was also resistance to tetracycline and chloramphenicol. The timeline for developing AMR is seen in Table 11.1. Timeline of the discovery, introduction and year of resistance of antibiotics).

How AMR is acquired

There are more than 15 classes of antibiotics whose targets are essential for the growth and development of bacteria, and not one has escaped a resistance mechanism. The target of penicillin and its relatives is the inhibition of the synthesis of the protective bacterial cell wall, the tetracyclines inhibit bacterial protein synthesis by binding to ribo-

Table 11.1 Timeline of the discovery, introduction and year of resistance of antibiotics

Antibiotic class; example	Year of discovery	Year of introduction	Year resistance observed
Sulfa drugs: prontosil	1932	1936	1942
β-lactams: penicillin	1928	1938	1948
Aminoglycosides; streptomycin	1943	1946	1946
Chloramphenicols; chloramphenicol	1946	1948	1950
Macrolides; erythromycin	1948	1951	1955
Tetracyclines; chlortetractcline	1944	1952	1950
Rifamycins; rifampicin	1957	1958	1962
Glycopeptides; vancomycin	1953	1958	1960
Quinolones; ciprofloxacin	1961	1968	1968
Streptogramins; streptogramin B	1963	1998	1964
Oxazolidinones; linezolid	1955	2000	2001
Lipopetides; daptomycin	1986	2003	1987
Fidaxomicin; (targeting Clostridium difficile)	1948	2011	1977
Diarylquinolines; bedaquiline	1977	2012	2006

Modified from Lewis, K. Nature Revs. Drug Discovery May 2012.

somes, and the sulfonamides and fluoroquinolones inhibit bacterial DNA synthesis. The mechanisms by which the bacteria achieve resistance is also variable. In some cases the bacterium resistance gene codes for an efflux pump that ejects the drug from the bacterial cell, so that lethal levels are not achieved within the body of the bacterium; in other cases the bacterium overproduces the target (enzyme) so that the concentration of antibiotic is too low to overwhelm it, and in other cases the target is modified in such a way that binding of the antibiotic is reduced. And in some cases the resistance genes give rise to enzymes that degrade the antibiotic or chemically alter to render it functionless.

Drug resistance is mobile; that is, the genes for resistance traits can be transferred among different kinds of bacteria by mobile genetic elements. Chromosomal genes can be acquired by one bacterium through the uptake of naked DNA from another microbe, a process called transformation. It was by transformation that penicillin-resistant *Streptococcus pneumoniae* (cause of bacterial pneumonia and blood poisoning) arose by acquiring the genes from a naturally occurring, penicillin-resistant, commensal viridans group streptococcus.

Chromosomal mutants of *Staphylococcus aureus* (causing blood poisoning, wound infections, and pneumonia) developed intermediate resistance to vancomycin soon after its introduction in 1972, and higher levels were acquired by a transposon (a movable bit of DNA that can change the genetic makeup of a cell) from other gut bacteria.

Although resistance genes are generally directed against a single type of antibiotic, it is possible to have multiple genes, each with a single resistance trait, accumulate through mobile genetic elements such as plasmids (extrachromosomal loops of DNA) and transposons, as well as viruses (bacteriophages). These generally carry higher levels of resistance. Sequential mutations in the bacterial chromosome were responsible for the initial emergence of penicillin- and tetracycline-resistant *N. gonorrhoeae,* but later on it was through transposons that higher levels of resistance were achieved. It is a regrettable twist of fate for us that many bacteria play host to specialized transposons, called integrons, that act like flypaper in capturing new genes. These integrons can consist of several resistance genes that are passed to other bacteria as whole regiments of antibiotic-defying guerrillas.

What can be done about AMR?

There are four core actions that will help fight AMR. First, prevent infections and prevent the spread of resistance by avoiding infections in the first place; this reduces the amounts of antibiotics that have to be used and reduces the likelihood that resistance will develop. Prevention can be achieved through immunization, safe food preparation, hand washing, and using antibiotics only when necessary. Second, track resistant bacteria so that specific strategies can be developed to prevent resistant bacteria from spreading. Third, improve the use of today's antibiotics by slowing down the development and spread of AMR infections by stopping inappropriate and unnecessary use of antibiotics in people and animals. And fourth, promote the development of new antibiotics and develop new diagnostic tests for resistant bacteria.

Coda

The 19th century ushered in measures for protection from and treatment of plagues. Anesthesia, antisepsis, the germ theory of disease, and the discovery of "magic bullets" worked together to alter the way disease was looked upon, brought about a new view of how hygienic practices could be harnessed to improve the public health, and provided a stimulus for the rational development of therapeutic measures. By the 20th century, "wonder drugs" such as the antibiotics, as well as chemotherapeutic agents, were available. In concert with improvements in immunization, these plague protectors—imperfect as they were and as they continue to be—benefited the health of many, but regrettably not all of humankind. There is much work yet to be done, but our greatest hope is that the tragedies of the coming plagues can be mitigated by the rational use of drugs and vaccines.

doi: 10.1128/9781683670018.ch12

12
The Great Pox, Syphilis

It was the fall of 1932, and syphilis was rampant in small pockets of the American South. The U.S. Public Health Service began a study of the disease and enlisted 399 poor, black sharecroppers living in Macon County, Alabama, all with latent syphilis. Cooperation was obtained by offering financial incentives such as free burial service, on the condition that they agreed to an autopsy; the men were also given free physical exams, and a local county health nurse, Eunice Rivers, provided them with incidental medications such as "spring tonics" and aspirin whenever needed. The men (and their families) were not told they had syphilis; instead, they were told they had "bad blood," and annually a government doctor would take their blood pressure, listen to their hearts, obtain a blood sample, and advise them on their diet so that they could be helped with their "bad blood." These men were not told, however, that they would be deprived of treatment for their syphilis, and they were never provided with enough information to make anything like an informed decision. The men enrolled in the Tuskegee Syphilis Study (as it was formally called) were denied access to treatment for syphilis even after penicillin came into use (in 1947). They were left to degenerate under the ravages of tertiary syphilis. By the time the study was made public, largely through James Jones's book *Bad Blood* and the play *Miss Evers' Boys*, 28 men had died of syphilis, 100 others were dead of related complications, at least 40 wives had been infected, and 19 children had contracted the disease at birth (Fig. 12.1).

The Tuskegee Study was designed to document the natural history of syphilis, but for some it came to symbolize racism in medicine, ethical misconduct in medical research, paternalism by physicians, and government abuse of society's most vulnerable—the poor and uneducated. On May 16, 1997, the surviving participants in the study were invited to a White House ceremony, at which President Bill Clinton said: "The United States government did something that was wrong—deeply, profoundly,

Figure 12.1 (Left) African Americans enrolled in the Tuskegee Syphilis Study. Many would have joined the "Great Migration" to northern cities. Courtesy of Everett Historical, Shutterstock.com.

morally wrong. It was an outrage to our commitment to integrity and equality for all our citizens. Today all we can do is apologize, but you have the power. Only you have the power to forgive. Your presence here shows ... you have not withheld the power to forgive. I hope today and tomorrow every American will remember your lesson and live by it."

A Look Back

In 1996, England "celebrated" the 500th anniversary of the arrival of syphilis there. Syphilis—"the Great Pox," as the English called it—was a disease that from 1493 onwards swept over Europe and the rest of the world, including China, India, and Japan. The claim was made that this new disease was brought to Naples by the Spanish troops sent to support Alphonso II of Naples against the French king, Charles VIII. Charles VIII launched an invasion of Italy in 1494 and besieged the city of Naples in 1495. During the siege his troops, consisting of 30,000 mercenaries from Germany, Switzerland, England, Hungary, Poland, and Spain, as well as France, fell ill with the Great Pox, which forced their withdrawal. It is generally believed that with the disbanding and dispersal of the soldiers of Charles VIII, who themselves had been infected by the Neapolitan women, the pox spread rapidly through Europe. In the spring of 1496 some of these mercenaries joined Perkin Warbeck in Scotland, and with the support of James IV invaded England. The pox was evident in the invading troops. Within 5 years of its arrival in Europe the disease was epidemic: it was in Hungary and Russia by 1497 and in Africa and the Middle East a year later. The Portuguese carried it around the Cape of Good Hope in 1498 with the voyages to India of Vasco de Gama. It was in China by 1505, in Australia by 1515, and in Japan by 1569. European sailors carried the Great Pox to every continent save for Antarctica.

The French called this pox "the disease of Naples," blaming the Italians for it; the Italians called it "the French disease." The basis for these various names is that the undisciplined troops of Charles VIII of France, during their retreat from Italy, carried the disease back to their homelands in many parts of Europe, and shortly thereafter it came to be called after the national origins of those people who were disliked and considered unclean—the Russians called it "the Polish disease," in Japan it was called "the Chinese disease," and in England the "French pox" as well as "the Spanish disease."

Victims suffered with fevers, open sores, disfiguring scars, and disabling pains in the joints and experienced gruesome deaths, leading Joseph Grünbeck (1473-1532) of Germany to write in the late 15th century: "In recent times I have seen scourges, horrible sicknesses and many infirmities affect mankind from all corners of the earth. Amongst them has crept in, from the western shores of Gaul, a disease which is so cruel, so distressing, so appalling that until now nothing so horrifying, nothing more terrible or disgusting, has ever been known on this earth." In 1496, Albrecht Dürer made a woodcut of a mercenary whose skin presented with multiple chancres (Fig. 12.2). But if the story of the outbreak of this pox in the army of Charles VIII is accurate, how did his men contract this disease?

The principal hypothesis is that the disease came to the Old World from the New World (the Columbian hypothesis). Christopher Columbus (1451-1506) visited the Americas, and on October 12, 1492, arrived in San Salvador. He set sail for home 3 months later, on January 16, 1493, and arrived in Spain in March of 1493, carrying with him several natives of the West Indies. The crew of 44, upon arrival in Spain, disbanded, and some of these joined the army of Charles VIII. The first mention of the disease occurs in an edict by Emperor Maximilian of the Holy Roman Empire at the Diet of Worms in 1495, where it is referred to as "the evil pox." Twenty-five years later Francisco López de Villalobos, in a book published in Venice, claimed that syphilis had been imported into Europe from the Americas. Favoring this idea was the severity of the outbreak—indicative of a new import. Indeed, from 1494 to 1516 the first signs were described as genital

Figure 12.2 Albrecht Durer's 1496 illustration of the syphilitic man.

ulcers, followed by a rash, and then the disease spread throughout the body, affecting the gums, palate, uvula, jaw, and tonsils, and eventually destroying these organs. The victims suffered pains in the muscles, and there was early death—clearly an acute disease. From 1516 to 1526 two new symptoms were added to the list: bone inflammation and hard pustules. Between 1526 and 1560 the severity of symptoms diminished, and thereafter its lethal effects continued to decline, but from 1560 to 1610 there was another sign: ringing in the ears. By the 1600s the Great Pox was an extremely dangerous infection, but those who were afflicted did not suffer the acute attacks that had been seen in the 1500s.

In the 1700s syphilis was a dangerous but not an explosive infection. By the end of the 1800s both its virulence and the number of cases declined. Even so, the numbers were by no means trivial: by the end of the 19th century it was estimated that 10% of the population of Europe was infected, and by the early 20th century mental institutions noted that a third of all patients could trace their neurological symptoms to syphilis.

Clearly either the people were developing an increased resistance or the disease's pathogenicity was changing.

What produced this dramatic outbreak of syphilis? According to the Columbian hypothesis, syphilis was a disease acquired in the New World and introduced into a naive Old World population, and with the increased rate of transmission by sexual means, it was transformed from what once had been a milder disease into a highly virulent one (such as happened with HIV). What is the evidence for this hypothesis? Perhaps the best evidence for the origin of syphilis lies in the bones and teeth: bone lesions—scrimshaw patterns and saber thickenings on the lower limbs of adults and notched teeth in children—are diagnostic of syphilis and have been found in the skeletal remains of Amerindians. No such characteristic bone lesions were found in skeletons in China that were dated to be older than the 1500s, or in Egyptian mummies. Hippocrates and Galen also make no mention of the disease, and so it appears that syphilis was not present in ancient Greece or in the Roman Empire.

In the 20th century, however, some researchers hypothesized that syphilis may have existed in Europe long before the Neapolitan outbreak (the pre-Columbian hypothesis) but that it was not recognized as a distinct disease until the Renaissance, and prior to 1495 it was confused with other infections such as leprosy. Indeed, in support of the pre-Columbian hypothesis, there have been a steady trickle of reports describing Old World skeletons with bone lesions that predate Columbus's New World voyages. In 2011 that hypothesis was questioned, however, by the paleoanthropologist Kristin N. Harper and colleagues, who examined 54 published reports of syphilis in pre-Columbian Old World skeletons and did not find a single case that met stringent and standardized criteria for syphilis. They contended that many of these older reports used nonspeciifc indicators to diagnose syphilis, that the photographs of the lesions were not of high quality, and the dating of the specimens was suspect. In summary, they noted that "it appears that solid evidence supporting an Old World origin for the disease (syphilis) remains absent." A precise answer to the question of whether the origin of the Great Pox was pre-Columbian or Columbian may come only when the bones of Columbus and members of his crew are found and can be shown to contain, by DNA analysis, the genetic signature of the germ of syphilis, but for the time being the evidence seems to favor a Columbian origin.

It has been claimed by those supporting the Columbian hypothesis that the disease we know today as syphilis originated in Africa in the form of yaws, a chronic disfiguring disease that affects the skin, bone, and cartilage transmitted by skin-to-skin contact with an infective lesion, allowing the bacterium to enter through a preexisting cut, bite, or scratch. Yaws then passed through Asia to North America, spinning off a mutation in the form of bejel, a chronic disease that causes mouth sores and destructive lumps in the bone, again with transmission by direct contact, via broken skin or contaminated hands, or indirectly by sharing drinking vessels and eating utensils. Bejel also passed through Asia into North America; it was in North America, however, that 8,000 years later another mutation took place, creating the causative agent of venereal syphilis.

Spirochete Discovered

What is the causative agent of the disease syphilis? Early observers believed syphilis was God's punishment for human sexual excesses. Public bathhouses were closed, and distrust arose between friends and lovers. One result was that Platonic love emerged as a social cult. Others believed that syphilis had an astrological basis. In 1484, Mars, Jupiter, and Saturn were in conjunction with Scorpio, the constellation most commonly associated with sexuality. As early as 1530, though, Girolamo Fracastoro (1483-1553) recognized that this disease was contagious, calling the disease "syphilis" after a fictitious shepherd, named Syphilis, who got it by cursing the gods. Fracastoro described the earliest stages of syphilis as small ulcers on the genitals followed by a skin rash; the pustules ulcerated, and the person suffered with a severe cough that eroded the palate. Sometimes the lips and eyes were eaten away, or ulcerated swellings appeared, and there were pains in the joints and muscles. Fracastoro theorized that syphilis was a result of "seeds of contagion," but for 400 years no one saw the "seeds" that caused the symptoms of the Great Pox. Then, in 1905, Fritz Schaudinn (1871-1906) and Erich Hoffmann (1869-1959), working in Germany, identified a slender, corkscrew-shaped bacterium, a spirochete (from *spiro*, meaning "coiled," and *chete*, meaning "hair"), in the syphilitic chancres (Fig. 12.3). When Hideyo Noguchi isolated the same bacterium from the brains of patients suffering with neurosyphilis paralysis (paresis) and insanity, it became clear that all the stages of the disease were linked to one kind of spirochete. Because of its shape, Schaudinn and Hoffmann called the microbe *Treponema* (*trep*, meaning "corkscrew," and *nema*, meaning "thread" in Latin), and because it stained so poorly, they named its kind *pallidum* (from the Latin *pallidus*, meaning "pale") (Fig. 12.3). The bacterium *T. pallidum* produces no spores, and it cannot be cultured on bacteriologic media. It can, however, be grown in experimental animals such as rabbits and guinea pigs, where it divides slowly, about once every 24 h. Human beings are the only natural host for *T. pallidum*. The spirochete is quite fragile and requires a moist environment. Other spirochetes related to *Treponema* cause relapsing fever and Lyme disease (*Borrelia burgdorferi*) (see p. 441).

One of the reasons it took so long for microbe hunters to identify the cause of syphilis was that it was confused (and associated) with another venereal disease, gonorrhea. Even the great anatomist and physician John Hunter of St. George's Hospital in London could not solve the puzzle.

It was 1748, and London was the center of the universe … they talked of … Voltaire, Blackstone was a judge, Chippendale a furniture maker. Samuel Johnson wrote … of the permanent and certain characteristics of the mind. The pox was sexually transmitted. The pox, if you were unlucky, could rot the organ of

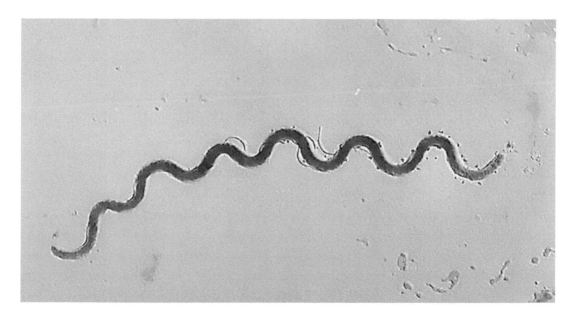

Figure 12.3 A photomicrograph of the spirochete *Treponema pallidum*. Courtesy CDC Public Health Image Library (14969); Susan Lindsley, 1972.

manhood. Some said the pox was one disease others said it was really two similar diseases. In John Hunter's mind … it wasn't transmitted by miasmas at all. The pox must be caused by … a putrid liquid of some sort. John Hunter had the pox all figured out. In his opinion there were two forms. If it began as a pimple on the penis, it took one specific course. The other form was seen when … the interior of the urethra was affected … and produced a liquid discharge. A patient with the pox stood before him … a tiny drop of yellowish liquid hung from the tip of his penis … it was a classic case … of the pox. To prove that the pox was a single disease all he needed to show was that the pus from this patient, a "wet" case, could produce a chancre, or a "dry" case in another penis. John only had one healthy penis handy, so he used it. While the patient stood before him, he took the droplet on a lancet and transferred it to the glans of his own penis. Then, through the droplet, he stabbed himself with the lancet … and again. He squeezed the small cuts open, so that the liquid … of the pox could take hold. He gave the man a bit of mercury … to rub … into his thighs. That Sunday he confided in his notebook that his penis began tingling … there was also a redness where the lancet had pierced the skin. By the following Tuesday there were two pimplelike chancres … A runny disease had produced a dry disease, proving, once and for all … the pox was but one disease. He rubbed mercury into his thighs and the chancres disappeared. Three months later he got a skin rash and … he rubbed a massive dose of mercury in his thighs and the symptoms, he wrote in his notebook, disappeared for good. The inevitable happened in 1793 … he died. John Hunter had been wrong. But it would be a century before microscopes focused on the gonococcus bacterium and syphilis spirochete. Only then did it become clear that a patient could have both at once and pass on both … or only one. They would discover that the form of the pox that seemed the worst to John Hunter, the wet one … was really less dangerous. It was gonorrhea. The other one that produced the chancre was syphilis. They would learn that syphilis had not two stages, but three. In its third stage it could attack a variety of organs, including the heart, brain, spinal cord—and aorta. John Hunter had died of tertiary syphilis—and by his own hand.

The Disease Syphilis

In the past, diseases such as syphilis have been called venereal diseases, or VD—pertaining to the Roman goddess of love, Venus—because under most circumstances they are transmitted by sexual contact. Since love and sex are not equivalent, however, the classic venereal diseases—gonorrhea, genital herpes, lymphogranuloma inguinale (caused by *Chlamydia trachomatis*), HIV, and syphilis—are now referred to as sexually transmitted diseases, or STDs.

Since there is enormous variation in the disease symptoms, syphilis has been called "the Great Imitator." How do we know about the clinical course of a syphilitic infection? Between 1890 and 1910 the Oslo study examined 1,404 untreated Swedish patients. Seventy-nine percent of those with late-stage syphilis developed neurosyphilis, 16% had destructive ulcers (gummas), 35% had cardiovascular disease, and in 15% of men and 8% of women syphilis was the primary cause of death. In the Rosahn and Black-Schaffer autopsy study of cadavers between 1917 and 1940 at the Yale University School of Medicine, 380 out of 3,907 individuals showed evidence of syphilitic lesions, an astounding 1 in 10! Ninety of 156 individuals with syphilitic lesions died primarily as a result of the disease, and the authors concluded that syphilis significantly reduced their life spans. Between 1932 and 1972 the now infamous Tuskegee Syphilis Study was designed to track untreated syphilis in poor, rural, and uneducated black males in Macon County, Alabama, where the overall infection rate was 36%. The study involved 399 men with latent syphilis and a control group of 201 healthy men. Signs of cardiovascular disease were found in 46% of the syphilitics, and 24% had an inflammation of the aorta, whereas in healthy controls the figures were 24 and 5%, respectively. Bone disease was found in 13% of the syphilitics but in only 5% of the controls. The greatest differences were seen in the central nervous system: in the syphilitics 8% had signs of disease, compared to 2% in the controls. After 12 years the death rate in the syphilitics was 25%, versus 14% in controls; at 20 years it was 39 versus 26%; and at 30 years it was 59 versus 45%.

In the early (primary) stage of infection, ~21 days (range, 3 to 90 days) after the initial contact with *T. pallidum*, a painless, pea-sized ulcer appears at the site of spirochete inoculation—a chancre. The chancre is a local tissue reaction and can occur on the lips, fingers, or genitals. If untreated, the chancre usually disappears within 4 to 8 weeks, leaving a small, inconspicuous scar. The individual frequently does not notice either the chancre or the scar, although there may be lymphadenopathy. (Because the initial lesion is larger than that of smallpox, syphilis has been called the Great Pox.) At the early stage the infection can be spread by kissing or touching a person with active lesions on the lips, genitalia, or breasts and through breast milk.

The secondary, or disseminated, stage usually develops 2 to 12 weeks (mean, 6 weeks) after the chancre; this stage, however, may be delayed for more than a year. *T. pallidum* is present in all the tissues, but especially the blood, where there is a high level of syphilis antigen. Serologic tests (such as the Wassermann test, VDRL test, and rapid plasma reagin [RPR]) are positive. There is now a general tissue reaction—headache, sore throat, a mild fever, and in 90% of the cases a skin rash. The skin rash may be mistaken for measles or smallpox or some other skin disease. The highly infectious secondary stage does not last very long. Then the patient enters the early latent stage, in which he or she appears to be symptom free; i.e., there are no clinical signs. Indeed, the most dangerous time is during the early latent stage, for the infected but apparently disease-free individual can transmit the infection to others. Transmission occurs by blood transfusion only rarely, because of the low incidence of spirochetes in the blood and because they do not survive longer than 24 to 48 h under blood bank storage conditions.

Untreated, the infection continues to progress, and after about 2 years the late latent, or tertiary, stage develops. In tertiary syphilis there are still spirochetes present in the body, but the individual is no longer infectious. For all intents and purposes such individuals are not infectious through sexual contact 4 years after initial contact. Tertiary syphilis develops in a third of untreated individuals 10 to 25 years after initial infection. The disease then becomes chronic. In about one-fifth of the cases, destructive ulcers (gummas) appear in the skin, muscles, liver, lungs, and eyes. In a tenth of the cases, the heart is damaged and the aorta is inflamed. In severe cases, the aorta may rupture, causing death, as it did in the case of John Hunter. In 40% of untreated cases, the spinal cord and brain become involved, causing incomplete paralysis (paresis), complete paralysis, and/or insanity, accompanied by headaches, pains in the joints, impotence, and epileptic seizures. The remaining cases have asymptomatic neurosyphilis. Most untreated patients die within 5 years after showing the first signs of paresis and insanity.

"Catching" Syphilis

At the primary (chancre) stage, syphilis can be spread by kissing or touching a person with active lesions on the lips, genitalia, and breasts. During the secondary stage, which usually does not last very long, the skin lesions render the individual infectious. In the early latent stage, there are no clinical signs, though the individual remains infectious.

Syphilis can be transmitted from the mother to the developing fetus via the placental blood supply, resulting in congenital syphilis; this is most likely to occur when the mother is in an active stage of infection. If the mother is treated during the first 4 months of pregnancy, the fetus will not become infected. Fetal death or miscarriage usually does not occur until after the 4th month of pregnancy at the earliest. Repeated miscarriages after the 4th month are strongly suggestive, but not unequivocal proof, of syphilis. A diseased surviving child may go through the same symptoms as the mother, or there may be deformities, deafness, and blindness. Of particular significance is that some offspring who are congenitally infected may have Hutchinson's triad: deafness, impaired vision, and a telltale groove across peg-shaped teeth, first described in 1861 by the London physician Jonathan Hutchinson.

Syphilis and Its Social Context

The highly pathogenic form of *T. pallidum* caused devastating effects from the 1500s to the 1700s. Over a 200- to 300-year period, however, it changed from an acute, lethal infection to a chronic, corrosive disease. Cerebral damage due to syphilis caused the deaths or affected the lives of Charles VIII and Francis I, Pope Alexander Borgia, Benvenuto Cellini, Henri de Toulouse-Lautrec, Heinrich Heine, Franz Schubert, Peter the Great, Catherine the Great, Guy de Maupassant, and Al Capone. Randolph Churchill, the father of Winston, was a syphilitic, prompting the question: would Winston Churchill have been so driven had he not felt compelled to revive and restore his father's name? Ivan the Terrible (1530-1584), the first czar of Russia, was syphilitic, and his son Fedor was a congenital syphilitic. Syphilis may have also caused Ludwig von Beethoven's deafness.

Ivan, Grand Duke of Muscovy, was born in 1530 and ascended the throne upon the death of his father. For years he ruled wisely and humanely. In 1552 his wife Anastasia gave birth to a son, Dimitri, who died at age 6 months, probably of congenital syphilis. Nine months later Anastasia gave birth to another son, Ivan, and in 1558 to a third son, Fedor. It is suspected that Ivan was infected with syphilis prior to his marriage, when he was a notorious womanizer. When Anastasia died in 1560, Ivan drowned his grief in drink and debauchery. Ivan remarried in 1561, and his wife gave birth to a son, Vassili, who lived for only 5 weeks. There is evidence that by 1564 Ivan was suffering from neurosyphilis. From 1565 to 1584 Ivan engaged in a reign of terror. He tortured, flogged, burned, and boiled those he considered to be his enemies. Claiming a conspiracy, he had thousands of citizens of Novgorod flogged to death or roasted alive or drowned under the ice; public executions were also held in Moscow. Ivan and his son raped the widow and daughter of Prince Viskavati, whom they had hanged, and in 1581, during a fit of rage, he stabbed to death his own son, the Tsarevitch Ivan. Ivan died a gibbering idiot at age 54. This left the throne to the congenital idiot Fedor, who was incapable of ruling, and so the throne fell to Boris Godunov. After the death of Godunov in 1605, chaos gripped Russia, and order was not restored until the election of the first Romanov in 1613.

Henry VIII of England (1509-1547) is suspected of being a syphilitic. His first wife, Catherine of Aragon, gave birth to a child who died within a few days, and she had three more stillborns, all in late pregnancy. Anne Boleyn miscarried one child at 6 months and another at three and half months. Jane Seymour had one son, Edward, born in 1537, who died at age 15, possibly of tuberculosis or poisoning, though he also had a suspicious skin rash shortly before his death. Elizabeth I, born of Anne Boleyn, lived to age 69 but was shortsighted, and so was Mary Tudor, who was also deaf and had a rather large, flat nose, which discharged foul-smelling pus. Henry became sterile or impotent in his late forties, suggesting that he may have been syphilitic. In 1527, Henry's character began to change, as did that of Ivan the Terrible. He suffered from headaches, insomnia, sore throats, and an ulcer on his leg. He had a gumma on his nose. Any person who slandered his marriage to Anne Boleyn was guilty of treason and was sentenced to death by hanging and quartering while still alive. He was bent on terror and slaughtered the Lollards, Lutherans, Anabaptists, and Catholics. He executed the prior of the Charterhouse and all his monks, and in 1535 beheaded Thomas More and Bishop John Fisher. Although he could have divorced Anne, he preferred to put her to death and to declare her daughter a bastard. He dissolved the monasteries and hung monks and abbots who resisted or delayed implementation of his policies. Henry never decayed mentally but became profoundly obese. His failure to produce a male heir was the beginning of the end of the Tudor line. The firm and efficient Tudor dynasty gave way, perhaps as a result of syphilis, to the absolutism of the much weaker Stuarts, and England was wracked by civil war. Did Henry VIII actually have syphilis? We do not know for sure, but syphilis was certainly common during his reign, and the disabilities of his offspring and his behavior make it likely that he did not escape infection. The historian William McNeill has aptly noted, "The inability of royal and aristocratic leaders to give birth to healthy children accelerated social mobility making more room at the top than there would have been otherwise"—thanks to syphilis!

Diagnosing Syphilis

In the primary (chancre) stage, syphilis is rarely diagnosed by growth and isolation of *T. pallidum* itself. Instead, examination of fluid from the lesion by dark-field microscopy and clinical findings are the basis for disease determination. The Wassermann test is a serologic reaction that originally used extracts of tissues infected with *T. pallidum*; when, however, it was found that uninfected tissues reacted positively with syphilitic sera, these antibodies (called reaginic) were considered nonspecific. The nonspecific reaction appears to be due to the external waxy layer of the spirochete. Despite this, these cheap, easy-to-use, and quick serologic tests continue to play a role in screening patients, especially those in high-risk populations, as well as in evaluating the adequacy of treatment. The most commonly used tests are the VDRL and the RPR card test. Both become negative 1 year after successful treatment for primary-stage syphilis and 2 years after successful treatment of secondary-stage syphilis. Patients who are infected with both HIV and syphilis commonly have a serologic response with very high titers. Fluorescent antibody absorbed tests and hemagglutination assays are sometimes used to confirm a positive nontreponemal, i.e., a nonspecific, test.

Treating Syphilis

In 1909, Paul Ehrlich (1854-1915) developed an arsenical derivative, salvarsan, that was effective in reducing the severity of the disease. Salvarsan was not without its problems—it was toxic, required years of treatment, and was difficult to administer. Only 25% of patients ever completed the therapy. In her book *Out of Africa,* Isak Dinesen tells how she was diagnosed with syphilis, contracted from her husband, and then treated (apparently successfully) with salvarsan. Salvarsan remained the drug of choice until 1943, when penicillin was introduced. Penicillin G benzathine is currently the recommended treatment for syphilis and is the only recommended treatment for pregnant women infected or exposed to syphilis.

Syphilis and the Social Reformers

Although the advent of drug treatment has helped to significantly reduce the incidence of syphilis, the disease has by no means been eradicated. A great deal of the failure can be attributed to social attitudes toward syphilis in particular and to sexually transmitted diseases in general. When salvarsan treatment became available in the early 1900s, syphilis became the focus of social reformers, since to them the disease was a clear indicator of the breakdown of the home, morality, and marital fidelity, and could be ascribed to promiscuity, prostitution, and immigration. Something had to be done, the social reformers said, about ridding society of the disease and its carriers.

The social reformers focused on the impact of sexually transmitted diseases on the family and the person who was infected—generally a man or woman who had strayed from the path of moral virtue and consorted with prostitutes or people of ill repute. Such individuals in turn infected the innocent spouse and the children. The moral code at the time, i.e., during the Victorian era, held that only sex within marriage was socially sanctioned, and therefore the victims of syphilis were those who "deserved" the disease and had brought it upon themselves by immoral behavior. The disease was considered a punishment. This is the litany sometimes expressed about those suffering with AIDS: those infected deserve it because of their sexual activity. In the early 1900s a punitive attitude led public health officials to spread the idea that simple contact was sufficient to spread syphilis, that one could get it from a drinking cup; from touching doorknobs, pencils, or pens; or from sitting on a toilet seat. This was of course not true, but saying that so-called "innocent" transmission was possible allowed people to be socially "pure" even if they had contracted the disease. Physicians and politicians also used such information as a further caution to keep individuals from engaging in socially unacceptable behavior, and as a means of reducing immigration into the United States, because immigrants were supposedly spreading syphilis in America, either via prostitution or through casual contacts.

By the time of World War I concern about syphilis had reached an unprecedented high. The military draft and consequent medical examinations had revealed that 13% of those drafted were infected with either syphilis or gonorrhea, and this touched off a vigorous anti-VD campaign. The focus of the campaign was on prostitutes, who were seen as the source of infection of military personnel. Posters were used by the Army (Fig. 12.4), and the idea was that, as one federal official stated, "to drain a red-light district and thereby to destroy the breeding ground of syphilis and gonorrhea is as logical as to drain a swamp thereby destroying the breeding place of malaria and yellow fever." As a result, thousands of women living and working near military training sites were quarantined during the war.

Those in charge of the VD campaign wanted not only to stop the spread of disease but also to change sexual and social behavior. Soldiers should not just be cured of syphilis or not just adopt precautions to avoid getting it, whether they wanted to have sex or not. They were supposed to live a morally pure life, to avoid extramarital sex under any circumstances, and in doing so they would avoid contracting a sexual-

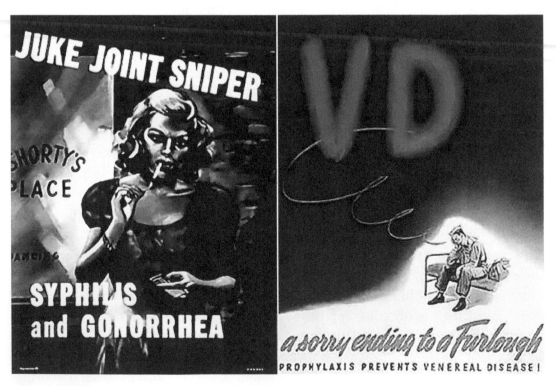

Figure 12.4 World War II posters warning of the perils of syphilis.

ly transmitted disease. Latex condoms were available during the time of World War I, but the military refused to provide them to the troops because it was believed this would contribute to their moral decline and encourage extramarital sexual behavior. The treatment for syphilis was made painful, and this was intended as a deterrent to contracting the disease. The U.S. Army also ruled that sexually transmitted diseases were injuries not incurred "in the line of duty," and so those who became infected lost their pay. The whole attitude was punitive: the infected soldier or sailor or marine had done something wrong, and the next best thing to telling them not to do it in the first place was to punish them afterwards.

As one might expect, the VD campaign was unsuccessful. Rates of venereal disease continued to stay at high levels or rose during and after WWI, and efforts to control syphilis lagged during the 1920s. In the 1930s, President Franklin D. Roosevelt appointed Thomas Parran as Surgeon General. Parran avoided taking a moral stance on the disease and instead concentrated on how to eradicate syphilis—an achievable goal, but one that was unpopular in some quarters. Parran developed a five-point program based on successful programs in Scandinavia. (i) Find the cases. To this end, diagnostic centers were established where there would be confidentiality and free testing to identify cases. (ii) Treat infected individuals promptly to prevent disease spread and to reduce virulence. (iii) Break the chain of transmission by tracing and treating sexual contacts. (iv) Institute mandatory premarital testing using the Wassermann test. (Wassermann testing for premarital screening turned out not to be effective because of false positives, and even a positive test did not indicate an active infection. Further, the tests were directed at the lowest-risk group, and this requirement assumed that those who wanted to marry would not do so if they tested positive, and that after marriage no one would acquire an infection. By the 1980s mandatory premarital Wassermann testing was dropped.) (v) Institute a massive public education campaign to inform people about syphilis, how it is contracted, its symptoms, and where to get treatment when infected. The heart of the campaign was provision of information: people were not told what to do or what not to do, and a moral position was not taken. The plan met with some opposition from the public and private sectors, since VD was something polite people were not supposed to discuss in public. Indeed, in 1934, when Parran was to make a radio broadcast on CBS about sexually transmitted diseases, he was allowed to do so only if he did not use the word "syphilis" or "gonorrhea." His will did eventually prevail, and the National Venereal Disease Control Act was passed in 1938. Congress allocated money for VD clinics, and testing and treatment were provided to those who could not afford it. During World War II, anti-syphilis efforts once again intensified when it was found that ~5% of the recruits had syphilis. Unlike in World War I, troops were provided with condoms, education, prophylaxis, and rapid treatment. If a soldier or sailor did not report the disease, however, it was a court-martial offense.

Distribution and Incidence of Syphilis

In the United States, after an outbreak in the 1990s, primary- and secondary-stage syphilis dropped to 2.1 cases per 100,000 people, the lowest rate since 1941. But that trend was beginning to reverse by 2002. According to the Centers for Disease Control and Prevention (CDC), the rate of primary- and secondary-stage syphilis increased by 15.1% from 2013 to 2014 to 6.3 cases per 100,000 people. The rate of congenital syphilis increased by 27.5% to 11.6 cases per 100,000 births. Researchers are still trying to work out why these increases are happening, but the 2014 CDC Sexually Transmitted Disease Surveillance Report offers a few clues. What is known is that some populations are especially at risk. Unlike the outbreak in the early 1990s that affected mainly heterosexual patients, the majority of syphilis cases in 2014 appeared in men who have sex with men (though syphilis rates also rose among women by 22.7%). It has been suggested that people might be less careful now that the threat of HIV/AIDS is less immediate than it was in the 1990s, or that partners might use strategies to prevent HIV transmission that aren't as effective for other STDs. Condoms, for instance, are a good precaution but not a reliable prevention method for syphilis, as the infectious sore might be on an area that remains uncovered.

Worldwide, in 1999 there were 12 million people infected with *T. pallidum*, with 90% of the cases in the developing world. In 2010 syphilis caused 113,000 deaths worldwide, and between 700,000 and 1.6 million pregancies ended in spontaneous abortions, stillbirths, and congenital defects because of syphilis.

Vaccines against Syphilis

What about a vaccine against syphilis? Some contend that the best hope for the control of syphilis is the development of a vaccine that prevents both disease and transmission. Although there have been many attempts to produce a successful syphilis vaccine by immunizing rabbits with whole killed or attenuated *T. pallidum*, only one immunization study, using multiple intravenous doses of gamma-irradiated *T. pallidum*, demonstrated complete protection against infectious challenges in the rabbit model. The protocol was very cumbersome and expensive, and impractical to test in humans. Although it has been shown that immunization with recombinant *T. pallidum* antigens can stimulate production of an immune response in the rabbit model, only partial protection was achieved, and there was no evidence of sterile immunity. The recent discovery of antigenic variation in *T. pallidum* makes the development of a protective vaccine even more formidable, however; studies are under way to test the ability of a cocktail of conserved regions of *T. pallidum* antigens to confer protection in the rabbit model.

Coda

Syphilis is more than a corrosive infectious disease. The history of syphilis is replete with discrimination against those with differing lifestyles and those who are marginalized in society—the urban poor, the uneducated, immigrants crowded into the cities, and those with alternative sexual orientations or other lifestyle factors. The terror of syphilis promoted the search for new drugs. Drug treatments, even those that have been shown to be successful, however, cannot be relied upon for complete eradication of STDs such as syphilis and gonorrhea. Syphilis has galvanized public health authorities to stem its spread through education, treatment, and avoidance of moral judgments and by attempts to blunt stigmatization. The incidence of syphilis, however, is a reflection of multiple factors: cultural beliefs and practices as well as economic and political forces. Its control will require comprehensive programs that are able to fit within the social and cultural dynamics of this sexually transmitted disease.

doi: 10.1128/9781683670018.ch13

13
The People's Plague: Tuberculosis

Disease and death are represented by Violetta in Giuseppe Verdi's opera *La Traviata* (1853) and by Mimi in Giacomo Puccini's opera *La Bohème* (1895). The young heroines are tall, thin, and pale-faced with cherry-red lips and flushed cheeks, and their voices are like those of the nightingale. But Mimi and Violetta are also mysteriously ill with a disease called consumption (from the Latin con, meaning "completely," and sumere, meaning "to take up"). To those living in the 19th century it seemed natural to link artistic talent to consumption, and Verdi and Puccini were well acquainted with this connection. Other composers and writers such as Keats, Shelley, the Brontës, Chopin, and Schiller were also consumptives. Consumption was characterized in an 1853 medical text as inducing the following features: nostalgia, depression, and excessive sexual indulgence. Indeed, at the time, it was believed that mental activity and artistic talent were stimulated by the poisons of this wasting disease.

Currently, consumption is more commonly called tuberculosis, or TB, but the pallor and emaciation of those afflicted gave it another name: the "white plague." In *Illness as a Metaphor* (1978), Susan Sontag wrote: "TB was—still is—thought to produce spells of euphoria, increased appetite, exacerbated sexual desire … Having TB was imagined to be an aphrodisiac, and to confer extraordinary powers of seduction." In the 1800s, when epidemic TB reached its peak in Western Europe, persons with TB were considered both beautiful and erotic: extreme thinness, long neck and hands, shining eyes, pale skin, and red cheeks. Yet such "beauty" had its price: a painful death by drowning in one's own blood. Because it was neither recognized nor understood that TB was a chronic infectious disease, it was romanticized.

The operas *La Traviata* and *La Bohème* were both based on Alexandre Dumas' 1848 semiautobiographical novel *The Lady of the Camellias* (made into the classic 1936 movie *Camille*, with Greta Garbo and Robert Taylor; Fig. 13.1). In the Dumas work the heroine (a courtesan) coughs blood onto her white handkerchief, which recalls the red-and-white colors of the camellia flower and symbolically represents

Figure 13.1 (Left) Movie poster for the 1936 tragic romance *Camille*.

her sexual availability at various times during the menstrual cycle. (The spitting up of blood was sanitized in the operatic versions.) A more accurate and less romantic description of the consumptive noted the incessant coughing, which made talking and eating almost impossible and breathing painful; loss of weight, which prevented walking; and the necessity to kill the pain, which required opium and whisky. By the time of death, emaciation was so complete that the individual resembled a cadaver. The romantic notion of TB in *La Traviata* (literally "The Fallen Woman") can be explained in part by the opera's composition before the cause of TB was discovered, in 1882. The themes of artistic genius and eroticism persist in *La Bohème* despite its later date, largely because "consumptive decline" was considered to result from a hereditary predisposition or specific living habits such as the debauched bohemian lifestyle, poverty, or sexual promiscuity. Although today it is known that this romanticized and morbid fascination with consumption was entirely without scientific basis, the linkage between disease, creativity, and eroticism persists: the 1996 rock musical *Rent* transplants Puccini's beloved bohemians to New York City, and despite giving them AIDS, rats, and roaches, composer Jonathan Larson passionately describes their hopes, losses, striving, death, and climactic resurrection.

A Look Back

TB is an ancient disease that has plagued humans throughout recorded history and even before. TB of the lungs (called pulmonary TB) is the form of the disease we are most familiar with, giving rise to the slang word "lunger." When localized to the lungs, TB can run an acute course, causing extensive destruction in a few months—so-called galloping consumption. It can also wax and wane, with periods of remission—mistaken in some cases for chronic bronchitis, which is also accompanied by the spitting up of blood. In 1839, Harriet Webster, the daughter of Noah Webster (the author of Webster's dictionary), wrote: "I began to cough and the first mouthful I knew from the look and feeling was blood ... I concluded to lie still and try what perfect quiet could do—swallowed two mouthfuls of blood and became convinced that if I could keep from further coughing I should be able to wait until morning without disturbing anyone. As soon as morning arrived, I looked at the contents of my cup. Alas, my fears were realized." She died 5 years later.

TB can affect organs other than the lungs, including the intestine and larynx, and sometimes the lymph nodes in the neck are affected, producing a swelling called scrofula (Fig. 13.2). ("Scrofula" comes from the Latin word for "pig," because the shape of the swollen neck looks like a little pig.) TB can also produce the fusion of the vertebrae and deformation of the spine (Fig. 13.3), called Pott's disease after Sir Percival Pott, who in 1779 described the condition. This may lead to a hunchback, and it may also affect the skin (when it is called lupus vulgaris) and the kidneys. TB of the adrenal cortex destroys adrenal function and results in Addison's disease, which probably killed Jane Austen.

The microbes that cause TB (and also leprosy) are called mycobacteria. Their free-living relatives inhabit the soil and water, where they fix nitrogen and degrade organic materials. Mycobacteria have a protective cell wall that is rich in unusual waxy lipids, such as mycolic acid, and polysaccharides, such as lipoarabomannan and arabinogalactan. Three mycobacteria, *Mycobacterium tuberculosis*, *Mycobacterium leprae*, and *Mycobacterium avium*, are human pathogens that respectively cause TB,

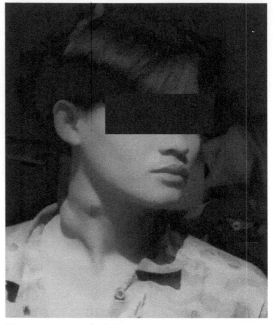

Figure 13.2 Scrofula in a young man. The enlarged lymph glands in the neck resemble a piglet and 'scrofula' in Latin means 'pig.'

leprosy, and a pulmonary disease with swollen glands in the neck. *M. avium* is an opportunistic infection found in immunocompromised people (e.g., those with AIDS) with fewer than 50 T4 cells/mm^3; its symptoms can include weight loss, fevers, chills, night sweats, abdominal pains, diarrhea, and overall weakness. Closely related to *M. tuberculosis* is a parasite of cattle, *Mycobacterium bovis*. Although *M. bovis* can infect people, it does so infrequently and with great difficulty. *M. bovis* grows under conditions where oxygen levels are low. When it does infect people, it is not associated with lung disease, whereas *M. tuberculosis* grows best when oxygen is plentiful and is associated with pulmonary TB, probably because the lung has

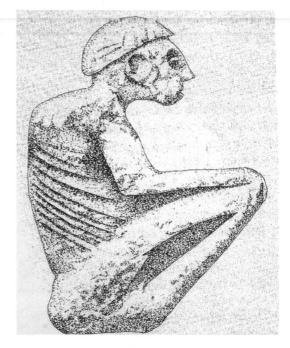

Figure 13.3 Tuberculosis of the spine (Potts disease) in a reconstruction of an Egyptian mummy.

high levels of oxygen. TB of the spine (Pott's disease), on the other hand, is associated with *M. bovis* and results from a blood infection that spreads to the spine via the lymph vessels. It has been hypothesized that *M. bovis* arose from soil bacteria and that humans first became infected with *M. bovis* by drinking milk. *M. tuberculosis*, on the other hand, is specific to humans and spreads from person to person through droplets of saliva and mucus. This airborne mode of transmission was favored, clearly, when humans settled down and established towns and cities. Genetically, *M. bovis* and *M. tuberculosis* have been shown to be >99.5% identical, and so the differences in their pathogenic nature are still to be explained.

Evidence of TB is found in bony remains that predate human writing. Pott's disease has been described from Egyptian mummies dating from 3700 to 1000 b.c. One of these mummies, from the Twenty-first Dynasty (1000 B.C.), discovered near the city of Thebes in 1891, is that of a priest who died with extensive destruction of the bones of the spine (Fig. 13.3). And there is DNA evidence of TB in vertebral samples from an Egyptian necropolis of Thebes dating from about 1450 to 500 B.C. Curiously, there is a paucity of evidence for pulmonary TB in any of the mummies from this period; in the tomb of a high priest of Ramses II, however, the body of a small boy has been found whose lungs had been stored, and these show the telltale signs of pulmonary TB. The burial site has been dated to 1000 to 400 B.C. Thus, it appears that tubercular disease of the lung is more recent than that of the bones. Historians of disease use this evidence to suggest that *M. tuberculosis* evolved from *M. bovis* after cattle were domesticated, between 8000 and 4000 B.C. Before this time the epidemic form of TB as we know it today did not occur. TB, it is believed, then spread to the Middle East, Greece, and India via nomadic tribes (Indo-Europeans) who were milk-drinking herdsmen who migrated, around 1500 B.C., from the forests of central and eastern Europe. A clay tablet in the library of the Assyrian king Ashurbanipal (ruled 668-626 B.C.) describes the disease: the patient coughs frequently; his sputum is thick and sometimes contains blood. His breathing is like a flute. His skin is cold. The Greek physician Hippocrates (460-400 B.C.) called the disease phthisis, meaning "to waste," and noted that the individual was emaciated and debilitated, had red cheeks, and that the condition was a cause of great suffering and death. Although Hippocrates believed that the disease was caused by evil air, he did not consider it contagious. Aristotle (384-322 B.C.), however, suggested that it might be contagious and the outcome of "bad and heavy breath." By the time of Galen (A.D. 129-200), however, the theory of contagion of phthisis came to be accepted in the Roman Empire, though the contagious agent could not be found.

During the Middle Ages (A.D. 500-1500) a feudal system developed in Europe in which a small elite (the nobility) ruled the rest of society, their subjects. Royalty claimed that their rights to rule and their talents were of divine origin, and they publicized this through claims of royal supernatural powers to heal disease, specifically scrofula. Kings and queens were able to heal those afflicted with scrofula by a simple touching (Fig. 13.4). Clovis of France (r. 481-511) and Edward the Confessor of England (r. 1042-1066) were the first kings endowed with this gift. Edward I of England (r. 1272-1307) touched 533 of his subjects in 1 month, Philip VI of Valois (r. 1328-1350) touched 1,500 in a single ceremony, and Charles II of England (r. 1660-1685) touched 92,102 people during his reign. On June 14, 1775, Louis XVI of France "ritually touched 2400 stinking sufferers from scrofula."

In Shakespeare's Macbeth (Act IV, Scene 3), an accurate description of the ceremony is given:

'Tis called the evil:
A most miraculous work in this good king;
Which often, since my here-remain in England,
I have seen him do. How he solicits heaven,

Figure 13.4 The royal touching to cure scrofula. Queen Anne is touching young Samuel Johnson to cure him of the 'king's evil.' Courtesy of Wikimedia Commons.

Himself best knows: but strangely visited people,
All swoln, and ulcerous, pitiful to the eye,
The mere despair of surgery, he cures,
Hanging a golden stamp about their necks,
Put on with holy prayers; and 'tis spoken,
To the succeeding royalty he leaves
The healing benediction. With this strange virtue,
He hath a heavenly gift of prophecy,
And sundry blessings hang about his throne,
That speak him full of grace.

During the ritual ceremony the king or queen touched the sufferer, made the sign of the cross, and provided the afflicted with a gold coin. In England this practice was known as the "king's touch" or the "royal touch," and it persisted until the early 18th century. One of the last sufferers of scrofula to be touched was the English writer, critic, and lexicographer Dr. Samuel Johnson, who was infected with TB by his wet nurse. Johnson was brought to Queen Anne (r. 1702-1714) at age 2 and touched; from all evidence it appears he was not cured by the touching (Fig. 13.4). But the crowd of sufferers anxious to be cured rushed to be touched, and several were trampled to death on this last occasion. Today, we know that royal touching had nothing whatsoever to do with curing scrofula; cure occurred in some cases because of the natural remission of the disease.

The word "tuberculosis" is of recent vintage and refers to the characteristic small knots or nodules, called "tubercles," found in the diseased lung. These were first described by Franciscus Sylvius in 1679; he also described their evolution into what he called lung ulcers. Almost all the great pathologists of his time, however, believed the disease was brought about by tumors or abnormal glands rather than by an infection. The first credible speculation on the infectious nature of TB was that of Benjamin Marten, who in 1772 proposed that the cause was an "animalcule or their seed" transmitted by the "Breath he [a consumptive] emits from his Lungs that may be caught by a sound Person."

In the 16th century the present epidemic wave of TB began in England, reaching its peak about 1780. At this time it was estimated that 20% of all deaths in England could be attributed to consumption. From there it spread to the rest of Western Europe, reaching a peak in the 1800s when La Traviata and La Bohème were written. Peaks of TB occurred in Eastern Europe between 1875 and 1880, and by 1900 it had reached North America.

The cause for this rise in TB may have been a demographic shift from rural to urban living, as well as the creation of "town dairies." These dairies were wooden buildings in the town center that housed dairy cows that formerly had been pastured;

now the tubercular cows were within the town, which provided ideal circumstances for animal-to-animal transmission as well as animal-to-human (zoonotic) transmission of TB. The result: a sharp rise of scrofula in the 17th century. Later, with a resurgence of trade, the walled towns of England continued to provide the means for human-to-human transmission. This was especially so when the textile industry became mechanized; this development led to a shift from a rural cottage industry to the more urban riverside sites where waterpower was available. Towns grew to become cities; as peasants streamed into these urban centers, people were more and more crowded together. These conditions are vividly shown in the paintings of Pieter Bruegel the Elder (1525-1569) and William Hogarth (1697-1764). Further, the practice in England of taxing a building partly based on its number of windows tended to affect building design by minimizing the number of windows, which enhanced the rebreathing of exhaled air of those living and working in such crowded and airless rooms. The increased density of people provided ideal conditions for the aerial transmission of *M. tuberculosis* and pulmonary TB. By the 19th century epidemic TB had raged for more than 2 centuries, and it was feared by some that TB might bring about the collapse of industrialized Europe as well as the end of all civilization.

John Bunyan wrote: "The Captain of all these men of death that came against him to take him away, was the consumption; for it was that that brought him down to the grave."

And an early-20th-century journalist described it: "Tuberculosis is a Plague in disguise. Its ravages are insidious, slow. They have never roused people to great, sweeping action. The Black Plague in London is ever remembered with horror. It lived one year; it killed fifty thousand. The Plague Consumption kills this year in Europe over a million; and this has been going on not for one year but for centuries. It is the Plague of all plagues—both in age and power—insidious, steady, unceasing."

To those living during the Victorian age (1837-1901), when the British Empire was at its height, TB was not only romantic but also attractive since the disease produced no obvious repulsive lesions, as did leprosy, its mycobacterial cousin. To the Victorians the blood in the sputum blended metaphorically with menstrual blood— and so in a strange way sickness and death were blended with eroticism and procreation. There is an eminent gallery of victims of TB, including Baruch Spinoza (1632-1677), Johann Wolfgang von Goethe (1749-1832), Friedrich Schiller (1759-1805), Fyodor Dostoevsky (1821-1881), Anton Chekhov (1860-1904), Sir Walter Scott (1771-1832), D. H. Lawrence (1885-1930), Percy Bysshe Shelley (1792-1822), John Keats (1795-1821), Alexander Pope (1688-1744), Samuel Johnson (1709-1784), Jean-Antoine Watteau (1684-1721), Niccolò Paganini (1782-1840), Elizabeth Barrett Browning (1806-1861), Igor Stravinsky (1882-1971), Robert Louis Stevenson (1850-1894), Edgar Allan

Poe (1809-1849), Franz Kafka (1883-1924), Amedeo Modigliani (1884-1920), Frederic Chopin (1810-1849), Henry David Thoreau (1817-1862), George Orwell (1903-1950), and Vivien Leigh (1913-1967). Keats died at age 25, and all six children of Reverend Brontë and his wife Maria (including Emily and Charlotte), as well as Percy Bysshe Shelley, died of "galloping consumption" before the age of 40. Some observers, having looked at this list of writers, composers, and artists, suggested that TB sparked genius, and one wrote: "Tuberculosis patients particularly young talented individuals ... display enormous intellectual capacity of the creative kind. Especially is this to be noted in those who are of artistic temperament, or who have a talent for imaginative writing. They are in a constant state of nervous irritability, but despite the fact that it hurts their physical condition, they keep on working and produce their best works." There is little scientific evidence, however, to show that TB had any real effect on the brain or on creativity.

The deaths of Vivien Leigh and Eleanor Roosevelt are less romantic and more tragic because they occurred after the disease agent had been identified and drugs for the cure were available. The actress Vivien Leigh developed TB in 1945, was hospitalized for a brief time, and recovered, but her physician recommended that she seek further treatment at a sanitarium. She refused. Although streptomycin and isoniazid became available to treat TB in the 1940s and 1950s, she again refused to avail herself of these cures. She died of TB at age 53. Eleanor Roosevelt was infected as a young woman and developed active TB at age 12, and then the disease went into remission. In the last 2 years of her life her health began to deteriorate, and she developed miliary TB (so called because the small tubercles in the lungs look like millet seeds and these spread throughout the body via the bloodstream). It was too late for treatment, and she died from disseminated TB in 1962 at age 78.

Some contend that TB was brought to the Americas by European explorers and settlers, though this appears not be true. An Inca mummy of an 8-year-old boy who lived in about A.D. 700—more than 700 years before Columbus arrived in the Americas—shows clear evidence of Pott's disease, and the lesions in the spine contain mycobacteria, most probably *M. bovis*. Domestic herds of cattle, as well as wild herbivores, probably served as the source of infection. In 1999 investigators found *M. bovis* in bone tissue removed from a 17,000-year-old bison that had fallen to its death in the Natural Trap Cave of Wyoming, providing clear evidence that TB was already present in prehistoric America, waiting for new human hosts. With urbanization, immigration of infected individuals, higher population densities, and poor hygiene, the conditions favored the spread of TB in the New World. Its first peak was in the early 19th century, and TB was especially prevalent in the cities along the Atlantic coast. Mark Caldwell, in his book The Last Crusade, described it: "Through its crowds and bad air made the city a crucible of TB, it was also the place where the poor congregated, where oppressive economic conditions prevailed, where filth and ugliness constantly assaulted the senses." In 1804 in New York City, a quarter of all deaths were due to consumption, and between 1812 and 1821, Boston had a similar fatality rate. TB did not affect all segments of the U.S. population equally. In 1850, African-Americans in Baltimore and New York City had higher death rates than did whites. In Baltimore, after the age of 15, the number of female deaths was twice that of males, but in New York and London it was the reverse. The reasons for this are not clear. TB was not strictly an urban disease; it was also present in rural areas. The critical element was found to be not the total population but the size of the household. In colonial America in the 18th century, irrespective of the size of the town or city, a typical dwelling had 7 to 10 inhabitants. Such crowding facilitated household transmission. Inefficient heating of the home usually led to a sealing of the windows and doors in winter to keep out the cold, and so transmission was enhanced. Further, behavioral patterns also contributed to the spread of pulmonary TB: caretakers of the sick frequently slept in the same bed as their patients, and physicians advised against the opening of windows in rooms where there were those sick with consumption.

Inadequate ventilation was also a contributing factor in the spread of urban TB. For example, during the 19th century most tenements were built with little concern for proper ventilation. As more and more impoverished immigrants arrived in the United States (especially during the 1830s and 1840s), they were forced to live crowded together in these miserable tenements. In 1800 a Boston apartment typically had an average of 8 people, and by 1845 the figure was over 10. The generally filthy conditions, as well as lack of ventilation characteristic of tenement housing, surely played a key role in the rise in the incidence of pulmonary TB in the United States.

In the urban centers of New York and Boston, consumption came to be regarded as "a Jewish disease" or the "tailors' disease " because so many young immigrant Jews who were consumptives earned their living in the garment industry, cutting, sewing, and stitching. Though TB was peculiar neither to Jews nor to those who worked in tailoring, they continued to be stigmatized as carriers of TB. During the 1920s, American nativists became concerned over the millions of emaciated and undersized Jewish immigrants who were lacking in the physical vitality so characteristic of the sturdy and robust Anglo-American stock, and who were now pouring into the cities along the Atlantic seaboard. They contended that because Jews were highly susceptible to TB, not only were these sickly Jews racially inferior but they would soon become public charges. Data gathered by the Jewish community refuted the claim that Jews were especially prone to TB. It was a fact, however, that TB did flourish among Jews (as it did among non-Jews) who lived in crowded and unhealthy conditions in the poorest parts of the city. Although the rates of consumption among Jews in both Europe and the United States were actually found to be lower than those of non-Jews, this did little to change the public perception. In effect, TB was used as a tool of anti-Semitism that justified the ostracism and persecution of Jews.

In the 1900s poorer people tended to have a higher mortality; the greatest number of deaths occurred between the ages of 15 and 45, but there was also a minor peak between 5 and 10 years of age. Native Americans, who were highly susceptible to TB, were virtually killed off by being herded together on reservations. Indeed, between 1911 and 1920, 26 to 35% of all deaths occurring among Native Americans were due to TB, and an examination of 600,000 Native Americans during 1920 showed that 36% had TB. TB was virtually unknown in sub-Saharan Africa until the beginning of the 20th century and was not found where there had been little or no European immigration.

Some believed TB to be an act of God afflicting both rich and poor, against which there was no defense. Others were convinced that it was a result of bad air that was present in the crowded and dirty cities. Many consumptives sought refuge in warmer or milder climates. In the United States there was a move to the Sunbelt of the West (Arizona, southern California, New Mexico, and Texas), while in Europe it was to the Mediterranean or South Africa or the Alps. Ocean voyages were also recommended for those with TB because of the slow tempo and the clean air, especially on cruise ships. Living at high altitudes where the air was unpolluted and crisp was considered beneficial, and so health seekers founded communities in Colorado. Some recovered, but others did not, and so clearly a change in scene did not cure. Though TB was not an environmental disease that could be cured by sunshine or fresh air, this view remained pervasive. Indeed, it was a critical factor in establishing the political map of Africa. Cecil Rhodes (1853-1902), who did more than any man of his time to enlarge the British Empire, was an ambitious and ruthless colonialist. He created the state of Rhodesia (now Zimbabwe and Zambia) by forcing the annexation of Bechuanaland (now Malawi and Botswana) and by acquiring the lands of the Matebele tribe by devious means. As the leading official of the British South Africa Company, he consolidated all the diamond mines in the Kimberley area into the De Beers Mines. This created a monopoly over the supply of diamonds, and in the process Rhodes became extremely wealthy. After his death, part of his fortune was used to establish the Rhodes scholarship at Oxford University. All this was because Rhodes, at age 16, was stricken with TB while attending Oxford University. To recuperate, he joined his brother, a cotton grower in Natal, South Africa. Later, the two became involved in the diamond mining business in an inland area now known as Kimberley. Upon regaining his strength, he returned to England and received his degree from Oxford in 1873. But back at Oxford he suffered a relapse and was told he had only 6 months to live. Rhodes was unable to resist the lure of Africa's riches and climate, and so he returned. In 1888 he tricked the Matebele ruler (Lobengula) into an agreement whereby Rhodes secured additional diamond mining concessions and was soon able to get complete control of the territory. Through the British South African Company he secured a monopoly of the Kimberley diamond production and created the De Beers Consolidated Mines.

In 1881, Rhodes became a member of the Parliament of the Cape Colony, a seat he held for the remainder of his life. In Parliament he favored a policy of containing the northward expansion of the Dutch Boers (Afrikaners) in the Transvaal Republic. In 1885, largely due to his persuasion, Great Britain established a protectorate over Bechuanaland. By 1890, Rhodes had become the prime minister and a virtual dictator of the British Cape Colony. He planned and promoted a Cape-to-Cairo railroad—from the Cape of Good Hope to Cairo, Egypt—dreaming that one day the map of Africa would be "painted red," which meant controlled by Great Britain (at the time mapmakers used the color red to show British territories). At first the Transvaal was of little interest to anyone, but when gold was discovered there in 1887, the prospect of wealth lured tens of thousands of non-Boers, many of them English, into the Transvaal to seek their fortunes. Transvaal's president, Paul Kruger (after whom the Krugerrand, the gold bullion coin, is named), refused to grant political rights to these foreigners. Rhodes's personal and business sympathies were with the British, who were living in and being discriminated against in the Transvaal, and in 1895 he saw this as an opportunity to overthrow the Boer-dominated government. He organized a raid on the Transvaal by Rhodesian troops, with the aim of triggering an insurrection against the Kruger government, but the raid (called the Jameson Raid after its leader, Leander Jameson) was poorly planned and executed, and it failed. As a result, Rhodes was censured by the British government and forced to resign as prime minister (1896). Relations between the British and Dutch, however, continued to deteriorate, and by 1899 their animosity resulted in the outbreak of the Boer War. Rhodes was in Kimberley at the time and was trapped there during a 4-month siege of the town by 5,000 Boers. Here he played an important role in directing the war and in building morale for those defending Kimberley, but 2 months before the war ended he had a heart attack and died at age 48. At the time of his death Rhodes had been instrumental in bringing almost 1 million square miles of Africa under British domination. The country, called Rhodesia, declared its independence from Britain in 1965; northern Rhodesia is present-day Zambia, and southern Rhodesia is now Zimbabwe.

334 | The Power of Plagues

The Germ of TB

In 1865 a French military physician, Jean-Antoine Villemin (1827-1892), succeeded in transmitting TB to rabbits. Villemin recovered pus from the lung cavity of a tuberculous patient who had died 33 h earlier and injected this under the skin of two healthy rabbits. Two other rabbits served as controls and were injected with tissue fluid from a burn blister. Some months later, when the rabbits were examined, Villemin found TB only in the lungs and the lymph nodes of those rabbits that had received the pus. Although it would seem that Villemin's demonstration of transmission would be proof enough of the contagious nature of TB, his announcement before the prestigious French Academy of Medicine was greeted with derision. His severest critic was Hermann Pidoux, who said that consumption in the poor was due to conditions of poverty, including overwork, malnutrition, unsanitary housing, and other deprivations, and when consumption developed among the rich Pidoux claimed it was brought about by their overindulgence in their wealth, laziness, flabbiness, overeating, excessive ambition, and habits of luxury. In both cases the result was, in Pidoux's words, "organic depletion." Such blind rejection of the infectious nature of TB remained until March 24, 1882, when the German physician Robert Koch (1843-1910) made a presentation before a skeptical audience of Germany's most prestigious scientific group, the Berlin Physiological Society, in which he claimed to have discovered the microbe of TB. Koch's announcement not only resolved the contagion argument but also resulted in a shift of scientific preeminence from France (and Louis Pasteur) to Germany. This shift took place for several reasons, including differences in the style of science. French scientists relied on intellectual deduction and tended to proclaim general laws from a limited set of experiments, whereas German scientists were more methodical, their approach was grounded more in reality, and they relied on repetitive experiments. In addition, after the defeat of the French in the Franco-Prussian War (1870-1872), the German economy and its science flourished, while post-Napoleonic France declined both in its wealth and in its scientific glory.

Koch's identification of *M. tuberculosis* was not so simple: the microscopic bacillus is colorless and unusually difficult to stain because of its waxy cell wall, and therefore it cannot be easily seen with the light microscope. To make it visible required heating and a special aniline dye (methylene blue). Later, the staining technique was improved upon by Paul Ehrlich, who had heard Koch speak in 1882; Ehrlich found that the bacilli would retain the dye color if they were first stained with red fuchsin followed by an acid wash; these stained bacilli were, in other words, acid fast (Fig. 13.5a). Koch also devised a method for growing the bacillus outside the body (by culturing it in test tubes containing coagulated serum as a nutrient source), and was able to use such pure cultures of tubercle bacilli to infect guinea pigs. It was methodical Koch,

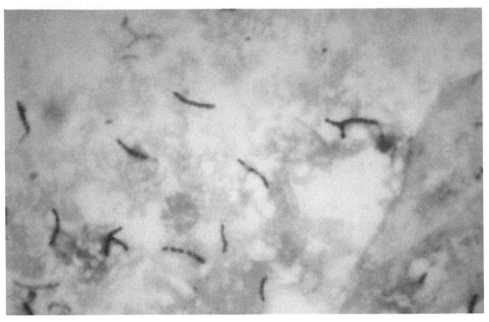

Figure 13.5a Stained tubercle bacilli in sputum as seen with the light microscope. Courtesy CDC Public Health Image Library (5789), Dr George Kubica, 1979.

the German, who, though he validated the imaginative approach of the Frenchman Villemin, ignored his findings and thus relegated him to obscurity.

In 1890, under pressure from the Prussian government and its leader, Otto von Bismarck, Koch forsook careful experimentation and prematurely announced that he had discovered a protective substance made from an extract of the bacillus called tuberculin (known today as PPD [purified protein derivative]). When injected into animals, tuberculin produced fever, malaise, and signs of illness. Koch believed that tuberculin sensitized the animal and effected cure. It was soon found that it did not. Indeed, instead of bringing about a cure, tuberculin turned out to be dangerous and sometimes lethal. Tubercle bacilli sensitize the body to tuberculin, so that when it is injected in sufficient quantities into a tuberculous individual, it can kill via a delayed hypersensitivity reaction. Yet Koch's tuberculin did serve a practical purpose as a diagnostic test. Even today it remains one of the most useful methods for determining previous exposure to TB).

Koch was also able to show, as had Villemin a decade earlier, that bovine TB (caused by *M. bovis*) could be infectious for humans, principally via the milk. At this time, when infected milk was recognized as a source of human disease, the tuberculin skin test became useful as a means for screening infected cows. Cows that had a positive test were killed. This meant that valuable animals would be taken out of production, and the losses could be devastating to farmers who had more than a few tuberculin-positive animals. Slaughter was unnecessary, however, since pasteurization of milk rendered it noninfective and protected those who drank such milk. Opposition to this slaughter-based control was vehement, and sometimes protection in the form of armed troops was needed to provide for the safety of the veterinarians who were involved in the testing in rural dairies where pasteurization was not employed. In 1932 there was a "Cow War" in Iowa when 400 angry farmers destroyed the cars of the veterinarians; ultimately martial law had to be declared. By the end of the 1930s more than 95% of the counties in the United States were declared to be free of infected cattle, and the threat of milk-borne TB faded away. The problem has not disappeared completely, though, since bovine TB continues in other countries, including Mexico.

The Disease of TB

Charles Dickens, in his 1879 novel Nicholas Nickelby, described TB:

> There is a dread disease, which so prepares its victims, as it were, for death ... a dread disease. In which the struggle between soul and body is so gradual, quiet, solemn, and the results so sure, that day by day, and grain by grain, the mortal part wastes and withers away, so that the spirit grows light ... a disease in which death and life are so strangely blended that death takes the glow and hue of life, and life the gaunt and grisly form of death—a disease which medicine never cured, wealth warded off, or poverty could boast exemption from—which sometimes moves in giant strides, or sometimes at a tardy sluggish pace, but slow or quick, is ever sure and certain.

Dickens could characterize TB, but he did not know that the principal risk behavior for contracting it was breathing. Persons with pulmonary TB may infect others through airborne transmission: coughing, sneezing, and speaking. A sneeze might contain an aerosol with a million microscopic (~10-µm) drops that evaporate slowly, producing floating droplet nuclei. TB bacilli, slightly bent microscopic rods ~2 to 4 µm in length and 0.3 µm in width (see Fig. 13.5b) and enclosed within droplet nuclei, move from place to place by riding on the gentle air currents. The bacteria-laden droplets can easily be inhaled. Tubercle bacilli are rather robust and can survive in moist sputum for 6 to 8 months. Inhalation is the major route of infection. Oral infection, through eating or drinking, is less efficient because the bacteria rarely survive passage through the acid-containing stomach. Infection may result if as few as 5 bacteria reach the grape-like clusters of the thin-walled air sacs (alveoli) of the lung, where the exchange of oxygen for carbon dioxide takes place. In the United States the lung is the primary site of infection in 80 to 85% of the cases.

Once in the alveolus, macrophages engulf the bacteria to initiate an infection. From here the bacilli within the macrophage are transported to other parts of the body by the lymph channels. For the first few weeks within the tissues of a susceptible individual (one who has not been tuberculous before), the bacilli multiply slowly, dividing once every 15 to 24 h. At first there is little damage or reaction, but after several more weeks of microbial multiplication, either at the initial site or at a more distant site, there is an inflammatory response; this becomes more intense, and fluid (lymph) leaks into the region. The site becomes infiltrated with fiber-secreting cells called fibroblasts that surround the free and macrophage-enclosed bacilli in an attempt to wall them off. A microscopic tubercle results. The tubercle grows in size, pushing aside normal tissue and producing the characteristic TB lesion—larger visible tubercles in the lung—and the tuberculin test is positive.

Primary TB is a self-limiting infection that often goes unnoticed—it appears to be merely a cold, and usually there is little impairment of lung function. If there is a protective immune response, which occurs in 85 to 90% of the cases, the disease may progress no further, calcification of the tubercles may take place, and the tuberculin test remains positive. In ~5% of cases, however, such a latent infection may become

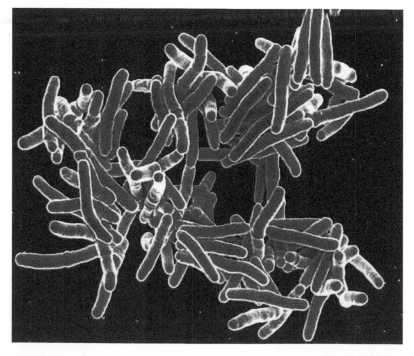

Figure 13.5b Digitally colorized scanning electron microscope (SEM) image of rod-shaped *Mycobacterium tuberculosis*. Courtesy NIAID, 2010.

active within 2 years of the primary infection, and another 5% will have active disease at a later point in time. This reactivation of infection occurs because some of the tubercle bacilli, even those within macrophages, are not killed. The tubercles can also break open to disseminate the bacilli via the blood; small alveolar blood vessels are eroded and rupture, hemorrhages occur, the bronchi can become irritated, there is coughing, the sputum becomes streaked with blood, fluid fills the lung, and breathing becomes more difficult. Now acid-fast M. tuberculosis can be found in the sputum. Reactivation of the infection results in the destruction of more lung tissue, producing a cheese-like consistency in which the bacteria survive but cannot multiply because of low levels of oxygen and acidity. Coughing, pallor, spitting of blood, night sweats, and painful breathing signify the spread of the disease. Chest X rays show tubercles, and because of fluid in the lung, a stethoscope placed on the chest reveals sounds of gurgling and slush. Continued inflammation, unfortunately, results in liquefaction of the lung tissue, and this oxygen-rich environment provides a rich growth medium; in some cases there can be >10 billion bacilli/ml. (It is a bitter irony that the tubercle bacillus flourishes in the apex of the lung, where oxygen is plentiful, and that fresh air, rich in oxygen, was believed to be curative for consumptives.) Over time the softened and liquefied lung contents are forced out of the lesion into the blood vessels to become disseminated throughout the body, and this is how new disease spreads to other parts of the body.

In TB it is important to distinguish between infection and disease. In the vast majority of cases in which the TB bacteria are inhaled, the bacilli are killed by macrophages or they are localized and grow slowly within tubercles. Although M. tuberculosis can grow and divide outside of the cell, it survives predominantly within macrophages; it can live within the macrophage because its waxy lipid coat makes it impervious to the killing mechanisms of the macrophage. Despite the formation of antibodies during a TB infection, these are unable to limit the disease, because the bacilli remain hidden within the macrophage. Cell-mediated immunity (see p. 261), especially that involving CD4 T helper cells (lymphocytes), is critical for disease arrest. The CD4 T cells produce cytokines, especially gamma interferon and tumor necrosis factor, that activate macrophages to kill or limit M. tuberculosis growth. The lesion becomes infiltrated with lymphocytes and macrophages, and a delayed hypersensitivity reaction, similar to that experienced with a bee sting, occurs. In TB, however, cell-mediated immunity is a two-edged sword—it is required for protection but is also involved in tissue damage—disease. In addition to the CD4 T helper lymphocytes, there are CD8 T cells (killer lymphocytes) that participate in the immune response; these produce gamma interferon and tumor necrosis factor, which activate macrophages, and in addition these T cells punch holes in the membrane of the macrophage by release of a molecule called perforin. Many macrophages are also killed by tuberculin-like products, causing them to release reactive oxygen species (hydrogen peroxide, hydroxyl radicals, and superoxide) and protein-degrading enzymes (proteases) that are detrimental to the host tissues. In addition, the products of the surviving tubercle bacilli activate suppressor T cells that depress the delayed hypersensitivity reaction as well as cell-mediated macrophage killing, and so the tubercle bacilli are able to spread to other tissues, and the inflammatory cycle and tissue destruction may begin anew. It is ironic that the macrophages that are able to deal with some of the tubercle bacilli by using their cell-killing mechanisms can damage the very lung tissues they are designed to protect.

Today's Diagnosis of TB

TB is a corrosive disease, and much of what had been described as its classic symptoms—pallor, coughing and spitting up of blood, weakness, and emaciation—were indicative of an advanced state of the disease. Much of the improvement in our understanding of prevention and treatment has involved diagnosis and the development of methods to limit the spread of the organism from both within and without the body.

Tuberculin, a substance isolated from tubercle bacilli by Koch in 1890, turned out not to be a cure for TB; in the control of TB, however, it serves as the gold standard for diagnosis. In the tuberculin test a small amount of tuberculin (or PPD) is injected under the skin of the forearm, and within 48 to 72 h after injection an inflamed area develops if the person has been exposed to TB. The redness persists for up to 7 days. A positive skin test does not indicate an active infectious case of TB but merely that the person has been exposed.

Although *M. tuberculosis* is Gram positive, it is weakly stained by the Gram method because of its waxy cell wall. The method of staining—red fuchsin followed by treatment with nitric acid—devised in 1882 by Paul Ehrlich is still used today (Fig. 13.5a) and makes identification much easier than the method originally described by Koch. Parenthetically, when Paul Ehrlich in 1887 showed the clinical symptoms of TB, he stained his own sputum and demonstrated that he in fact was infected. He had a mild infection, went to Egypt to rest, and recovered.

In 1895, Wilhelm Roentgen (1845-1923) discovered X rays. X-ray photography or radiography made visible the tubercular lesions caused by the disease long before its symptoms became noticeable, and this allowed for treatment of the disease at a much earlier stage. In actual fact, the X-ray photograph did not become a reliable diagnostic tool until the 1920s, but even at that time treatments for TB remained less than satisfactory.

Another tool, the stethoscope, was developed in 1816 by the French physician René Laennec. Laennec was asked to examine a rather overweight and buxom woman. Realizing he could not directly tap on her chest to determine whether there was fluid in her lungs, he recalled from his boyhood that one could hear the scratch of a pin when the ear was placed in contact with one end of wooden cylinder and the other end scratched. In an imaginative stroke, he took a paper notebook, rolled it up tight, applied it to the woman's chest, and listened. Much to his surprise, he clearly heard the sounds of the woman's heart and lungs. Further experimentation eventually led him to develop the first stethoscope, a hollow wooden tube about a foot in length. Laennec's stethoscope has been improved upon, but it continues to provide the physician with an acoustic picture of the conditions within the lung. It cannot, however, identify the specific cause of the abnormal chest sounds. Laennec theorized, on the basis of his clinical experience, that phthisis and scrofula as well as miliary TB (that spreads as seeds throughout the body) were different forms of the same disease; this idea was opposed by German scientists, who were convinced that a specific clinical picture had to be the result of a single infectious agent. When Villemin demonstrated transmission of TB to rabbits and guinea pigs, however, it became clear that Laennec was correct. Laennec, who devoted his entire life to the study of TB, died from it in 1826.

Heroin and TB

In the late 1800s painful respiratory diseases such as pneumonia and TB were the leading causes of death, and in the days before antibiotics the only remedy was for physicians to prescribe narcotics to alleviate patient suffering. In 1874, while seeking a less addictive alternative to morphine, the chemist Charles Alder Wright, working at St. Mary's Hospital Medical School in London, synthesized a diacetyl derivative of morphine by boiling anhydrous morphine with acetic anhydride and acetic acid over a stove for several hours. After running a few experiments with it on animals, though, he abandoned his work on the drug. Twenty-three years later Felix Hoffmann, working at Bayer in Germany, took an interest in acetylating morphine (as he had done with salicylic acid to produce aspirin) and managed independently to synthesize diacetylmorphine. Diacetylmorphine was found to be significantly more potent than morphine, and so Heinrich Dreser, head of the pharmacological laboratory at Bayer, decided that they should move forward with it rather than with the aspirin they had recently created. Dreser was apparently well aware of Wright's having synthesized diacetylmorphine years before, but despite this, he claimed Bayer's synthesis was an original product, and by early 1898, Bayer began animal testing, testing it primarily on rabbits and frogs. They next moved on to testing it on people, primarily workers at Bayer, including Dreser himself. After successful trials Bayer named the drug heroin, from the German adjective heroisch (meaning "heroic"), since 19th-century doctors called such a medicine powerful.

In 1898, Dreser presented heroin to the Congress of German Naturalists and Physicians as more or less a miracle drug that was "ten times" more effective than codeine as a cough, chest, and lung medicine and worked even better than morphine as a painkiller. He also claimed that it had almost no toxic effects, and even that it was completely nonaddictive. Dreser particularly pushed the potent and faster-acting heroin as the drug of choice for treating asthma, bronchitis, and especially TB. In early 1898, G. Strube of the Medical University Clinic in Berlin tested oral doses of 5 and 10 mg of heroin in 50 TB patients to relieve coughing and to induce sleep. He noted no unpleasant reactions and that most patients continued to take the drug after he no longer prescribed it. Because heroin worked well as a sedative and respiration depressor, it did indeed work extremely well as a type of cough medicine and allowed people affected by debilitating coughs to be able finally to get some proper rest, free from coughing fits. Further, because it was marketed as nonaddictive, unlike morphine or codeine, it was initially seen as a major medical breakthrough and was quickly adopted by the medical profession in Europe and the United States.

Just 1 year after its release heroin became a worldwide hit, despite its not actually being marketed directly to the public, but rather solely to physicians. Heroin was soon sold in a variety of forms: mixed in cough syrup, made into tablets, mixed in a glycerin solution as an elixir, and put into water-soluble heroin salts, among others. At the end of this first year it was popularly sold in more than 23 countries, and Bayer produced around a ton of it in that year.

It quickly became apparent, however, that Bayer's claims that the drug was not addictive were completely false, with reports popping up within months of its wide-spread release. Despite this, it continued to sell well in the medical field. The United States was the country in which heroin addiction first became a serious problem. By the late 19th century countries such as Britain and Germany had enacted pharmacy laws to control dangerous drugs, but under the U.S. Constitution individual states were responsible for medical regulation. Late in the century some state laws required morphine or cocaine to be prescribed by physicians, but drugs could still be obtained from bordering states with looser regulations. This era, moreover, was the peak of a craze for over-the-counter "patent" medicines that were still permitted to contain these drugs. At the turn of the century it is believed that over a quarter of a million Americans (out of a population of 76 million) were addicted to opium, morphine, or cocaine.

After years of resistance, American patent medicine manufacturers were required by the Federal Pure Food and Drug Act of 1906 to label the contents of their products accurately. These included "soothing syrups" for bawling babies and "cures" for chronic ills such as consumption or even drug addiction, which previously had not declared (and had sometimes denied) their content of opium or cocaine. Consumers by this time were becoming fearful of addictive drugs, so the newly labeled medicines either declined in popularity or removed their drug ingredients. (The preeminent survivor from this era was a beverage from Atlanta called Coca-Cola.) In 1914, President Woodrow Wilson signed the Harrison Narcotics Act, which exploited the federal government's power to tax as a mechanism for enabling, finally, federal regulation of medical transactions in opium derivatives or cocaine. And in 1924 the U.S. Congress banned all domestic manufacture of heroin. Under federal law and the laws of all 50 states heroin possession is now a crime.

Controlling Consumption

Prior to 1940, and as long as TB was diagnosed in its late phase, relief of symptoms became the prime method of treatment. There were many fads, and many of these did no good. Early treatments for TB included creosote, carbolic acid, gold, iodoform, arsenic, and menthol oil, administered either orally or as a nasal spray. Some physicians prescribed enemas of sulfur gases and the drinking of papaya juice. None worked to cure TB. During the late 19th century surgical techniques were used to "rest" the lung; this included pneumothorax or collapsed lung treatments that required removal of several ribs and reduction in the size of the thoracic cavity. In some cases, lung resections, i.e., the removal of infected lung tissue, were performed. Despite a lack of success in curing the disease, these measures were continued well into the 1940s.

More than 2,000 years ago Hippocrates recommended a change in climate for phthisis, and this notion of the benefits of "clean air and sunshine" lasted well into the 20th century. Particularly in the 1800s there developed a movement both in Europe and in North America to treat TB patients in open-air hospitals—sanitaria (from sana, meaning "cure" in Latin). Life in the sanitarium was a reaction to the stuffy, overheated rooms of the patient's home and workplace. Patients were made to take outdoor treatment whatever the weather. Wide-open windows and outside balconies were the places where the patients took the cure, whether in the hot summer or the freezing winter. The art of wrapping oneself in a blanket became an essential ritual. Fresh air was taken with a vengeance. Sanitaria were considered to be indispensable for cure, and patients were provided with brilliant sunshine, fresh air, quiet, rest, and good nutrition—but no anti-TB medicines, since none were available. Life in a luxury sanitarium (in Davos, Switzerland) is immortalized in Thomas Mann's novel *The Magic Mountain*, and one of the most famous sanitaria in the United States was the Trudeau Institute in the Adirondack Mountains at Saranac Lake, New York, established by Dr. Edward Livingstone Trudeau.

In 1873, Trudeau came down with TB after taking care of his consumptive brother, who had died 8 years earlier. Trudeau suffered from fever and weakness and began to cough up blood. Believing the end to be near, he decided to retire to the Adirondack Mountains, where he hiked, hunted, fished, swam, and anticipated death. Within a few months of this outdoor regimen, however, he had gained weight and recovered his energy, and his fever had abated. In 1882, after Trudeau read an account of the benefits of a sanitarium in Europe, he raised money through donations and used his own funds to establish the first sanitarium in the United States at Saranac Lake. The sole industry of the town of Saranac Lake became the sanitarium.

The sanitarium movement soon went into high gear. In 1900 the United States had 34 sanitaria with 445 beds; 25 years later there were 356 sanitaria with 73,338 beds. Sanitaria, however, did not cure TB. In the pre-chemotherapy era (1938-1945), for patients in sanitaria with advanced TB the death rate was 69%, whereas for those with minimal disease it was 13%—about the same as in the general population. The significant aspect of sanitaria is that they isolated the contagious individuals and that physicians maintained complete control of their patients.

Soon after Koch's discovery of the tubercle bacillus, anti-TB campaigns began, first in Europe and then in North America. In 1889-1890, Hermann Biggs of the New York Department of Public Health issued an education leaflet that contained information on how to prevent the spread of consumption. Included were:

1. A campaign of education using newspapers and circulars calling attention to the dangers of TB, possibilities for treatment, and precautions to prevent it.

2. Compulsory reporting of TB by all public institutions.

3. Assignment of inspectors to visit the homes of patients in order to enforce sanitary regulations regarding the disposal of sputum and to arrange for disinfection.

4. Provision of separate wards in hospitals for TB patients with pulmonary disease.

5. Provision of special consumptive hospitals to be used exclusively for the treatment of TB.

6. Provision of laboratory facilities for the bacterial examination of sputum.<

Biggs's Riverside Hospital was established to confine, voluntarily or not, tubercular individuals whose "dissipated and vicious habits" endangered the health of the community. It was more a prison than a hospital, and in many respects this approach to TB was reminiscent of the establishment of colonies and hospitals for those suffering from leprosy. Biggs's idea was that the protection of the public health was more important than individual freedom. He was the chief of the health police. Much of this attitude toward public health was based on a misguided interpretation of Darwinian evolution: TB was no longer a romantic disease, it was a sign of corruption, and pruning out the unfit would benefit mankind. Public health authorities impulsively attributed illness to the environment, how you lived and where you lived, which were under your control. If you chose to live in filth, lacked ambition, or were too lazy to work yourself out of the tenements, or were indolent and dozed by the fire instead of taking brisk walks in the fresh air and sunshine, then you brought TB on yourself. At all costs, it was reasoned, the infected should be prevented from infecting others in society. Now TB, once regarded as beautiful and a source of creative inspiration, was marked as a hereditary weakness. Succumbing to TB, some contended, was in a person's genes.

The United States declared a "War on Consumption" beginning in the early 1900s. This consisted of vigorous anti-TB campaigns, the most famous being Christmas Seals of the National Tuberculosis Foundation, which later became the American Lung Association. Early TB campaigns blamed capitalism for the disease, because it was found in the crowded cities full of impoverished factory workers. Other times TB victims were stigmatized by their "choice" of living as paupers; some claimed that those with TB were lazy and unambitious and that such individuals were inferior. It is ironic that the whole TB eradication movement rose, flourished, and vanished without ever establishing its effectiveness against the disease. The greatest benefits of the "War on Consumption," which was in full swing by 1915, were education, loss of stigma, abandonment of TB as a spiritualizing and romantic force, a greater understanding of contagion, the initiation of attempts to clean up tenements, and improvements in medical diagnosis and care, but not to quarantine.

Selman, Schatz, and Streptomycin

Penicillin (see p. 290) was a "wonder drug"; it was, however, useless against the germ of TB. The search for an antibiotic that would inhibit *M. tuberculosis* began not in a pharmaceutical company but in the laboratory of a soil microbiologist, Selman Waksman (1888-1973), at the Rutgers Agricultural College in New Jersey. Waksman was born in the Ukrainian town of Novoya Priluka and received his early education from private tutors. He obtained a diploma from the Fifth Gymnasium in Odessa in 1910 and immediately thereafter immigrated to the United States. Installed in the New Jersey home of a cousin, a chicken farmer, he enrolled at Rutgers College, intending initially to study medicine. He changed direction, though, after taking agriculture courses and working on his fourth-year research project, in which he carried out assays of bacteria in culture samples from soil. It was at this time that he developed an abiding interest in filamentous bacteria, the actinomycetes. Actinomycetes became the research focus of his master's degree at Rutgers (1916) and his doctorate at the University of California (1918). He then returned to Rutgers, where he continued to study soil bacteria. In 1924 he took a 6-month sabbatical in Europe, visiting a variety of laboratories and making contact with the luminaries of microbiology. During the return voyage he met a young French biologist, René Dubos (1901-1982), who was immigrating to the United States. Selman offered Dubos a place in his Rutgers laboratory, where the younger scientist worked on cellulose decomposition by soil bacteria and received the Ph.D. in 1927. Dubos was then on the staff of the Rockefeller Institute for Medical Research, working with Oswald Avery (of transforming principle fame) and looking for something that would attack the capsular polysaccharide of the pathogen *Streptococcus pneumoniae*. Dubos isolated a bacterium from soil samples in 1939, *Bacillus brevis*, that killed the pneumococci, and with the help of Rollin Hotchkiss isolated and purified the killing agents, gramicidin and tyrothricin. Excited, Dubos reported to Waksman that he had used a soil-derived culture of bacteria to combat an infectious disease. This stimulated Waksman to turn his principal focus of research from soil bacteria interactions to a search for antibacterial organisms in soil samples and to the isolation of antibacterial substances, that is, antibiotics. Waksman and his students did their screening by looking for growth inhibition zones around single colonies of pathogenic bacteria, as Dubos had done and as had been reported in 1928 by Alexander Fleming with the mold *Penicillium*. Despite a lack of government funding between 1939 and 1943, the Waksman group described four natural inhibiting substances, mainly from actinomycetes—actinomycin, streptothricin, clavacin, and fumigacin. Unfortunately, all were toxic to animals, as had been the case with Dubos' gramicidin and tyrothricin. Toward the end of this period a Ph.D. student, Albert Schatz (1920-2005), was being supervised by Waksman. In October of 1943, Schatz isolated two strains of *Streptomyces griseus* (identified first by Waksman in 1919). One strain, 18-6, came from soil, and the other, D-1, came from a culture from the throat of a sick chicken that had been eating soil. Elizabeth Bugie, Waksman's technician, tested each strain for a suitable medium to produce a putative antibiotic and found meat broth to be best. Schatz developed methods to grow large amounts of *Streptomyces*, and then he isolated the antibiotic, named it streptomycin, and tested it against *M. tuberculosis* in soil. In 1944 the results, which were a part of Schatz's doctoral dissertation, were published in the *Proceedings of the Society for Experimental Biology and Medicine*, under the authorship of Schatz, Bugie, and Waksman: "Streptomycin, a Substance Exhibiting Antibiotic Activity against Gram Positive and Gram Negative Bacteria."

Learning of the action of streptomycin on tubercle bacilli from Rutgers, William Feldman and Corwin Hinshaw of the Mayo Clinic requested and obtained the antibiotic for use in guinea pigs infected with a virulent strain of M. tuberculosis. By June of 1944 it was concluded that streptomycin worked against TB in animals and was not toxic to them. The drug, however, did not destroy all the bacteria, but it was bacteriostatic; that is, it stopped the infection from spreading. The work of Hinshaw and Feldman was published on November 15, 1944, in the *Proceedings of the Staff Meetings of the Mayo Clinic*, and then Hinshaw and Feldman carried out clinical trials in humans. Following this lead, the American Trudeau Society and the National Tuberculosis Association took an active part in further investigations. Within 3 years after the announcement of the isolation of streptomycin came the almost complete elucidation of its chemistry, its first derivative (dihydrostreptomycin) was obtained, and the first 1,000 clinical cases were reported.

During this period Waksman, who had a long-standing relationship with Merck (going back to 1939), which had developed methods for producing penicillin, entered into an agreement to give exclusive drug development rights to Merck, and in 1945 a patent was taken out in the names of Waksman and Schatz. Merck broke ground on a $3.5 million plant in Elkton, Virginia, employing 400 workers to produce the therapeutic agent streptomycin. The royalty would be 2.5% to the Rutgers Research and Endowment Foundation, and the Rutgers Foundation provided 20% of its royalties to Waksman. Waksman, however, did not tell Schatz of the Merck deal, and Schatz received no royalties. By 1942, Waksman had received more than $140,000 in royalties—more than 10 times his salary. Schatz received a few thousand dollars from Waksman as "gifts." But in 1949, when Schatz discovered that he had been deprived of royalties, he hired a lawyer, and there ensued a contentious legal battle. On December 20, 1950, after much wrangling, the case was settled. Schatz received $125,000 and was given 3% of the royalties, which amounted to $15,000 annually for several years. Waksman received 10% of the royalties, half that amount going to his Foundation for Microbiology (established in 1949) for research and publications in microbiology. Waksman's work on streptomycin was recognized by numerous scientific societies in the United States, France, Denmark, Sweden, Israel, Spain, and Turkey. In 1952, Selman Waksman alone received the Nobel Prize for Physiology or Medicine for "ingenious, systematic and successful studies of soil microbes that have led to the discovery of streptomycin, the first antibiotic remedy against tuberculosis." In his acceptance speech Selman spoke of streptomycin but said not a word about Schatz's contribution.

Despite the Schatz lawsuit, the public relations department of Rutgers University continued to laud only Waksman, and the academic community rallied to Waksman's defense. Schatz was blackballed from positions in academe. Waksman continued to oversee the work on antibiotics at Rutgers and published 500 scientific papers and wrote or edited some 28 books. He wrote an autobiography, *My Life With Microbes,* and in 1954, Rutgers University opened a $3.5 million neo-Georgian structure, the Institute of Microbiology, funded principally by royalties from streptomycin paid to the Rutgers Foundation.

Schatz's contribution to the discovery of streptomycin went unnoticed until 1990, when Milton Wainwright (Sheffield University) wrote about it (*Miracle Cure: The Story of Penicillin and the Golden Age of Antibiotics*), and it was recognized again in 2012 in Peter Pringle's book *Experiment Eleven: Dark Secrets Behind the Discovery of a Wonder Drug*). In 1994, on the 50th anniversary of the discovery of streptomycin, Rutgers University awarded Schatz the Rutgers Medal, and the National Museum of American History exhibited two test tubes containing *Streptomyces*. Next to the tubes was a picture of Schatz in a white lab coat and the caption "Dr. Schatz discovered streptomycin in 1943 when he was a 23 year old graduate student working with Dr. Selman Waksman at Rutgers University." Schatz died on January 27, 2005, at age 84 of pancreatic cancer.

So was Schatz deserving to share in the Nobel Prize? And did he deserve to be vilified by the academic community? Albert Sabin (who also was never awarded a Nobel Prize) said that Schatz had behaved like an ungrateful, spoiled, immature child and would regret what he had done. (Schatz never did.) Sabin went on to say the Nobel Prize is awarded for the discovery of important new principles that open up new fields of research; this Waksman achieved. In this achievement Schatz made no contribution.

In the matter of isolating the streptomycin-producing *Streptomyces* there is little doubt that Schatz should have received credit and that Waksman should not have excluded him from royalties or recognition in the public arena, but as Dubos has written, "the cruel law of scientific life … [is that] credit goes to the man who convinces the world, not to the man to whom the idea first occurs."

Drug Resistance

The mode of action of streptomycin was discovered later: it inhibits the synthesis of the waxy cell wall of the bacillus, thus leaving the tubercle bacilli naked and unprotected from the onslaught of the killing machinery of the macrophage. But the antibiotic victory over TB was short-lived. Soon there were signs of streptomycin-resistant TB bacilli. Once drug-resistant strains of *M. tuberculosis* appeared, there was a need to increase the initial streptomycin dose by 1,000 times to stop the growth of bacteria. In the 1940s, Jörgen Lehmann in Sweden was inspired by a paper in *Science* magazine that reported increased growth of *M. tuberculosis* in the presence of salicyclic acid (the main ingredient in aspirin). Lehmann theorized that an analog of salicylic acid might have the opposite effect, i.e., inhibit growth. The new chemical was para-aminosalicylic acid (PAS). It was not as effective as streptomycin, but by 1949, Lehmann found that combining PAS with streptomycin worked better than one drug on its own. In 1952, Gerhard Domagk, who had isolated Prontosil (see p. 285), found that a new class of thiosemicarbazones also had growth-inhibitory effects on *M. tuberculosis* in the test tube, and in testing one derivative, isoniazid, a drug first synthesized from coal tar in 1912, Domagk found it to be a potent anti-TB agent. By 1952 isoniazid had become the mainstay in the treatment of drug-resistant TB. Isoniazid also blocks the synthesis of the mycolic acids that are a main constituent of the waxy wall of *M. tuberculosis*.

Streptomycin and other drugs did not eradicate TB; they did, however, effectively eliminate sanitaria. In 1954 the Trudeau Institute was closed, and by the 1960s almost all sanitaria had vanished. In 1963 rifampin (derived from another Streptomyces species, *S. mediterranei*), an inhibitor of the synthesis of tubercle bacillus RNA, was introduced for treatment. These days, the initial treatment for TB involves taking several drugs for 6 to 9 months. There are 10 drugs approved by the U.S. Food and Drug Administration for TB treatment, the first-line drugs being isoniazid, rifampin, streptomycin, and ethambutol and pyrazinamide.

Drug Resistance Redux

To minimize the emergence of drug resistance, TB patients are treated with a drug cocktail (multidrug therapy), i.e., isonaizid plus rifampin or a combination of fluoroquinolones and injectables such as amikacin or capreomycin for 20 to 30 months.

How does drug resistance come about, and why is it prevented by a drug cocktail? Drug resistance develops this way: an average TB patient may have a billion tubercle bacilli, and 1 in 10,000 of these carries a mutation, a genetic change, that allows that bacillus to evade the lethal effects of the anti-TB drug. Once the patient is treated with that drug, only the mutant survives, and these are the drug-resistant bacilli. The result is that there are now ~10,000 tubercle bacilli that can reproduce in the presence of the drug and increase their numbers, and now almost the entire population is resistant. But let us assume that a second, and equally effective, drug is added along with the first drug to the billion tubercle bacilli. Again, only 1 in 10,000 is resistant to this drug. When the drugs are added together, though, there is less than a single resistant bacillus. A third drug added even further reduces the possibility of a surviving bacterium. Thus, multidrug therapy makes it possible to delay or even prevent the emergence of drug resistance.

In the absence of drug resistance, isoniazid and rifampin administered for 9 months is curative. If rifampin is not used, 18 months is the minimum duration of therapy for cure. If pyrazinamide is added to isoniazid and rifampin for the first 2 months, the treatment period can be shortened to 6 months. Currently, the most commonly used regimen consists of isoniazid, rifampin, and pyrazinamide administered daily for 8 weeks, followed by isoniazid and rifampin given daily, twice a week, or three times a week for 16 weeks. More than 85% of patients who receive both isoniazid and rifampin have negative sputum cultures within 2 months after treatment has begun.

The emergence of multidrug-resistant strains of *M. tuberculosis* has provoked great concern because the disease caused by such strains is often fatal. These strains often develop in people who begin treatment but lapse after a few weeks, allowing larger numbers of mutant bacteria to survive; these bacteria overcome and resist the drugs. In the late 1980s it became clear that a substantial number of patients with TB were not completing their treatment. To address this problem, the Centers for Disease Control and Prevention (CDC) recommended that direct observation of therapy (DOT) by a trained health care worker be considered for all patients to ensure that they swallow each and every pill. Maintaining DOT is not easy, however, since it requires trained medical staff, infrastructure, and money. In practice, one-quarter of TB cases are not diagnosed before they become infectious, and many who begin treatment do not stay the course.

There is evidence from studies of twins that susceptibility to TB has a genetic basis, but no major TB susceptibility gene has ever been identified. Susceptibility to TB is dramatically enhanced in HIV-infected persons, and the spread of AIDS has been paralleled by a resurgence in TB cases. HIV-infected individuals who adhere to the standard regimen for TB do not have an increased risk of treatment failure or relapse. The use of protease inhibitors and reverse transcriptase inhibitors for HIV treatment has complicated the treatment of TB in HIV-infected individuals. The administration of these drugs with rifampin can result in lower levels of antiviral drugs and toxic levels of rifampin. Rifabutin may be substituted for rifampin, which leads to fewer side effects with some antiviral drugs but not with others.

Vaccination against TB

Why is there no vaccine against TB? There is. In the 1920s two French bacteriologists, Albert Calmette (1863-1933) and Camille Guérin (1872-1961), used a technique first used in 1882 by Louis Pasteur—they attenuated bovine TB (M. bovis). They grew the bacteria on a nutritive medium containing beef bile for 231 generations over a period of 13 years, during which time the strain became weakened. This non-disease-producing bacillus, which could elicit immunity, became the vaccine called BCG (bacillus Calmette-Guérin), which was cross-reactive with human TB. When BCG was first introduced in a trial in Lübeck, Germany, in 1930, there were 76 deaths of 249 babies inoculated. These fatalities were the result not of the inoculation with BCG itself but rather of live, virulent *M. tuberculosis* stored in the same incubator; the adverse publicity, however, so poisoned the public's interest that for decades it was difficult to achieve popular support for BCG vaccination. This accident may have had one beneficial effect: it encouraged the search for anti-TB drugs.

Injecting BCG produces a mild infection and induces immunity, and has never resulted in a virulent infection. It does, however, have some adverse reactions and is not 100% effective. (Estimates of protective efficacy range from 30 to 80%.) Though the mortality from TB in countries not using BCG has declined as markedly as it did in countries that have used BCG since 1950 (Fig.. 13.6), most believe that BCG has contributed to the decline in TB, but there is a reluctance on the part of many in the United States to use a live vaccine. (Without BCG vaccination, the number of infections in the United States has declined, as is shown in Fig. 13.7.) Further, vaccination with BCG renders the tuberculin test positive, and its use as a vaccine would be of little benefit to those already infected.

Incidence of TB

By the 1940s—before antibiotics were introduced and strict public health measures

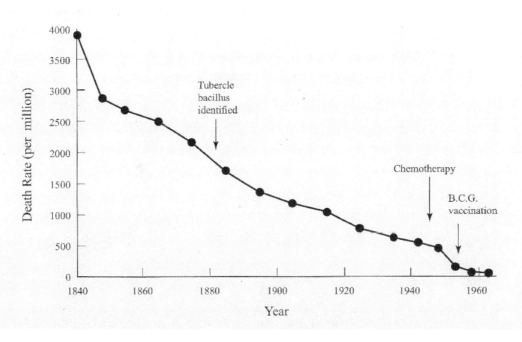

Figure 13.6 The death rate from tuberculosis in England and Wales. From T. McKeowan, The Role of Medicine, Princeton University Press, 1976, p.9.

were instituted—the lethal incidence of epidemic TB was declining. Why? Some suggested that it was due to the emergence of less virulent strains of the mycobacterium, whereas others proposed that it was due to an increase in the resistance of the human population. Still others contended that public health measures such as the segregation of the infected in sanatoria, as well as forbidding the practice of spitting in public, led to a decline in TB. Higher standards of hygiene, including cleaner cities, destruction of tubercular cows, pasteurization of milk, and hand washing, all led to reduced transmission, and better nutrition and a higher standard of living have also contributed to the decline. Although the incidence of TB continues to decline, it has not disappeared altogether. Since the World Health Organization (WHO) established a global reporting system 20 years ago, it has received reports of 78 million TB cases, 66 million of which were treated successfully. According to the WHO, 9.6 million people were estimated to have fallen ill with TB in 2014: 5.4 million men, 3.2 million women, and 1 million children. Globally, 12% of the 9.6 million new TB cases were HIV positive. Of the 418,000 cases of multidrug-resistant TB that were estimated to have occurred in 2014, only about one-fourth of these—123,000—were detected and reported. Of 9.6 million people infected with TB, 58% were in Southeast Asia and the Western Pacific region, with India, Indonesia, and China having the largest numbers. In the United States in 2014 the CDC reported a total of 9,421 cases (a rate of 2.96 cases per 100,000 people). But in 2015 there were 9,563 cases. Approximately

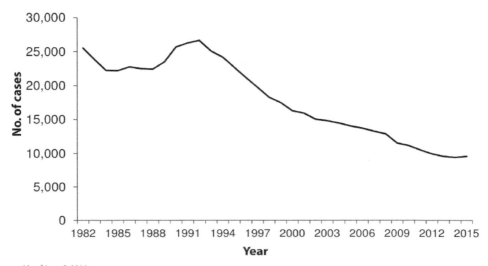

Figure 13.7 Reported Tuberculosis (TB) cases in the US 1982-2015

half of these cases were in California, Texas, Florida, and New York. In 2014, 66% of the cases of TB occurred in foreign-born persons, with a case rate of 15.4 cases per 100,000. In 2013 there were 555 deaths from TB in the United States. While the active form of TB is contagious, there are about 11 million people in the United States with latent infections—those that don't show symptoms and are not contagious. In the United States the rates for TB differ by race and ethnicity: Native Americans have 5.0 cases per 100,000, Asians 17.8 per 100,000, African-Americans 5.1 per 100,000, Native Hawaiian and other Pacific Islanders 16.9 per 100,000, Hispanics or Latinos 5.0 per 100,000, and Caucasians 0.6 per 100,000. In the United States the total number of multidrug-resistant TB cases decreased slightly from 1.4% (96 cases) in 2013 to 13.3% (91 cases) in 2014; the proportion of multidrug-resistant cases of TB occurring among foreign-born persons, however, increased from 31% in 1993 to 88% in 2014.

Coda

When humans were hunter-gatherers and lived in small roving bands, TB was not a serious threat, but with the development of agriculture and animal husbandry the numbers of people increased, as did exposure to new pathogens. A soil bacterium infected cattle or game and became *M. bovis*, and perhaps by crossing the species barrier it evolved into *M. tuberculosis*. Coincident with urbanization were epidemics of TB. The poor—undernourished, crowded, and living under unhygienic conditions—became the breeding ground for the "white plague." Today, it is recognized that TB is an infectious and a societal disease. Understanding this contagious illness demands that social and economic factors be considered as much as the way in which the tubercle bacillus causes damage to the human body, how it manages to evade the immune system, and how it is able to overcome the most potent anti-TB drugs. In the past, fear of the disease, fueled by preexisting prejudices, led to public and institutional reactions including mandatory testing and isolation and tended to stigmatize immigrants and those with different lifestyles. TB has and will continue to affect the emotional and intellectual climate of human populations throughout the world. Now, more than a century after the discovery of the cause of TB, it must be realized that for the "People's Plague" to be eradicated the subtle interplay between disease and society must be fully appreciated. Until that time, TB remains a disease that could reemerge to threaten us once again.

Bene dic
deo et
morere

doi: 10.1128/9781683670018.ch14

14
Leprosy, the Striking Hand of God

There is a morbid interest in and dread of leprosy, largely stemming from frequent references in the Bible. To the readers of the Old Testament, leprosy was an abomination. Job says, "Pity me, pity me, and pity me, you, my friends, for the hand of God has struck me." Although the Bible does not specify that Job suffered from leprosy, it was assumed that he did, and for a time he was considered the patron of lepers. Indeed, in medieval art Job is always depicted as being covered with black spots, and body spots are also seen in illustrations of lepers (Fig. 14.1). In Leviticus this account of the disease is given: "When a man shall have the skin of his flesh a rising, a scab or bright spot, and it be in the skin of his flesh like the plague of leprosy; then he shall be brought unto Aaron the priest or unto one of his sons the priests."

The mention of leprosy in both the Old Testament (Exodus, Numbers, Deuteronomy II, Samuel II, and Kings II) and the New Testament (Matthew, Mark, and Luke) has surely contributed to the fear of the sores associated with leprosy, and the notion that what is blemished is unclean and is also displeasing to God because it is defiled. Christ curing lepers became a metaphor for divine salvation. The fictional account of Ben-Hur, as well as the parables of Christ and his encounter with the 10 lepers on the road to Galilee, gave rise to a massive literature on the imagery of leprosy as a disease of the soul and one that was highly contagious.

In Job 19:13-21 (Jerusalem version) we read:

> My brothers stand aloof from me, and my relations take care to avoid me, my kindred and my friends have all gone away, and the guests in my house have forgotten me. The serving maids look on me as a foreigner, a stranger, never seen before. My servant does not answer when I call him; I am reduced to entreating him. To my wife my breath is unbearable, for my brothers I am a thing corrupt. Even the children look down on me ever ready with jibe when I appear. All my dearest friends recoil from me in horror; those I loved best have turned against me.

Figure 14.1 (Left) Job stricken with a plague as depicted in a 17th century woodcut, "Feldt-buch der Wundarzney" by Hans von Gersdorff, INTERFOTO / Alamy Stock Photo.

Often, we think of ourselves as being more enlightened, and above such attitudes toward the diseased, but when one considers the hysteria associated with the outbreak of AIDS, and the draconian measures that were initially suggested for controlling its spread, our approach to disease may not be so very different than the way leprosy was viewed in the Bible and the Middle Ages.

A Look Back

Leprosy probably arose in the Far East, about 1400 B.C., for there are accurate descriptions in the sacred Hindu writings of the Veda, and there are also descriptions in the Chinese literature.

The earliest accounts of leprosy occur in an Indian text, *Charaka Samhita*, written between 600 and 400 B.C. The *Nei Chang*, a textbook of medicine, describes a disease wherein the patient has stiff joints, the eyebrows and beard fall off, the flesh becomes nodular and ulcerates, numbness results, and finally the bridge of the nose changes color and rots. The *Nei Chang* was written between 250 and 230 B.C.

It has been claimed that leprosy was brought from India to Greece in the 4th century B.C. by the soldiers of Alexander the Great, but it is also possible that leprosy spread from the Far East to the west along the trade routes arriving in the Mediterranean about the time of Christ. Certainly Hippocrates, who lived from about 460 to 377 B.C., did not describe leprosy, because he never saw it. Rather, the best description of leprosy in Europe comes from a contemporary of Galen, Aractus, in A.D. 150. Leprosy then spread further west, and in the graveyards of Britain and France there is evidence of the disease in bones dating from 500 to 700. The emperor Constantine (274-337) suffered from leprosy, and pagan priests believed bathing in the blood of sacrificed children would cure him. It did not!

The Hebrew word *saraath* was used to describe many skin conditions, and it has been variously translated as "defiled" or "accursed" or simply "scaly." When the Bible was translated into Greek some 300 years before Christ, *saraath* became *lepros*. St. Jerome Latinized *lepros* into *lepra*, and in the first English translation of the Bible in 1384, John Wycliffe (1329-1384) translated *lepra* as "leprosy." And leprosy in the Middle Ages, and even before this time, was considered a disease of the soul. The association of the disease with sinfulness stigmatized the leper. Medieval medical authors refer to lepers as being crafty, irascible, and suspicious and above all having a burning desire for lustful sex. Sometimes the disease was known as satyriasis—an insatiable sexual appetite (from the satyr of Greek mythology, a forest deity with the ears and tail of a horse given to merriment). The belief that venereal disease was due to immorality was "logically" extended to leprosy; in this way, leprosy became accepted as divine punishment for the sins of the flesh. By the 16th century there was also confusion between syphilis and leprosy.

The disfigurement of the face and hands (Fig. 14.2a and b) contributed to the alienation of the leper, and the sores on the body led to the belief that leprosy was contagious. Lepers were not considered "nice people," and they were cast out of society. In 1179 the Third Lateran Council issued a decree urging the segregation of lepers

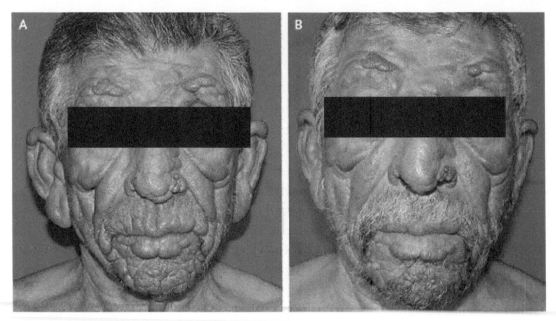

Figure 14.2a Disfigurement of the face. ©2012 Massachusetts Medical Society.

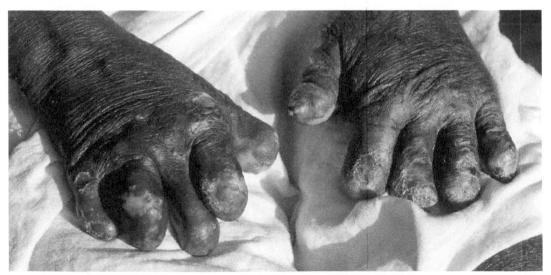

Figure 14.2b Deformity of the hands. Courtesy of Wikimedia and licensed under creative commons licensing: CC BY-SA 3.0.

Figure 14.3 Leper announcing his arrival in the city with the ringing of a bell. Watercolor by R.Cooper. Courtesy Wellcome Library, London, CC-BY 4.0.

from society: the leper was told not to mix with the crowds, to use his own container in drawing water, and not to touch anything unless he paid for it first. He was to wear a distinctive garment and announce his presence with a bell or clappers (Fig. 14.3).

At times the lepers were treated with kindness, but more frequently, as in the time of Henry II of England (1133-1189) and Philip V of France (1293-1322), the leper was strapped to a post and burned alive. On the other hand, Edward I of England (1239-1307) was considered kinder—he buried them alive. Some good Christian people did sympathize with the plight of the leper, and almshouses or refuges were established for them. Many were called lazar houses or lazarets after Lazarus of Bethany, who, it was believed, suffered from leprosy and who replaced Job as the patron saint of the leper. St. Francis of Assisi (1181-1226) was notable for his devotion to taking care of lepers, embracing and kissing them. He never contracted leprosy, leading to speculation that he may have had tuberculosis, which conferred immunity.

The pandemic of leprosy reached epidemic proportions in the 12th century and had its peak in Europe in the 13th and 14th centuries; records showing the construction of 19,000 lazarets document this continuous increase in leprosy in Europe. By the 16th century, though, most of the lazarets had been closed, and by the 18th century they had all but disappeared. The historian William McNeill has suggested that leprosy may have retreated with the rising incidence of pulmonary tuberculosis in the 1300s, and this provoked greater resistance to leprosy, but scientific evidence for this is scanty. In addition, it has been hypothesized that the increased supplies of woolen textiles for clothing may have broken the chain of skin-to-skin contact needed for the spread of the disease, and at the same time contributed to the spread of typhus. This, too, is only conjecture.

Knights participating in the Crusades contracted leprosy, and in 1048 they formed their own spiritual order, the Order of Lazarus. The return of these knights to Europe was probably a contributing factor to the spread of leprosy throughout Europe. Leprosy spread from Spain and Africa to the Americas in the 16th and 17th centuries. It has been claimed that the last leprosy patient in Great Britain died in Scotland in 1798. But as late as the mid-19th century there were thousands of cases in Scandinavia, and with the importation of Chinese laborers into the Pacific islands by the colonizing Europeans, there was a further spread of the disease. The first reference to leprosy in Hawaii was in 1823, and two generations later 5% of Hawaiians were infected. In the early 1900s it was estimated that ~1% of the deaths in Fiji were due to leprosy.

In the early 1860s there was a "rediscovery" of leprosy, with the point being made that it was a disease of those of foreign birth. Indeed, leprosy had by this time reached epidemic proportions on the Hawaiian Islands. In an effort to halt its spread, the Hawaiian Kingdom followed a practice of quarantine using isolation and segregation. The Americans and Europeans who lived in Hawaii feared the disease, and the Western imperialistic attitude served to strengthen the negative image of those who were diseased. Officials rounded up the lepers and loaded them onto ships bound for a settlement at Kalawao, on an isolated peninsula of Molokai's north shore. Many of those transported were shoved off the ships hundreds of yards from the land and forced to swim to shore. Some of the weaker ones never made it. In 1865 a leper colony was established on Molokai, holding a total of 142 leprous individuals. Some of those banished to the island, however, probably had other diseases. Although the Chinese were singled out as the source of leprosy, 97% of the lepers on Molokai were Hawaiian. The prevailing view was that leprosy was a very contagious disease and that the Hawaiians lacked social utility. Thus, their isolation on Molokai seemed totally justified.

In 1873, Father Damien, a Roman Catholic priest from Belgium, joined the colony as its resident priest. His assignment was to be for 3 months, but he stayed for 16 years. The determined Father Damien cleaned and bandaged sores and built homes, a hospital, a reservoir, and a plumbing system; and for the next decade and a half he buried many hundreds of leprosy victims. He provided comfort and compassion to those who had been shunned by their friends and families because they were considered to be unclean and sinners. Father Damien did not avoid contact with the lepers of Molokai, and when he died in 1889 at the age of 49, the story was spread that he had contracted leprosy during his residency. He came to be called the "Martyr of Molokai" and was beatified in 1995. (Since it was known earlier, however—perhaps in 1876—that he already showed signs of the disease, he may have contracted the disease before coming to Molokai.) Father Damien's leprosy caused much fear and reinforced the belief that the disease was highly infectious. Although Hawaii became a territory of the United States in 1900, the involuntary isolation of those with leprosy at Molokai continued until 1974.

In the United States proper there was also a fear of the Chinese—a condition that increased after 1850, and so exclusionary laws were enacted to prevent them from bringing their diseases to the United States. By 1870 leprosy was considered one of the diseases specifically associated with the Chinese. Calls for action arose, and in 1894 the state of Louisiana established the Louisiana Home for Lepers, located 85 miles from New Orleans. Run by the Sisters of Charity, it was a neglected asylum. The quarantine of lepers at the Louisiana Home served to reinforce the stigma of the disease.

By 1917, at the height of U.S. racial discrimination and xenophobia, a leprosy bill was passed to establish a leprosarium. The old Louisiana Home for Lepers became the American National Leprosarium at Carville, and in 1921 the U.S. flag was raised over the institution. It was surrounded by a barbed-wire fence with a No Trespassing sign. The rules and regulations were more like those of a prison than a hospital. Patients were sent there by special leper trains, some arriving in shackles and accompanied by armed guards, and the patients' mail both ingoing and outgoing was disinfected. Babies born to the patients were given up for adoption. Yet the only crime of these children was that they might have contracted leprosy. Patients were not permitted to marry until 1952, despite the fact that drugs to control the disease such as the sulfone dapsone were introduced in the late 1940s. The leprosarium was purposely placed between the men's and women's penitentiaries as a means of preventing mingling of the sexes, since the fear of contracting leprosy by passing through this "zone of contagion" served to scare the inmates from considering escape for conjugal visits. By 1956 it became a voluntary hospital under the Public Health Service and was renamed the Gillis W. Long Hansen's Disease Center. Up until the 1960s strict public health laws forbade those with leprosy to use public transport, to fly over certain states, to use public restrooms, or to live freely in society. The basic freedoms guaranteed to most Americans were denied them. Although over the years some deregulation did take place, Carville continued to be a shameful reminder of the stigma and the stereotype of leprosy in the United States. In 1997 the last 135 people hospitalized (of the maximum number of 450) were "set free" as President Bill Clinton transferred the facility back to the state of Louisiana, where it has become a school and training center for at-risk youth. The former Carville patients will receive $33,000 annually, will have lifelong medical care for their disease, and can reside anywhere they desire.

The Disease of Leprosy

What is leprosy, and what causes it? The first objective and scientific appraisal of leprosy came in 1873 with the discovery of the leprosy bacillus, *Mycobacterium leprae*, by Gerhard Armauer Hansen (1841-1912) in Norway. (Indeed, the preferred name for the disease today is Hansen's disease.) *M. leprae*, like its cousin *Mycobacterium tuberculosis*, is an acid-fast bacillus (so called because after staining and rinsing with acid the stain remains), but it differs in several significant ways—it cannot be grown in tissue culture, and it grows very slowly and only in humans, mice, and nine-banded armadillos, taking about 2 weeks to divide. (Tuberculosis bacilli divide every 12 h.) In the mouse the bacteria grow best in the footpad, where the temperature is 30°C and as few as 1 to 10 bacilli can initiate infection, but to get 1 million bacilli it takes 5 to 6 months. In the armadillo the disease becomes disseminated, and one can get 1 billion bacilli (~200 g) in 15 months. *M. leprae* has been found in 10% of the armadillos in certain regions of Louisiana.

There are no known vectors or reservoir hosts, and there is no satisfactory way of detecting past or preventing unapparent infections. Therefore, what we know of the disease and its spread is derived from clinical cases. Leprosy is a disease of low infectivity, and it is also a spectral disease, that is, one showing different manifestations: tuberculoid leprosy, localized in skin nodules where the bacteria may be abundant; and disseminated or lepromatous leprosy, with bacteria in macrophages and the skin. Most patients show the tuberculoid type, which develops a year or two after exposure, whereas the lepromatous type requires longer. The manifestations appear to depend on the immune status of the host. Tuberculoid leprosy is associated with severe nerve damage. The disease involves cell-mediated immunity (see p. 261), with T helper cells and interleukin-2 secretion in tuberculoid leprosy, but since the bacteria multiply within the Schwann cells that insulate the nerve, there is damage to the nerves, and anesthesia results. In lepromatous leprosy the T helper cells do not respond to the bacilli—gamma interferon is not produced, macrophages are not activated, and as a consequence the bacteria multiply within the macrophages, and the disease spreads with multiple organ involvement, leading to facial deformity (Fig. 14.2a) and blindness. There is osteoporosis and shortening of the digits, with deformity of the hands (Fig. 14.2b). The testes are damaged, and there is no natural remission. Death results most commonly from renal failure, pneumonia, and tuberculosis. (Lepromatous leprosy patients do have antibody to *M. leprae* antigens, but this not protective.) Natural resistance to leprosy is highest in African blacks and lowest amongst Caucasians.

It is believed that after the bacteria enter the body (via the nose or through open wounds) they somehow find their way to the Schwann cells. The affinity for Schwann cells by *M. leprae* is the result of the high affinity of the bacterium for a specific region of the molecule laminin found on the outer surface of these cells, and once they adhere to the laminin using a bacterial surface protein called H1p, they then invade. The mechanism of invasion is unknown, but once inside the Schwann cells the bacilli are temporarily protected from the host immune system. Over time, however, the immune system attacks the infected Schwann cells, destroying nerves in the process. Accompanying this loss in sensation are secondary infections that, according to Douglas Young and Brian Robertson, "whip up the immune system into a frenzy … and under siege, soft tissues, even bones become degraded, particularly in the limbs and digits."

Where Leprosy Is

There has been a dramatic decrease in the global burden of leprosy: from 5.2 million people infected in 1985 to 753,000 in 1999 and 175,554 at the end of 2014. During that same year 213,899 cases were reported to the World Health Organization (WHO).

Most cases are found in the tropics. Pockets of endemicity remain in Angola, Brazil, the Central African Republic, India, Madagascar, Nepal, and Tanzania. In 2014, 175 new cases of leprosy were detected in the United States. Approximately 85% of the U.S. cases are found in immigrants, although small endemic foci occur in Florida, Louisiana, Texas, and along the Gulf of Mexico in Mexican and Asian California populations, and in Spanish Americans in New York City. Leprosy may be a zoonosis in the southern United States because armadillos are a large reservoir for the disease.

Even today leprosy leads to a lifetime of social ostracism and misery. The fear of contagion, coupled with the possibility of disfigurement, has contributed to an acceptance of terrible cruelty toward lepers. The higher the standard of living, the lower the incidence of leprosy. Leprosy is often an endemic disease, one usually acquired in childhood, but infection can occur even at age 70. Many who are exposed develop no disease. In many parts of the world males are more frequently infected than are females (2:1), but the reasons for this remain unclear.

Leprosy Today

Because leprosy bacilli cannot be cultured in the laboratory, it is difficult to assess the onset of the disease. *M. leprae* can be grown in nude (immunodeficient) mice and nine-banded armadillos, and this provides for microbiological analysis and studies on immunology and treatment. The diagnosis of leprosy, however, has not changed substantially in 100 years and relies on microscopic examination and evaluation of the response of the patient to a pinprick or heat. To date, examination of individuals with tuberculoid lesions stained for acid-fast bacilli is the diagnostic method, the one used by Hansen in 1873. Newer molecular methods such as PCR are being tested to detect the presence of *M. leprae* nucleic acids.

Leprosy can be cured and disability averted if treatment begins early. Dapsone, developed in the 1940s, controls the growth of the bacilli. Its mode of action is by interference with the synthesis of bacterial nucleic acids through blocking the conversion of *p*-aminobenzoic acid to folic acid; because in humans folic acid is obtained in the diet, dapsone does not affect the cells of the human body. In 1991 the WHO recommended multidrug therapy using dapsone, clofazimine, and rifampicin for lepromatous leprosy and rifampicin and dapsone for tuberculoid leprosy (also called paucibacillary leprosy because of the smaller numbers of bacilli in the skin lesions), with treatment lasting 6 to 24 months before the cure is complete. Since 1995 treatment has been made available free of charge by the WHO. It has been estimated by the WHO that over the past 20 years more than 16 million people have been treated.

In 1998, India announced approval of a leprosy vaccine called Leprovac. It contains a heat-killed, fast-growing, nonpathogenic mycobacterium called *Mycobacterium w* (*M. w*), which shows the closest antigenicity of all mycobacteria to *M. leprae*. The vaccine stimulates the immune system of leprosy patients, disrupting immune tolerance and provoking an immune response that kills and clears *M. leprae* from the body. In this manner, *M. w* resembles the action of the tuberculosis vaccine, BCG, which also stimulates clearance of leprosy bacilli. Because it has been shown that the cytokines gamma interferon and interleukin-2 induce clearance of bacilli in leprosy patients, both vaccines may act by triggering the release of cytokines that then serve to enhance the killing capacity of macrophages. Leprosy patients must be given eight doses of Leprovac at 3-month intervals in order to effect clearance; for ethical reasons, however, the patients are also given multidrug therapy.

"Catching" Leprosy

How is leprosy spread? The bacilli are shed with the skin, but most bacilli are found in nasal secretions. Some leprosy patients shed 100 viable *M. leprae* bacilli each day from the nose. There remains a question as to whether leprosy can be spread by mosquitoes. The route of entry into the body is still uncertain, but it may be through inhalation, since the bacteria cannot penetrate the skin directly. Leprosy seems to occur in clusters of families, which suggests that susceptibility depends on genetic factors, but since few of the members of a family ever contract the disease, leprosy is considered a disease of low virulence. Leprosy patients have no greater susceptibility to tuberculosis than do those without infections. There appears to be no cross-protection between tuberculosis and leprosy, although in some cases BCG can protect against both diseases.

Coda

Leprosy has been a dreaded and disfiguring disease since time immemorial. Although today we know that for transmission to occur there must be prolonged contact with the infected individual and that the disease is curable, we still do not know the precise mode of transmission or how nerves are destroyed. But most importantly, as with AIDS, we still do not know how to completely remove its stigma from our consciousness so that discrimination of those afflicted is abandoned.

doi: 10.1128/9781683670018.ch15

15
Six Plagues of Africa

Africa, with 12 million square miles, contains nearly one-fifth of the world's land surface and is 3 times the size of the United States. The real Africa, what some historians have called "Black Africa," stretches from about 15°N to the Limpopo River Valley in the south. Prior to the 19th century Black Africa had 80 million people. It was virtually unknown to the rest of the world, and for those who knew about it the knowledge was scant or ill informed. The reason for this was that there was little communication with the rest of the world except for brief incursions by the Phoenicians, the Romans on the Nile, and a few Arab caravans in search of slaves. Black Africa also appeared to lack the climatic attractions and plunder of the New World, and it was difficult to penetrate for several reasons. First, there were few deep bays or gulfs for the shelter of ships. Second, most of the rivers were not navigable because of sandbars at the mouth or rapids a short distance upstream. Third, beyond the shore lay miles of impassable mangrove swamps, and further inland was a tropical forest that obliterated the paths of traders, explorers, and natives. Fourth, for the Europeans the climate was oppressive, and there were savage animals, as well as hostile natives, and finally there were parasitic diseases.

The early humans in Africa initially acquired their parasites from the forest animals. Over the millennia these became the endemic tropical infections that included malaria, yellow fever, sleeping sickness, river blindness, Guinea worm, and blood fluke disease (see p. 45). The African people acquired some immunity to these diseases, and though debilitated, they survived. This was the pattern for tens of thousands of years. The Africans of Black Africa were not subject to "crowd diseases" such as smallpox, tuberculosis, or measles; however, all this changed as more and more Europeans began to intrude.

Figure 15.1 (Left) Lady Africa and her ladies.

Slavery and European Exploration

Henry the Navigator of Portugal (1394-1460) inspired the passion for exploration of Africa and Asia. However, until the mid-1400s, Europeans could not sail down the west coast and return home because of the strong winds from the north all year long. However, by 1440, when better sailing ships had been developed, ships could return to Europe from the west coast, and this made possible further exploration of Africa. In 1442 two of Prince Henry's ships brought back with them a dozen Africans whom they had captured on the west coast in the course of an unprovoked attack on an African village. From this time onward, contact of Africans with the remainder of the world increased, and the health of the African was disrupted more and more.

In 1460, Captain Alvise da Cadamosto, sailing in the service of Prince Henry the Navigator, described the Gambian coast of West Africa in the following way: "there are tree-fringed estuaries, inlets and animals such as elephants, hippopotami, monkeys, and in the villages I saw Muslim traders." Initially, the Portuguese had little inclination to venture further into the interior, because there seemed to be little of commercial value, but after 1482, Portugal sent an expedition to obtain gold that was available in the villages that lined the Bight of Benin. By the 16th century 10% of the world's gold supply was being sent to Lisbon. In the early 17th century the English attempted to obtain the riches of Africa's interior, but their efforts failed. But with the settling of the Americas, a demand was created for another African commodity, one more valuable than gold—slaves.

By the mid-1400s the Portuguese had begun buying small numbers of slaves for sale in Europe; however, this trade in human cargo accelerated once America was discovered. Slaves became an important part of the American economies, which were heavily dependent on crops such as cotton, sugarcane, and tobacco. Slaves were also needed because the local labor force, the Amerindians, had died from European diseases, and the Europeans themselves were highly susceptible to the tropical diseases endemic to the Caribbean. Africans, on the other hand, were relatively immune to tropical diseases such as yellow fever and malaria, so they could work in the Americas. As a consequence, Europeans bought slaves for work in the American colonies, and this trade increased in the 1500s when sugar plantations in Brazil and the Caribbean were established. The Spanish kings initially employed the Portuguese as intermediaries for obtaining their slaves, but as their conquests in the West Indies and mainland America increased, so too did the demand for slaves, and so they inaugurated a system of special contracts under which they bestowed a monopoly upon chosen foreign nations, corporations, and individuals to supply slaves for their American possessions. Soon other European countries including Britain, Holland, and France were in competition with one another for these sources of human flesh. In 1562 the first British contractor, John Hawkins, with Queen Elizabeth as a "silent" partner, embarked on a career of murder and brigandage using his ship *Jesus*. Ten years later Elizabeth knighted Hawkins and commended him for "going every day on shore to take the inhabitants with burning and spoiling their towns." Toward the middle of the 17th century the British became direct exporters of slaves from the west coast through "the African Company" and from the Mediterranean coast of Morocco through "the Barbary Merchants," among whose directors were the Earls of Warwick and Leicester. Though the French were also one of the players in the slave trade, when the Napoleonic Wars began (in 1792) and Britain blockaded France and other trading nations, it was primarily the British ships that moved as many as 50,000 slaves per year. The Atlantic slave trade reached its peak between 1740 and 1790, and it is estimated that from 1515 to 1870 Europeans brought 11 million people from Africa to the Americas on slave ships.

The slave trade also encouraged the local African chiefs to sell their prisoners for European products such as cloth, guns, and iron—materials that the Africans found easier to buy than to develop on their own. By 1780, Arabs and Africans began a slave trade on the east coast of Africa. East coast African slaves were shipped to Zanzibar and other parts of eastern Africa and then to countries around the Red Sea and the Persian Gulf. Slavery uprooted Africans from their homes, and disease spread among them with the greatly increased population movement. The incursions of slavers led to dispersal of rural clans, resulted in tribal warfare, and also contributed to the ever-wider distribution of disease.

The River Niger fascinated historians and explorers since the time of Herodotus in 450 B.C. The prevalent hypothesis in the 18th century was that the Niger rose near Bambuk and flowed eastwards to the sea. In 1788 the Association for Promoting the Discovery of the Interior Parts of Africa (the African Association) was formed in London by 12 English gentlemen. Their aim was to open up trade with the gold-rich lands of the interior, particularly Tellem, a city on the Niger that was said to be built entirely of gold. They picked Mungo Park (1771-1805), a medically trained Scot who was a surgeon with the East India Company, to lead the first expedition to establish communication between the Niger and Gambia Rivers. In 1795, Park and his party of 30 sailed down the west coast of Africa to the mouth of the Gambia River, where the English had already established a fort. After traveling up the Gambia, they moved into the dense jungle of the interior, and a year later they reached their destination (Sego) on the Niger some 750 miles from the coast. The city of Tellem was not found, and by that time Park's team had run out of money, so they returned to England (1797) via the coast. On his return he published *Travels in the Interior Districts of Africa* (1799) and raised sufficient funds for a second expedition to find Tellem and the legendary city of Timbuktu. In April 1805, Park departed with 44 men to penetrate and navigate the Niger; during the 16-week trek to reach the Niger River they suffered from attacks of dysentery, black vomit (yellow fever), and ague (malaria). Three-quarters of the men had died by the time they reached the river, and after 8 months only five were left to follow the Niger's course. Park and the remainder of his party disappeared into the jungle on November 19, 1805. Of the men who started with him, not a single one returned to England, yet only six were killed by something other than malaria or dysentery. In *Travels*, Park recorded some of the diseases he encountered. He observed that although malaria was severe in his party, it was mild among the African adults. He also described elephantiasis, yaws, and leprosy. He saw many cases of gonorrhea but not a single case of syphilis, and he also saw individuals with smallpox, but no measles.

When news of Park's disappearance reached England, the director of the African Society, Joseph Langley, mounted an expedition to find him. The team followed Park's route up the Gambia and crossed through the jungle to reach the Niger River, and then they paddled their canoes toward Tellem. Langley wrote in his 1808 account of the trip, *Dark River*, that Tellem was not a city of gold but instead consisted of a small village of mud huts populated by cannibals. Returning to the river, the Langley team and their canoes were carried downstream; eventually they reached a friendly village, were picked up by a British ship, and returned to England to tell their tale.

The lower course and termination of the Niger River remained unknown in the early 1800s. Richard and John Lander were commissioned by the British government to fill in this gap. Leaving England in January 1830, the two brothers arrived on the Guinea Coast (presently the region between Senegal and Angola) in March, reached the banks of the Niger in June, and then, using a compass as their only scientific instrument, began their river journey by canoe. After 18 months of hardship and adventure they brought back definitive news that the Niger emptied into the Bight of Benin. During their expedition the Landers found traces of Mungo Park near Bussa Falls, and although Richard Lander died in 1832 (during a subsequent Niger expedition to Africa under Major Macgregor Laird with the purpose of founding commercial settlements on the river), he and his brother informed the government of their discoveries in their *Journal of an Expedition to Explore the Course and Termination of the Niger* (1832). John Lander died in England on November 16, 1839, of a disease he contracted in Africa. Following the report of the Lander brothers, two British ships set out northwards on a tributary of the Niger in 1832 under the command of Major Laird. Among the 48 Europeans who set out, only 10 survived the fever. In 1841 the British sent a larger, second British Niger expedition under Captain H. D. Trotter, which moved from the Niger delta with three steamboats, 145 Britons, and 158 locally recruited Africans with the mission to preach the end of slavery and cannibalism. When the steamers reached a point about 100 miles from the sea, fever broke out. As they pushed on further upstream, the number of sick was so great that the expedition had to be terminated and they returned to the coast. After 9 weeks on the river 130 Britons had come down with fever, and of these, 50 died; but of the Africans, only 11 developed fever and none died.

The Congo River is the second-largest river in Africa; it drains an area about as large as Europe without Russia and flows for 2,000 miles through dark forests. Although its mouth had been discovered by a Portuguese explorer (Diego Cão) in 1482, its interior course was uncharted. In February 1816, Commander James Tuckey of the British Royal Navy pioneered a river route up the Congo River; however, the ship could only proceed as far as the rapids, so he landed his party and they moved farther along its banks on foot. By September he and his 44 men were attacked by "remittent fever" and black vomit, and 35 died.

The human losses on the Niger and the Congo provoked horror as well as an intense interest in malaria on the part of the British. The disease agent responsible for malaria was unknown and would not be discovered until 1880, but despite this, a breakthrough in treatment—a triumph of empiricism—occurred in 1640 with the discovery of the fever bark, called by the indians of Peru *kin-kina*. The drug was brought to Europe by the Jesuits. However, despite its successes in curing malaria, there were failures because samples of bark varied greatly in potency. But in 1820, when the cinchona alkaloids (which were the active principle) were isolated and crystallized by two French chemists, Pierre Pelletier (1788-1842) and Joseph Caventou (1795-1877) (and named quinine), then it could be prescribed in a powder form of known strength, and its reliability increased. By 1854 quinine powder was generally available. Quinine was not without its disadvantages. It has a bitter taste and causes ringing in the ears, nausea and vomiting, and in some cases deafness. It also is eliminated from the body rapidly and so had to be taken daily.

The dramatic effect of quinine was seen in 1854, on the third Niger expedition, when not a single man died of malaria. However, its commander, Dr. William Baikie, did not realize that malaria could relapse, and so he stopped taking quinine on leaving the African coast. He died from a relapse of malaria on his way home to England.

Livingstone and Stanley

Dr. David Livingstone (1813-1873) was born in Balantyre in the lowlands of Scotland to a poor religious family. At age 10 he was sent to work as a "piecer" in the cotton factory, but even so he continued to read and study. Young David probably read Mungo Park's *Travels*, and from that time forward Africa became for him a land whose secrets would be revealed only to those who were bold and brave.

Although the 1833 Act of Abolition abolished slavery in the British Empire, the practice did continue in other nations. In response, there came into being an active church-sponsored British missionary movement. The Missionary Society for the Extinction of the Slave Trade and for the Civilization of Africa was founded in London in 1839-1840. Unlike the African Association, the missionary movement concerned itself with the salvation of the souls of the African people rather than the accumulation of geographic and scientific information. By age 20 Livingstone had decided to become a doctor and a lay missionary; he attended medical college at Glasgow University and then gained admission to the London Missionary Society. At age 27 he embarked for Africa, and 13 weeks later, after arrival at Cape Town, he made his way by ox wagon to the northernmost mission station at Kuruman. It was from here that he ventured north into the jungle, and in 1848, accompanied by two big-game hunters, William Oswell and Mungo Murray, he discovered Lake Ngami. This discovery made his reputation as an explorer and convinced him that his primary inclinations were in exploration rather than as a sedentary preacher. In 1850, Livingstone, now married and with a family of three, set out to revisit Lake Ngami and to find a water highway capable of being quickly traversed by boats to a large section of the well-peopled interior territory in order to establish missions and to drive out slavery, then being carried out by the Portuguese, who were swapping arms for people at the rate of one gun for one youth. At Ngami, Livingstone and his children had their first experience with malaria, but he was already well aware of the efficacy of quinine and so he administered from his kit a large dose that effected relief from the fever. The "Livingstone prescription" consisted of 3 g calomel (mercurous chloride), 3 g quinine hydrochloride, and 10 g rhubarb, plus grape or cinnamon flavoring (to reduce the bitterness of the quinine) mixed with a bit of alcohol. In 1851, Livingstone and Oswell ventured forth again and this time discovered the lower Zambezi River. Although this glorious, magnificent, and beautiful river was "discovered" by Livingstone (and others over the next 20 years), it was well known to the Africans living in the interior, and on most expeditions they served as his guides. However, it was Livingstone and other European explorers who provided the observations of latitude, longitude, and altitude that enabled accurate maps to be prepared. It is for this reason that their finds have been called discoveries.

Concern for the medical dangers and the welfare of his family prompted Livingstone to dispatch them to England from Cape Town. This was not an act of pure altruism on his part; indeed, it freed him to continue his explorations unencumbered by wife and children. On June 8, 1852, he left Cape Town by wagon and began his 4-year epic trans-African journey, which covered 6,000 miles. Usually Livingstone traveled only 6 miles a day, and his equipment for the journey consisted of two guns, three muskets, biscuits, a few pounds of tea and sugar, 20 lb of coffee, a bag of spare clothing, medicines, a box of books, a slide projector, a chronometer, a thermometer, a compass, a horse blanket for a bed, a gypsy tent, and beads for bartering with the natives. Throughout the trek—partly on foot, partly by oxen, and partly by canoe—malaria, prolapsed piles, fatigue, attacks by hostile tribes, and torrential rains hampered him. Livingstone was aware that the scourge of Africa was malaria and that "it was destined to preserve inter-tropical Africa for the black races of mankind." He also recognized that eradication of malaria would be critical if missionary and commercial penetration of Africa were to succeed and remarked that the inventor of the mosquito net was deserving of a statue in Westminster Abbey. Although Livingstone was aware of the correlation between the myriads of mosquitoes and his persistent bouts of malaria, he never pursued the connection, believing that the fevers were due to exposure to an the east wind arising from the swamps, to drinking of milk in the evening, or to drinking water that flowed over granite. Livingstone was the first to recognize the relationship between the tsetse fly and nagana (an animal form of sleeping sickness) in cattle; he also described the river blindness worm (*Onchocerca*) in the anterior chamber of the eye; and when he reported on the bloody urine he saw in his porters, he provided evidence for the presence of blood fluke disease.

Between August 1852 and May 1853 he traveled from Kuruman to Linyanti, and from November 1853 to May 1854 he traveled down the Zambezi from Linyanti to the west coast city of Luanda. After 4 months of rest he turned inland for an 11-month trek back to Linyanti, and en route to Quilimane (which he reached in May 1856), on the Indian Ocean, he discovered the great waterfall called "the Smoke That Thunders" and named it Victoria Falls after his queen. Livingstone returned to Britain just before Christmas 1856, where he was celebrated as a folk hero and a great explorer, and he transcribed his African journals into an 1857 bestselling book, *Missionary Travels*. Anxious to return to Africa, Livingstone convinced the British government and the Royal Geographic Society to sponsor a Zambezi River expedition (1858-1864) to explore the river and its tributaries. He also accepted appointment as consul in the District of Quilimane and the Eastern Coast of Africa. Together with six assistants, Livingstone and his family left England in March 1858, arrived in Cape Town in April, and a month later departed for the mouth of the Zambezi. Recognizing the perils of malaria, the staff were instructed to take daily doses of quinine in a glass of sherry. The Zambezi proved not to be navigable due to impassable rapids, but forays on foot led to the discovery of Lake Nyasa (September 1859), now known as Lake Malawi. During the second half of 1860 he joined the exploration party of Bishop Charles Mackenzie's Universities' Mission to the Shire Highlands. Despite quinine prophylaxis, malaria struck the Mackenzie party within a year. This included Livingstone's wife, and she and several other members died of the fever.

After the Zambezi expedition Livingstone returned to London (July 1864), but this time he did not receive a hero's welcome, since the expedition was considered an expensive failure. In addition, the discoveries of Richard Burton and John Hanning Speke and their search for the source of the Nile diminished his accomplishments. Yet despite this, he wrote a second book, *The Zambezi and Its Tributaries*, and convinced the Royal Geographic Society that there was still much to be discovered about the lakes and the central African watershed. In March 1866, at age 53, Livingstone embarked on his last journey into the interior of Africa. The plan was to start out in Zanzibar, march northwards from Lake Nyasa to Lake Tanganyika, and proceed along its east coast to Ujiji, where the great caravan route to the interior ended. He reached Ujiji in March 1869, but the travel was exhausting and he was plagued with malaria, bleeding piles, and tropical ulcers. On this venture he became acquainted with "earth-eating," or "sufura," finding it prevalent among the slaves. This condition, known today as geophagy, remains a sign of hookworm disease, although Livingstone did not make the connection. During the following 2 years he traced the lakes and the rivers with which they communicated, but little of this was known in England. Not hearing of Livingstone's whereabouts and fearing for his safety, the *New York Herald* sent its star reporter, Henry Morton Stanley, to seek news of his health. Stanley arrived in Zanzibar in January 1871. Starting out with 192 men and 6 tons of equipment, Stanley met Livingstone at Ujiji on November 10, 1871, with the now famous line: "Dr. Livingstone, I presume." The two remained together for 4 months, and although Stanley encouraged Livingstone to return to England, he refused, believing he was on the verge of a great discovery. Stanley retraced his path to the coast and arrived in England to a hero's welcome. Livingstone remained in Africa, where he died on May 4, 1873. His heart was buried in Africa, but his mummified body was returned to England, where it is interred in Westminster Abbey.

Stanley and the scramble for Africa

David Livingstone was broad-shouldered, of medium height, with a determined demeanor but gentleness; he was also a manic-depressive. Courageous and enthusiastic, he marched across Africa, with Bible in hand, condemning slavery, speaking the native languages, dispensing medicine, and loving the African continent and its people. When Stanley met him at Ujiji, he was emaciated and in ill health, yet he was dressed for the occasion: he wore a gold-braided consular cap, a woolen jersey, gray tweed trousers, and tight shoes. Stanley, on the other hand, rode on horseback, wearing white flannel trousers, a white-powdered pith helmet with a neatly folded headband, and shiny Wellington boots.

Henry Morton Stanley (1841-1904), born John Rowlands in Denbigh, Wales, of unmarried parents, was brought up in crushing poverty. As a child, he was placed in a workhouse, where he was exposed to a sadistic master. He escaped, came to the United States as a cabin boy, jumped ship, and was adopted by a New Orleans merchant named Henry Hope Stanley, shortly thereafter taking the name of his adopted father and changing his middle name from Hope to Morton. Stanley fought in the U.S. Civil War, worked as a deckhand and a clerk, and, after discovering he had a way with words, became a newspaper reporter. Stanley was described by Queen Victoria as as "a determined, ugly little man with a strong American twang." His face was red and round, with high cheekbones, a wispy moustache, and penetrating gray eyes. In contrast to Livingstone, Stanley was tough and intent on gaining fame, not caring who got hurt in the process; for his entire life he was overly sensitive to criticism, perhaps because of the stigma of his illegitimate birth. Stanley was not above stretching the truth. He was also a harsh leader who drove his African porters and was arrogant in his belief in his own racial superiority. Whereas Livingstone loved Africa and its people, Stanley disliked both, saying: "I do not think I was made for an African explorer for I detest the land most heartily." But Stanley's efficient and powerful manner contributed to several successful explorations, and in the minds of many he remains one of Africa's greatest explorers.

When Stanley heard of Livingstone's death, he promised to return to Africa to carry on his work. Three years later Stanley was commissioned by the *New York Herald* and London's *Daily Telegraph* to continue the explorations begun by Livingstone. With three British companions and 300 Africans, he left Zanzibar in 1874, circumnavigated Lakes Tanganyika and Victoria (1878), navigated down the Lualaba and Congo Rivers, and made the trip to the Atlantic (1879) via the Congo River. All the other whites and 150 Africans died during the expedition.

Returning to Europe, Stanley tried to interest the British government in further exploration and development of the Congo Basin but met with no success. His expeditions did, however, attract another European monarch, Belgium's King Leopold II, a man of boundless energy and ambition. King Leopold was a devoted supporter of the Roman Catholic church and very rich. He was a promoter of scientific research and a patron of missionary efforts. Leopold II had founded (in 1878) the International Association of the Congo, financed by a consortium of bankers, and he recruited Stanley to help him explore the Congo under the auspices of this association. Stanley arrived at the mouth of the Congo River in 1879 and began the journey upstream. He founded Vivi, the first capital, across the river from present-day Matadi and then moved farther upriver, reaching a widening he named Stanley Pool (now Pool de Malebo) in mid-1881. There he founded a trading station and the settlement of Léopoldville (now Kinshasa) on the south bank. The north bank of the river had been claimed in 1880 by Pierre de Brazza for France, leading ultimately to the creation of the colony of French Congo. Stanley claimed for Belgium that part of the Congo that had not been claimed for France.

The road from the coast to Vivi was completed by the end of 1881, and Stanley returned to Europe. He was back in Africa by December 1882 and sailed up the Congo River to Stanleyville (now Kisangani), signing more than 450 treaties on behalf of Leopold II with persons described as local chieftains who had agreed to cede their rights of sovereignty over much of the Congo Basin. Stanley's series of journeys and treaties throughout the Congo became the Congo Free State (1885-1908), with its capital Leopoldville (now Kinshasa) and its other major city Elizabethville (now Lubumbashi). From 1908 to 1960 the country was called the Belgian Congo, from 1960 to 1994 Zaire, and since 1994 the Democratic Republic of the Congo. Bordering on the east was the French Middle Congo, with its capital Brazzaville.

The period of European colonial expansion, termed by some writers as "the scramble for Africa," began in the 1880s. European powers were able to pick up huge parcels of land cheaply and with little risk from armed resistance since they now possessed two critical resources: quinine and firearms (repeating rifles and machine guns). The power differential between Europe and Black Africa was enormous and would remain unchanged for the next 70 years. In 1884, Germany laid claim to Togo, Cameroon, Tanganyika, and Namibia. The British, who already had a presence in West Africa (Gambia and Nigeria), secured land around Mombasa and Berbera to ensure a safe sea route to India, and then Germany proclaimed sovereignty over Dar es Salaam. Great Britain formalized occupation of the Guinea Coast (an old geographic designation for the west coast of Africa that now stretches from Senegal to Angola), and the French established themselves on the right bank of the Congo. In this way the delicate balance of power in Europe was preserved, but power could be expressed by European nations in the less sensitive area of colonial lands. A gain by one country was countered by another in a kind of overtrumping in a card game.

The Conference of Berlin, held in 1884-1885, was convened to settle disputes among the European nations in Africa. By international agreement at the Berlin Conference all signatories had to be notified of a power's intent to annex. The term "spheres of power" entered into the treaty language, and in 25 years almost all of Africa was partitioned in a manner that had no reference to history, geography, or ethnic considerations. To Europeans these were simply lines traced on a map that indicated a possession. By 1914 nearly 11 million square miles of Africa was in the hands of European nations. Only 613,000 square miles was independent. At the outbreak of World War II there were only three independent African states: Egypt, South Africa, and Liberia.

Separately, the Berlin Conference recognized Leopold II's International Association of the Congo, which had already adopted its own flag, as an independent entity. Shortly afterward the association became the Congo Free State. By the General Act of Berlin, signed at the conclusion of the conference in 1885, the powers also agreed that activities in the Congo Basin should be governed by certain principles, including freedom of trade and navigation, neutrality in the event of war, suppression of the slave traffic, and improvement of the condition of the indigenous population. The conference recognized Leopold II as sovereign of the new state.

Shortly thereafter, in order to meet the conference's legal requirement of "effective occupation," Leopold II proceeded to transform the Congo Free State into an instrument of colonial hegemony. Indigenous conscripts were promptly recruited into his nascent army, the Force Publique, manned by European officers. A corps of European administrators was hastily assembled, which by 1906 numbered 1,500 people, and a skeletal transportation grid was eventually assembled to provide the necessary links between the coast and the interior. The cost of the enterprise proved far higher than had been anticipated, however, as the penetration of the vast hinterland could not be achieved except at the price of numerous military campaigns. Some of these campaigns resulted in the suppression or expulsion of the previously powerful Afro-Arab slave traders and wary merchants. Only through the ruthless and massive suppression of opposition and exploitation of African labor could Leopold II hold and exploit his personal fiefdom. Though he was respected as a diplomat and philanthropist, promoting research to prevent sleeping sickness and proclaiming how much he loved the African, his actions suggest that the real motive was greed. The Belgian goal was to teach Africans not to work in the "childish" pursuits of their own culture but to be organized in rational routines of productive wage labor in the European manner for European employers. Such labor was considered a civilizing influence, and a profitable by-product was cheap labor. Agricultural programs required the Africans to raise designated crops—cotton for export and food for the towns and mines within the colony, neither of which threatened European interests, nor did they ensure the health and well-being of the native population. Leopold II, through his thousands of European civil servants and officers, established a system of state socialism and issued decrees by which all land, all rubber, and all ivory were the property of the state—himself. He made it illegal for the natives to sell these to other Europeans or for other Europeans to buy from the natives. His decree was to "neglect no means for exploiting the forests." Vast profits were accumulated by Leopold in a simple scheme: each village was ordered to bring in a specific quota of rubber, and the workers were to neglect all other activities until this quota was met. If they failed, the women were taken away as hostages or became parts of harems, and other native troops were sent into the villages to brutalize, kill, and spread terror. To prove that they had killed the villagers, they had to bring in one right hand for every cartridge used. Due to this, and to the accelerated spread of sleeping sickness by the movement of hostages from one province to another, in 15 years the native population was reduced from 20 million to 9 million.

The increased population movement and the punitive expeditions to quell uprisings brought European diseases to the virgin territories of Africa. Forced labor, a feature of colonial rule, also spread disease: for example, when rubber workers were recruited from the Congo Free State, thousands left their villages and moved up the tributaries of the Congo; these new settlers brought their diseases far across the continent. The men of Mozambique, Zimbabwe, and Malawi took their diseases to the gold mines of Johannesburg and brought back with them to their homelands new strains of disease organisms. Those who lived in the highlands of Kenya had little experience with malaria, so when they were recruited to labor on harbor works in Mombasa, they died of malaria at an appalling rate.

The imposition of efficient administrations by the Europeans on their colonies tended to concentrate people in larger villages and also forced them to grow crops that were highly remunerative, such as coffee, tea, and groundnuts, which removed acreage needed by Africans for food production. The movement of people between village and forest spread disease, and the roads and railways provided increased opportunities for disease movement. Different communities with different backgrounds, resistance patterns, and diseases became mixed, and this had disastrous consequences for the Africans. And finally, a different kind of destabilizer entered the picture—firearms were brought to Africa in large numbers. Some of those guns were obsolete firearms that had been replaced by breech-loading rifles in the American Civil War, but they were snatched up by the local tribal chiefs. There was tribal warfare, and with an expansion of tribal domains, there was more movement and more exchange of diseases. Civil wars, social upheavals, economic disruptions, and endemic diseases, as well as diseases introduced by the European colonists, including tuberculosis, whooping cough, diphtheria, plague, typhoid fever, and cholera, became more and more a part of Africa's decline in health. When Livingstone traveled in Africa between 1841 and 1873, he noted that the Africans were not a sickly people, but after 1880 they were described very differently by another writer: they were "a hive of gangrenous germs."

The Europeans were quite aware that the native Africans were comparatively free of the indigenous diseases. The nature of their immunity was unknown, and some of the European whites feared that if it was survival of the fittest they would eventually be replaced by the better-endowed blacks. Such a thought was not very comforting to the whites, so they concocted fanciful explanations. One was that the Africans had a greater facility for sweating and this threw off the noxious vapors and humors in the blood. But this theory was abandoned when a group of ex-slaves from

America were settled in Freetown in Sierra Leone and came down with the same fatal diseases as the Europeans.

The Europeans eventually overcame the "immunity problem" by chemotherapy (and force). With quinine it was possible for Europeans to enter Black Africa and live without fear of death. Quinine therapy combined with the introduction of the breech-loaded rifle gave the Europeans increased confidence, and exploration moved into high gear with a move toward the Rift Valley. East Africa was not only strategically placed but it had great plains, immense game, and snow-capped mountains. The British of India were encouraged to spend their leaves there, and a new brand of intrepid explorers was born. These were of the safari type, and their adventures could be recorded by sketches and camera. Big-game hunting and ivory trading became the rage. Missionaries in increasing numbers came to Africa, and they opened the continent even more. There were handbooks on how to travel and how to avoid or cure disease, and the number of European visitors increased. The terror of Africa for Europeans soon disappeared.

Endemic Diseases of Africa

Sleeping sickness

One of the deadliest of African diseases is sleeping sickness. It is undoubtedly of African origin and today remains so. The disease, in humans and cattle, occurs in the fly belt of Africa between 20°N and 20°S, extending over 10 million square miles, an area larger than the continental United States. The disease precludes farmers from keeping cattle and small ruminants and accounts for low livestock productivity. Indeed, the production of this region of Africa is 1/70 that of the animal protein produced in Europe on the same amount of land. The losses in milk, meat, hides, manure, and tractive power result in an estimated annual loss of $5 billion, and 30% of the 150 million cattle are at risk.

The story of human sleeping sickness begins with fishes and frogs. In 1841, Professor G. Valentin of Berne, Switzerland, gave a very brief description of a protozoan he found in the blood of a trout, and he noted that it moved by means of its undulating membrane. Two years later David Gruby, working in Paris, discovered a similar organism in frog blood, and because its swimming motion suggested the action of a corkscrew, he called it *Trypanosoma sanguinis*. The name *Trypanosoma* is derived from two Greek words, *trypano*, meaning "auger" or "screw-like," and *soma*, meaning "body"; the species name *sanguinis* comes from the Latin word for blood. Trypanosomes were also described from the blood of field mice, voles, and rats, but they were considered mere curiosities and of no economic, medical, or veterinary importance. All that changed in 1880 when Griffith Evans, an English veterinarian serving in Punjab, India, found trypanosomes in the blood of horses, mules, and camels suffering from a fatal wasting disease called surra. Inoculation of blood containing trypanosomes into healthy animals produced surra. Evans was convinced that the trypanosome (later named *Trypanosoma evansi*) was a parasite, but he did not discover how horses, mules, and camels became infected. (Leonard Rogers discovered the vector, a biting stable fly, in 1899.) In 1892, Alphonse Laveran, who had discovered the malaria parasite in 1880 and had then become interested in other blood parasites, summarized all that was known about trypanosomes in 11 pages and 18 references!

In the early 1890s the British colonial farmers of Zululand were faced with the decimation of their European breeds of cattle by a wasting disease called nagana, a word meaning in Zulu "in low or depressed spirits." At the farmers' urging, in 1894 the governor of Zululand requested that Surgeon-Captain David Bruce (1895-1931) of the British Army carry out an investigation to determine the cause of nagana. Bruce was trained as a bacteriologist and initially believed that the disease was caused by a bacterium, but none could be cultured. However, when he examined the blood of diseased cattle, he discovered the cause and he described it: a rapidly vibrating body, lashing about among the red blood corpuscles, probably a trypanosome (Fig. 15.2).

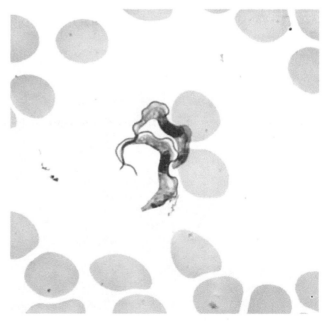

(In 1899 that trypanosome was named by Plimmer and Bradford *Trypanosoma brucei* after its discoverer.) Bruce then went on to establish Koch's postulates for nagana: if he injected blood from cattle suffering with this disease into dogs, severe wasting symptoms resulted, and abundant trypanosomes were found in the dogs' blood. He wrote: "the clinical features of nagana are defined by the constant occurrence in the blood of a protozoan parasite." In 1895, Bruce discovered the vector for nagana: the bloodsucking tsetse fly (genus *Glossina*). But where did the tsetse fly acquire the trypanosome? Bruce hypoth-

Figure 15.2 *Trypanosoma* the causative agent of African sleeping sickness. Courtesy CDC Public Health Image Library #1182.

esized, and then proved, that wild game—buffalo, wildebeest, and bushbuck—were the source of infection and that transmission was the result of the bite of the tsetse flies that infested the area in which these game animals lived. Bruce's work on the "fly disease" nagana set the stage for a fuller understanding of the human disease known as African sleeping sickness.

Between 1375 and 1406 an Arab historian had reported human sleeping sickness (Fig. 15.3) in Mali. It went largely unnoticed. During the late 1700s increased numbers of British missionaries and explorers went to Africa, where they established settlements and colonies. In 1702 an English naval surgeon, John Atkins, described a disease in Africans living along the Guinea Coast. Atkins believed its cause was the accumulation of phlegm around the nerves, which impaired their function, along with a poor diet and a lack of mental exercise, which led to the symptoms of the "sleepy distemper." But in 1803, Thomas Winterbottom, a physician working in the colony of Sierra Leone, published an account of the "African lethargy." He offered no cause, but recognized a telltale clinical characteristic: swelling of the cervical lymph nodes (Fig. 15.4). This swelling, called "Winterbottom's sign," was recognized by

Figure 15.3 Victims of African sleeping sickness. Courtesy Wellcome Library, London, CC-BY 4.0.

slave traders as an indicator that the African would suffer from lethargy; they either would not buy such slaves or, if they could, would rid themselves of such as soon as the sign appeared. In hindsight, it can be recognized that Winterbottom had described a clinical condition of individuals suffering from Gambian sleeping sickness, but before 1901 there was no record of human disease caused by a trypanosome. In May 1901 a 42-year-old Englishman who worked on the steamships plying the Gambia River came down with a fever. He was admitted to the hospital and was treated with quinine for malaria without success. When his blood was examined, no malaria parasites could be found; however, there were trypanosomes. In 1902 this trypanosome of humans was named by Joseph Everett Dutton of the Liverpool School of Tropical Medicine *Trypanosoma gambiense*. This was the first demonstration of trypanosomes in a European patient suffering from a disease resembling sleeping sickness. To carry out further work on African trypanosomes, Dutton traveled from England to Gambia. There he found that a very small proportion of the local natives also carried the parasite in their blood but suffered no ill effects; he suggested that they probably served as a reservoir to infect the Europeans. Dutton's death in 1903 (of relapsing fever) cut short his brilliant research career and prevented the completion of studies on transmission of Gambian sleeping sickness.

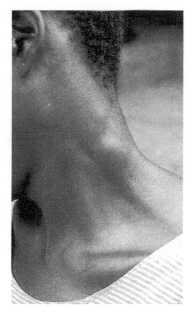

Figure 15.4 Swelling of the lymph node in the neck known as Winterbottom's sign. Courtesy Wallace Peters and Peter Janssen.

In 1901 a severe epidemic of sleeping sickness broke out in Uganda, and so the Royal Society of London sponsored a three-person scientific commission headed by Bruce to investigate its cause. By 1902 the commission had discovered that the distribution of individuals with Winterbottom's sign corresponded with the distribution of sleeping sickness. Further, examination of the cerebrospinal fluid from a case with sleeping sickness by one of the members of the commission, Dr. Aldo Castellani, showed trypanosomes. However, Castellani, who was supposed to investigate the evidence for a bacterial cause, became convinced that the disease was due to a streptococcus, not the trypanosome, and he did not appreciate his own finding until much later. Castellani refused to tell Bruce of his discovery unless he agreed that the discovery be published under his name alone, and that no credit or information be given to the other member of the commission, D. N. Nabarro. Bruce agreed. Then he and Castellani were able to show that in the cerebrospinal fluid from 34 other cases 20 had trypanosomes, whereas in 12 control cases none were found. Castellani published the findings on his own and took full credit for himself. Bruce honored his agreement with Castellani but wrote in a report: "For the sake of a future historian, at the time of the arrival of the commission Dr. Castellani did not consider … that this trypanosome had any causal relationship to the disease but thought it was an accidental concomitant like Filaria. As Dr. Castellani has not entered into any detail respecting these matters in his reports, it is thought advisable to supplement his account with the above, as the history of the discovery of the cause of such an important disease must always be of interest." Bruce suspected that the tsetse fly was involved in the transmission of Gambian sleeping sickness. Later, when it was found that monkeys were susceptible to the disease, and that a tsetse fly fed on a human case could transmit the disease to monkeys, Bruce concluded that sleeping sickness was a human tsetse fly disease.

In 1910, J. W. Stephens of the Liverpool School of Tropical Medicine discovered a new species of trypanosomes: it was from a patient with sleeping sickness who had acquired the disease in 1909 in Rhodesia, an area where *T. gambiense* and its vector (*Glossina palpalis*) did not occur. When Stephens inoculated this patient's blood into rats, the trypanosome caused a more acute disease and had a slightly different morphology than the more benign *T. gambiense*. Stephens, in collaboration with H. B. Fantham, described and named the new species *Trypanosoma rhodesiense*. In 1912, Bruce headed another sleeping sickness commission in an area near Lake Nyasa. He found *T. rhodesiense* in the blood of one-third of the 180 game animals examined. Bruce compared the morphology of *T. rhodesiense* with *T. brucei* from cases of nagana and found them to be identical. He concluded: "*T. rhodesiense* is neither more nor less than *T. brucei*, and the human trypanosome disease in Nyasaland is nagana." The vector was suspected to be the most abundant tsetse, *Glossina morsitans*. Thus, early in the 20th century the causative agent, the vectors, and the pathogenesis of African sleeping sickness came to be discovered.

T. gambiense gives rise to a milder chronic infection, is found in West and Central Africa, and is transmitted by riverine species of *Glossina* (*palpalis* and *tachinoides*) that are associated with human habitation. Although occasionally pigs can serve as a reservoir host, the tendency is for parasitemias to be low in the human, and so transmission to other animals is infrequent. Asymptomatic individuals may harbor parasites in the blood for long periods of time and could be a source of infection for the vector. The disease is an anthroponosis: fly to human to fly to human. *T. rhodesiense* is the East African form and is the more virulent subspecies, producing an acute infection. It may also be the subspecies that is more recently evolved, i.e., within the past 100 years. *T. rhodesiense* is transmitted by the dry savannah bush flies *G. morsi-*

tans, *Glossina swynnertoni*, and *Glossina pallidipes*. In wild game animals (bushbuck, gazelles, and wildebeest) the disease is mild, but in humans it is highly pathogenic, killing in 4 to 6 months if untreated. The disease is an anthropozoonosis: fly bites infected ungulate, fly becomes infected, and then fly transmits disease to human.

The spread of sleeping sickness was increased in about 1890 because of another infection brought into Africa by cattle from outside the continent: rinderpest, a virus of cattle and game with high mortality. The kill-off by rinderpest upset the balance of the eastern savannah, where there was a delicate balance between cattle, game, humans, and tsetses. When rinderpest killed the cattle and game, then the tsetse sought an alternative host: humans. And when the game was gone, the amount of vegetation also increased because grazing was reduced. Now the flies had new breeding grounds, and so sleeping sickness extended its grip. Rinderpest also decimated the cattle herds of the pastoral Africans, so that protein deficiency became more severe and human susceptibility to other diseases increased.

Control of sleeping sickness is yet to be achieved. Since the vector of *T. gambiense* thrives in hot, humid regions and along swift-running streams, clearing of vegetation has worked to a limited degree, but *T. rhodesiense* has a reservoir in game animals where it causes little harm, and it also has a tsetse vector that lives in the dry savannah, which is impossible to clear. Today there are reported to be ~30,000 human cases per year, with 50 million people at risk; however, the true number of infections may be 10 to 15 times greater. Treatment of sleeping sickness is of limited value and depends on early use of rather toxic drugs administered intravenously. Pentamidine, melarsoprol, and suramin are all used.

Figure 15.5 Young girl leading a blind man suffering with river blindness. It is not uncommon in parts of Africa to see blind adults being led to the fields by children who have not yet lost their sight. Courtesy WHO.

What about a protective vaccine against sleeping sickness? All efforts to date have failed, largely because the trypanosomes in the blood are covered with a surface coat that is shed periodically, thus providing a constantly changing target for the immune system. It is estimated that there may be several thousand variants, expressed sequentially. So every time the immune system mounts an attack against one variant surface coat, the trypanosomes stop making that variant and switch to a new surface coat, which has to be reacted to by the immune system anew. Absent a vaccine and with poor drugs for treatment, sleeping sickness and nagana continue to plague Africa.

River blindness

In the region of Africa called the Upper Volta, and in Sierra Leone, Liberia, northern Ghana, and Central Africa, it is not uncommon to see blind older men being led by the young to the fields (Fig. 15.5). The cause of their blindness is the worm named *Onchocerca volvulus*. In some African localities some 80 to 100% of individuals are infected with *Onchocerca*. This parasite is also found in the highlands of Guatemala and southwestern Mexico in the coffee-growing areas. River blindness is probably an African disease, brought to the Americas with the importation of slave laborers from Africa to cultivate sugarcane and coffee. However, it was not described from the Americas until 1915.

River blindness is transmitted by a blackfly called *Simulium damnosum*, which carries the microscopic infective juvenile stages (called microfilariae) of *Onchocerca* (Fig. 15.6). The disease is commonest in rich, oxygenated water in rapidly flowing rivers where the blackflies breed—the same areas where people congregate for bathing, fishing, drawing water, and defecating. Thus, the river is an excellent place for the fly to meet its human host. The infected flies

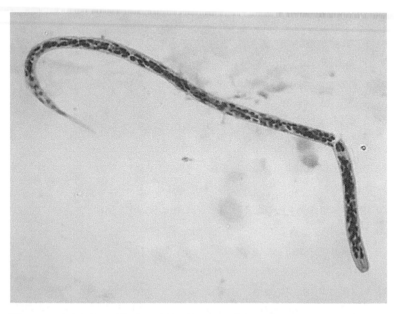

Figure 15.6 Microfilaria of *Onchocerca*. The microscopic worms are about 300 microns in length or ~20 times the diameter of a white blood cell. Courtesy CDC Public Health Image Library, #1147.

bite, and the infective stages migrate out of the mouthparts and enter the wound. The worms become mature at the site of the bite and become enclosed in a subcutaneous connective tissue nodule (Fig. 15.7a and b). Here they are trapped, the worms mate, and the adult female worms give rise to live microfilariae, which migrate throughout

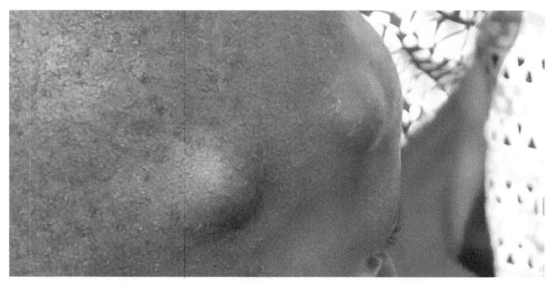

Figure 15.7a Young boy with two nodules containing *Onchocerca* adults. Allen JE, Adjei O, Bain O, Hoerauf A, Hoffmann WH, et al. 2008. Of Mice, Cattle, and Humans: The Immunology and Treatment of River Blindness. PLOS Neglected Tropical Diseases 2(4): e217; CC-BY 4.0.

Figure 15.7b Three male *Onchocerca* and one adult female removed from nodule. Courtesy P. Soboslay.

the body and are a source of infection for the biting blackflies. Development in the fly may take 1 to 3 months, and an adult female (which can live for up to 2 years) can produce 1,500 microfilaria each day.

Why is it called river blindness? The microfilaria lives in association with a bacterium (*Wolbachia*), and in concert they cause the pathology. The circulating microfilariae migrate into the eye, invading the cornea, retina, and choroid and creating mechanical damage, and due to an inflammatory response to the bacterium there is a partial loss of vision. Ultimately the optic nerve itself is affected and total blindness results (Fig. 15.8). It is the repeated exposure of humans to the bite of the infected river-dwelling blackflies that ultimately results in blindness.

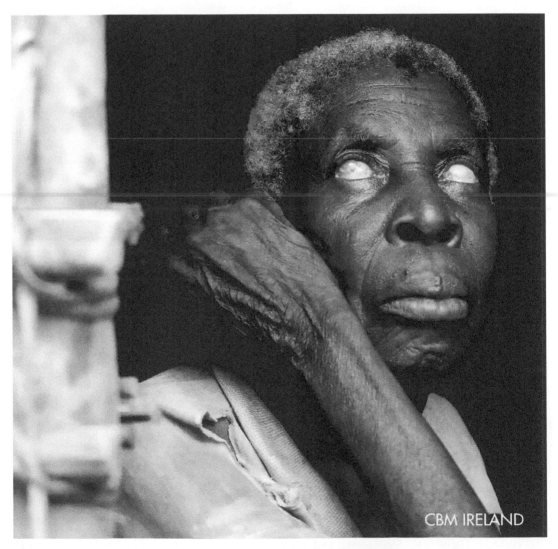

Figure 15.8 An adult woman blind from *Onchocera*. Courtesy CBM.

The search for and identification of the cause of river blindness took 70 years. In 1875, John O'Neill, a British naval surgeon attached to the HMS *Decoy* at Cape Coast Castle in the Gold Coast (Ghana), became interested in a scabies-like skin disease (known as "craw-craw"). After examining material from the skin nodules in a drop of water under a microscope, he observed thread-like undulating worms, the microfilariae, and concluded that these were "the immediate cause of the complaint." In 1890, Rudolph Leuckart, using samples collected and sent by missionaries from this same region, described the morphology of the adult female and male worms. The adult worms were coiled together to form a knotted ball, with female worms 30-50cm in length and the males 15-45mm (Fig. 15.7b). In 1901, William T. Prout described and named the worms *Onchocerca* (for "hooked tail") *volvulus* (for "roll about"). Despite these findings, the life cycle of the parasite as well as its transmission from person to person remained a mystery. Then, in 1915, Rodolfo Robles discovered the presence of onchocerciasis among coffee plantation workers in Guatemala; he removed a nodule from the forehead of a boy and found within it the worms described earlier by Leuckart. Robles ruled out that transmission took place via water, and noted that the laborers who lived outside the disease zones and entered areas of endemicity only during the day to work became infected with onchocerciasis. He hypothesized that the vector of the disease was a day-biting insect, and, more specifically, black-flies that were found in the areas of endemicity. Robles, however, never verified his hypothesis. Eleven years later, in 1926, Donald Blacklock of the Liverpool School of Tropical Medicine, working in Sierra Leone, picked up where Robles had left off. Blacklock found that the microfilariae were found not in the blood but in the skin, and he postulated that the vector must be an insect that can damage the skin and dislodge the juveniles in order to feed on blood. He discovered a biting, blood-feeding blackfly, *S. damnosum*, near streams where the disease was endemic. Blacklock collected several wild blackflies and proceeded to infect them by allowing them to bite infected individuals. He then proceeded to trace the development of the parasite in the gut, thorax, head, and proboscis of the flies and in this way provided evidence for the transmission of the disease. This work was later confirmed by other workers in other parts of Africa, and in 1946, Jean Hissette was able to link onchocerciasis with blindness in Ghanaians living along the Volta River.

Incidence

Worldwide, there are more than 120 million people at risk of contracting onchocerciasis, with some 18 million people infected. More than 99% of the disease burden is in Africa; however, river blindness is not limited to Africa. It occurs in 13 foci in six countries of the Americas (Brazil, Colombia, Mexico, Guatemala, Ecuador, and Venezuela), where it was introduced through the slave trade. Starting in the early

16th century, slaves from areas of West Africa where the disease is heavily endemic were brought to Central and South America, bringing with them *O. volvulus*. The slaves then migrated within the colonies, including coffee plantations, and carried with them the parasite. Suitable *Simulium* species were present in the Americas, and so the parasites were transmitted to the indigenous American population and were able to spread further through migration, including among certain contiguous border countries such as Guatemala and Mexico. Through migration and the presence of suitable vectors, onchocerciasis moved into relatively small, circumscribed areas of Ecuador, Colombia, Guatemala, Mexico, and northern Venezuela. The economic and social consequences of river blindness are profound, and surely much of the local history is tied to the progressive depletion of human resources. In some regions the people became so accustomed to blindness that they could not understand why all people did not become blind. In both Africa and the Americas those coming in close contact with blackflies could look forward to a working expectancy of no more than a few years before suffering from photophobia and then being condemned to darkness for the remainder of their lives.

Diagnosis and treatment

How is river blindness diagnosed? Small snips of skin are removed from the shoulders or head of infected individuals; these are placed in a drop of salt solution, and the emergence of microfilariae is observed under the microscope. Other methods use slit lamps to detect microfilariae in the cornea and anterior chamber of the eye. More recently, detection has been accomplished by screening for parasite DNA and parasite antigens.

How can river blindness be prevented or, better yet, eliminated? (i) Destroy the blackfly breeding sites. This is sometimes especially difficult to effect with rapid-flowing rivers and streams. (ii) Kill the blackfly vector by spraying a biological control agent such as *Bacillus thuringiensis* (called Bt). (iii) Surgically remove the nodules to decrease the parasite burden and the source of microfilariae. (iv) Treat humans with the drug ivermectin (Mectizan), which kills the microfilariae. Since 1987, Merck has donated Mectizan, "as much as necessary for as long as necessary," through local and international nonprofit organizations, national governments, their ministries of health, the World Health Organization (WHO), and the World Bank. A total of 11,069,285 ivermectin treatments were administered in the Americas during 1993-2012. By the end of 2012 transmission of the infection had been interrupted or eliminated in 4 of 6 countries and halted in 11 of 13 foci. In 2013, Colombia became the first country to eliminate river blindness, and Ecuador and Mexico followed in Colombia's steps in 2014 and 2015, respectively. Since 1995 no new blindness has been attributed to onchocerciasis in the Americas.

The Carter Center (in Atlanta), in conjunction with the ministries of health, has broken river blindness transmission in Uganda and Sudan by providing twice-per-year Mectizan treatments. In Uganda, 2.7 million residents are now free of the disease, and in Abu Hamad, Sudan, the most isolated focus in the world, more than 100,000 people are no longer at risk. These successes have spurred elimination projects in Ethiopia and Nigeria. The work to eliminate river blindness through education and mass treatment goes on.

Guinea worm

The Guinea worm has been known since antiquity, particularly in the Middle East, Africa (Cameroon and Senegal), and India. In Numbers 21:4-8 it is written: "And they journeyed from Mount Har by Way of the Red Sea to … the land of Edom … and the Lord sent fiery serpents among the people; and much people of Israel died … And the Lord said unto Moses 'Make thee a fiery serpent and set it upon a pole and it shall come to pass that everyone … when he looketh upon it shall live.'" It is believed that this biblical reference is to Guinea worm and that the caduceus—the Aesculpian staff (from Aesculpius, the Roman god of medicine)—the symbol of modern medicine, is not of Greek origin but refers to the Jews in Egypt who were able to remove the worms

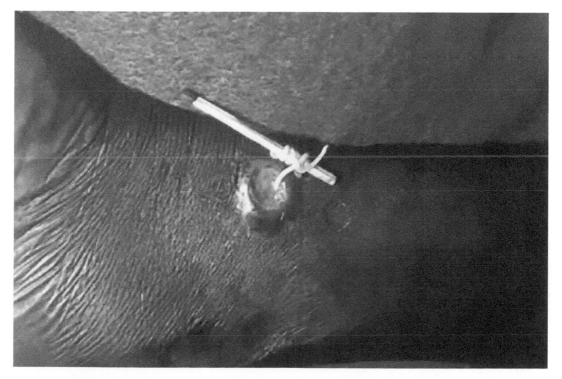

Figure 15.9a An adult Guinea worm being removed by rolling onto a matchstick. Courtesy CDC Public Health Image Library #1342, 1968.

by wrapping them around a wooden stick (Fig. 15.9a). The existence of Guinea worm in Egypt in the time of the pharaohs has been confirmed by the recent finding of calcified worms in mummies from the period of the New Kingdom in the 15th century B.C.

In 1758, Linnaeus recognized the Guinea worm as a roundworm (nematode), naming it *Gordius medinensis* after the city of Medina, where the condition was prevalent. In 1868 the Englishman Henry Bastian provided a detailed description of the worm and called it *Dracunculus*, after the Latin word *draco*, meaning "snake" or "serpent." The adult female Guinea worm is a meter in length and 2 mm in diameter, whereas the male is ~2 cm long (Fig. 15.9b). Despite the female worm's being so long and active under the skin—resembling a wriggling varicose vein—the entire life cycle was not demonstrated until 1905 by Robert Leiper. Both adult worms live in the skin, and the female is fertilized by the male, who dies soon thereafter. The female discharges between 1 million and 3 million juvenile worms through a blister in the skin; this discharge is facilitated by contact with cool water. After release of the offspring, the parent worm shrivels and dies and the tissues absorb the remnants. The blistering of the skin is due to an intense allergic reaction and can be painful; hence the name "fiery serpent." The juvenile worms are fed upon by a small one-eyed crus-

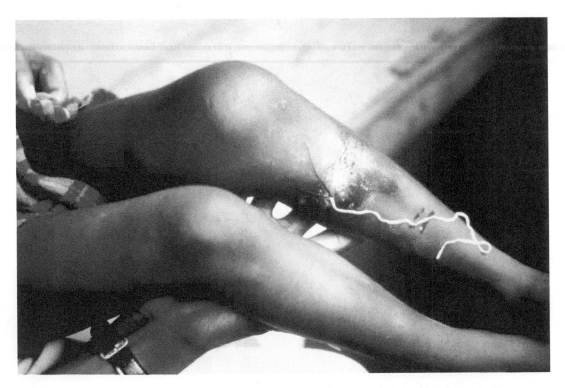

Figure 15.9b An adult female Guinea worm emerging from a painful and enlarged ulcer. Courtesy Biophoto Associates/Science Source.

tacean (*Cyclops*), commonly called a "water flea" (Fig. 15.10), where over a period of 1 to 3 weeks they develop further, becoming infectious juvenile stages. (The role of the water flea in the infection was described by the Russian Alexei Fedchenko in 1870, working in central Asia.) Humans drinking raw contaminated water ingest the infected water fleas. The water flea is digested by gastric juice in the stomach, the juvenile worms are liberated, and within 2 weeks they migrate from the intestinal tract through the abdominal cavity and then to the skin and joints. Maturation to an adult worm may take up to a year.

Incidence

In 1986 Guinea worm afflicted an estimated 3.5 million people a year in 21 countries in Africa and Asia. Although infection with Guinea worm is not a cause of mortality, it is a burden in terms of morbidity, incapacity, and debilitation. Infected humans suffer from allergic reactions, rash, nausea, and dizziness. The blisters, which can be quite painful, often become infected with bacteria, leading to abscesses, arthritis, buboes, and tissue damage. Persons with Guinea worm may be unable to work during planting and harvesting, may have difficulty in caring for their children, and may suffer permanent crippling.

Treatment

There are no drugs or vaccines. Traditional treatment typically involves rolling the emerging adult female around a stick or surgery. Under the best of circumstances, after worm extraction, the person may be disabled for 2 to 4 weeks, but more often incapacitation may last for several months due to the presence of many worms, sensitivity of the area in which the worms reside, and secondary infections of ulcerated areas.

Eradication programs for Guinea worm were begun in the early 1920s and focused on water sanitation in order to eliminate the *Cyclops*. These efforts have continued with great success in 13 sub-Saharan countries where the disease remains endemic, since control can be accomplished by passing the raw drinking water through a fine mesh screen to remove infected water fleas. The Carter Center spearheads the international Guinea worm eradication campaign in partnership with the WHO, the Centers for Disease Control and Prevention, UNICEF, and many other partners to provide education to

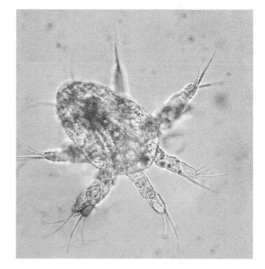

Figure 15.10 The nauplius larva of a cyclops copepod. Courtesy Wikipedia Commons.

change behavior, such as teaching people to filter all drinking water and preventing transmission by preventing anyone with an emerging worm from entering water sources. As a consequence, the number of cases has dropped 99% to 22 cases in 2015. As a a result of these efforts, 80 million cases of the disease have been averted. The Carter Center estimates that Guinea worm disease will be the second disease to be eradicated by human intervention.

Plagues Out of Africa

Yellow fever, the Louisiana Purchase, and the Panama Canal

Two of the most significant African diseases to establish themselves in the New World were malaria and yellow fever. Both were important factors in determining the pattern of human settlement and survival in the Americas. Before Christopher Columbus and his crew arrived, it appears that neither malaria nor yellow fever existed in the Western Hemisphere, and some contend that yellow fever was successfully transferred from West Africa to the Caribbean for the first time in 1648 when epidemics broke out in Yucatan and Cuba. What delayed the establishment of yellow fever was that the disease could not become epidemic in the New World until a specialized mosquito, *Aedes aegypti*, could find a suitable niche in the Americas.

Aedes is a highly domesticated mosquito that breeds in small bodies of still water—water casks or cisterns are excellent sites for the female to lay her eggs. The *Aedes* mosquito crossed the ocean aboard ships and established itself in areas where the temperature always stayed above 72° F. This peculiarity of mosquito and human association meant that mosquitoes carrying yellow fever could remain on shipboard for weeks and months. Yellow fever is usually a fatal disease, and a voyage lasting several months could decimate a crew—no one knew who would be next. The British feared the "Yellow Jack"—the flag symbolizing an infectious disease onboard ship.

Yellow fever was called "Bronze John on his saffron steed" and the "saffron scourge" because of its symptoms: jaundice, fever, nausea, black vomit due to hemorrhages in the stomach, and as well as headaches and muscular pain. Coma and death result from kidney and liver damage, but recovery can begin in 2 weeks, and convalescence may require several more weeks.

Yellow fever was called the "scourge of the American South" because it was restricted geographically due to climate and slavery. The climate of the American South, which made possible the cultivation of tobacco, cotton, and sugarcane, also provided a favorable environment for the breeding of *A. aegypti*. Yellow fever came to the Americas by ship during the slave trade. The fear of "Bronze John" drove many people from the coastal communities, and those who could afford to left the South during what was called the summer "sickly season." Yellow fever can only become endemic where the climate is warm enough to permit year-round mosquito activity. In parts of the American South where there was cold and frost during the winter, transmission ceased, but each summer it was reintroduced from Latin America and Africa.

The first recorded yellow fever epidemic in the Western Hemisphere was in 1648 in Yucatan, Cuba, and the Barbados. By the 1690s it was in North America, especially Charleston, New York, and Philadelphia—port cities with dense populations and good opportunities for transmission. There was an outbreak in Boston in 1693 after the arrival of the British fleet from the Barbados. Epidemics broke out in Philadelphia and Charleston in 1699 and in New York in 1702. Between 1763 and 1792 there was no yellow fever in the British colonies, but in 1793 there was an epidemic in Philadelphia after a ship arrived from Tobago, which caused 5,000 deaths in a total population of 60,000.

During the 1800s there were 39 epidemics in New Orleans, a city called "Bronze John's port of entry." One of the last outbreaks in the United States occurred in 1878 in Memphis, Tennessee. After the 1820s the disease was confined to areas south of the Mason-Dixon Line. The North was no longer troubled with yellow fever about the time that slavery was eliminated. However, in New Orleans, the major port of entry from Latin America, almost every summer between 1790 and the Civil War there were epidemics. Because its cause and means of transmission were not understood, it was accepted as a way of life in the American South and for many it was considered to be due to a miasma.

In New Orleans yellow fever had an interesting pattern. (i) It was confined to cities, since *Aedes* is a cosmopolitan species and in cities population densities were higher. (ii) It affected newly arrived immigrants and northerners because they lacked the immunity of the local residents. (iii) Blacks were rarely affected, demonstrating a natural immunity. As a consequence, yellow fever had its major impact on the urban South, not the rural South.

The peak of yellow fever epidemics in the United States took place in 1850, when there was an intense debate over slavery. Some feared an uprising of the immune blacks. Some people suggested that yellow fever was God's punishment of the whites who engaged in slavery, whereas others claimed that slavery was good for the southern plantation system since the deaths due to yellow fever were less catastrophic for black laborers.

The epidemic nature of yellow fever, with its high mortality and dramatic symptoms, attracted more attention than did other diseases, and it had great impact in the northern press. Consequently, the negative impressions of the South were reinforced, leading to a lack of migration from North to South and increased levels of absenteeism in the South during the epidemic season of the summer. Yellow fever was bizarre, exotic, and dreadful, and it was in the South. It is no wonder the American South became insular.

It was not until 1880 that Patrick Manson discovered mosquitoes to be transmitters of elephantiasis. The mosquito vector for yellow fever, however, would not be discovered until 1900, and the actual cause of the disease would not be described until 1927. Thus, it is not surprising that until 1905 no real control of yellow fever took place, and that was 250 years after the disease had arrived in the United States.

Yellow fever, caused by an arthropod-borne virus (arbovirus), was isolated by A. H. Mehaffey and J. Bauer in 1927 from the blood of a mild human case from Ghana and used to infect rhesus monkeys. For natural transmission to occur, the *Aedes* mosquito must feed on the yellow fever victim within the first 3 to 4 days of illness, and then the virus must be incubated for 10 to 12 days in the mosquito before the insect is infective. The mosquito is then infective for the remainder of her life. After an infected mosquito has bitten a person, it takes 3 to 6 days before fever occurs.

A New Orleans physician described yellow fever in this way: "Often I have met and shook hands with some blooming handsome young man today and later I have been called to see him in black vomit, with profuse hemorrhages from the mouth, nose, ears, eyes and even the toes; the eyes, prominent, glistening yellow and staring; the face discolored orange." Three to six days after infection, fever and headache, dizziness, and muscle ache occur. In some cases the disease goes no further, but in others the fever rises sharply, the gums and stomach bleed, and there can be delirium and coma. The skin turns yellow—jaundice—because the yellow fever virus attacks the liver; as a result, the liver cannot function properly and yellow bile pigments accumulate in the skin. Although only 5% of those infected commonly die of yellow fever, the death rate can be much higher (20 to 50%) during an epidemic. Once recovered, the individual is immune. In 1937, Max Theiler, a South African working in the Rockefeller Foundation laboratories in New York, came upon an attenuated strain of the yellow fever virus (called 17D). This live attenuated virus conferred immunity without pathogenicity and thus became a practical vaccine against yellow fever. The yellow fever vaccine (YFV) is probably the safest of any human vaccine. Approximately 500 million doses of this vaccine have been administered, with <50 documented vaccine-associated fatalities.

Yellow fever and the Louisiana Purchase

In 1803 the United States purchased Louisiana from France for about $15 million. The area was 827,987 square miles and included all the land between the Mississippi River and the Rocky Mountains, and from Canada to the Gulf of Mexico. Today, it includes 15 states. This is how the United States acquired it. In 1682 the French explorer Robert La Salle led 50 men from the Great Lakes region down the Mississippi River and claimed the entire Mississippi Valley for France. He named the region Louisiana after his king, Louis XIV. Louisiana became a French colony, and by 1718, New Orleans was its capital. The French were disappointed in the revenues from the colony, so in 1762 they ceded the trading rights to Spain. After several skirmishes between the French and Spanish settlers, Spain took firm control in 1769. Sugar planting and processing began in 1795, and by 1800 it was the major cash crop. Louisiana prospered with sugarcane.

During the American Revolutionary War, Spain allowed the Continental Congress to use New Orleans as a supply base for moving goods up the Mississippi River to supply the struggling colonies. In 1800, France secretly persuaded Spain to return Louisiana. This brought concern to the new president, Thomas Jefferson, since Spain had allowed the Americans to deposit their goods duty free in New Orleans and then permitted them to export them to Europe and elsewhere. At about this time there was a rumor that Spain also planned to give up parts of its colonies to France, including Florida. If this were carried out, the fear of President Jefferson was that the French would cut off American access to the Gulf of Mexico, especially New Orleans. In 1801, Jefferson's secretary of state, James Madison, learned of the possible deal between Spain and France, but exactly how much land was to be transferred was not known. The United States tried to block the transfer, but Napoleon Bonaparte turned the U.S. down. Then, in 1802, the Spanish governor, probably under pressure from Napoleon, suspended the right of deposit for American goods in New Orleans. This put the United States on the brink of war with France, and the U.S. threatened to send 50,000 troops to New Orleans to take the port by force. Instead, a land deal was effected. The reason for averting a war: yellow fever.

In 1697 the French established a colony on the western side of Haiti, with the Spanish taking the eastern side. Haiti was a major sugar producer largely because this labor-intensive crop used slave labor imported from Africa. In 1791 and 1794 there were slave revolts, and one of the leaders, Francois Toussaint L'Ouverture—himself the son of slave parents—declared himself governor of Haiti. It was clear to Napoleon that he had to assert his authority, but Haiti was critical for another reason—it would be a staging area for Napoleon's invasion of Louisiana and ultimately the establishment of Napoleon's North American Empire. Were Napoleon to succeed, this would interfere with the British interests in North America. In late 1801, Napoleon dispatched 20,000 well-equipped soldiers to Haiti to subdue the rebellion under the command of his brother-in-law General Victor Emmanuel Le Clerc. The French landed in Haiti in early 1802, and after a few battles they were in control. But there were still minor skirmishes with rebel forces, and Le Clerc was concerned by the number of sick—there were already 1,200 in the hospital, and by April the numbers had increased. The cause of the illness was yellow fever, and soon one-third of the French force was ill. By June the French soldiers were dying at a rate of 30 to 50 per day. In October, Le Clerc himself died of yellow fever. By 1803 the disease had killed 20,000 additional replacements and the French gave up. Of all the French dispatched to Haiti, only 3,000 were left alive, and it is estimated that 50,000 of their comrades died. On April 11, 1803, Napoleon told his finance minister, Charles Maurice de Talleyrand, that he could ill afford an American campaign and he feared the British might seize the territory or that they would form an Anglo-American alliance against him. Napoleon renounced all claim to Louisiana, and the Louisiana Purchase treaty was signed May 2, 1803.

The colony of Haiti had been one of France's richest, and in 1798 it had 520,000 inhabitants, but by 1804 not a single white person was left, and the total population consisted of 10,000 mulattoes and 230,000 blacks. Why was Napoleon's army decimated by yellow fever? (i) They were a naive population. (ii) The indigenous Haitian population was immune and served as a reservoir for the disease. (iii) The climate was favorable for mosquito transmission—1802-1803 was particularly rainy. (iv) Most of the cities were burned to the ground, and so hospitals and medical supplies were unavailable. (v) The hot, humid climate stressed Napoleon's soldiers. (vi) The French army bivouacked in the low-lying areas where the mosquitoes were more abundant.

Yellow fever and the Panama Canal

The British Empire existed between the reigns of two great queens, Elizabeth (1558-1603) and Victoria (1837-1901). British domination was based in part on its ability to rule the waves by controlling strategic points around the globe. These served not only for provisioning, coaling, and repair of ships but for military ventures as well. The first of these strategic bases was Gibraltar, but later they included Cape Town, Hong Kong, Singapore, Ceylon, the Falklands, Aden, and Suez. Britain held control of the Mediterranean, Red Sea, and Indian Ocean but was forced to let go of the Isthmus of Panama, though not because of fortresses or cannons. The reason: yellow fever.

The success of Ferdinand de Lesseps in building the Suez Canal (1850-1869) encouraged the French to attempt to build a Panama canal. (The Suez Canal—connecting the Mediterranean and the Red Sea—shortened the route between England and India by 6,000 miles, yet the canal itself is only 100 miles long.) The French acquired the land rights to build the Panama Canal from Colombia, and also gained title to the Panama Railroad (which had been built in 1855 by a group of New York executives at a cost of $5 million) for $20 million. The French began digging in 1882, but by 1889, after removing 76 million cubic yards of earth, they gave up due to mismanagement, poor skills, theft, and disease. More than 22,000 workers died during that period. The DeLesseps Panama Canal Co. went bankrupt in 1889.

The California gold rush stirred U.S. interests in a canal to connect the Atlantic and the Pacific as early as 1849, and a canal through the Isthmus of Panama became even more enticing during the Spanish-American War of 1898 because of the difficulty of sending ships from San Francisco to Cuba to reinforce the Atlantic fleet. (The distance from New York to San Francisco with the canal is 5,200 miles; without it, it is 13,000 miles, because ships have to go around South America.) In 1899, Congress authorized a commission to negotiate a land deal and the French sold their rights and the railroad for $40 million. However, Colombia refused to sign a treaty involving a payment of $10 million outright and $250,000 annually, because the figure was considered to be too low. A group of Panamanians feared that Panama would lose the commercial benefits of a canal, and the French worried about losing its sale to the United States, so with the backing of the French, the Panamanians, encouraged by the U.S., began a revolution against Colombia. The United States, in accord with its treaty of 1864 with Colombia to protect the Panama Raiload, sent in troops. The Marines landed at Colon and prevented the Colombian troops from marching to Panama City, the center of the revolution. On November 6, 1903, the United States recognized the Republic of Panama, and 2 weeks later the Hay-Bunau-Varilla Treaty was signed, giving the U.S. permanent and exclusive use of the Canal Zone (10 miles wide and 50 miles long). Dealing directly with Panama, the United States paid $10 million outright plus $250,000 annually beginning in 1913. The U.S. also guaranteed Panama's independence.

Construction began in 1907, and the canal was completed in 1914. The project removed 211 million cubic yards of earth. More than 43,400 persons worked on the canal, most of whom were blacks from the British West Indies. The cost of construction was $380 million. In 1936 the U.S. payments to Panama increased to $430,000; by 1955 the U.S. payments to Panama increased to $1,930,000, and in the 1970s to $2,328,000. In the 1960s there were riots, and other concessions were granted. Panama continued to exercise greater and greater control over the canal and took complete control on December 31, 1999.

The successful construction of the Panama Canal is related to the effective control of disease. In 1880, Carlos Finlay, a Cuban physician, showed that *Aedes* mosquitoes could transmit yellow fever; however, not everyone believed his results with human volunteers. Of course, few at that time believed in mosquito transmission, most claiming that yellow fever was carried by filth. In 1900, U.S. Army major Walter Reed carried out the definitive transmission experiments, and he reported his findings in 1901. Reed used human volunteers and screening to keep out mosquitoes.

The volunteers were paid $100 for participating and another $100 bonus for getting yellow fever. Several volunteers died. The experiments of Reed clearly showed that the *Aedes* was capable of transmitting the disease.

In March 1901, Major William C. Gorgas, the sanitary engineer on the canal project, initiated measures to eliminate mosquitoes and their breeding sites. During construction of the canal all patients with yellow fever were kept in screened rooms. Workers were also housed in copper-screened houses. Drainage and kerosene spraying were the mainstays of mosquito control. The magnitude of the accomplishment can be best appreciated by the number of deaths. During the de Lesseps canal days the death rate was 176/1,000. When the canal was completed, the death rate from all causes was 6/1,000. The last fatal case of yellow fever in Panama occurred in 1906.

Recent outbreaks

Cases of yellow fever continue to occur mainly in Africa and some South American countries. In a 1993-1994 epidemic there were >70 cases in Brazil, with the principal reason for the outbreak being a lack of vaccination. In 1995 an epidemic in Peru involved >800 cases, with a fatality rate of 38%. In 2016 the worst outbreak of yellow fever in decades occurred in Africa. On February 12, 2016, the WHO announced an outbreak in Luanda, Angola. There were at least 3,867 suspected and confirmed cases reported nationally, including 369 deaths. From Angola the virus spread to the Congo. In April the WHO was notified of 21 deaths from yellow fever in the Democratic Republic of Congo in a province that borders Angola. At the end of May, 700 suspected cases and 63 deaths were been reported to the WHO through the national surveillance system. Fifty-two cases were laboratory confirmed by the National Institute for Biomedical Research in Kinshasa and the Pasteur Institute in Dakar. Forty-six of the 52 cases were imported from Angola. Local transmission continued in urban areas in Angola and Congo. On May 30 a mass vaccination campaign was launched in Congo, but as of June 22 vaccine was in short supply. On June 20 the health minister declared the outbreak of yellow fever in three provinces, including Kinshasa. Transmission within Kinshasa was of concern because of the large and densely packed population. As of July 8 the WHO was notified of 1,798 suspected cases, with 68 confirmed cases (59 imported from Angola) and 85 deaths. Although the WHO quickly dispatched >6 million doses of yellow fever vaccine, >1 million doses disappeared in Angola. Thousands more vaccinations were delayed when syringes got waylaid and ice packs to keep the vaccine potent went missing. With limited yellow fever vaccine in the stockpile, the WHO recommended that doses be diluted. By August the WHO reported that the number of yellow fever cases was declining. More than 17 million additional people were expected to be vaccinated in massive campaigns scheduled to take place in both countries before the rainy season began in this part of Africa.

Annually, about 1,000 cases of yellow fever are reported worldwide, but the actual number may be 200 times greater. Yellow fever can be maintained in a forest transmission cycle between monkeys and mosquitoes, and humans can acquire this jungle yellow fever. It can also be transmitted from human to human even in an urban setting, since *A. aegypti*, the domestic vector, can be present at high density due to its capacity for breeding in artificial containers used to store water. N.B.: The Aedes mosquito is also a vector for the emerging Zika virus (see p. 436).

Malaria and the American South

There is no convincing evidence that malaria existed in the Western Hemisphere before the arrival of Columbus and the conquistadors. Malaria was probably first introduced into Mexico and Central America by the Spanish explorers, as evidenced by the writings of a physician with Cortez, who in 1542 clearly described the disease in the soldiers. Although malaria did not exist in pre-Columbian North America, its mosquito vector, *Anopheles,* did. In short, before the colonists and explorers arrived in the Americas, there were anopheles mosquitoes but no malaria.

During the 16th and 17th centuries malaria was introduced into the United States by early settlers coming from England, especially from the fen country as well as Kent, where endemic malaria was called the "Kentish disorder" or the "lowlands disease." In the American colonies both the landed gentry and indentured laborers provided the seedbed for infection of *Anopheles* in Maryland, Virginia, and the Carolinas. The early French and Spanish explorers and settlers along the coast of the Gulf of Mexico were also a source of malaria for those living in the American South and the lower Mississippi Valley. But it was only after the introduction of slaves from Africa—to work in the rice fields and on the sugar plantations—that malaria became a prominent feature of the American South, and this condition persisted until the 20th century.

By the Revolutionary War malaria was endemic from Georgia to Pennsylvania, and it was then was carried up the coastal rivers and across the Appalachians into western New York and Pennsylvania and then down the Ohio River and its tributaries into Kentucky and Ohio. As the pioneers moved westward, they carried with them their earthly possessions and their diseases, and so it was that malaria moved into Indiana and Illinois and crossed the Mississippi River into Missouri. Moving relentlessly, the fever plague of malaria reached New England, southern Michigan, and Wisconsin and as far north as Minneapolis. French settlers sailed down the Mississippi River, bringing malaria as far south as the Mississippi Delta and into the Tennessee Valley. As people migrated west in search of new places to settle, and chose the fertile areas of river bottoms and alongside streams and creeks, malaria became established in Iowa, Nebraska, eastern Texas, the Dakotas, the San Joaquin and Sacramento Valleys of California, and even in the Willamette Valley of Oregon, because these were also the areas where anopheline mosquitoes flourished. As a consequence, by 1850, with increased numbers of settlers and the opening up of more and more land for agriculture, malaria became endemic in much of the United States and was hyperendemic in the Ohio and Illinois River valleys, but especially in the Mississippi valley from St. Louis to New Orleans.

In the American South, malaria was transmitted principally by *Anopheles quadrimaculatus*, a mosquito well adapted to warm standing pools of water, flooded rice fields, and tree holes, and with a distinct preference for human blood. In the valleys of California, *Anopheles freeborni*, a mosquito associated with irrigation agriculture, that prefers animal blood, and that is restricted by the dry hot summers, became the transmitter of "miner's fever." Mosquitoes have a flight range of about 1 mile and live for several weeks. Blood is required for egg laying. Depending on the temperature and humidity, a new generation of mosquitoes can develop every 10 to 14 days, but mosquito populations do rise and fall seasonally. Incidence usually increases in the late spring and early summer, then declines slightly, and rises again in late August and early September. By the end of the 19th century, with few exceptions, malaria was no longer found in the North. This was because the land was cleared and drained, and this reduced mosquito breeding sites; the introduction of cattle provided a source of blood other than that from humans; and the cold winters killed off infected mosquitoes. In the South, however, with its warmth, humidity, and breeding sites (ponds, bottomlands, and cultivated fields of rice, sugar, tobacco, and cotton), mosquitoes were present year-round, and so was malaria, which became endemic. Due to the slave trade, there was a continuous introduction of infections from Africa. In the American South, malaria struck down newcomers, contributed to high infant mortality, and reduced the vitality of the inhabitants. It also made the southerner more susceptible to other disorders. Endemic malaria gave rise to the classic picture of a thin, sallow-skinned, dull, lazy southerner, an image that remained until the 20th century.

Individuals who survive an initial attack of malaria can continue to harbor the parasites in the blood for months to years, but because of immunity, they may show no signs of clinical disease. However, such individuals can serve as reservoirs for infecting mosquitoes and thus contribute to the spread of disease. Most adult southerners—black and white—had some measure of acquired immunity, but blacks had a distinct advantage because 90% of those from West Africa were naturally resistant to vivax malaria, and 10 to 20% were endowed with sickle cell trait and, as a result, were resistant to falciparum malaria (see p. 153). However, outbreaks of malaria did occur amongst the slaves and their descendants because resistance was not absolute; that is, protection from one type of malaria by a specific genetic trait did not confer resistance to the other type of malaria. When slaves were transferred to the South and

West from older plantations, they could also develop malaria because their acquired immunity was specific, and thus they were susceptible to the newer strains of malaria.

When federal troops returned from the Mexican War (1848-1850), they brought malaria to Texas, New Mexico, Arkansas, and the Atlantic states. During the Civil War malaria was present, especially in the Union troops from the North who were garrisoned in the South. In the period from 1861 to 1866, malaria represented 25% of all reported diseases among the Union troops. There were 1,213,814 cases of malaria and 12,199 deaths among the Union forces. Though malaria was more prevalent among the Confederate troops, it was less fatal, probably due to a degree of acquired immunity. When the Civil War ended and the Union troops returned home, there was an increase in malaria in the northeastern United States, but this was considerably lower than that in the South. In the South during the Reconstruction period mosquitoes flourished in the fallow agricultural lands because there were fewer individuals to work in the fields—the labor force having been depleted by death and disease during the Civil War. Although the northern border of the malaria belt began to retreat between 1890 and 1920, it did not fall back as significantly in the South. Indeed, the mortality in Missouri in 1860 was 112 (per 100,000 people), and by 1910 it was 17, but in Illinois and Wisconsin, where it was 57 and 37, respectively, it had declined to <1 by 1920.

It was recognized as early as 1900 (after the discovery of the parasite by Laveran and its vector by Ronald Ross and Giovanni Battista Grassi) that to reduce the incidence of malaria one would have to block transmission. This required reductions in both the numbers of mosquitoes and human infections. Killing mosquitoes, however, was found to be labor-intensive and expensive. In addition, it was soon realized that there needed to be an integrated program involving the screening of beds and houses in malarious areas, the confinement of malaria patients under screens, the treatment of humans with quinine, the elimination of breeding sites by drainage and filling, the use of mosquito larvicides such as oil and Paris green (arsenic trioxide), and education. In 1917-1918 the U.S. Army pursued mosquito control in the South, where troops were stationed and trained, but this was of limited success. It was clear that ridding the entire United States of malaria was too costly and too inefficient to be handled by local governments. From 1933-1935 (and during the height of the Depression) federal resources were committed to an anti-malaria campaign using the Civil Works Administration, the Emergency Relief Agency, and the Works Project Administration. An Office of Malaria Control (the forerunner of the Centers for Disease Control) was established in 1942. Despite this, malaria remained entrenched in the South because southern leaders would not accept the hypothesis of mosquito transmission, and endemic poverty also limited local funding for mosquito control. Even the meager efforts at control, such as they were, were interrupted by World

War II. After the war the attempt to eradicate malaria became a national program. By 1946 there was widespread use of DDT (dichlorodiphenyltrichloroethane) and "flit guns" loaded with pyrethrum to kill adult mosquitoes. The Tennessee Valley Authority set in place mosquito abatement programs to avoid the problems associated with damming streams and rivers for flood control and hydroelectric power as well as recreational and navigational purposes. In a 10-year period (1942-1952) malaria in the United States was reduced to negligible proportions, at a cost of $100 million.

Of the three chronic disorders associated with the American South—hookworm, pellagra (a dietary deficiency disease), and malaria—malaria had the most profound effect, since it struck young and old, rich and poor, and, though not invariably fatal, it was always debilitating. Malaria was eliminated from the American South by a combination of factors: public health measures, diversified farming, shifts in population (for example, cotton growing moved to the Southwest), mechanization of farming, and an improved standard of living. As malaria receded from the South, so too did the distinctive image of a fever-ridden area populated by poverty-stricken, lazy sharecroppers growing cotton and tobacco.

George Washington's ally

The American Revolution had its battle heroes: Washington, Lafayette, Rochambeau, de Grasse, Pulaski, John Paul Jones, and Nathan Hale. But one hero hardly ever mentioned was a small, fickle female named *Anopheles*. Here is how she became Washington's ally and Britain's foe. The Revolutionary War began on April 19, 1775, when a group of colonists fought British soldiers at Lexington, Massachusetts. Ill feelings between the British government and its colonies culminated in Britain's attempt to force the colonies to depend on it for manufactured goods, as well as the restriction of the colonists' rich trade in rum, slaves, gold, and molasses in Africa and the West Indies. From 1776 to 1778 both sides gained victories. New York fell to the invading British, but after the British general John Burgoyne surrendered at Saratoga, the French entered on the side of the United States. In April 1778, Sir Henry Clinton was in command of the British army in the American colonies. He was under orders to take Philadelphia, evacuate the city, and then concentrate his forces in New York City in order to meet the Continental Army under General George Washington. By summer Clinton began to move his army to New York City by land instead of by sea. This was a serious blunder on his part, for in the march northward to their strategic destination Clinton's men had moved through the mosquito-infested lowlands of New Jersey. By the time the men were garrisoned in New York, in early August, 7 out of 100 soldiers had died of malaria and many more were ill with fever. One of Clinton's lieutenants wrote: "the troops more sickly than usual and the same for the inhabitants."

On August 29 six British warships were anchored off the coast of New York to reinforce Clinton's troops, and 2,000 men disembarked—all sick with scurvy (a dietary deficiency disease). To add further to Clinton's problems, 10,000 of his men were relocated to the West Indies, Georgia, and Florida. As a consolation for this he was promised reinforcements at a later date. Clinton's strategy was to attack and destroy Washington's Continental Army at West Point. However, because the available troops in the West Indies, Florida, and Georgia were now ill with malaria and typhus (see p. 113), Clinton was forced to redraw his battle plans: he would subjugate the South by landing at Charleston, South Carolina, and then move inland to capture Virginia. Clinton embarked on December 26, 1778, with 8,500 men. Major General Charles Cornwallis was his second in command, and by May 1779, Charleston and most of South Carolina were under British control. The situation appeared to be so well in hand that Clinton returned to New York City and left Cornwallis in command of this southern contingent. Cornwallis moved his forces northward to Virginia, but by July 1779 two-thirds of his officers and men were unable to march due to malaria fever. Over the next few months they continued to suffer from relapses. Still they pushed on. Fever plagued the men and Cornwallis himself as they marched toward Virginia. Although battles, maneuvers, retreats, and advances characterized the next 2 years of the Revolutionary War, it was the particularly severe malaria among the British troops that would make the difference in the future course of the war. Between January and April 1781, Cornwallis's army had dwindled from 3,224 to 1,723 men, and by May he had only 1,435 men fit and ready for battle. Because Clinton expected Washington to attack New York City, the question was whether Cornwallis should adopt a defensive posture and remain in Charleston or move north to assist him. If he took the latter course, then a port in Virginia would be critical to resupplying his depleted army by ship. Cornwallis chose to remain in Virginia at the port of Yorktown on the River York, but did so unhappily. He complained of being surrounded by acres of unhealthy swamps and felt that he would soon become prey for the Continental Army. The plan was for Washington's army to march south and surround Yorktown, and have the French blockade the harbor, preventing Cornwallis's escape. On August 14, 1781, Washington received word that the French fleet (which had arrived in the Caribbean) was heading for the Chesapeake Bay. Six days later Washington's army crossed the Hudson River. Clinton expected Washington to attack him at Staten Island, but instead Washington and his troops headed south for Yorktown, where they were to be met by eight French ships sailing from Rhode Island. By September 5 the French ships had arrived in the Chesapeake Bay and defeated the British fleet, and now they controlled the mouth of the York River. Cornwallis was bottled up. Clinton advised Cornwallis that Washington was not going to attack New York City but instead was moving troops against Yorktown. Clinton also told Cornwallis he was dispatching 4,000 troops from New York City to assist him. Such help never came, because the men who had been garrisoned in New York were decimated by malaria acquired during the previous two autumns. On September 17 and again on the 29th, Cornwallis wrote to Clinton: "This place is in no state of defense … if relief doesn't come soon you must prepare to hear the worst … medicines are wanted." By medicines he meant pulverized fever bark. Cornwallis, with one-third of his besieged forces too sick from malaria and no relief in sight, surrendered at Yorktown on October 19, 1781. After 7 years the Revolutionary War was over. Great Britain gave up all hope of conquering the colonies, and on September 3, 1783, it recognized the new republic in the Treaty of Paris. It may have been the superior tactical strength of the combined American and French forces that brought victory to the Continental Army and independence for the United States, but Washington's army had a secret ally: the malaria-carrying *Anopheles* mosquito.

Hookworm and the American South

Yellow fever and malaria are microparasites, whereas hookworm is a macroparasite. Both the adult male and female hookworms live in the small intestine, where they attach to the wall, abrade it, and suck blood (Fig. 15.11). After the eggs of the thread-like female worms (~10 mm in length and 0.4 mm in diameter) are fertilized by the smaller male worms (~7 mm in length and 0.3 mm in diameter), they pass out with the feces, for the juvenile worms to hatch out of the egg requires wet, warm, and sandy soil. When the hatched-out juvenile worm comes in contact with the skin, it burrows into it, enters the bloodstream, and finally gets into the digestive tract, where it matures, feeds, and reproduces. The adult worms can live in the small intestine for up to 10 years, and a female can produce ~9,000 eggs each day.

In Europe hookworm infections were considered an occupational disease of miners. In Hungary in 1786, 1,200 coal miners were attacked by it, and in 1802 and again in 1820 there were outbreaks in the French coal mines. Similar outbreaks occurred in Germany and in the mines of Cornwall, England. In Cornwall it was called "miner's bunches" due to a telltale sign of infection: inflammation at the site where the juvenile worms had penetrated the skin. In 1880, during the construction of the St. Gotthard tunnel between Switzerland and Italy, 10,000 workers developed severe hookworm anemia and there was an exceedingly high fatality rate, especially among the Italians. Disease was the result of infected miners passing eggs in their feces and then having skin contact with the rungs of ladders soiled with mud and feces.

In the 18th century the clinical picture of hookworm infection was described among African slaves in the West Indies. It was unknown among the early European settlers in the Americas; however, it was commonly seen in their African slaves. By 1808 it was observed that anemia and geophagy (literally "dirt-eating") were common in both slaves and poor whites living in the cultivated lowlands of the American South where the soil was sandy and the climate humid and warm. Because the blood-feeding hookworm produces anemia, it was referred to as the "vampire of the American South."

The ancient Chinese medical literature described hookworm disease as the "able to eat but lazy to work yellow disease." The Ebers papyrus of ~1550 B.C. refers to the characteristics of infections in ancient Egypt, and the Greek physician Hippocrates in 450 B.C. described it thus: "the skin is yellow, the intestine disturbed and the person has an appetite for eating clay." Depending on the particular species, an adult hookworm can cause a daily blood loss of 0.03 to 0.15 ml. In a typical infection blood loss can be as large as a cupful of blood. Victims of hookworm disease show a characteristic set of symptoms: anemia, diarrhea, fever, and sometimes an insatiable appetite to eat clay or dirt. (It has been claimed that clay eaters are attempting to replace their iron stores, since red clays have a high iron content.) The individual is lethargic, the complexion is sallow, and physical development is slowed, as is mental acuity. In areas of endemicity hookworm disease is particularly prevalent in children, where in addition to severe

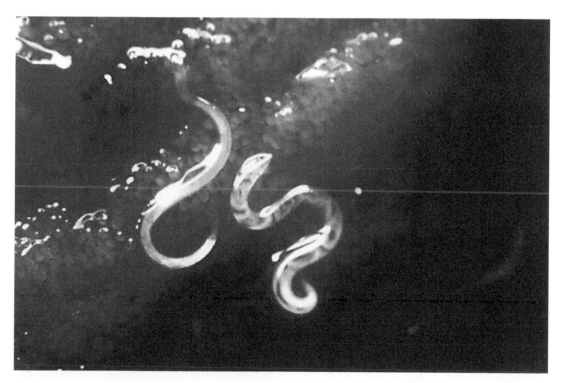

Figure 15.11 This enlargement shows hookworms, *Ancylostoma* attached to the intestinal mucosa. Barely visible larvae penetrate the skin (often through bare feet), are carried to the lungs, go through the respiratory tract to the mouth, are swallowed, and eventually reach the small intestine. This journey takes about a week. Courtesy of CDC, 1982.

anemia it may produce grotesque deformities.

Humans can be infected by two kinds of hookworms: *Ancylostoma duodenale* (Fig. 15.12a) and *Necator americanus* (Fig. 15.12b). Hookworm infections probably

existed among the ancient Egyptians, and before 1838 the disease had been described from Italy, Arabia, and Brazil. In 1838 the Italian physician Angelo Dubini discovered and named the minute whitish worm he found in the small intestine of a woman patient dying in the hospital, *A. duodenale*; however, he did not associate the presence of the worm with a specific disease. In 1843, when he found many worms in the intestine of a man dying with pneumonia, as well as in 20% of the cadavers he autopsied, he suspected that this worm was parasitic. Dubini named the worm *Ancylostoma* (*ancylo*, "hooked," and *stoma*, "mouth") because the worm

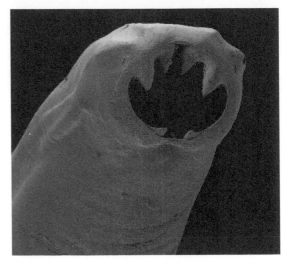

Figure 15.12a Color-enhanced Scanning Electron Micrograph (SEM) of the Old World hookworm *Ancylostoma duodenale*, an intestinal parasite. ©PhotoResearchers 2017, David Scharf.

had teeth and *duodenale* because of its location in the duodenum of the small intestine. In 1847, Franz Pruner, a German working as a personal physician to the Abbas Pasha of Egypt, described the same worm. However, it wasn't until 1853 that Wilhelm Griesinger and Theodor Bilharz associated the presence of these minute roundworms with anemia. In 1878, Giovanni Battista Grassi and his associates in Italy identified hookworm eggs in the feces of infected patients. Up until the 19th century it was believed that the disease was spread by ingestion of eggs through fecal contamination, since human volunteers could be infected by feeding them fully developed eggs but not hatched-

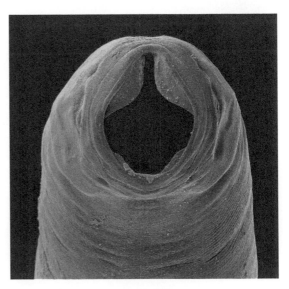

Figure 15.12b Scanning electron microscope view of the razor-sharp teeth of *Necator,* Credit: ©PhotoResearchers 2017, David Scharf.

out juvenile worms. Indeed, it was found that the infected volunteers were passing eggs in a month's time. But in 1893, when Arthur Looss accidentally dropped some juvenile worms onto his hand and then developed itching and redness, he began to suspect that the juvenile hookworm could directly penetrate the skin. This hypothesis was proven experimentally in 1902 by C. A. Bentley, a medical officer working on a tea plantation in India. Bentley took two samples of fecal-contaminated soil: one that was associated with "ground itch" due to the presence of hookworm juveniles, and the other comprising soil contaminated with feces but without worms. A portion of each soil sample was heated and dried for 8 h, and the remainder was left unheated. Each soil sample was applied to the wrists and held in place with a bandage. "Ground itch" developed only in the unheated soil sample containing living hookworm juveniles.

For hookworm to be an important parasite requires several conditions: lack of sanitary facilities for disposal of feces; bare feet; and the presence of a moist, warm climate and sandy soil. All were present in the American South, and not in the North. Initially it was assumed that the hookworm in the United States was *A. duodenale*; however, in 1902, Charles W. Stiles identified differences in the parasite and named it *N. americanus*. By 1902 one health worker boldly claimed that hookworm was an endemic disease of the American South. Indeed, this was substantiated by further work. Although Stiles was ridiculed as the discoverer of the "germ of laziness," he persisted in calling for a public health program to control the disease. By 1910 the Rockefeller Sanitary Commission for the Eradication of Hookworm Disease, with a grant of $1 million, began its work. It was dedicated to the proposition that sanitary privies, wearing of shoes, and health education would eliminate the disease, and by 1914, together with chemotherapy, it succeeded in significantly reducing the incidence. However, the greatest reduction in the number of infections in the American South required persistent and multifaceted control programs and did not occur until 1950.

Although eradicated from the United States and most of Europe, hookworm remains a worldwide problem, with an estimated prevalence of 630 million people, largely due to social customs, inertia, and economic constraints. Hookworm is most prevalent in rural, poorly nourished individuals with low resistance to infection. In the well-nourished individual there is resistance, sometimes called "self-cure," evidenced by lower worm numbers upon reinfection and lower egg production by the worms.

Where did hookworm come from? It is suspected that humans first acquired *A. duodenale* at the time of the domestication of the dog (ca. 3000 B.C.) in the Fertile Crescent. From here it spread with the migration of the human population across the continents. *Ancylostoma* originally infected people in the warmer, nonarid areas north of the Tropic of Cancer. Mediterranean, Middle Eastern, Indian, and Asian immigrants brought the parasite to the Americas. *N. americanus* had as its original range south of the Tropic of Cancer in the South Pacific islands, Indonesia, southern Asia, and Africa. It was probably introduced into the Western Hemisphere with the importation of slaves from Africa. Today, *Ancylostoma* is found in southern Europe, North Africa, and northern Asia, whereas *Necator* predominates in tropical areas in the Western Hemisphere and in equatorial Africa.

If clinical descriptions of the disease stem from 1742 in the West Indies, and from the American South 100 years later, why did it take nearly 200 years to eliminate the disease from the southern United States? Hookworm eggs are microscopic, so the cause of the infection may be missed. But this cannot explain the situation completely, since the adult worms, minute though they are, were seen in an autopsy as early as 1840. A more plausible reason is that the disease was not epidemic and hookworm anemia was masked by the abundance of malaria as the primary cause of anemia in humans living in the American South. More importantly, however, there was no motive for eradication until 1902. Around that time the United States was developing a distinct national pride and a vision for itself. This resulted in two kinds of proposals: physical exclusion of the deviants or their integration into the society. It was the latter approach that sought to bring the poor "white crackers" and the dirt eaters as well as blacks—the so-called aberrant and distinctive groups—into the American nationality, and this probably provided the medical impetus to look for the cause of the group's distinctiveness, as well as the discovery of the cause and cure for hookworm. In addition, social reform played a role in the public health measures that eventually made the South truly American.

doi: 10.1128/9781683670018.ch16

16
Emerging and Reemerging Plagues

"A creature without memory cannot discover the past; one without expectation cannot conceive of a future." "Those who do not remember the past are condemned to repeat it."
–George Santayana

Diseases caused by infectious agents can affect the course of human events. They have in the past, and it is certain they will do so in the future. Great plagues, such as the bubonic plague or influenza, can happen again. Plagues are natural and almost predictable phenomena. Although remarkable scientific advances have been made in controlling diseases through sanitation, chemotherapy, antisepsis, antibiotics, improved nutrition, and immunization, we continue to live in evolutionary competition with microbes, and there is no guarantee that we can always beat them at their own game. Lurking out there are germs and worms that may spread to our domestic animals, our domesticated plants, and us. These are the seeds of coming plagues.

Many factors lead to the emergence of infectious diseases, some of which are obvious: a change in a parasite's virulence, a breakdown in public health and surveillance measures, a change in human behavior, crowding, alterations in the environment, technological advances, economic downturns, poverty, and even something as innocent as foreign travel for a vacation. The emergence of infectious disease is a complex and dynamic process, and most new infections are not caused by a never-before-encountered pathogen. New parasites are actually old pathogens in an altered form. Some of these arise because of breaches in the species-specific barrier, but the most potent factors that drive the emergence of disease are human activities—social, economic, political, technological, environmental, and behavioral. As the cartoon character Pogo said: "We have met the enemy and he is us."

Paradoxically, the current global situation favors rather than discourages the spread of epidemic disease. The reason for this is that for most of human history

Figure 16.1 (Left) *The Scream* by Edvard Munch 1893, Courtesy Wikipedia

populations were relatively isolated and only in recent centuries has there been extensive contact between peoples of different continents. Much of the world's population now lives in urban areas, and there is a move toward "megacities." These areas are able to support the persistence of some infections and allow for the emergence of others. And some of these will be in tropical regions—places that serve as the most effective cradles for the emergence of infectious diseases. Human behavior and close contact with animals, as well as monoculture (growing purebred strains) in agriculture, also provide the means for the spread of these diseases and may contribute to a collapse in the economy and social order.

Blame It on the Rodents

A "new" disease, Argentine hemorrhagic (Junin) fever, developed because of a change in agricultural practices. When Argentine farmers turned from growing mixed crops and began clearing land to plant corn, they provided a new niche for a little mouse named *Calomys musculinis*. Herbicides were applied to increase corn production; this killed off much of the native plant life and reduced the diversity of the rodent species. Consequently, *Calomys* increased, because this mouse thrives on corn, but it also carries the Junin virus. Plant more corn, and the result is more mice, more virus, and also more human infections. Now there are more than 600 cases of this hemorrhagic fever annually.

HIV is not the only animal virus that can infect humans. In the summer of 1993 a severe respiratory disease appeared in the Four Corners area of Arizona, New Mexico, Colorado, and Utah. Typically the disease was characterized by rapid pulmonary failure. A 31-year-old woman arrived at the hospital with a 4-day history of fever and sore throat followed by headache, body pains, diarrhea, and vomiting; her breathing was shallow and she complained of chest pains when she breathed deeply. Initially her temperature was 38.5°C, but within a day it had risen to 40.5°C. It was believed that she was suffering from gastritis and dehydration, and so she was placed on morphine and given intravenous fluids. Chest X rays showed her lungs to be filled with fluid. Because she did not respond to antibiotics, the preliminary diagnosis was mycoplasma or viral pneumonia. Although by the third day her blood pressure had dropped to 70 mm Hg, she made a remarkable recovery, and on her own she began to resolve the infection. Although she continued to complain of headache, cough, general achiness, and diarrhea, she was sufficiently improved in a week's time that she was discharged from the hospital. She was lucky. Of the 160 cases in 26 states, 30% of patients died. When state and county health department officials conducted an investigation near the patient's home, they caught 17 rodents. Fifteen were deer mice, and one of them was positive for hantavirus, a hemorrhagic fever that is endemic in the area stretching from Western Europe to far eastern Asia. Its appearance in the United States was unique. The disease was called Sin Nombre, and it is caused by a virus transmitted via the urine of deer mice, whose range includes most of the 48 states. The disease is more prevalent in men than in women, probably as a result of the greater occupational exposure of men to rodents. This particular outbreak was related to a large crop of pinyon nuts that was being harvested by the men on the Navajo reservation, and this in turn led to their inhalation of dried rodent urine that was virus-laden. Between 1993 and 2011 a total of 587 cases of hantavirus disease were confirmed. In 2012 hantavirus infected 10 people who stayed overnight in tents in Yosemite National Park; 3 died. As a result, the signature tents had to be closed. Human-to-human transmission did not occur, but one wonders what would have been the result if this virus had acquired capacity for pulmonary transmission, as happened with pneumonic plague.

Lassa fever is another case study. In 1960 diamond deposits were discovered in Sierra Leone, and this encouraged human migration. More humans produced more food and garbage. Occupying the same area was the rodent *Mastomys natalensis*, a natural reservoir for the Lassa fever virus. With the increase in humans there was an increase in virus-infected *Mastomys*, which can be considered a domesticated species because it lives in close association with human habitation. As with hantavirus, this virus is shed in the urine, the urine dries, and the virus-laden dust is carried into the air, whence humans can inhale it. With an incubation period of 1 or 2 weeks, it soon causes hemorrhages and death. Outbreaks first occurred in 1969. Between August 2015 and 17 May 2016, the World Health Organization (WHO) reported the possibility of 273 cases of Lassa fever, including 149 deaths in Nigeria. Of these, 165 cases and 89 deaths have been confirmed through laboratory testing.

Since August 2015, 10 health care workers have been infected with Lassa fever virus, 2 of whom have died. Of these 10 cases, 4 were nosocomial infections. As of 17 May 2016, eight states are currently reporting Lassa fever cases (suspected, probable, and confirmed) or deaths or have reported that they are following contacts for the maximum 21-day incubation period.

In 2002 there was an outbreak of SARS (severe acute respiratory syndrome)—a pandemic that ultimately affected 32 countries, with more than 8,500 cases, 800 deaths, and nearly $100 billion in losses due to cancelled travel, reduced trade, and decreased investment in Southeast Asia. SARS appears to have been the result of an animal virus (Figure 16.2) of the masked palm civet (a species eaten as a delicacy in China) that crossed the species-specific barrier in the Dongmen market of Hong Kong, where hundreds of vendors sell live animals and seafood and where the animals and food handlers come in contact with geese, ducks, chickens, pigeons, doves,

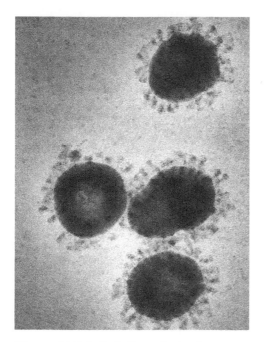

Figure 16.2 A digitally- colorized transmission electron microscopic image of a coronavirus similar to that causing SARS. Courtesy CDC Public Health Image Library #15523/Fred Murphy and Sylvia Whitfield, 1975.

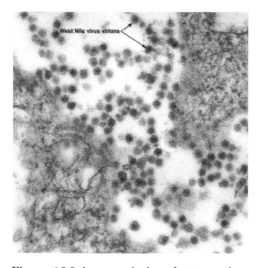

Figure 16.3 A transmission electron microscopic image of the West Nile viruses. Courtesy CDC Public Health Image Library #10700/Cynthia Goldsmith and P. E. Rollin.

wild birds, raccoon dogs, civet cats, and other exotic species.

In 2003 there was an outbreak of monkeypox (a cousin of smallpox) in 4 people and possibly 30 others that was traced to a "pet swap meet" in northern Wisconsin. The source was an infected Gambian giant rat from Africa. The rat virus was transmitted to pet prairie dogs that then bit children. Clearly, the handling of exotic animals transported from their usual habitats may provide the seeds of epidemics. We were lucky in this case, since human-to-human transmission did not occur, but in other cases involving birds, mosquitoes, and our food, human infections have occurred.

On the Wings of Birds

In the United States the most commonly diagnosed arbovirus (arthropod-borne virus) disease is West Nile virus (WNV) (Figure 16.3). WNV was first isolated in 1937 from the blood of an infected woman in the West Nile country of Uganda. The woman had a rapid onset of fever, headache, backache, muscle pains, and anorexia. She had a rash on her face and chest and swollen glands in her neck. She recovered without any treatment. Twenty-two years later the same virus, which is antigenically related to St. Louis encephalitis virus and Japanese encephalitis virus, caused a more severe illness—meningitis, encephalitis, and paralysis in 59 people in New York City. During 2000-2001 the disease spread through the South and Midwest, and by 2002 WNV was in six western states (Arizona, Utah, Nevada, Oregon, Alaska, and Hawaii), where in 2003 it triggered the largest encephalitis epidemic in recent U.S. history, with 4,000 cases and 284 deaths. Since 2004 the incidence of reported WNV illnesses in the United States and Canada has stabilized, and the number of reported cases in the United States has remained relatively stable or has decreased, suggesting a more endemic pattern, the exception being 2012, which witnessed the largest outbreak of WNV in the United States since 2003. As of 2014 nearly 50,000 human cases of WNV illness and 1,760 deaths had been reported in the United States, with the highest incidence of disease in the western and Mountain West regions.

In Southern California, WNV has been a transitory epidemic, peaking every 4 years in response to the rise and fall of its primary host, birds. Since 2014, however, the rates of disease transmission have run mostly unchecked year after year. In Los Angeles County in 2014, 218 cases were reported, with 31 fatalities, and in 2015 there were 300 cases, with 53 deaths.

The incidence of WNV encephalitis and death increases 20-fold among those older than 50, and the death rate is ~10% from meningitis or encephalitis, because of inflammation in the brain and spinal cord, leading to small hemorrhages and degeneration of nerve cells in the brain stem. Aside from supportive measures, there is no established treatment, and there is no FDA-approved vaccine.

How did WNV get from Africa to America? In 1957 there was a WNV outbreak in an Israeli nursing home, and subsequent outbreaks occurred in Israel and Africa in the 1960s and 1970s. These were mild, however, and the symptoms were flu-like. But in the mid-1990s outbreaks in Russia, Romania, and Israel had other characteristics—patients suffered from meningoencephalitis—and WNV disease was now a serious neurological disorder. The most significant factor in all of the outbreaks was the involvement of the common house mosquito *Culex* as the vector. *Culex* had never before been an important vector, but environmental conditions helped to make it so. The first epidemics reported from urban areas, in August 1999, occurred at a time when rainfall had been lower than normal; such dry conditions probably increased the number of favorable breeding sites for *Culex*, which readily lays its eggs in stagnant and polluted waters.

The female *Culex* mosquito feeds at dawn and dusk. She flies in search of a blood meal, which is necessary to obtain proteins for the nourishment of her eggs. Receptors on the mosquito antennae are sensitive to the carbon dioxide, alcohols, and fatty acids that waft off the skin, and to the warmth of a bird or human. Landing on the skin, the mosquito pushes its long proboscis into the skin. Four needle-like stylets cut and retract the flesh, and a fifth lubricates the site with saliva, which is both an anesthetic and an anticoagulant. A sixth stylet probes deeper through the layer of fat into the network of arterioles and venules, each slightly larger than a capillary. Like a skilled phlebotomist, the mosquito pulls back slightly with the first taste of blood and then begins pumping blood into its abdomen. Once full—having ingested more than twice its body weight—it withdraws the stylets and takes flight. In the process of feeding, the infected *Culex* mosquito has introduced soccer ball-shaped viruses into the bloodstream of the host. In the human the virus favors specific cells, especially those of the nervous system. WNV can break through the meninges that protect the brain and spinal cord and in its replication destroys the neurons. About 80% of people infected with WNV never know they have the infection. Their bodies fight off the infection. The rest develop fever, and in <1% of the cases the symptoms become more dire—encephalitis, meningitis, tremors, and paralysis.

In recent years there has also appeared to be a change in the virulence of this arbovirus. In New York City, concurrent with human infections, there were an unusual number of bird deaths, primarily wild crows and some exotic captive birds in the Bronx Zoo. More than 14,000 cases of WNV occurred in 2003 in horses, with a mortality rate of 35%. The most common way humans (and horses) acquire the infection is by the bite of a *Culex* mosquito that has fed on an infected bird, especially crows and blue jays. Although WNV was initially spread from Africa by migratory birds, now person-to-person infections can occur through blood products, organ transplantation, breast milk, and intrauterine infection. WNV is not transmitted from person to person by mosquitoes, and thus people and horses are figuratively and literally "dead-end" hosts. "The future epidemiologic pattern of WNV remains unclear. Whereas WNV infection in the United States appears to have reached an endemic pattern, the possibility of large future outbreaks remains, and re-emergence of WNV in European countries raises the possibility of significant WNV activity globally."

Prevention of WNV can be accomplished by reducing contact between humans and the *Culex* mosquito; this means maintaining window and door screens, wearing long-sleeved shirts and long pants when outdoors, and applying insect repellents such as DEET (diethyltoluamide).

Anthrax

Three days after the shock of the jetliner assaults of September 11, 2001, anthrax-containing letters were received at NBC News, the *New York Post*, and the Florida-based publisher of the supermarket tabloids the *Sun* and the *National Enquirer*. Three weeks later two more envelopes containing anthrax spores arrived at the Senate offices of Tom Daschle and Patrick Leahy. Before this anthrax attack ended, 17 people would fall ill and 5 would die—a photo editor at the *Sun*, two postal employees in Washington, DC, a hospital stockroom clerk in New York City, and a 94-year-old widow in Connecticut whose mail had probably crossed paths with the anthrax envelopes somewhere in the labyrinths of the postal system. Government mail service was shut down, and police were swamped with calls from citizens fearful of opening their mail. There was pressure on law enforcement to find the perpetrator, who initially was believed to be an American passing as an Islamic jihadist. Suspicion soon centered on a limited number of American scientists familiar with anthrax. One of these was Steven Hatfill, an M.D. who had worked at the U.S. Army's elite Medical Research Institute of Infectious Diseases at Fort Detrick, Maryland, where anthrax stocks were held for the development of biogical warfare weapons. Hatfill was hounded by the Department of Justice for years, but after it produced no witnesses, and no firm proof to show that Hatfill had even touched anthrax, let alone killed anyone with it, he was removed from the list of suspects in 2007. Instead, the FBI turned its attention to another individual who had complete access to anthrax: senior microbiologist Bruce E. Ivins. Ivins was given the same thorough harassment by the Department of Justice as Hatfill. Ivins was placed under 24-hour surveillance, and the FBI dug into his past and told him he was a murder suspect. Ivins was banned from the Fort Detrick laboratory, where he had worked for 28 years. In July 2008, following a voluntary 2-week stay in a psychiatric clinic for treatment of depression and severe anxiety, Ivins, then 62 years of age, checked himself out of the clinic, went home, and downed a fatal dose of Tylenol. Two weeks later the Department of Justice officially exonerated Hatfill, and in 2008 they apologized and quietly settled with Hatfill for $5.82 million. In 2009 the Department of Justice officially ended its investigation and issued a 96-page report concluding that Ivins hatched the anthrax-by-mail scheme in the hopes of creating a scare that would rescue what he considered to be his greatest achievement, an anthrax vaccine program that he helped create but which in 2001 was in danger of failing.

A look back

Anthrax may be one of the latest weapons of bioterrorism, but the disease has been killing humans for more than a century. In 1847, after mohair and alpaca wool were imported into England from Peru and Asia Minor for use as textile fibers, there were recurring deaths among those who worked in sorting the wool. The peculiar, rapid, and fatal occupational illness became known as "woolsorters' disease." As early as 1849, Franz Pollender had observed large, rod-shaped bacteria in the blood of anthrax-infected sheep, but he did not publish his findings until 1855. In 1863, Casimir-Joseph Davaine published similar observations and found identical bodies in human skin ulcers. But none of this evidence was compelling proof that these specific microbes were the cause of anthrax. Indeed, it could have just as easily been concluded that the bacteria were the result of the disease or innocent bystanders.

Robert Koch

In 1870, Robert Koch (1843-1910) became a physician in the small town of Wollstein, where he established a clinical practice and was the district medical officer. There he began to investigate anthrax—a disease that in 4 years had killed 528 people and 56,000 livestock and was a concern not only in Germany but to farmers all over Europe. For his 28th birthday Koch's wife gave him a microscope. He started using the microscope to examine everything he could get his hands on, until he examined a drop of blackened blood from anthrax-infected sheep. There he discovered what others had already seen—tangled threads that were alive. Like Davaine, he transferred the blood into rabbits, and from the infected rabbits to mice. He was able to grow the bacteria outside the body in the fluid taken from the eyeballs of cattle, and this too could be introduced into clean mice, which then became infected. Koch infected mice in a series of mouse-to-mouse inoculations, and even after 20 passages the bacteria remained true to form and their capacity to cause disease undiminished. These experiments ruled out the possibility that a poison from the original animal caused the disease in mice, since only an agent capable of multiplication within the body could produce such a result after being diluted 20-fold.

Koch devised a simple but elegant method to study the bacteria, a hanging drop. On a glass microscope slide Koch placed a drop of fluid scooped out from the center of the eye of a freshly butchered calf, and into it he put a small piece of spleen from a mouse that had died from anthrax. Over this he carefully placed a thicker glass slide with its center scooped out, and then he ringed the surface with petroleum jelly. When the two were placed together, the slides stuck together to form an airtight "sandwich." By quickly turning the glass sandwich upside-down, Koch was able to suspend the drop over the space of the scooped-out thicker slide. This preparation allowed him to sit for hours examining with his microscope the contents of the drop and not have it dry out. He saw the anthrax bacteria divide and observed the transformation of the motile rod (that is, the bacillus) into a bead-like, dormant spore (Figure 16.4). When fresh medium was added to the spores, they transformed into active bacilli. The extreme hardiness of the spores, Koch explained, was the reason for the persistence of the anthrax germ in the soil in contaminated pastures, and on the wool itself to cause "woolsorters' disease." Understanding the natural history of the disease, Koch suggested, was the means for control of anthrax: proper disposal of the contaminated carcasses.

Convinced that he had solved the riddle of anthrax, Koch visited Professor Ferdinand Cohn, Germany's leading microbe hunter, at the University of Breslau and demonstrated his experiments. Finding the results convincing, Cohn sponsored for publication Koch's paper "The Etiology of Anthrax. Based on the Life Cycle of *Bacil-*

lus anthracis." Koch contended that it was only the spores that produce the specific effect when they gain access to the circulation by way of the skin or lungs. The lung infection, or "woolsorters' disease," results from inhalation of anthrax spores in the dust from the products (mostly the wool) of the dead animals. The lung disease begins with a chill or slight shivering; the tongue is coated and thirst is present, with some nausea; vomiting is common; there is a feeling of tightness in the chest, with difficulty in breathing; there is a cough, which may continue for days; the skin is moist, bathed with perspiration; and there is fever. Massive lesions soon form in the lungs and the brain. A few thousand bacilli

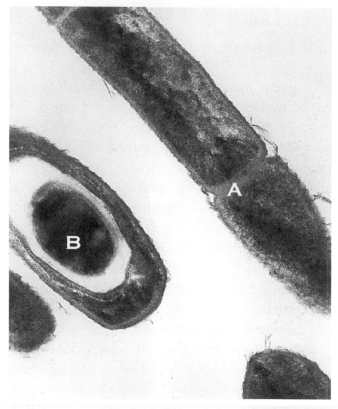

Figure 16.4 A transmission electron microscopic image of *Bacillus anthracis* showing a dividing bacillus (A) and the spore (B), Courtesy CDC Public Health Image Library #1813/ Dr. Sherif Zaki and Elizabeth White.

propagate within days into literally trillions. The final stages before death are excruciatingly painful. As their minds disntegrate, victims literally drown in their own fluids. In the skin, when the anthrax spores enter a wound, they cause localized tissue death, a painless black sore, and this blackness gives it the species name *anthracis*, meaning "coal" in Greek. Unlike the lung infection, where the fatality rate may be as high as 50%, the skin lesions resolve spontaneously in 80 to 90% of cases.

Louis Pasteur

Louis Pasteur (1822-1895), the son of a tanner from the French village of Arbois, was a diligent but unexceptional student who abandoned art to devote himself to science. During his time as a doctoral student in chemistry, Pasteur used his microscope and discovered that the tiny crystals of tartaric acid existed as mirror images—right and left—much like a pair of hands. Later, when he was appointed professor and dean at the Faculty of Sciences in Lille, he was called upon to help the beet-sugar distillers, who were having problems with their fermentations. Applying the microscopic technique used in studying crystals, Pasteur observed that in the healthy vats where sugar was fermented into alcohol there were tiny globules—yeasts—whereas in the sick vats there were no yeasts but a tangled mass of rods. These rod-like bacteria, called "bacilli," from the Latin *baculum*, meaning "rod," always produced the acid characteristic of sour milk. Pasteur concluded that fermentation to alcohol was due to live yeast and that the bacilli fermented the sugar to acid. Pasteur wrote: "It is those yeasts that my microscope showed me in the healthy beet vats, it is those yeasts that turn sugar into alcohol—it is undoubtedly yeasts that make beer from barley and it is certainly yeasts that ferment grapes into wine—I haven't proved it yet, but I know it." Later, Pasteur would again use his microscope (as well as growth in a nutritious broth) to determine the kinds of bacteria that cause butter to become rancid, milk to go sour, and wine to turn ropy. Pasteur's conclusions flew in the face of those who contended that fermentation was a chemical process and that "living ferments" such as yeast were its product, not its cause. In addition, and of equal importance for us even today, Pasteur devised a heating-cooling method to control the offending microbe: pasteurization.

In 1865, at the request of his former teacher the chemist Jean-Baptiste Dumas, now Minister of Agriculture, Pasteur became involved in studies of a disease of silkworms, called pebrine because the sick worms were covered with black spots resembling pepper, which was devastating France's silkworm industry. For 6 years he struggled with pebrine disease; using the microscope, he discovered that the spots contained a microbe, and experimentally he was able to pass the disease to healthy silkworms. He triumphantly announced: "The little globules are alive—they are parasites! The little corpuscles are not only the sign of the disease, they are its cause." Pasteur now traveled the countryside showing the farmers how to keep their healthy silkworms away from all contact with mulberry leaves that had been soiled by the sick worms. He had no sooner settled the problem of pebrine than another sickness of silkworms appeared. Now prepared for the problem, he was able to find the microbe of this disease much more quickly. At the age of 45, Pasteur basked in the glory of having saved the silkworm industry of his beloved country, and he boldly imagined a day when it would be in the power of humans to make infectious diseases disappear from the face of the earth.

Pasteur's research on silkworms provided a transition between studies of fermentation and the diseases of animals. Pasteur was small in stature and sturdily built, and his face appeared to be carved from granite—a high forehead, grayish green eyes with a deep look—and he seemed both serious and sad. The atmosphere in the Pasteur laboratory was severe; smoking was not permitted; and his assistants had strict hours, were treated more like servants than colleagues, and all worked in silence. Pasteur spent hours in silent observation, his eye glued to the microscope. Indeed, as one of his colleagues remarked, "Nothing escaped his myopic eye and jokingly we said he saw the microbes grow in the broth." He was authoritarian and demanding, and he would focus on only one thing at a time, becoming completely absorbed by the subject he had chosen to study. He trusted only himself, and he alone kept the records of the experiments by writing down the information provided by his assistants.

The rivalry between Pasteur and Koch

When Koch's paper on the identification of *B. anthracis* appeared in 1876, Pasteur was 54 and widely known for his work on fermentation and the germ theory of disease. As Pasteur turned his attention to anthrax, he came into direct competition with Koch. In 1878, Pasteur spoke before the Academy of Sciences and described his experiments with the anthrax bacillus, never once mentioning the name of Robert Koch. In 1880 he once again presented to the members of the Academy his results with a chicken cholera vaccine. In that talk he casually mentioned that Jean Joseph Henri Toussaint, a professor at the Veterinary School in Toulouse, was the first to isolate the microbe that caused chicken cholera. For 9 months Pasteur kept secret how he attenuated the virulence of the cholera microbe. Finally, he disclosed that attenuation of the cholera microbe involved simply culturing it for 2 to 3 months, exposed to atmospheric oxygen. He then went on to describe how a "live attenuated" cholera vaccine was able to protect fowl against a lethal challenge. Pasteur theorized that the live attenuated vaccine protected by "depleting the host of a certain element that healing does not reconstitute and that the absence of which hinders the development of the microbe when re-inoculated a second, third or fourth time." Pasteur, ever bold, prophesied that it would be possible to attenuate the virulence of all microbes simply by growing them under special conditions.

Flush with success at having produced a cholera vaccine, Pasteur began to develop a vaccine for anthrax. In this his competitor was not his archrival, the German Koch, but a fellow Frenchman. On July 12, 1880, Henri Bouley (a veterinarian) read before the Academy a report from his friend Toussaint that described his anthrax vaccine, produced by killing the bacteria by heating them for 10 min at 55°C, and that had been used on 8 dogs and 11 sheep. Of the 8 dogs, 4 injected with the killed vaccine survived a series of four successive injections of virulent live anthrax. All unvaccinated dogs died after the first injection. Similar results occurred with the sheep. When Pasteur heard of Toussaint's results from Bouley, he was astonished. He wrote, "it overturns all the ideas I had on viruses, vaccines etc. I no longer understand anything … it makes me want to confirm it to my own satisfaction." Pasteur could not, would not, accept with grace that he was wrong in his concept of immunity: namely, that the biological activity of a living, if attenuated, microbe was critical to deplete the host of essential nutrients.

By August of 1880, Toussaint had changed the method for killing the anthrax bacteria, and instead of heat he had resorted to carbolic acid (phenol), which Joseph Lister had used as an antiseptic (see p. 273). In February 1881, Pasteur announced that he had an attenuated anthrax vaccine, produced by the same technique he had used for attenuation of the cholera microbe, i.e., atmospheric oxygen. A month later, when Pasteur reported successful results of his anthrax vaccine in sheep, a veterinarian, Hippolyte Rossignol from Pouilly-le-Fort, a farm town 40 miles outside of Paris, challenged him. Rossignol prodded Pasteur to test the vaccine at his farm. Pasteur had to accept the challenge, or else his claim of priority over Toussaint would have been damaged. Further, there were rumors that Pasteur was seeking to profit financially from his "secret remedies" against livestock diseases. The flamboyant Pasteur "welcomed" the challenge, and a public trial was conducted at Pouilly-le-Fort in June of 1881. There were more than 200 observers, including government officials, politicians, veterinarians, farmers, cavalry officers, and newspaper reporters. Of 50 sheep in the trial, half were vaccinated on May 5 and May 17, and the remainder were left unvaccinated. On May 31 all the sheep were challenged with a virulent culture of anthrax bacilli. By June 3 all the unvaccinated sheep were dead. The London correspondent of the Paris *Journal des Debats* wrote: 'When Pasteur spoke, when his name was mentioned, a thunder of applause rose from all benches from all nations. An indefatigable worker, a sagacious seeker, a precise and brilliant experimentalist, an implacable logician and an enthusiastic apostle, he has produced an invincible effect on every mind." Pasteur's fame spread throughout France, Europe, and beyond.

There is, however, another side to the dramatic success of Pasteur's attenuated anthrax vaccine. Charles Chamberland, in Pasteur's laboratory, had been experimenting with a "dead" vaccine prepared by chemical treatment with potassium dichromate. In small-scale tests it worked. Only Pasteur and his collaborators knew the nature of the vaccine used for his famous trial at Pouilly-le-Fort. Indeed, it is now known from an examination of Pasteur's laboratory notebooks that Pasteur used Chamberland's "dead" vaccine, not the live attenuated one he emphasized was critical to success with chicken cholera. At the time, however, Pasteur, not Toussaint, received credit for developing the first successful vaccine against anthrax. Toussaint subsequently published only two more scientific papers before his death in 1890 at age 43, after suffering a mental breakdown. In France, Pasteur was recorded as the inventor of the vaccine against anthrax.

Pasteur and Koch first met in the summer of 1881 at the 7th International Congress of Medicine, held in London. Pasteur, 59, was at the height of his fame, having published on fermentation and having solved the problems of pebrine disease, soured milk, and spoiled wine. Pasteur could be irritating. He had a flamboyant style in speeches, was haughty, and was often contemptuous of the work of others. At the time, he had carried out preliminary studies on anthrax transmission and was in the process of producing vaccines for chicken cholera and anthrax through attenuation. At the Congress, Koch, 38, demonstrated his plate culture technique that led him to identify the causative agent of anthrax, *B. anthracis*. For the most part Pasteur ignored Koch's work. Koch attacked Pasteur's work on vaccines. Koch's attack may partly have resulted from an upstart's feeling resentful at being ignored by the grand master and partly have been a legitimate reflection of differences in techniques, but nationalism also played an important role in their bitter rivalry. France had been humiliated in losing the Franco-Prussian War (1870-1871), a war in which Koch was exempt because of nearsightedness; he had volunteered anyway, becoming a physician in a field hospital, where he witnessed severe battle wounds and the devastating effects of typhoid fever. In 1868, Pasteur had suffered a massive cerebral hemorrhage. Initially he lost the ability to speak or to use the limbs on the left side of his body, but gradually his speech and powers of thought returned. He never regained the use of his left hand, however, and had to rely on assistants to carry out the experiments. The construction of Pasteur's Paris laboratory was interrupted by the Franco-Prussian War, and his only son, Jean-Baptiste, had enlisted and was gravely ill with typhoid. During the siege of Paris the Pasteur family was forced to flee, first to Geneva and then to Lyon. The war affected everything Pasteur held dear—family, country, and science—yet he was physically unable to fight back. He became obsessed with a hatred of Germans and their nation. He swore: "each of my studies will bear the epigraph: Hatred to Prussia, vengeance!"

Pasteur answered Koch in detail when both were in attendance at the 4th International Congress of Hygiene, held in Geneva in 1882. Koch, who was at the height of his fame after his discovery of the bacterium responsible for tuberculosis a few months earlier, was unexpectedly aggressive on hearing Pasteur's work on attenuation. His attack on Pasteur was due in part to a problem in translation. Neither spoke nor understood the other's language. Pasteur referred to Koch's work as *"recueil allemand,"* meaning a "collection of German works," but this was mistranslated by Koch's colleague as *"orgeuil allemand,"* meaning "German arrogance." Koch was insulted and angered, and a written response to Pasteur was published in a paper ("On Inoculation against Anthrax. In a Reply to Pasteur's Lecture in Geneva") in which Koch attacked Pasteur in a highly insulting manner, writing: "all what we heard was completely useless data … he (Pasteur) is not even a physician … and all material served as a vehicle for a violent polemic directed against me." Pasteur responded in a long emotional letter and expressed surprise at the virulence of Koch's attack. There was, however, a more fundamental difference than language between the two. Koch believed in the permanence of virulence by a particular species and could not countenance how attenuation could be possible. Pasteur, on the other hand, believed (correctly!) that virulence was a property that could be lost and recovered. Pasteur suggested that such variations in virulence could explain how epidemics might arise because of the temporary or short-term increases in virulence by a particular microbe. Further, such changes in virulence might arise when a particular microbe acquired virulence that enabled it to infect a previously unsusceptible animal species. These remarks by Pasteur made decades ago continue to be relevant to HIV/AIDS and other emergent diseases.

Pasteur and Koch were different from one another in significant ways: Pasteur, trained as a chemist and more interested in broad philosophical questions and larger scientific issues than in specific medical problems, established microbes as agents of disease (the germ theory); and Koch, educated as a physician, unequivocally assigned a specific microbe to a particular medical malady. Despite their bitter antagonism for one another, the studies that Koch and Pasteur carried out were complementary. Koch would make his major contributions in hygiene and public health measures, and through his associates in protection through chemotherapy (Paul Ehrlich) and serum therapy (Emil von Behring), while Pasteur would make primary medical contributions to immunization through microbe attenuation.

Currently, because anthrax is considered to be an appealing biological warfare weapon, there is ongoing research into the production of a protective anthrax vaccine. The currently licensed anthrax vaccine in the United States is not based on a live attenuated vaccine, as proposed by Pasteur, nor is it an inactivated preparation first developed by Toussaint and recognized only in 1998 by the French government; rather, it consists of a protein, called protective antigen, a central component of the anthrax toxin, adsorbed to aluminum hydroxide with small amounts of formaldehyde and benzethonium chloride added as preservatives. Prepared from attenuated cultures of a capsule-free strain and named Biothrax, it is licensed and recommended for a small population of mill workers, veterinarians, laboratory scientists, and others with a risk of occupational exposure to anthrax. Parenthetically, those infected with anthrax can be treated presently with the antibiotic ciprofloxin (Cipro).

Madness and the Infectious Protein

The sign in the window of McDonald's read: "We no longer serve British beef." The year was 1996, and London was declaring itself "hamburger free." Sales of beef dropped, schools stopped serving beef, and there was widespread concern about eating beef, or for that matter any meat product. This was not an economic boycott on British beef. Its cause was an outbreak of "mad cow disease." Fear and panic gripped the country because of the appearance of Creutzfeldt-Jakob disease (CJD), a rare and fatal neurological disorder. The victims were, unlike most CJD cases, young, and the link to their deaths was that all had eaten British beef. It was called variant CJD, or vCJD. The British government responded by forbidding the sale of beef, and more than 250,000 cattle were slaughtered to stem the mini-epidemic. Over the next several years the cattle industry lost $40 million a year, and trade in beef with Britain was virtually nil. Beef infected with mad cow disease, or technically bovine spongiform encephalopathy (BSE), had unexpectedly jumped the species-specific barrier, and in doing so entered the human food supply. Some suspect that industrial cannibalism may explain the 1990s outbreak of BSE. Twenty years earlier there had been an Arab oil embargo that had caused an increase in the cost of energy needed to heat the rendering vats to process brain, spinal cord, and lymphoid tissue from cattle into lard. There was also an increase in the price of organic solvents, and during this same period people began to turn away from lard in cooking. As a result, there was a surplus of tallow, and its price fell. The British government, in order to accommodate these economic changes, allowed the processors to reduce the temperature of the rendering vats and to omit organic solvents. As a result, the offal that was cooked yielded a high fat protein fraction; this was then milled into bone and meat meal, and when this was fed to livestock, the animals became infected with the causative agent of BSE. The origin of the causative agent of BSE remains unclear; it is possible, however, that it arose spontaneously in cattle, as does sporadic CJD in people, or more likely that it came from scrapie (see below) in sheep. Whatever the source, 10 or more years passed before the infectious agent that survived the rendering process began to produce a recognizable disease in cattle. Since the human disease was reported in 1996, more than 200 teenagers and young adults developed vCJD as a result of BSE.

The story of the linkage of CJD with BSE began not in Britain in 1996 but in the Highlands of New Guinea, with studies by an American pediatrician, Carlton Gajdusek (1923-2008), who had been asked to look into several cases of encephalitis among children of a hunter-gatherer tribe, the Fore. The disease was called kuru, and it was characterized by uncontrolled tremors, blurred speech, and pathologic laughter. In 1957 he described it as "a classical advancing Parkinsonism involving every age, overwhelmingly females although many boys and a few men also have had it, is a mighty strange syndrome … well-nourished, healthy young adults dancing about, with … tremors which look far more hysterical than organic, is a real sight." Gajdusek had always wanted to be a microbe hunter. As a teenager growing up in Yonkers, New York, he had read Paul de Kruif's book *Microbe Hunters*, and his heroes became Louis Pasteur, who devised the first laboratory-produced vaccines; Robert Koch, who identified the causative agents of anthrax, cholera, and tuberculosis; Antonie van Leeuwenhoek, who was the first human to see a parasite (*Giardia*) by looking through his magnifying lenses; and Ronald Ross and Walter Reed, who flagged the *Anopheles* and *Aedes* mosquitoes as the vectors for malaria and yellow fever, respectively. Gajdusek had been well prepared for his work on kuru; he had been trained as an M.D. at Harvard Medical School and then went on to study the growth of poliovirus in kidney cells with John Enders at Harvard, immunology and hepatitis with Frank Macfarlane Burnet in Australia, and protein chemistry with Linus Pauling at the California Institute of Technology (Caltech).

The disease Gajdusek investigated in New Guinea was a mysterious one. Upon autopsy the brains of kuru victims were found to have extensive damage—there were holes and microscopic knots of protein called amyloid plaques—but remarkably there was no sign of inflammation. The condition resembled another disease that had been described in 1913 in Breslau, Germany, by the neurologist Hans Creutzfeldt. A woman, an elderly mental patient, died 3 days after Creutzfeldt had examined her. Before her death he noted that she had facial tics; made jerky and twitching movements of her arms and legs; had difficulty swallowing; and had epileptic attacks, screaming fits, and "unmotivated outbursts of laughter." Creutzfeldt's work on this neurological disorder was interrupted by World War I, but at its conclusion he wrote up his findings and as a courtesy sent the manuscript to a colleague, Alfons Jakob of Hamburg, who had observed four other patients with the same deadly symptoms. Stained slices of brain tissue from these deceased patients were examined by light microscopy. Creutzfeldt and Jakob described the microscopic findings—"it was full of holes as in a sponge"—and they called the condition "a spongioform change." (Figure 16.5). Although the brains of these patients were severely damaged, there were no signs of an inflammatory response. The published their findings in 1920, and the disease they described was named after them, Creutzfeldt-Jakob disease.

Upon hearing of Gajdusek's findings in New Guinea, a pathologist colleague remembered the work of Creutzfeldt and Jakob and realized that the brains of kuru victims and those of CJD patients were quite similar. There was a critical difference in the age of the victims, however: the CJD patients were older, whereas those with kuru were younger women and children. Gajdusek began to search for the causes of kuru and the reasons for its pathology. In detective-like fashion he examined the food the Fore ate; calculated the incidence of kuru in the Fore tribe; determined the contact the Fore would have with mosquitoes, flies, and ticks; and evaluated whether it was a familial disease. In his examination of the distribution of kuru among the Fore, Gajdusek

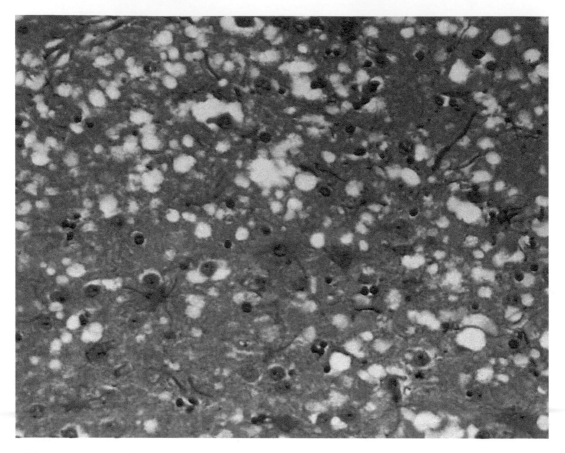

Figure 16.5 A stained section of cow brain tissue with BSE showing the presence of "holes" giving it a spongelike appearance. Courtesy CDC Public Health Image Library #5435/Al Jenny, USDA-APHIS, 2003.

carried out an epidemiologic study similar to that first conducted by John Snow for cholera a century earlier (see p. 181). No correlations were found for a food toxin, and vector or person-to-person contact also showed no pattern consistent with the occurrence of kuru. The disease also seemed not to be inherited. Then in 1961-1963, during a kuru epidemic, a husband-and-wife team of Australian anthropologists made a critical finding: the members of the Fore tribe engaged in ritual cannibalism. Not all members, however, were ancestor eaters. Only women and children partook in the practice, and the brains of the deceased relatives were considered a special delicacy. Cannibalism was not an ancient practice among the Fore. It is thought to have started in the 19th century, with the first remembered case in about 1920. At its peak, kuru was estimated to have killed 1% of the population annually. With the pathology and transmission of kuru now known, Gajdusek prepared a photographic exhibit of its features, and this traveled to medical museums throughout the world. One visitor to such an exhibit in

London, William Hadlow, was a veterinarian. After seeing the photographs, Hadlow wrote to the medical journal *Lancet* that the pathology seen in the kuru brains resembled a disease of sheep called scrapie, with which he was quite familiar.

Scrapie is a neurological disease, and sheep itch so intensely that they tend to scrape the wool off their hindquarters by rubbing themselves against fences, trees, and walls. In addition to the scraping syndrome, the sheep stagger, have tremors, go blind, become paralyzed, and die. Where scrapie originated is unclear, but epidemics were recorded in the 19th century in parts of Europe, and by the 20th century it was endemic in Britain. In the 1930s, French and English veterinarians unintentionally demonstrated that scrapie was an infectious disease. Both groups attempted to produce a vaccine in the fashion of Louis Pasteur: brains of scrapie sheep were homogenized, formalin was added to kill the pathogen, and this vaccine was injected into the brains of healthy animals. After many months, and before the animals could be challenged, all the "immunized" sheep came down with scrapie. Because of the long delay before symptoms appeared—5 years in sheep—a pathologist in Iceland, Bjorn Sigurdsson, proposed that the infectious agent of scrapie was a "slow virus." (This is in contrast to "quick" or acute virus diseases such as smallpox, polio, yellow fever, and measles.) Gajdusek, the pediatrician trained in virology, embraced the name and used it to explain kuru's cause. When, after autopsy, the brains of the "slow virus"-infected scrapie sheep were examined, spongioform degeneration was evident. The question for Gajdusek then became: how is this "slow virus" transmitted in nature? He embarked on a series of experiments designed to satisfy Koch's postulates (see p. 169). Brains from kuru victims were homogenized and injected into the brains of healthy chimpanzees. In 10 months the monkeys came down with kuru-like symptoms, and all died shortly thereafter. It was also possible to infect other chimpanzees from these kuru-infected monkeys. The brains were examined by light microscopy, and they were spongioform. In Gajdusek's mind, Koch's postulates were satisfied, and now he had joined his boyhood heroes as a full-fledged microbe hunter. Then, using the 1878 dye methods of Paul Ehrlich, the brain tissue was stained with the aniline dye Congo red; when viewed with polarizing light, the amyloid plaques (with their aggregated and knotted fibrils) stood out with a greenish glow. In 1978 homogenized brain tissue from scrapie-infected sheep was concentrated and examined by Beatrice Merz using the electron microscope; she saw filaments that were twisted and tangled, but there was nothing that looked like a virus particle.

Although sheep farmers tended to believe that scrapie was a hereditary disease, there was mounting evidence that transmission was by sheep eating the afterbirth of an infected animal or grazing in a pasture where infected animals had been kept. Scrapie and kuru seemed to follow the same transmission path: eating. Proof of kuru transmission by cannibalism was clearly shown by its decline among the Fore after the practice was stopped. It ceased among 4-to-9-year-olds by 1968, among 10-to-14-year-olds by 1972, and among 15-to-19-year-olds by 1973. No child born after the Fore abandoned cannibalism developed kuru. Deaths from kuru among the Fore tribe dropped dramatically. In 1976, Carlton Gajdusek received the Nobel Prize for "discoveries concerning new mechanisms for the origin and dissemination of infectious diseases."

In 1974 a woman who had previously received a corneal transplant came down with CJD. Examination of the donor record showed that he had died of pneumonia but had a history of memory loss and tremors. It was later believed that the donor had suffered from CJD. Now there was the frightening possibility that CJD was transmissible by transplantation. This suspicion was tragically confirmed when other transplant recipients came down with CJD, as did those who received injections of growth hormone prepared from pituitary glands from human cadavers. Prior to the development of recombinant DNA methods for the synthesis of human growth hormone, pituitary pools were used as a source of hormone in the treatment of those of small stature. Between 1963 and 1985, ~500,000 glands were used in the preparation of hormone and used to treat 10,000 children; the result was an alarming number (several hundred) of cases of CJD. Clearly, the transmissible infectious agent responsible for the epidemic in cattle was also capable of infecting humans.

What was the infectious agent responsible for scrapie, kuru, and CJD? In 1976, Stanley Prusiner (b. 1942) at the University of California at San Francisco began carrying out biochemical analyses of scrapie and found that, unlike any virus, it contained no nucleic acid but was made entirely of protein. He called the agent responsible for the neurodegenerative diseases a "prion" (pronounced *pree-on*), for proteinaceous infectious particle, thereby eliminating Gajdusek's "slow virus" as the culprit in kuru, scrapie, and CJD. How can a protein-only particle, the prion, multiply without having any self-replicating genetic material such as DNA or RNA? In 1982, Prusiner and his colleagues analyzed the amino acid sequence of the scrapie protein, abbreviated PrP^{sc} (protein of prion, scrapie). This amino acid sequence was then used to work backwards—amino acid sequence → RNA nucleotide sequence → DNA nucleotide sequence—to find the DNA code (or gene) that specified the prion protein (now abbreviated PrP). Surprisingly, the gene for PrP^{sc} from the brains of scrapie-infected animals was identical to the amino acid sequence of PrP from normal brains. The only difference between the two was that PrP could be digested with a proteolytic enzyme (such as that found in the pancreas, which we use to digest proteins in our diet) while PrP^{sc} could not, despite its identical amino acid composition. Prusiner found that in PrP^{sc} the conformation of the molecule had been changed from an α-helix to a misfolded β-sheet, a form that could not be digested with proteolytic enzymes (just as the protein in your hair is virtually indigestible by proteolytic enzymes). This also explains, of course, how the prion is able to resist the action of stomach enzymes and why eating can be a means for transmission. Prusiner also found, further, that if one eliminated the gene for PrP (called a "knockout") in animals susceptible to scrapie, an infection did not develop after brain inoculation of PrP^{sc}. The obvious conclusion from these experiments was that a normal protein, PrP, was changed into the misfolded disease form, PrP^{sc}. How could PrP be reshaped into PrP^{sc} to become infectious?

The most plausible hypothesis—and it is only a hypothesis—is by nucleation. It may work as did ice-nine in Kurt Vonnegut's 1963 novel *Cat's Cradle*. In the novel a brilliant scientist invents a new kind of ice form with different properties from ordinary ice:

> "Now suppose," chortled Dr. Breed, enjoying himself, "that ... the sort of ice we skate upon and put into highballs—what we might call ice-one—is only one of several types of ice. Suppose water always froze as ice-one on Earth because it never had a seed to teach it how to form ice-two, ice-three, ice-four ... ? And suppose there were only one form, which we will call ice-nine—a crystal as hard as this desk—with a melting point of, let us say ... one hundred-and-thirty degrees. ... When the rain fell, it would freeze into hard little hobnails of ice-nine—and that would be the end of the world."

(And this is what happens in *Cat's Cradle* when the nucleating ice-nine is accidentally dropped into the ocean and all the water of the Earth's surface freezes, making life impossible.)

You can demonstrate the nucleation process for yourself by dissolving sugar in a glass of hot water until no more will dissolve; then cover the solution with a glass plate, turn off the heat, and let it cool. In cooling the sugar solution is supersaturated, but sugar crystals do not form because they have not been "instructed" as to the form they should take. Add but a single sugar crystal, and the solution crystallizes. The crystal "seed" has supplied a template—structural information—that the solution needs to organize itself. Similarly, in scrapie, kuru, and CJD, the nucleating form is PrP^{sc}, which, when it gets into a cell, especially nerve cells, results in the conversion of PrP into PrP^{sc}—the abnormal form. And with more abnormal forms there are more nucleating centers, and finally the cell, to use Vonnegut's words, "freezes into hard little hobnails" that are neurotoxic. Prusiner's group has also produced in the laboratory synthetic prions—consisting of 55 amino acids—and by treating them with acetonitrile has been able to produce the infectious β-sheet conformation; when this is inoculated into mice, a year later they manifest neurological changes. The transmissible and infectious nature of this synthetic prion was demonstrated by serial passage in mice. Stanley Prusiner received the Nobel Prize in 1997 for his discovery of "prions—a new biological principle of infection."

What is the function of the PrP, this precursor to neurological pathology? It is hypothesized that in normal cells prion-like proteins might form or stabilize long-term memories by creating long-lived protein clusters at the synaptic junctions in the nervous system. Prusiner believes that through a random process, which most of the time is a dead-end pathway, small numbers of prions are cleared via protein quality control pathways; when, however, a sufficient quantity of prions are formed and reaches a threshold level, there is self-propagation, leading to nervous system dysfunction.

Demostrating that prions are involved in certain neurodegenerative diseases may lead to better diagnostics and effective therapies. Prusiner suggests that "early diagnosis will require identification of prions long before symptoms appear. Meaningful treatment will probably require cocktails of drugs to diminish precursor protein, intefere with its conversion into prions and enhance clearance."

Ebola

In March 2014 the WHO declared there was an outbreak of Ebola across the African country Guinea. The disease claimed dozens of lives and was on its way to neighboring countries … it went on to kill 10,000 people … in the 12 months that followed and the … menace has left a permanent mark on the world. As the crisis deepened Sierra Leone, Liberia and Guinea shut their schools. A whole generation of children missed some six months of education … The Global Business Coalition for Education suggested up to 5 million children were denied classes. And many are unlikely to return … Margaret Chan, the Director-General of the WHO, said, "The world including the WHO was too slow to see what was unfolding before us."

David Quammen, the author of the book *Spillover*, wrote: "The story of the 2014 outbreak was so rivetingly awful that it competed for headline space with contemporaneous events in Syria, Ukraine and the Gaza strip." The *New York Times* mobilized dozens of reporters, photographers, and video journalists, producing more than 400 articles, including 50 front-page articles. The World Bank estimated that the three countries will have lost $2.2 billion in foreign economic growth as a result of this epidemic. About 200,000 people were having trouble getting food. Doctors Without Borders sent 700 doctors and aid workers from around the world to Ebola-stricken countries. The governors of New York and New Jersey announced mandatory quarantines for medical workers returning from West Africa. Five U.S. airports screened travelers from West Africa. The World Bank mobilized $1.62 billion for Ebola response and recovery to support the countries hardest hit by Ebola, to pay for essential supplies and drugs, personal protective equipment and infection prevention control materials, health worker training, hazard pay and death benefits to Ebola health workers and volunteers, contact tracing, vehicles, data management equipment, door-to-door public health education outreach, and, in Liberia, psychosocial support for those affected by Ebola. To support a surge of foreign health workers to the three countries and to provide budget support to help the governments of Guinea, Liberia, and Sierra Leone to cope with the economic impact of the outbreak, and financing the scale-up of social safety net programs for people in the three countries.

Ebola is a newly emerging disease, a zoonosis. A zoonosis is an animal infection that is transmissible to humans. Bubonic plague, malaria, AIDS, West Nile virus, influenza, rabies, and Lyme disease are some other zoonotic infections. As with these diseases, Ebola is a disease that came about by an interspecies leap. It might surprise you to learn that the jump of a pathogen from one species to another is not such a rare occurrence. Indeed, about 60% of all infectious diseases currently known either cross routinely or have recently crossed from other animals to us. So what causes Ebola disease, and where did it come from?

Ebola disease

Ebola disease is caused by a filovirus, so called because when viewed with an electron microscope the viral particles are shaped like sinewy filaments (Figure 16.6).

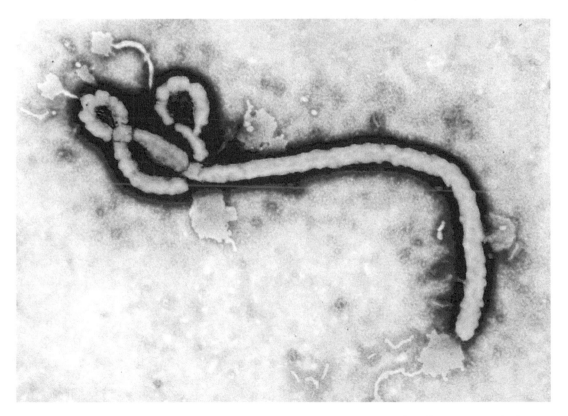

Figure 16.6 A colorized transmission electron microscopic image of the filamentous Ebola virus. Courtesy CDC Public Health Image Library #10815/Frederick A. Murphy.

Filoviruses, the most lethal human pathogens in the world, include two genera: Ebola virus (EBOV) and Marburg virus (MARV). (MARV was the first known filovirus, identified in 1967 during an outbreak of hemorrhagic fever in Marburg and Frankfurt in Germany and Belgrade in the former Yugoslavia from the importation of infected monkeys from Uganda.) Filoviruses cause severe hemorrhagic fever in humans and nonhuman primates, killing between 40 to 90% of infected individuals within 2 or 3 weeks after the onset of disease.

In 1994, Richard Preston described Ebola disease in the most horrific way in his bestselling book *The Hot Zone*. The virus "jumped from bed to bed, killing patients left and right," causing them to bleed profusely and liquefying their organs until "people were dissolving in their beds." Virtually every part of the body is digested into a slime of virus particles. "Droplets of blood stand out on the eyelids. You may weep blood … the blood refuses to coagulate." In actual fact, the early symptoms—headache and fever, aches and pains, and vomiting—resemble a case of the flu and begin about 8 to 10 days after exposure to the virus but can appear as late as 21 days. Then in about half the cases the victims may vomit blood or pass it into the urine or bleed under the skin or from the eyes and mouth. But the bleeding is not usually what kills the patient. Rather, blood vessels deep within the body begin leaking fluid; as a result, the blood pressure plummets and the heart, kidneys, liver, and other organs begin to fail.

The Ebola virus

The virus that causes Ebola disease consists of a single strand of RNA that codes for seven structural proteins: nucleoprotein (NP); virion proteins (VP) 24, 30, 35, and 40; polymerase protein (PP); and glycoprotein (GP), which forms spikes on the surface of the virus particle and functions in attachment and allows entry into the host cell. VP24 and VP40 are matrix proteins and VP30 is essential for transcription and replication once the virus is within the cell. EBOV pathogenicity is related to GP as well as VP24 and VP35. (Mutations in VP24 and NP are primarily responsible for the acquisition of high virulence.) With internalization of the virus, there is a marked replication of virus, as well as impairment of the immune system due to a loss of CD8 T cells as well as plasma cells, concomitant with the inhibition of production by the host of the antiviral interferon. There is also vascular dysregulation, leading to hypotension, coagulation disorders, and bleeding, which finally results in the sudden onset of severe shock.

Disease emergence

The first appearance of Ebola disease was in June and July of 1976 in Nzara, a small town in southern Sudan bordering on the African rainforest. The outbreak started with a single individual who was working in a cotton factory. The disease spread rapidly from him to close relatives; then it moved on to neighboring areas, and high levels of transmission occurred in the hospital. The outbreak lasted until November, at which time the total number of infected individuals was 284, with 151 deaths, for a fatality rate of 53%. In 1976 there was another outbreak, with a case fatality rate of 88% (318 cases with 280 deaths), that occurred in Zaire (today the Democratic Republic of Congo), and it was presumed that this initial infection was acquired in the Mission Hospital. Virus was isolated from patients of both outbreaks and named Ebola virus after a small river in Zaire. Although initially it was believed that the two outbreaks were directly linked, it has since been discovered that the outbreaks were caused by two distinct species of filovirus, Sudan ebolavirus and Zaire ebolavirus.

To everyone's surprise, EBOV next emerged in the United States in 1989. It was detected in cynomolgus monkeys (*Macaca fascicularis*) that were imported from the Philippines into a primate facility in Reston, Virginia. The 1989 monkey outbreak spread through the cages and rooms of the animal colony via small particle aerosols. This distinct species of EBOV, known as the Reston ebolavirus, had a lower human pathogenciity than previous isolates.

Since the first description of EBOV, it has reemerged several times in Central Africa, where the disease is considered to be endemic. In 1992 and 1994 there were Ebola-associated deaths in a troop of chimpanzees in the Tai National Park in Ivory Coast. A 34-year-old woman who performed the necropsy on the chimps came down with Ebola disease. This disease was caused by a distinct species, Ivory Coast ebolavirus. Although it was once thought that EBOV originated in gorillas because human outbreaks began after people ate gorilla meat, now it is believed that bats are the natural reservoir for the virus and that apes and humans catch it from eating food that bats have drooled or defecated on, or by coming into contact with surfaces infected by bat droppings and then touching the eyes or mouth. Ebola can also be introduced into human populations through close contact with blood, secretions, organs, and bodily fluids of infected animals such as chimpanzees, gorillas, fruit bats, monkeys, forest antelopes, and porcupines that are found ill or dead in the rainforest.

The world paid little attention to Ebola disease despite Preston's dramatic description of EBOV in the 1994 book *The Hot Zone*. But in 1995 EBOV reemerged in Kikwit in the Democratic Republic of the Congo. Subsequently there were outbreaks in Gabon and Uganda. The largest-ever epidemic of Ebola disease, since the first case occurred in a 2-year-old in December 2013, claimed thousands of lives in West Africa in 2015—more than 10,000 across Guinea, Sierra Leone, and Liberia. This outbreak was cause by Zaire ebolavirus. Ebola virtually shut down clinics and public health infrastructure in many areas. Visits for routine health services dropped precipitously. In Guinea alone, health facilities treated an estimated 74,000 fewer malaria cases (compared with previous years) during this outbreak. Immunization levels dropped across all three countries, so that should there be a regional oubreak of measles, there could be thousands of cases and potentially more deaths than would be caused by EBOV. Craig A. Spencer, a worker with Doctors Without Borders and Director of Global Health in Emergency Medicine at New York Presbyterian-Columbia University Medical Center, wrote in the August 15, 2015, *New York Times*: It will be years before these West African countries are able to train nurses, develop and implement a sustainable medical education model, and supply an adequate number of homegrown health care workers. Without sustained assistance from the international community, the nations of West Africa face a losing war of attrition with the epidemic. Although the epidemic reached its peak in late 2014, and had decreased significantly since then, as of today (2016) approximately 11,302 people have succumbed to the disease.

Transmission

EBOV relies on intimate social interactions for transmission. It has an R_0 of 1.5 to 2.5, similar to that of other aerosol droplet infections such as SARS and influenza (but far lower than measles and polio, with R_0s of 12 to 18 and 5 to 7, respectively). The virus is passed on through close contact with the bodily fluids of infected individuals. A cough from a sick person could infect someone who has been sprayed with saliva. The virus can be present on a person's skin after symptoms develop, and according to the Centers for Disease Control and Prevention (CDC), the virus can survive for a few hours on the dry surface of doorknobs and countertops and may survive for several days in puddles or collections of bodily fluids. People remain infectious so long as the blood contains the virus. Bleach solutions can kill the virus.

The most vulnerable to Ebola disease are those who care for the sick or engage in traditional burial ceremonies, such as washing the body of the deceased. At the height of the 2013-2016 outbreak, entire communities were quarantined. Traditional burial ceremonies were banned. The most simple human touch—a handshake or a hug—was discouraged. Ebola response teams turned up in full spacesuit gear, armed with bleach spray, to take bodies away and to treat the sick. Since the start of the epidemic nearly 7% of health care workers in Sierra Leone and more than 8% in Liberia have died from Ebola disease. At least 24 cases were treated in Europe and the United States, many of whom were health workers who had contracted the disease in West Africa and had been transported back to their home countries for treatment.

Treatment

Treatment of Ebola disease involves supportive care rehydration with oral and intravenous fluids; treatment of specific symptoms improves survival. There is as yet no proven treatment available; a range of potential treatments, however, including blood products, immune therapies, and drug therapies, are being evaluated. No licensed vaccines are available, but several potential vaccines are under development.

Diagnosis

The differentiation of Ebola disease from other infectious diseases such as malaria, typhoid fever, and meningitis can be difficult. Confirmation that symptoms are caused by EBOV infection are made using the following investigations: antibody-capture enzyme-linked immunosorbent assay (ELISA), serum neutralization test, reverse transcriptase PCR (RT-PCR) assay, electron microscopy, and virus isolation by cell culture.

Prevention and control

The WHO recommends reducing the risk of wildlife-to-human transmission by eliminating contact with infected fruit bats or monkeys/apes and the consumption of their raw meat. The WHO also recommends that animals be handled with gloves and other appropriate protective clothing, and that animal products (blood and meat) should be thoroughly cooked before consumption. Further WHO recommendations are that the risk of human-to-human transmission be reduced by minimizing direct or close contact with people with Ebola symptoms, particularly with their bodily fluids, and that gloves and appropriate personal protective equipment be worn when taking care of ill patients at home. Regular handwashing is required after visiting patients in hospital, as well as after taking care of patients at home.

To reduce the risk of possible sexual transmission, the WHO recommends that male survivors of EBOV disease practice safe sex and hygiene for 12 months from the onset of symptoms or until their semen tests negative twice for EBOV. Contact with body fluids should be avoided, and washing with soap and water is recommended. The WHO does not recommend the isolation of male or female convalescent patients whose blood has tested negative for EBOV. Outbreak containment measures, including prompt and safe burial of the dead, identifying people who may have been in contact with someone infected with Ebola and monitoring their health for 21 days, separating the healthy from the sick to prevent further spread, and maintaining good hygiene and a clean environment, are all important.

Zika Disease

In 2015 worldwide attention was focused on the report from Brazil that thousands of infants born in the country were suffering with microcephaly, literally "small head."

The birth defect—a malformation of the brain leaving the afflicted children with varying degrees of shrunken heads—leads to neurological problems that include delay in development and cognitive impairment. Brazil usually sees a couple of hundred cases of microcephaly per year, some of which result from toxoplasmosis, genetic mutations, and even excessive alcohol consumption during early pregnancy, but between October 2015 and February 2016 well over 3,500 infants in Brazil were born with microcephaly. The outbreak could not have come at a worse time: in the midst of Brazil's worst recession and a political crisis in which the president was impeached. The increase in the number of cases of microcephaly coincided with the explosive spread of the Zika virus in the area. In February 2016 the WHO issued a global health emergency, and various governmental organizations advised pregnant women not to travel to countries where the Zika virus is endemic. As the WHO emergency declaration made clear, establishing the virus as the causative agent would be an important step, but public health responses did not wait for proof. Acts in the face of limited evidence are as old as epidemiology itself. Indeed, more than a century ago in 1854, John Snow, the "father" of modern epidemiology, had the Broad Street pump handle removed to protect the citizens of London from cholera (see p. 181), and this was long before Robert Koch identified *Vibrio cholerae* as the causal agent of the disease. That the Zika virus itself was the cause of the recent spike in cases of microcephaly has been proven by further studies.

Finding Zika virus

In the late 1930s and late 1940s, British and American scientists were housed in Entebbe, Uganda, primarily with the aim of isolating yellow fever viruses. To isolate these viruses, the researchers placed Asian rhesus monkeys (*Macaca mulatta*) in cages on platforms atop the forest canopy in the Zika forest near Entebbe—presumably a location where they could serve as potential hosts for the local mosquito population, known to be vectors of yellow fever. In 1947 blood from one of these sentinel monkeys (monkey 766) was injected into the brains of Swiss albino mice. This led to sickness; a number of viruses were isolated from a brain homogenate from the sick mice, and among these viruses was Zika virus 766 (Figure 16.7). This same virus was also isolated from mosquitoes found in the same location. It was suspected that the Zika virus was circulating and endemic in the indigenous African monkeys, including 13 species of *Colobus*. That the Zika virus was isolated from an Asian monkey (*M. mulatta*) implies that the African virus was able to adapt and overcome the species barrier. It is not known when the Zika virus crossed into the human population; blood samples taken from several individuals living in the area, however, were shown to have antibodies to Zika, which suggests that the virus had been circulating in the human population by the 1950s. Although Zika fever or Zika disease was known from

Uganda and Tanzania in the early 1950s, the virus was not isolated from a human (in Nigeria) until 1969. The Zika virus that originated in Africa then spread in two directions: within Africa, and then to Asia and the Pacific region. Until 2007, Zika infections were sporadic and without any major outbreak. Then in 2007 nearly 75% of the population on the Pacific island of Yap was infected. Another outbreak occurred in French Polynesia in 2013 and was traced to those who had attended a cultural event on Easter Island. In addition, there were cases reported from New Caledonia and the Cook Islands. The current epidemic started in Brazil with the first confirmed case of Zika disease in May 2015. No one knows how the virus got to Brazil, but the suspicion was that it arrived in 2014 carried in the blood of an individual attending the World Cup football competition. As many as 1.3 million persons have been affected in Brazil alone. Since then, the virus has spread to more than 20 countries and territories. Territories in the Caribbean have already reported local transmission, and by October 2016, according to the CDC, there had been 3,936 cases in the continental United States and 25,955 cases in the U.S. territories of Puerto Rico, the U.S. Virgin Islands, and American Samoa.

Transmission of Zika

The Zika virus is transmitted to humans through the bite of an infected female mosquito, notably *Aedes aegypti* and *Aedes albopictus*. The *Aedes* mosquito originated in Africa, but it is now found in almost all subtropical and tropical climes. The mosquito breeds in small containers of water close to or even inside homes. Some have speculated that El Niño has played a role in the spread of Zika in Latin America, since it creates ideal conditions for the proliferation of mosquitoes. Of great public health concern is that more than half the world's human population lives in areas where these mosquitoes are abundant. Because *Aedes* is active and aggressive during the daytime hours, but especially at dusk or dawn in

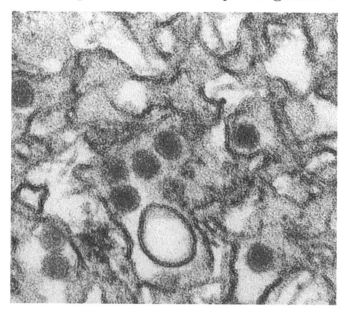

Figure 16.7 A digitally colorized transmission electron microscopic image of Zika viruses showing outer envelope and inner dense core. Courtesy CDC Public Health Image Library #20538/ Cynthia Goldsmith, 2016.

cloudy weather, and often indoors, control can be difficult.

Zika virus may be transmitted perinatally as well as sexually through semen. A patient who was infected in Senegal transmitted the infection to his wife by sexual contact. The Zika virus can also be transmitted by blood transfusion. The fact that Zika is mostly a mild disease may actually hinder control measures, since asymptomatic individuals may be free to spread the disease to others without being identified.

Clinical

About 20 to 25% of people infected with the Zika virus commonly develop mild symptoms that include a rash and painful joints; headache, vomiting, and edema occur in a minority of cases. A rare complication of Zika virus is Guillain-Barré syndrome, a condition in which one's own immune system damages nerve cells, leading to muscle weakness. If a pregnant woman develops a Zika infection, the virus can cross the placenta and infect the fetus. The impact on the fetus is more severe, especially if the infection is contracted early in pregancy. In the infected fetus there may be complications such as microcephaly, fetal growth restriction, central nervous system insufficiency, and death.

The most alarming feature of Zika infections is the threat to the fetus. Zika, like other viruses such as cytomegalovirus and varicella, "loves" the brain, where the virus can multiply and destroy cells during early embryonic development. Damage to the fetal brain is severe because the fetus has a minimal immune system, in contast to an adult with a fully functional immune system. In an autopsy of a brain from a microcephalic fetus, the brain was not only very small, but there was a complete absence of cerebral folds, the lateral ventricles were severely dilated, there were calcifications throughout the cerebral cortex, and there was hypoplasia of the brain stem and spinal cord. A large amount of viral genomic RNA was detected within the brain cells. Because Zika pathogenesis is difficult to study directly in humans, mouse models have been used to point to how a mother's infection harms the fetus. When pregnant mice were injected with Zika virus isolated from a person in French Polynesia, high levels of virus were found in the placenta—1,000 times more than in the blood. This suggests that at least one effect is for the virus to cut off the blood supply to the fetus. The virus also turned up in the head of the fetus. In another study, again in mice, the Zika virus was injected directly into the fetal brain, where damage was associated with neuronal stem cells. Here there was cell-cycle arrest and inhibition of differentiation that resulted in cortical thinning and microcephaly.

Prevention

With transmission of Zika virus by mosquitoes, most preventive measures involve controlling the insects using insect repellents, using mosquito nets at night, wearing light-colored clothing, and staying covered. Other control measures include removing stagnant water and plants that contain water that can serve as breeding grounds for *Aedes* and spraying where mosquito larvae are detected. Education of the public regarding both insect bite prevention and mosquito management is important. Because of Zika's pandemic nature, travel precautions may help to restrict transmission.

A novel approach has been employed in Brazil: genetically modified *Aedes*. The Oxitec company has created mosquitoes that contain an inserted gene producing a protein called tetracycline repressible activator variant, which ties up the cell's machinery and prevents the expression of other genes key to survival, leading to the early death of the insect. Population control is achieved by releasing male mosquitoes carrying the self-destructing gene, which will mate with the blood-feeding female mosquitoes. Offspring that inherit the gene die, causing the mosquito population to decline dramatically. In April 2015, Oxitec released these genetically engineered mosquitoes in a Brazilian city (Piracicaba) and reported an 82% reduction in wild larvae by the end of the year. Another approach used was to infect *Aedes* with the bacterium *Wolbachia*, which does not infect *Aedes* in nature. *Wolbachia*-infected mosquitoes do not pick up or transmit viruses. The *Wolbachia* is transmitted to the next generation via the eggs and can be sustained in the environment for a considerable period of time.

Conspiracy theories

As has happened with pandemics from the time of the Black Death (Chapter 4) to the era of AIDS (Chapter 5), the rapid appearance of Zika virus has spawned a plethora of conspiracy theories. Some Argentine doctors blamed microcephaly on the use of a larvicide and the contamination of the drinking water. The larvicide pyriproxyfen, however, was not used in areas where microcephaly occurred. Another group has blamed the release in 2011 of genetically modified mosquitoes for the control of dengue fever. Another popular misconception is that chickenpox and rubella vaccines are responsible for the surge in microcephaly. A bizarre rumor is that the Zika epidemic is a plot by the global elite to depopulate the world. The antidote to these unfounded theories about the Zika outbreak is to publicize authentic scientific explanations.

Vaccine

There is at present no specific antiviral agent or vaccine against the Zika virus. Previous vaccination against yellow fever did not protect laboratory workers against a Zika infection. More than a dozen companies, in addition to the National Institutes of Health and the Brazilian government, are attempting to produce an effective vaccine. How long it will take to produce such a safe and protective vaccine is unknown; researchers, however, are optimistic in saying that a Zika vaccine could be available by 2018.

Diagnosis

There is no gold-standard diagnostic tool for Zika disease. Antibodies are limited in usefulness, since they are frequently cross-reactive between flaviviruses. Viral culture is not routinely performed, and so far there is no antigenic detection test available. At present, diagnosis is mainly by molecular means—the detection of viral RNA using RT-PCR and a plaque reduction neutralization test.

Lyme Disease

Vilia G. was gardening outside her New England home. She was a biologist, so she simply plucked the bloodsucking tick from her skin and thought nothing more of the occurrence. But the next year brought agony: fatigue, fevers that would come and go, aching joints, and trouble lifting her arms or walking. "It was," she said, "a living hell. … Every day you wake up with less of your body working … you are desperately sick and then have to fight for care."

Vilia was suffering with Lyme arthritis or Lyme disease (LD), the most common vector-borne illness in the United States, and one that continues to grow as a public health problem. Since the CDC began tracking occurrences in 1981, the number of reported LD cases has soared to nearly 35,000 per year. Although this number seems large, it is estimated that the number of actual cases may be more than 10 times those that have been reported. The ticks that spread LD now live in almost half of U.S. counties and are spreading northward into Canada. LD affects all age groups, though children between the ages of 5 and 9 years have the highest incidence. Although LD was first recognized by concerned mothers in 1976 in Old Lyme, Connecticut (whence the disease gets its name), it is not restricted to the United States or Canada. It has also been reported from forested areas in Europe and Asia.

The "germ" of LD

When LD was first found in children, it was believed to be caused by juvenile rheumatoid arthritis; subsequent investigations, however, showed that the arthritis is a late manifestation of a tick-transmitted, motile, corkscrew-shaped bacterium, *Borrelia burgdorferi*. (Figure 16.8). The spirochete *B. burgdorferi* has a small, linear chromosome and 9 circular and 12 linear plasmids that encode a number of lipoproteins, including six outer surface proteins (Osp's, designated by the letters A through F) and another surface-exposed one, VlsE, that undergoes antigenic variation to allow resistance to the acquired immune response. Because the genome of *B. burgdorferi* encodes few proteins involved in biosynthetic activity, the bacterium depends on the host for most of its nutritional requirements. Unlike diphtheria (see p. 225), *B. burgdorferi* requires no iron in its environment for its growth and it produces no toxins; instead,

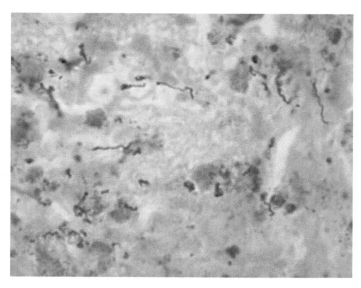

Figure 16.8 Silver-stained *Borrelia burgdorferi* spirochetes . Courtesy CDC Public Health Image Library #835 , Edwin P. Ewing Jr.

Figure 16.9 Adult female western black-footed tick *Ixodes pacificus* the vector of Lyme disease. Courtesy CDC Public Health Image Library #8686/James Gathany et al. , 2006.

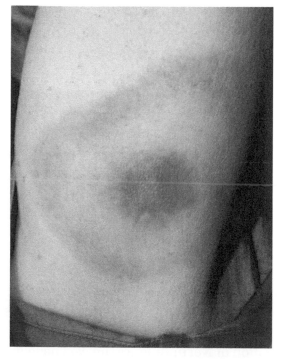

Figure 16.10 "Bull's eye" rash at the site of the tick bite in a Maryland woman who subsequently developed Lyme disease. Photo courtesy CDC Public Health Image Library #9875 James Gathany, 2007.

LD is the result of spirochete migration through tissues, adhesion to host cells, and evasion of immune clearance.

Transmission and pathogenesis

B. burgdorferi is transmitted primarily by the bite of two bloodsucking ticks: *Ixodes scapularis*, the black-footed tick, and *Ixodes pacificus*, the western black-footed tick (Figure 16.9). At the site of the bite there are infiltrates of lymphocytes and macrophages as part of the innate immune response. Macrophages engulf and kill the spirochetes, and inflammatory cells within the lesion produce inflammatory cytokines, including tumor necrosis factor and gamma interferon. Antibodies are also produced against OspC. In 70 to 80% of cases there is a slowly expanding skin lesion—the telltale bull's-eye skin rash (Figure 16.10)—and the sufferer has flu-like symptoms with malaise, fatigue, fever, muscle aches, and stiff joints. Days to weeks afterwards the spirochete disseminates throughout the body, including the blood, cerebrospinal fluid, brain, muscle, bone, and retina. Dissemination is achieved by the bacteria binding to fibronectin; fibrinogen on blood cells; heparan and dermatan sulfate on neurons; and collagen on the cells of the heart, nervous system, and joints. Although there is an active immune response, the spirochetes survive by changing their surface antigens and by inhibiting host immune responses. Although dissemination ceases within weeks to months because of antibody and innate immune mechanisms, the spirochetes may still survive in localized niches for several years. Months after the onset of illness, ~60% of untreated patients may experience intermittent attacks of arthritis, especially in the knees. Untreated LD can even spread to the heart, causing Lyme carditis, a rare but fatal complication. About 10% of patients have joint inflammation for months or years, even after treatment with antibiotics, and this may be due to autoimmunity, presumably because OspA is similar antigenically to human lymphocyte function-associated antigen 1, which is present on T and B lymphocytes as well as macrophages.

Reemergence of LD

Although the earliest cases of LD occurred on Cape Cod in the 1960s, *B. burgdorferi* DNA has been found in museum specimens of mice and black-footed ticks from the late 19th and early 20th centuries, suggesting that the infection has been present in North America for millennia. So why the reemergence? During the European colonization of North America the woodland in New England was cleared for farming, and deer were hunted almost to extinction. During the 20th century, however, farmland reverted to woodland, deer proliferated, and white mice were plentiful, and as a result, the deer tick population increased. Finally these areas became populated with both humans and deer as more rural woodlands became suburban woodlands, and as the deer, protected from hunting and without predators, multiplied. It is this increase in the interactions of ticks, deer, mice, and humans, as well as climate changes—warmer temperatures, increases in rainfall, and milder winters—that has led to the reemergence of Lyme disease.

In nature *B. burgdorferi* organisms live in an enzootic (endemic in animals) cycle in ticks and warm-blooded mammals, including deer, white-footed mice, and chipmunks. The blood-feeding ticks have a 2-year life cycle. In the first year the adult tick lays eggs in the spring and the larvae hatch in the summer, typically in August. The larvae feed on small mammals, particularly white-footed mice in the summer, when they acquire the LD infection. Once the tick is infected, it remains so for life. After feeding, the larvae become infective until the following spring, when they change into blood-feeding nymphs, which are smaller than the head of a pin. The nymphs feed on small rodents and can transmit LD to humans when they insert their mouthparts into a human's skin and release the bacteria while feeding. Most of this nymphal feeding occurs in late spring and summer, and as a result, most human infections are acquired between May and August. The nymphs molt into adult ticks (about the size of a sesame seed) in the fall and early spring, then feed on blood for 7 to 10 days and mate on larger mammals such as deer. On occasion the adult female ticks will also feed on humans, so that they too can cause an infection. The adult female lays eggs on the ground in the late spring, and in this way the 2-year tick cycle is completed.

Diagnosis, prevention, and treatment

The variable signs and symptoms of LD are nonspecific, making diagnosis difficult. The physician first looks for the characteristic rash and determines the patient's medical history and where the patient has been. Following this preliminary examination, laboratory tests including ELISA and Western blotting are conducted for immunological responses to *B. burgdorferi* antigens.

The risk of catching LD from ticks can be minimized by spraying skin and clothing with insect repellent, such as DEET or lemon permethrin. Long sleeves and pant legs tucked into socks may keep the ticks from contacting the skin, and wearing light-colored clothing may make it easier to spot the ticks. After being outdoors in a tick-infested area, one should place all clothes in a dryer for 20 min to kill ticks. Attempts to control infection by eradication of deer or widespread use of acaricides have had limited acceptance.

LD can be treated with the antibiotic doxycycline, either orally or intravenously, and if there is persistent joint inflammation after 2 months of oral antibiotic or 1 month of intravenous treatment, then nonsteroidal anti-inflammatory agents, antirheumatic drugs, or arthroscopic synovectomy may be employed. Although a recombinant OspA vaccine was developed in the 1990s and licensed commercially, it was withdrawn by the manufacturer in 2002 because of limited patient and physician acceptance. There was theoretical, though never-proven, concern that in rare cases vaccination might trigger autoimmune arthritis.

The Conquest of Plagues

The conquest of infectious disease involves both medicine and social and moral factors, and the reverse is also true; that is, severe outbreaks of disease—plagues—can influence the social, political, and economic structure. Here in a modified form (from Daniel Fox) are 10 generalizations on epidemic disease. (i) Never underestimate the severity of the problem when an epidemic occurs. (ii) Expect fear, anxiety, scapegoating, and attempts to segregate and quarantine. (iii) Enlist widespread support for a public health policy or program—include the medical and the social, the business community and the government. (iv) Even with an enlightened and powerful science, don't expect a quick fix—it takes time, money, and lots of scientific effort to develop "magic bullets" to control the spread of disease. (v) Education can contribute to the solution, but it must be used in conjunction with other measures. (vi) Voluntary public health programs work better than compulsory ones. (vii) Recognize the importance of the cultural climate. (viii) All infectious diseases will not be completely eradicated—we will have many more epidemics, but we should be optimistic that it may not require decades for the effects of an infectious epidemic disease to be blunted. (ix) Learn the unique natural history of the disease—the nature of the microbial agent and how it passes through a society. (x) Know that epidemic disease has influenced our history and will continue to do so in our future, but always keep in mind two questions. How will the coming plagues be able to impact our tomorrows? And how can we be best prepared for the emerging and reemerging plagues that lurk on the horizon?

Appendix

Cells and Viruses

In the living world, the fundamental unit is the cell, and all the processes of life occur within cells. A cell may be defined as the standard of biological activity, bounded by a membrane and able to reproduce itself independently of any other living system. All living organisms, large and small, plant and animal, human or microbe, fungus or fly, trypanosome or tapeworm, are made up of cells. All cells are basically similar to one another having many structural features in common. How can we recognize a cell and what do we mean when we say it is the standard unit upon which all living systems are based? We can more easily understand the concept of a cell by making an analogy and comparing living organism to buildings. The rooms of buildings are analogous to the cells of the organism.

Rooms and cells have boundaries with exits and entrances—rooms have walls, floors, ceilings, doors and windows, whereas cells have walls and membranes with pores of various sizes. Rooms and cells come in a variety of shapes and sizes, and each kind of room may have its own particular kind of use, function or specialty. Buildings may be composed of one room or many rooms. In the same way organisms may be composed of one cell, in which case we describe them as unicellular, or of many cells, in which case they are referred to as multicellular. By combining rooms of various kinds we can construct a variety of buildings—homes, apartments, offices, schools. And so on. Similarly different organisms are constructed of different cell types. Just as there are no buildings without rooms there is no life without cells.

Cells, are microscopic in size, and are separated from their environment by an interface or plasma membrane (Figure A-1a). In some cells the outer surface of the plasma membrane is wrapped around with a cell wall. Everything within the plasma membrane is referred to as the cytoplasm, a jellylike material containing various structures collectively known as organelles, or "little organs" Continuing our analogy of cells with rooms organelles might be thought of as items of furniture or equipment (such as stoves, sinks, refrigerators) that have a specialized function. Organelles are: nucleus, mitochondrion, ribosome, endoplasmic reticulum, Golgi apparatus (and in green plants the chloroplast).

On the basis of size, structure and nutritional habits we can generally divide organisms into plants (those that contain a cellulose wall, pigment, make their own food and do not move about), fungi (those that have cell walls, are colorless, absorb

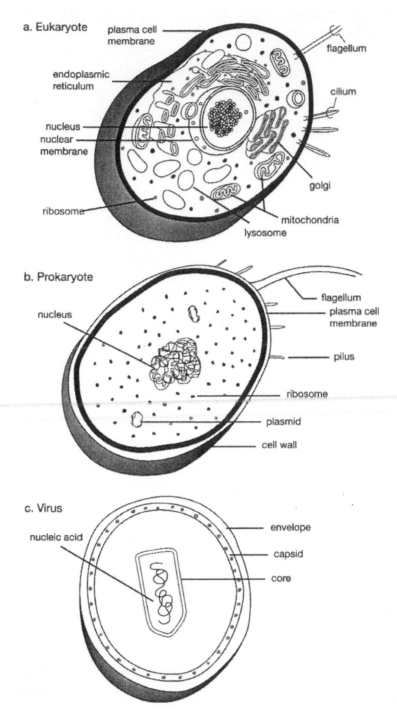

Figure A1. The appearance of cells and a virus. Although all are shown to be the same size, in actual fact they vary widely in size. The eukaryote cell (a) has been enlarged 1000 times (actual size 50um), the prokaryote cell (b) has been enlarged 25,000 times (actual size 2um) and the virus (c) has been enlarged 500,000 times (actual size, 0.1um).

their food and do not move about) and animals (colorless, motile organisms that cannot make their own food).

Those cells with their hereditary material (nucleic acids) enclosed in a membrane (nuclear membrane) are called eukaryotes and those with nucleic acid not enclosed in a membrane are called prokaryotes (Figure A1b). Examples of prokaryotes are bacteria, rickettsiae and blue-green algae. The cytoplasm of prokaryotes contains no mitochondria, endoplasmic reticulum or Golgi apparatus but is richly endowed with ribosomes; some prokaryotes can move about by means of whiplike structures called flagella. In eukaryotes the nucleic acid is combined with protein to form chromosomes, which are enclosed by the nuclear membrane; they may move by means of cilia, flagella or pseudopods, and usually have the full array of organelles.

Viruses, smaller than bacteria, come in a variety of shapes and sizes, but they all have the same basic structure: an outer protein coat and a central core of double-stranded nucleic acid, either RNA or DNA (Figure A1c). Viruses do not have a plasma membrane and their genetic code is incomplete—so although the DNA or RNA contains all the information for asembling a new virus particle, the information needed for reproduction is lacking. To reproduce, a virus must enter a living cell and use the machinery of the cell to produce its offspring. Because sviruses are not completely independent, they are not considered to be cells. Some biologists might even say they are not alive. However, viruses can be killed if their DNA or RNA is destroyed by irradiation or drugs, and they are able to change their genetic code by mutation. Viruses can be crystallized, just as table salt can be crystallized, and still remain infective when the crystals are dissolved and come in contact with a suitable host cell. All viruses are parasitic; however, in some cases there is no evidence of injury to the host. Most of the viruses we are aware of do cause ill effects that we call disease and therefore are regarded as pathogenic. Viruses are responsible for such maladies as AIDS, rabies, smallpox, polio, influenza, the common cold, fever blisters, genital herpes, mumps, measles, yellow fever, viral pneumonia, infectious hepatitis, SARS, Ebola and Junin fever, and Zika disease. Some cancers e.g. Rous sarcoma are caused by viruses. Viruses cannot be grown on non-living substances but must have a living cell (either prokaryotic or eukaryotic) to supply the raw materials and enzymatic machinery necessary for their reproduction. The virus nucleic acid provides the blueprint for making new virus particles, and directs the living host cell to perform the actual synthetic operation of viral replication. Viruses <0.1um in size, are too small to be seen with the light microscope but can be seen with the electron microscope.

Notes

In researching this book I relied on primary and secondary literature sources. These Chapter Notes consist of references as well as comments and are keyed to page numbers.

Chapter 1. The Nature of Plagues

Specific

1.   M. Chien et al., "The genomic sequence of the accidental pathogen *Legionella pneumophila*," *Science* 305 (2004): 1966–1968. Describes the genomic sequence of *Legionella* and has many references to the outbreak. In addition, there is a website on *Legionella*: http://www.cdc.gov/legionella/about/index.html.

2.   R. Hajjeh et al., "Toxic shock syndrome in the United States: surveillance update, 1979–1996," *Emerging Infectious Diseases* 5 (1999): 807–810; L. K. Altman, *Who Goes First? The Story of Self-Experimentation in Medicine* (Berkeley: University of California Press, 1998), 195–199; D. Davis et al., "Toxic shock syndrome: case report of a postpartum female and a literature review," *Journal of Emerging Medicine* 16 (1998): 607–614.

3-4.   K. Tsang et al., "A cluster of cases of severe acute respiratory syndrome in Hong Kong," *New England Journal of Medicine* 348 (2003): 1977–1985.

5-9.   L. S. Roberts and J. Janovy, Jr., *Foundations of Parasitology*, 6th ed. (New York: McGraw-Hill, 2000); R. Buckman, *Human Wildlife: The Life That Lives on Us* (Baltimore: Johns Hopkins University Press, 2003).

10-11.   J. W. Leavitt, *Typhoid Mary: Captive to the Public's Health* (Boston: Beacon Press, 1996).

11-13.   For an infection to persist in the population, each individual on average must transmit the infection to at least one other individual. The number of individuals each infected person infects at the beginning of an epidemic is given by R_0; this is the basic reproductive ratio of the disease or, more simply, the multiplier of the disease. The multiplier helps to predict how fast a disease will spread through the population. The simplest way of obtaining a value for R_0 is to multiply the average probability per contact per unit time (β) by the duration of infectiousness (D) by the number of contacts per unit time (c). See also R. E. Lenski and R. M. May, "The evolution of virulence in parasites and pathogens: reconciliation between two competing hypotheses," *Journal of Theoretical Biology* 169 (1994): 253–265; A. P. Dobson and E. R. Carper, "Infectious diseases and human population history," *Bioscience* 46 (1996): 115–126.

12-13.   O. Razum et al., "SARS, lay epidemiology, and fear," *The Lancet* 361 (2003): 1739–1740; C. Dye and N. Gay, "Modeling the SARS epidemic," *Science* 300 (2003): 1884–1885.

13-14.   M. B. Oldstone, *Viruses, Plagues, & History* (New York: Oxford University Press, 1998), 73–79.

14.   The calculation uses the formula $R_0 = \beta Dc$, where β is the average probability of contact, D is the duration of infectiousness, and c is the number of contacts per unit time. For HIV, $D = 0.5$, $c = 0.2$, and if $R_0 = 1$, $\beta = 10$.

15-16.   A. Schuchat et al., "Measles in the United States since the millennium: perils and progress in the postelimination era," in *Emerging Infections 10*, ed. W. M. Scheld et al. (Washington: ASM Press, 2016), 131–142; P. Willon and M. Mason, "California Gov. Jerry Brown signs new vaccination law, one of nation's toughest," *Los Angeles Times*, 17 October 2016.

16-17. R. M. Anderson and R. M. May, "Vaccination and herd immunity to infectious diseases," *Nature* 318 (1985): 323–329; Lenski and May.

17. W. H. McNeill, *Plagues and Peoples* (New York: Anchor Books, 1976), 9.

17-18. G. Dwyer et al., "A simulation model of the population dynamics and evolution of myxomatosis," *Ecological Monographs* 60 (1990): 423–447; D. Ebert and J. J. Bull, "Challenging the trade-off model for the evolution of virulence: is virulence management feasible? *Trends in Microbiology* 11 (2003):15–20.

18-19. P. W. Ewald, *Evolution of Infectious Disease* (New York: Oxford University Press, 1994); R. M. Nesse and G. C. Williams, *Why We Get Sick: The New Science of Darwinian Medicine* (New York: Random House, 1994).

Chapter 2. Plagues, the Price of Being Sedentary

General

J. C. Davis, *The Human Story: Our History, from the Stone Age to Today* (New York: HarperCollins, 2004).

R. Dunbar, *The Human Story: A New History of Mankind's Evolution* (London: Faber and Faber, 2004).

C. Lockwood and C. Stringer, *The Human Story: Where We Come From and How We Evolved* (London: Natural History Museum, 2014).

C. Stringer and P. Andrews, *The Complete World of Human Evolution*, 2nd ed. (London: Thames and Hudson, 2012).

E. F. Torrey and R. H. Yolken, *Beasts of the Earth: Animals, Humans, and Disease* (New Brunswick, N.J.: Rutgers University Press, 2005).

C. Zimmer, *Smithsonian Intimate Guide to Human Origins* (New York: Harper Perennial, 2007).

Specific

27. I. Tattersall, "Once we were not alone," *Scientific American* 282 (2003): 56–62; M. Leakey and A. Walker, "Early hominid fossils," *Scientific American* 276 (1997): 74–79; *Walking with Cavemen* (London: BBC Video, 2003), television documentary series.

28. D. Johanson and B. Edgar, *From Lucy to Language* (New York: Simon and Schuster, 1996); G. Burenhult, ed., *The Illustrated History of Humankind*, 5 vols. (London: HarperCollins, 1993); R. Leakey and R. Lewin, *Origins Reconsidered: In Search of What Makes Us Human* (New York: Doubleday, 1992).

30. R. Fiennes, *Zoonoses and the Origins and Ecology of Human Disease* (London: Academic Press, 1978); , "The African emergence and early Asian dispersal of the genus *Homo*," *American Scientist* 84 (1996): 538–551; R. A. Weiss, "The Leeuwenhoek Lecture 2001. Animal origins of human infectious disease," *Philosophical Transactions of the Royal Society of London. Series B, Biological Sciences* 356 (2001): 957–977.

31-32. S. Jones et al., ed., *The Cambridge Encyclopedia of Human Evolution* (Cambridge: Cambridge University Press, 1992); R. Lewin, *Human Evolution: An Illustrated Introduction*, 4th ed. (London: Blackwell, 1999).

28-30. E. Trinkaus and P. Shipman, *The Neandertals: Changing the Image of Mankind* (New York: Knopf, 1993); C. Stringer and C. Gamble, *In Search of the Neanderthals* (London: Thames and Hudson, 1993); I. Tattersall, *The Last Neanderthal: The Rise, Success, and Mysterious Extinction of Our Closest Human Relatives* (New York: Macmillan, 1995).

32. A. Coale, "The history of the human population," *Scientific American* 231 (1974): 41–51; E. S. Deevey, Jr., "The human population," *Scientific American* 203 (1960): 195–204.

31.   T. Malthus, *An Essay on the Principle of Population* (London: Dent and Sons, 1927–1928).

33-36. J. Diamond, *Guns, Germs, and Steel* (New York: Norton, 1997), 108, 111, 89, 90, 208.

38.   McNeill, 67.

38-41. This discussion is based on chapters 4–6 in Diamond.

Chapter 3. Six Plagues of Antiquity
Specific

45.   E. V. Hulse, "Joshua's curse and the abandonment of ancient Jericho," *Medical History* 15 (1971): 376–386.

46-47. McNeill, 39.

47-50. W. D. Foster, *A History of Parasitology* (Edinburgh: Livingston, 1965); Roberts and Janovy; J. Farley, *Bilharzia: A History of Imperial Tropical Medicine* (Cambridge: Cambridge University Press, 1992).

50-53. P. J. Hotez and A. Kamath, "Neglected tropical diseases in sub-Saharan Africa: review of their prevalence, distribution, and disease burden," *PLoS Neglected Tropical Diseases* 3 (2009): e412; P. J. Hotez et al., "Neglected tropical diseases of the Middle East and North Africa: review of their prevalence, distribution, and opportunities for control," *PLoS Neglected Tropical Diseases* 6 (2012): e1475.

53-55. F. F. Cartwright and M. Biddiss, *Disease and History*, 2nd ed. (Phoenix Mill, U.K.: Sutton, 2000), 5–7; M. Grmek, *Diseases in the Ancient Greek World* (Baltimore: Johns Hopkins University Press, 1989).

55.   See the website: http://www.archeology.org/online/news/kerameikos.html. It has been estimated that the population in Athens was 250,000 to 300,000, and the total number of deaths was between 65,000 and 78,000. See also P. E. Olsen et al., "The Thucydides syndrome: Ebola déjà vu? (or Ebola reemergent?)," *Emerging Infectious Diseases* 2 (1996): 155–156.

55-56. Thucydides, *Peloponnesian Wars*, Book 2, chapters 47–52.

56-59. A. Celli, *The History of Malaria on the Roman Campagna from Ancient Times* (London: Bale, 1933); R. Sallares, *Malaria and Rome: A History of Malaria in Ancient Italy* (Oxford: Oxford University Press, 2002). Before the Roman conquest of the Italian peninsula, the local population had a life expectancy of 28 to 42 years and 5 to 15% of the children died during the first 10 years of life. After the conquest, life expectancy declined to ~27 years and infant mortality increased to ~25%.

59-61. Cartwright and Biddiss, 19–21.

61-62. McNeill, 104–107, 108, 109–114; V. Nutton, "Portraits of science. Logic, learning, and experimental medicine," *Science* 295 (2000): 800–801.

64.   Most scientists accept that the plague of Justinian was bubonic plague; however, a recent book—S. Scott and C. J. Duncan, *Biology of Plagues: Evidence from Historical Populations* (Cambridge: Cambridge University Press, 2001)—using epidemiologic analyses contends that it was some other infectious disease, such as Ebola. See also J. W. Wood et al., "The temporal dynamics of the fourteenth-century Black Death: new evidence from English ecclesiastical records," *Human Biology* 75 (2003): 427–448.

        Bratton calculates that given an estimated population of ~290,000, ~115,000 would have contracted plague, and with a fatality rate of 20%, ~58,000 would have died. T. L. Bratton, "The identity of the plague of Justinian. (Part II)," *Transactions and Studies of the College of Physicians of Philadelphia* 3 (1981): 174–180.

65.   McNeill, 137.

Chapter 4. An Ancient Plague, the Black Death

General

R. S. Bray, *Armies of Pestilence: The Effects of Pandemics on History* (Cambridge: Butterworth, 1996), chapters 6–9.

C. B. Cunha and B. A. Cunha, "Impact of plague on human history," *Infectious Disease Clinics of North America* 20 (2006): 253–272, viii.

D. Defoe, *A Journal of the Plague Year* (New York: Oxford University Press, 1969).

R. S. Gottfried, *The Black Death: Natural and Human Disaster in Medieval Europe* (New York: Free Press, 1983).

C. T. Gregg, *Plague: An Ancient Disease in the Twentieth Century* (Albuquerque: University of New Mexico Press, 1985).

D. Herlihy, *The Black Death and the Transformation of the West* (Cambridge, Mass.: Harvard University Press, 1997).

J. Kelly, *The Great Mortality: An Intimate History of the Black Death, the Most Devastating Plague of All Time* (New York: HarperCollins, 2005).

C. McEvedy, "bp,"*entificerican*1988–

R. Porter, ed., *The Cambridge Illustrated History of Medicine* (Cambridge: Cambridge University Press, 1996).

M. B. Prentice and L. Rahalison, "Plague," *The Lancet* 369 (2007): 1196–1207.

P. Slack, *The Impact of Plague in Tudor and Stuart England* (London: Routledge Kegan & Paul, 1985).

B. W. Tuchman, *A Distant Mirror: The Calamitous 14th Century* (New York: Ballantine, 1978).

G. Twigg, *The Black Death: A Biological Reappraisal* (New York: Schocken, 1985).

P. Ziegler, *The Black Death* (New York: Harper, 1969).

Specific

67.   Herlihy, 4.

67-68.   G. Boccaccio, *The Decameron*, quoted in Ziegler, 46.

70-74.   Herlihy, 59–82.

74-76.   Ziegler, 84–109.

74-76.   Herlihy, 80.

76-77.   Ziegler, chapter 5.

77-79.   Herlihy, 68–72.

78-79.   Herlihy, 46–52.

79-80.   E. Bendiner, "Alexandre Yersin: pursuer of plague," *Hospital Practice (Office ed.)* 24 (1989): 121–128, 131–132, 135–138 passim.

81-82.   L. Gross, "How the plague bacillus and its transmission through fleas were discovered: reminiscences from my years at the Pasteur Institute in Paris," *Proceedings of the National Academy of Sciences of the United States of America* 92 (1995): 7609–7611.

82-83.   A. K. Hufthammer and L. Walløe, "Rats cannot have been intermediate hosts for *Yersinia pestis* during medieval plague epidemics in Northern Europe," *Journal of Archaeological Science* 40 (2013):1752–1759; B. V. Schmid et al., "Climate-driven introduction of the Black Death and successive plague reintroductions in Europe," *Proceedings of the National Academy of Sciences of the United States of America* 112 (2015): 3020–3025.

84-85.   E. Carniel, "Plague," in *Encyclopedia of Microbiology*, 2nd ed., vol. 3, ed. J. Lederberg (San Diego: Academic Press, 2000), 654–661; Ziegler, 18.

85-86.   S. P. "Yersinia virulence factors—a sophisticated arsenal for combating host defences," *F1000 Research* 5 (2016): 1370.

85.   R. D. Pechous et al., "Pneumonic plague: the darker side of *Yersinia pestis*," *Trends in Microbiology* 24 (2016): 190–197.

85-86.   M. Achtman et al., "*Yersinia pestis*, the cause of plague, is a recently emerged clone of *Yersinia pseudotuberculosis*," *Proceedings of the National Academy of Sciences of the United States of America* 96 (1999): 14043–14048.

86.   "Plague confirmed as cause of park biologist's death," *San Diego Union-Tribune*, 17 November 2007; J. Booth, "Woman in Los Angeles catches Black Death," *Los Angeles Times* Online, 9 January 2008.

87.   World Health Organization, "Plague around the world, 2010–2015," *Weekly Epidemiological Record* 91 (2016): 89–104.

87.   Food and Drug Administration, "FDA approves additional antibacterial treatment for plague," Food and Drug Administration, 8 May 2015, www.fda.gov/NewsEvents/Newsroom/PressAnnouncements/ucm446283.htm.

88.   http://www.jewishgen.org/Ukraine/Photo_Album/Stamps/haffkine.htm.

88.   V. A. Feodorova et al., "Assessment of live plague vaccine candidates," *Methods in Molecular Biology* 1403 (2016): 487–498; V. A. Feodorova and V. L. Motin, "Plague vaccines: current developments and future perspectives," *Emerging Microbes & Infections* 1 (2012): e36; S. K. Verma et al., "A recombinant trivalent fusion protein Fi-LcrV-HSP70(II) augments humoral and cellular immune responses and imparts full protection against *Yersinia pestis*," *Frontiers in Microbiology* 7 (2016): 1053.

89.   N. C. Stenseth et al., "Plague: past, present, and future," *PLoS Medicine* 5 (2008): e3.

89.   B. L. Ligon, "Plague: a review of its history and potential as a biological weapon," *Seminars in Pediatric Infectious Diseases* 17 (2006): 161–170; T. V. Inglesby et al., "Plague as a biological weapon: medical and public health management," *JAMA* 283 (2000): 2281–2290; S. Riedel, "Plague: from natural disease to bioterrorism," *Proceedings (Baylor University Medical Center)* 18 (2005): 116–124.

Chapter 5. A 21st-Century Plague, AIDS

General

H. Y. Fan et al., *AIDS: Science and Society*, 7th ed. (Sudbury, Mass.: Jones and Bartlett,)

G. Stine, *AIDS Update 2014* (San Francisco: Pearson Benjamin Cummings, 2014).

B. S. Weeks and I. E. Alcamo, *AIDS: The Biological Basis*, 5th ed. (Sudbury, Mass.:Jones and Bartlett, 2009).

Specific

91.   J. McGeary, "Death stalks a continent," *Time*, 12 February 2001, 37, http://content.time.com/time/world/article/0,8599,2056158,00.html.

91.   http://www.unaids.org/sites/default/files/media_asset/MDG6_ExecutiveSummary_en.pdf; http://www.unaids.org/en/resources/documents/2016/Global-AIDS-update-2016; http://www.cdc.gov/hiv/statistics/overview/ataglance.html.

92-93.   R. C. Gallo, "The AIDS virus," *Scientific American* 256 (1987): 46–56.

93.   G. G. Simpson et al., *Life: An Introduction to Biology* (San Francisco: Harcourt Brace, 1957), 18.

93-94. J. S. Henderson et al., *A Notable Career in Finding Out: Peyton Rous, 1879–1970* (New York: Rockefeller University Press, 1971); H. A. Lechevalier and M. Solotorovsky, *Three Centuries of Microbiology* (New York: Dover, 1974), 282.

94-95. G. M. Cooper et al., ed., *The DNA Provirus: Howard Temin's Scientific Legacy* (Washington: ASM Press, 1995).

95-97. P. Mohammadi et al., "24 hours in the life of HIV-1 in a T cell line," *PLoS Pathogens* 9 (2013): e1000361.

102-103. I. Molotsky, "U.S. approves drug to prolong lives of AIDS patients," *New York Times*, 21 March 1987.

103-105. M. E. Avery, *Gertrude B. Elion: 1918–1999*, vol. 78 of *Biographical Memoirs* (Washington: National Academy Press, 2000); G. B. Elion, "The purine path to chemotherapy," *Science* 244 (1989): 41–47.

105-106. E. J. Arts and D. J. Hazuda, "HIV-1 antiretroviral drug therapy," *Cold Spring Harbor Perspectives in Medicine* 2 (2012): a007161; M. E. Badowski et al., "New antiretroviral treatment for HIV," *Infectious Diseases and Therapy* 5 (2016): 329–352; https://aidsinfo.nih.gov/education-materials/fact-sheets/21/58/fda-approved-hiv-medicines; http://www.healthline.com/health/hiv-aids/medications-list.

106.   D. G. Carnathan et al., "Activated CD4+CCR5+ T cells in the rectum predict increased SIV acquisition in SIVGag/Tat-vaccinated rhesus macaques," *Proceedings of the National Academy of Sciences of the United States of America* 112 (2015): 518–523.

107-108. P. Aggleton et al., "Risking everything? Risk behavior, behavior change, and AIDS," *Science* 265 (1994): 341–345; S. M. Blower and A. R. McLean, "Prophylactic vaccines, risk behavior change, and the probability of eradicating HIV in San Francisco," *Science* 265 (1994): 1451 1454.

109-112. J. Goudsmit, *Viral Sex: The Nature of AIDS* (New York: Oxford University Press, 1997); J. Moore, "The puzzling origins of AIDS," *American Scientist* 92 (2004): 540–549; P. M. Sharp and B. H. Hahn, "The evolution of HIV-1 and the origin of AIDS," *Philosophical Transactions of the Royal Society of London. Series B, Biological Sciences* 365 (2010): 2487–2494; J. Pepin, *The Origins of AIDS* (Cambridge: Cambridge University Press, 2011).

Chapter 6. Typhus, A Fever Plague

Specific

113.   H. Zinsser, *Rats, Lice and History* (Boston: Little, Brown, 1935).

114.   G. Weissmann, "Rats, lice, and Zinsser," *Emerging Infectious Diseases* 11 (2005): 492–496.

114-117. Cartwright and Biddiss, chapter 4.

117-118. R. K. Peterson, "Insects, disease and military history," *American Entomologist*, , 147–160.

118-119. L. Gross, "How Charles Nicolle of the Pasteur Institute discovered that epidemic typhus is transmitted by lice: reminiscences from my years at the Pasteur Institute in Paris," *Proceedings of the National Academy of Sciences of the United States of America* 93 (1996): 10539–10540.

121-122. Lechevalier and Solotorovsky, 325–332.

123-124. J. W. Maunder, "The appreciation of lice," *Proceedings of the Royal Institution of Great Britain* 55 (1983): 1–31.

124-125. D. Raoult and V. Roux, "The body louse as a vector of reemerging human diseases," *Clinical Infectious Diseases* 29 (1999): 888–911.

125-129. H. Markell, *Quarantine! East European Jewish Immigrants and the New York City Epidemics of 1892* (Baltimore: Johns Hopkins University Press, 1997).

130. W. Szybalski, "Maintenance of human-fed live lice in the laboratory and production of Weigl's exanthematic typhus vaccine," in *Maintenance of Human, Animal, and Plant Pathogen Vectors*, ed. K. Maramorosch and F. Mahmood (Enfield, N.H.: Science Publishers, 1999), 161–180.

130-131. A. Allen, *The Fantastic Laboratory of Dr. Weigl: How Two Brave Scientists Battled Typhus and Sabotaged the Nazis* (New York: Norton, 2014).

131. Zinsser.

Chapter 7. Malaria, Another Fever Plague

Specific

133. R. Kapuscinski, *The Shadow of the Sun* (New York: Vintage Books, 2002).

134-135,
143-145. I. W. Sherman, "A brief history of malaria and discovery of the parasite's life cycle," in *Malaria: Parasite Biology, Pathogenesis, and Protection*, ed. I. W. (Washington: ASM Press, 1998), 3–10.

133-143. R. Ross, *Memoirs* (London: John Murray, 1923); E. R. Nye and M. E. Gibson, *Ronald Ross: Malariologist & Polymath, a Biography* (London: Macmillan, 1997); W. F. Bynum and C. Overy, *The Beast in the Mosquito: The Correspondence of Ronald Ross and Patrick Manson* (Amsterdam: Editions Rodopi, 1998); P. F. Russell, *Man's Mastery of Malaria* (London: Oxford University Press, 1955); G. A. Harrison, *Mosquitoes, Malaria and Man: A History of the Hostilities since 1880* (New York: Dutton, 1978).

145-152. I. W. Sherman, *Magic Bullets To Conquer Malaria: From Quinine to Qinghaosu* (Washington: ASM Press, 2010).

150. The 2015 Nobel Prize in Physiology or Medicine was awarded to Youyou Tu "for her discoveries concerning a novel therapy for malaria." Youyou Tu discovered artemisinin.

153-156. A. V. Hill and D. J. Weatherall, "Host genetic factors in resistance to malaria," in *Malaria: Parasite Biology, Pathogenesis, and Protection*, ed. I. W. Sherman (Washington: ASM Press, 1998), 445–456.

157. W. Trager and J. Jensen, "Human malaria parasites in continuous culture," *Science* 193 (1976): 673–675.

158. J. W. Barnwell and M. R. Galinski, "Invasion of vertebrate cells: erythrocytes," in *Malaria: Parasite Biology, Pathogenesis, and Protection*, ed. I. W. Sherman (Washington: ASM Press, 1998), 93–120.

158-160. I. W. Sherman, *Malaria Vaccines: The Continuing Quest* (London: World Scientific, 2016); R. A. Seder et al., "Protection against malaria by intravenous immunization with a nonreplicating sporozote vaccine," *Science* 341 (2013): 1359–1365; A. S. Ishizuka et al., "Protection against malaria at 1 year and immune correlates following PfSPZ vacination," *Nature Medicine* 22 (2016): 614–623; T. L. Richie et al., "Progress with *Plasmodium falciparum* sporozoite (PfSPZ)-based malaria vaccines," *Vaccine* 33 (2015): 7452–7461; S. L. Hoffman et al., "The march toward malaria vaccines," *American Journal of Preventive Medicine* 49 (2015): S319–S333.

Chapter 8. "King Cholera"

General

M. Echenberg, *Africa in the Time of Cholera: A History of Pandemics from 1817 to the Present* (New York: Cambridge University Press, 2011).

G. N. Grob, *The Deadly Truth: A History of Disease in America* (Cambridge, Mass.: Harvard University Press, 2002).

S. Hempel, *The Strange Case of the Broad Street Pump: John Snow and the Mystery of Cholera* (Berkeley: University of California Press, 2007).

S. Johnson, *The Ghost Map: The Story of London's Most Terrifying Epidemic—and How It Changed Science, Cities, and the Modern World* (New York: Riverhead, 2005).

C. E. Rosenberg, *The Cholera Years: The United States in 1832, 1849, and 1866* (Chicago: University of Chicago Press, 1962).

J. Snow, *On The Mode of Communication of Cholera* (London: John Churchill, 1855), http://www.ph.ucla.edu/epi/snow/snowbook.html.

Specific

163. J. Franklin and J. Sutherland, *Guinea Pig Doctors: The Drama of Medical Research through Self-Experimentation* (New York: Morrow, 1984), 181–182.

164. P. Vinten-Johansen et al., *Cholera, Chloroform, and the Science of Medicine: A Life of John Snow* (New York: Oxford University Press, 2003); Snow.

164. "Louis Pasteur (1822-1895)," *Microbes and Infection* 5 (2003): 553–560; P. Debré, *Louis Pasteur* (Baltimore: Johns Hopkins University Press, 1998); G. L. Geison, *The Private Science of Louis Pasteur* (Princeton, N.J.: Princeton University Press, 1995).

165. R. Koch, *Essays of Robert Koch*, trans. K. C. Carter (Westport, Conn.: Greenwood Press, 1987); T. D. Brock, *Robert Koch: A Life in Medicine and Bacteriology* (Washington: ASM Press, 1999); R. Münch, "obert,"*Microbes and Infection* 5 (2003): 69–74.

166-169. I. W. Sherman and V. G. Sherman, *Biology: A Human Approach*, 4th ed. (New York: Oxford University Press, 1983); P. de Kruif, *Microbe Hunters* (San Diego: Harcourt Brace Jovanovich, 1954), chapters 1–3.

169. G. G. Simpson.

170. McNeill, 231.

171. D. Barua, "History of cholera," in *Cholera*, ed. D. Barua and W. B. Greenough III (New York: Plenum, 1992), 1–36; G. C. Cook, "The Asiatic cholera: an historical determinant of human genomic and social structure," in *Cholera and the Ecology of* Vibrio cholerae, ed. B. S. Drasar and B. D. Forrest (London: Chapman & Hall, 1996), 18–53.

174. A. Zavis, "UN admits a role in cholera epidemic," *Los Angeles Times*, 19 August 2016.

176. R. R. Colwell, "Global climate and infectious disease: the cholera paradigm," *Science* 274 (1996): 2025–2031; C. Dold, "The cholera lesson," *Discover*, February 1999, 71–75.

177-181. Cartwright and Biddiss, 157–166; G. Gill et al., "Fear and frustration—the Liverpool cholera riots of 1832," *The Lancet* 358 (2001): 233–237; F. F. Cartwright, *A Social History of Medicine* (London: Longman, 1977); S. E. Finer, *The Life and Times of Edwin Chadwick* (London: Methuen, 1952); C. Hamlin, *Public Health and Social Justice in the Age of Chadwick: Britain, 1850-1854* (Cambridge: Cambridge University Press, 1997).

181-186. Snow.

188-190.   E. Huxley, *Florence Nightingale* (London: Weidenfeld and Nicholson, 1975). More-critical views of Florence Nightingale can be found in L. Strachey, *Eminent Victorians* (London: Continuum, 2002); H. Small, *Florence Nightingale: Avenging Angel* (London: Constable, 1998); and F. B. Smith, *Florence Nightingale: Reputation and Power* (New York: St. Martin's Press, 1982).

190-193.   http://www.entrenet.com/~groedmed/grosseisle/gi.html; http://ballinagree.freeservers.com/sumsorrow.html.

194.   K. Sévère et al., "Effectiveness of oral cholera vaccine in Haiti: 37-month follow-up," *American Journal of Tropical Medicine and Hygiene* 94 (2016): 1136–1142; F. J. Luquero and D. A. Sack, "Effectiveness of oral cholera vaccine in Haiti," *The Lancet. Global Health* 3 (2015): e120-e121; A. S. Azman et al., "The impact of a one-dose versus two-dose oral cholera vaccine regimen in outbreak settings: a modeling study," *PLoS Medicine* 12 (2015): e1001867.

Chapter 9. Smallpox, the Spotted Plague

General

I. Glynn and J. Glynn, *The Life and Death of Smallpox* (New York: Cambridge University Press, 2004).

G. Williams, *Angel of Death: The Story of Smallpox* (Basingstoke, U.K.: Palgrave MacMillan, 2010).

Specific

197.   J. L. Carrell, *The Speckled Monster: A Historical Tale of Battling Smallpox* (New York: Dutton, 2003); Cartwright and Biddiss, chapter 4; Diamond, chapter 3.

198.   J. N. Shurkin, *The Invisible Fire: The Story of Mankind's Victory over the Ancient Scourge of Smallpox* (New York: Putnam, 1979); H. J. Parish, *Victory with Vaccines: The Story of Immunization* (Edinburgh: Livingston, 1968).

199-201.   Oldstone, chapter 4.

199-202.   D. R. Hopkins, *Princes and Peasants: Smallpox in History* (Chicago: University of Chicago Press, 1983); C. W. Dixon, *Smallpox* (London: Churchill, 1962); Glynn and Glynn; F. Fenner et al., *Smallpox and Its Eradication* (Geneva: World Health Organization, 1988); J. M. Eyler, "Smallpox in history," *Journal of Laboratory and Clinical Medicine* 142 (2003): 216–220; D. Baxby, *Jenner's Smallpox Vaccine: The Riddle of Vaccinia Virus and Its Origin* (London: Heinemann, 1981); H. Bazin, *The Eradication of Smallpox: Edward Jenner and the First and Only Eradication of a Human Infectious Disease* (San Diego: Academic Press, 2000).

202-203.   R. Preston, *The Demon in the Freezer: A True Story* (New York: Random House, 2002).

205-207.   S. S. Bernstein, "Smallpox and variolation; their historical significance in the American colonies," *Journal of the Mount Sinai Hospital, New York* 18 (1951): 228–244.

209.   P. J. Pead, "Benjamin Jesty: new light in the dawn of vaccination," *The Lancet* 362 (2003): 2104–2109.

212-213.   P. Berche, "The threat of smallpox and bioterrorism," *Trends in Microbiology* 9 (2001): 15–18; C. S. McClain, "A new look at an old disease: smallpox and biotechnology," *Perspectives in Biology and Medicine* 38 (1994): 624–639; T. O'Toole, "Smallpox: an attack scenario," *Emerging Infectious Diseases* 5 (1999): 540–546.

213-215.   M. R. Albert et al., "The last smallpox epidemic in Boston and the vaccination controversy, 1901-1903," *New England Journal of Medicine* 344 (2001): 375–379; J. Farrell, *Invisible Enemies: Stories of Infectious Disease* (New York: Farrar, Straus & Giroux, 1998); J. Duffy, *Epidemics in Colonial America* (Baton Rouge: Louisiana State University Press, 1953); H. N. Simpson, *Invisible Armies: The Impact of Disease on American History* (Indianapolis, Ind.: Bobbs-Merrill, 1980).

Chapter 10. Preventing Plagues: Immunization

General

P. A. Offit, *Vaccinated: One Man's Quest to Defeat the World's Deadliest Diseases* (Washington: Smithsonian Books, 2008).

D. M. Oshinsky, *Polio: An American Story* (New York: Oxford University Press, 2005).

J. S. Smith, *Patenting the Sun: Polio and the Salk Vaccine* (New York: Morrow, 1990).

Specific

217-218. W. R. Clark, *At War Within: The Double-Edged Sword of Immunity* (New York: Oxford University Press, 1995).

218-221,
221-224,
224-226. L. Sompayrac, *How the Immune System Works*, 5th ed. (Chichester, U.K.: John Wiley & Sons, 2015); A. M. Silverstein, *A History of Immunology*, 2nd ed. (San Diego: Academic Press, 2009); D. J. Bibel, *Milestones in Immunology: A Historical Exploration* (Madison, Wis.: Science Tech Publishers, 1988); G. B. Pier et al., *Immunology, Infection, and Immunity* (Washington: ASM Press, 2004); A. K. Abbas et al., *Cellular and Molecular Immunology*, 8th ed. (Philadelphia: Saunders, 2014); K. Murphy and C. Weaver, *Janeway's Immunobiology*, 9th ed. (New York: Garland, 2016).

229-230. P. Ehrlich, in *Great Scientists Speak Again II*, ed. I. W. Sherman (unpublished); de Kruif, 326–350; M. Marquardt, *Paul Ehrlich* (New York: Schuman, 1951).

225-227,
231-232. http://www.cdc.gov/diphtheria/clinicians.html; P. G. Guilfoile, *Diphtheria* (New York: Chelsea House, 2009); Centers for Disease Control and Prevention (CDC), "Fatal respiratory diphtheria in a U.S. traveler to Haiti—Pennsylvania, 2003," *MMWR. Morbidity and Mortality Weekly Report* 52 (2004): 1285–1286, https://www.cdc.gov/mmwr/preview/mmwrhtml/mm5253a3.htm; F. J. Grundbacher, "Behring's discovery of diphtheria and tetanus antitoxins," *Immunology Today* 13 (1992): 188–189; F. Winau and R. Winau, "Emil von Behring and serum therapy," *Microbes and Infection* 4 (2002): 185–188.

232-233. M. Granström, "The history of pertussis vaccination: from whole-cell to subunit vaccines," in *History of Vaccine Development*, ed. S. A. Plotkin (New York: Springer, 2011), –

234-240. J. M. Barry, *The Great Influenza: The Epic Story of the 1918 Pandemic* (New York: Viking, 2004); A. W. Crosby, *America's Forgotten Pandemic: The Influenza of 1918*, 2nd ed. (Cambridge: Cambridge University Press, 2003).

240-243,
243-245. S. L. Katz, "The history of measles virus and the development and utilization of measles virus vaccines," in Plotkin, 199–206; M. R. Hilleman, "The development of a live attenuated mumps virus vaccine in historic perspective and its role in the evolution of combined measles-mumps-rubella," in Plotkin, 207–218.

247-248. Granström.

248-250. A. Gershon, "Vaccination against varicella and zoster: its development and progress," in Plotkin, 247–264; S. L. Katz et al., "The role of tissue culture in vaccine development," in Plotkin, 145–149.

254-261. C. D. Jacobs, *Jonas Salk: A Life* (New York: Oxford University Press, 2015); M. R. Jiménez, *Albert B. Sabin, 1906–1993* (Washington: National Academy of Sciences, 2014), http://www.nasonline.org/publications/biographical-memoirs/memoir-pdfs/sabin-albert.pdf.

Chapter 11. The Plague Protectors: Antiseptics and Antibiotics

Specific

265-266. A. S. Lyons and R. J. Petrucelli, *Medicine: An Illustrated History* (New York: Abradale, 1987).

266-268. http://www.barberpole.com/art of.htm; http://www.pbs.org/kqed/demonbarber/ bloodletting/the humours.html; http://www.cleavlandopera.org/four/educational/ncoc/ barbers.html; http://www.rcseng.ac.uk/wellcome/history_of_the_college/.

269-273. O. W. Holmes, *The Contagiousness of Puerperal Fever* (New York: P. F. Collier & Son, 1909-1914); M. Thompson, *The Cry and the Covenant* (New York: Doubleday, 1949); S. B. Nuland, *The Doctors' Plague: Germs, Childbed Fever, and the Strange Story of Ignac Semmelweis* (New York: Norton, 2003); F. G. Slaughter, *Immortal Magyar: Semmelweis, Conqueror of Childhood Fever* (New York: Schuman, 1950).

274. Porter, chapter 6.

274-280. V. S. Thatcher, *History of Anesthesia* (New York: Garland, 1984); J. M. Fenster, *Ether Day: The Strange Tale of America's Greatest Medical Discovery and the Haunted Men Who Made It* (New York: HarperCollins, 2001); L. Stratmann, *Chloroform: The Quest for Oblivion* (Phoenix Mill, U.K.: Sutton, 2003); R. J. Wolfe, *Tarnished Idol: William T. G. Morton and the Introduction of Surgical Anesthesia* (San Anselmo, Calif.: Norman Publishing, 2001).

280-281. J. J. Beer, *Emergence of the German Dye Industry* (Urbana: University of Illinois Press, 1959); W. M. Gardner, ed., *The British Coal Tar Industry* (Philadelphia: Lippincott, 1915); J. Drews, "Drug discovery: a historical perspective," *Science* 287 (2000): 1960–1964; Marquardt; A. S. Travis, *The Rainbow Makers: The Origins of the Synthetic Dyestuffs Industry in Western Europe* (Bethlehem, Pa.: Lehigh University Press, 1993).

283-285,
286-290. I. W. Sherman, 2010, 82–93.

290-299. C. Walsh, *Antibiotics: Actions, Origins, Resistance* (Washington: ASM Press, 2003); R. Hare, *The Birth of Penicillin and the Disarming of Microbes* (London: Allen & Unwin, 1970); E. Lax, *The Mold in Dr. Florey's Coat: The Story of the Penicillin Miracle* (New York: Henry Holt, 2004); G. Macfarlane, *Alexander Fleming: The Man and the Myth* (Cambridge, Mass.: Harvard University Press, 1984); G. Macfarlane, *Howard Florey: The Making of a Great Scientist* (Oxford: Oxford University Press, 1979).

299-302. S. B. Levy and B. Marshall, "Antibacterial resistance worldwide: causes, challenges and responses," *Nature Medicine* 10 (2004): S122–S129; C. L. Ventola, "The antibiotic crisis. Part 1: causes and threats," *Pharmacy & Therapeutics* 40 (2015): 277–283; C. L. Ventola, "The antibiotic crisis. Part 2: management strategies and new agents," *Pharmacy & Therapeutics* 40 (2015): 344–352; S. B. Levy, "The challenge of antibiotic resistance," *Scientific American* 278 (1998): 46–53; D. K. Byarugaba, "Mechanisms of antimicrobial resistance," in *Antimicrobial Resistance in Developing Countries*, ed. A. D. Sosa et al. (New York: Springer, 2009), 15–26; K. Lewis, "Platforms for antibiotic discovery," *Nature Reviews. Drug Discovery* 12 (2013): 371–387; E. D. Brown and G. D. Wright, "Antibacterial drug discovery in the resistance era," *Nature* 529 (2016): 336–343.

Chapter 12. The Great Pox, Syphilis

Specific

305-306. "The scientific environment of the Tuskegee Study of Syphilis, 1920–1960," *Perspectives in Biology and Medicine* 43 (1999): 1–30; J. H. Jones, *Bad Blood: The Tuskegee Syphilis Experiment* (New York: Free Press, 1981).

306-308.   D. S. Jones, "Syphilis, historical," in *Encyclopedia of Microbiology*, 2nd ed., vol. 4, ed. J. Lederberg (San Diego: Academic Press, 2000), 538–544; , "*Treponema pallidum* (syphilis)," in *Principles and Practice of Infectious Diseases*, 5th ed., ed. J. Mandell et al. (New York: Churchill Livingstone, 2000), 2474–2490; Cartwright and Biddiss, chapter 3; C. Quetel, *The History of Syphilis*, trans. J. Braddock and B. Pike (Baltimore: Johns Hopkins University Press, 1990); M. Tampa et al., "Brief history of syphilis," *Journal of Medicine and Life* 7 (2014): 4–10.

308-309.   K. N. Harper et al, "The origin and antiquity of syphilis revisited: an appraisal of Old World pre-Columbian evidence for treponemal infection," *American Journal of Physical Anthropology* 146 (2011): 99–133; K. N. Harper et al., "Syphilis: then and now," *The Scientist*, 1 February 2014.

310-311.   Franklin and Sutherland, 25–57.

312.   P. D. Rosahn and B. Black-Schaffer, "Studies in syphilis: III. Mortality and morbidity findings in the Yale autopsy series." *Yale Journal of Biology and Medicine* 15 (1943): 587–602.

311-313.   Tramont.

313.   T. Rosebury, *Microbes and Morals* (New York: Viking, 1971).

314-315.   Cartwright and Biddiss, chapter 3.

315.   McNeill, 227.

315.   Tramont.

316.   F. Winau et al., "Paul Ehrlich—in search of the magic bullet," *Microbes and Infection* 6 (2004): 786–789; de Kruif, chapter 12.

316-319.   Rosebury; T. Parran, *Shadow on the Land: Syphilis* (New York: Reynal and Hitchcock, 1937); D. S. Jones.

319.   N. Sharp, "The return of syphilis," *The Atlantic*, 3 December 2015; http://www.cdc.gov/std/syphilis/stats.htm.

319.   E. L. Ho and S. A. Lukehart, "Syphilis: using modern approaches to understand an old disease," *Journal of Clinical Investigation* 121 (2011): 4584–4592; K. V. Lithgow and C. E. Cameron, "Vaccine development for syphilis," *Expert Review of Vaccines* 16 (2017): 37–44; C. E. Cameron and S. A. Lukehart, "Current status of syphilis vaccine development: need, challenges, prospects," *Vaccine* 32 (2014): 1602–1609.

Chapter 13. The People's Plague: Tuberculosis

General

H. Bynum, *Spitting Blood: The History of Tuberculosis* (Oxford: Oxford University Press, 2012).

Cartwright and Biddiss, chapter 7.

T. M. Daniel, *Captain of Death: The Story of Tuberculosis* (Rochester, N.Y.: University of Rochester Press, 1997).

T. Dormandy, *The White Death: A History of Tuberculosis* (New York: New York University Press, 2000).

R. Dubos and J. Dubos, *The White Plague: Tuberculosis, Man, and Society* (Boston: Little, Brown, 1952).

M. Kato-Maeda and P. M. Small, "User's guide to tuberculosis resources on the Internet," *Clinical Infectious Diseases* 32 (2001):1580–1588.

The People's Plague: Tuberculosis in America (Arlington, Va.: PBS, 1995), video.

L. B. Reichman and E. S. Hershfield, ed., *Tuberculosis: A Comprehensive International Approach*, 2nd ed. (New York: Marcel Dekker, 2000).

Specific

323-324. L. Hutcheon and M. Hutcheon, *Opera: Desire, Disease, Death* (Lincoln: University of Nebraska Press, 1996).

324-326. S. Grzybowski and E. A. Allen, ","*The Lancet* 346 (1995): 1472–1474; M. Bloch, *The Royal Touch: Sacred Monarchy and Scrofula in England and France* (New York: Dorset, 1989); T. Daniel et al., "History of tuberculosis," in *Tuberculosis: Pathogenesis, Protection, and Control*, ed. B. R. Bloom (Washington: ASM Press, 1984), 13–24; , "The origins of *Mycobacterium tuberculosis* and the notion of its contagiousness, in *Tuberculosis*, ed. W. N. Rom and S. M. Garay (Boston: Little, Brown, 1996), 3–20; G. M. Taylor et al., "First report of *Mycobacterium bovis* DNA in human remains from the Iron Age," *Microbiology* 153 (2007): 1243–1249.

326-328. S. M. ,"Old Testament biblical references to tuberculosis," *Clinical Infectious Diseases* 29 (1999): 1557–1558; S. T. Cole et al., "Deciphering the biology of *Mycobacterium tuberculosis* from the complete genome sequence," *Nature* 393 (1998): 537–544; ,"The complete genome sequence of *Mycobacterium bovis*," *Proceedings of the National Academy of Sciences of the United States of America* 100 (2003): 7877–7882; A. R. Zink et al., "Molecular identification of human tuberculosis in recent and historic bone tissue samples: the role of molecular techniques for the study of historic tuberculosis," *American Journal of Physical Anthropology* 126 (2005): 32–47.

328-331. Grob; C. Zimmer, "The stories behind the bones," *Science* 286 (1999): 1071, 1073–1074; M. Caldwell, *The Last Crusade: The War on Consumption, 1862–1954* (New York: Atheneum, 1988).

329. D. E. Hammerschmidt, "Bovine tuberculosis," *Journal of Laboratory and Clinical Medicine* 141 (2003): 359–360.

331-332. A. M. Kraut, "Plagues and prejudice: nativism's construction of disease in the nineteenth- and twentieth-century New York City," in *Hives of Sickness: Public Health and Epidemics in New York City*, ed. D. (New Brunswick, N.J.: Rutgers University Press, 1995), 65–90; B. Bates, *Bargaining for Life: A Social History of Tuberculosis, 1876–1938* (Philadelphia: University of Pennsylvania Press, 1992); G. D. Feldberg, *Disease and Class: Tuberculosis and the Shaping of Modern North American Society* (New Brunswick, N.J.: Rutgers University Press, 1995); K. Ott, *Fevered Lives: Tuberculosis in American Culture since 1870* (Cambridge, Mass.: Harvard University Press, 1996).

332-333. Columbia University Press, "Cecil John Rhodes," *The Columbia Encyclopedia*, 6th ed. (New York: Columbia University Press, 2001).

334-336. Brock; R. Koch, "The aetiology of tuberculosis," in *Medicine and Western Civilization*, ed. D. J. Rothman et al. (New Brunswick, N.J.: Rutgers University Press, 1999), 319–329.

336-339. P. M. Small and P. I. Fujiwara, "Management of tuberculosis in the United States," *New England Journal of Medicine* 345 (2001): 189–200.

338-340. F. Ryan, *The Forgotten Plague: How the Battle against Tuberculosis Was Won—and Lost* (Boston: Little, Brown, 1993).

340-342. P. N. Jenkins, "Heroin addiction's fraught history," *The Atlantic*, 24 February 2014; R. Askwith, "How aspirin turned hero," *Sunday Times* (London), 13 September 1998; I. Scott, "Heroin: a hundred-year habit," *History Today* 48 (1998): 6–8.

342-344. Ryan.

344-347. S. A. Waksman, *The Conquest of Tuberculosis* (Berkeley: University of California Press, 1964); P. Pringle, "Notebooks shed light on antibiotic's contested discovery," *New York Times*, 12 June 2012; W. Kingston, "Streptomycin, Schatz v. Waksman, and the balance of credit for discovery," *Journal of the History of Medicine and Allied Sciences* 59 (2004): 441–462; H. B. Woodruff, "Selman A. Waksman, winner of the 1952 Nobel Prize for Physiology or Medicine," *Applied and Environmental Microbiology* 80 (2014): 2–8; R. D. Hotchkiss, *Selman Abraham Waksman: 1888–1973*, vol. 83 of *Biographical Memoirs* (Washington: National Academy Press, 2003).

348-349. emedicine.medscape.com/article/230802-treatment; http://www.cdc.gov/tb/topic/treatment/tbdisease.htm.

349-350. K. T. Chung and C. J. Biggers, "Albert Léon Charles Calmette (1863-1933) and the antituberculous BCG vaccination," *Perspectives in Biology and Medicine* 44 (2001): 379–389; D. S. Barnes, "Historical perspectives on the etiology of tuberculosis," *Microbes and Infection* 2 (2000): 431–440.

350-352. World Health Organization, *Global Tuberculosis Report 2016* (Geneva: World Health Organization, 2016), http://apps.who.int/iris/bitstream/10665/250441/1/9789241565394-eng.pdf?ua=1; L. Sun, "TB cases increase in US for first time in 23 years," *Washington Post*, 24 March 2016.

Chapter 14. Leprosy, The Striking Hand of God

General

M. S. Duthie et al., "Advances and hurdles on the way toward a leprosy vaccine," *Human Vaccines* 7 (2011): 1172–1183.

J. C. Lastória and M. A. Abreu, "Leprosy: a review of laboratory and therapeutic aspects—part 2," *Anais Brasileiros de Dermatologia* 89 (2014): 389–401.

J. H. Richardus and L. Oskam, "Protecting people against leprosy: chemoprophylaxis and immunoprophylaxis," *Clinics in Dermatology* 33 (2015): 19–25.

Specific

355. D. W. Beckett, "The striking hand of God: leprosy in history," *New Zealand Medical Journal* 100 (1988): 494–497. It has been suggested that the biblical references to leprosy are not due to that disease but are more likely to have described smallpox.

356-359. J. Troutman, "The history of leprosy," in *Leprosy*, 2nd ed., ed. R. C. Hastings (Edinburgh: Churchill Livingstone. 1994), 11–25.

359. McNeill, 155–156.

359-361. Z. Gussow, *Leprosy, Racism, and Public Health: Social Policy in Chronic Disease Control* (Boulder, Colo.: Westview, 1989).

362. "How does *Mycobacterium leprae* target the peripheral nervous system?" *Trends in Microbiology* 8 (2004): 23–28.

362-363. http://www.who.int/mediacentre/factsheets/fs101/en/; http://emedicine.medscape.com/article/220455-overview.

364. P. J. Brophy, "Subversion of Schwann cells and the leper's bell," *Science* 296 (2002): 862–863; W. J. Britton and D. N. Lockwood, "Leprosy," *The Lancet* 363 (2004): 1209–1219; , "Genomics: leprosy—a degenerative disease of the genome," *Current Biology* 11 (2001): R381–R383.

Chapter 15. Six Plagues of Africa

General

Duffy.

P. Farmer, *Infections and Inequalities: The Modern Plagues* (Berkeley: University of California Press, 1999).

C. Hibbert, *Africa Explored: Europeans in the Dark Continent, 1769-1889* (London: Allan Lane, 1982).

Hotez and Kamath.

T. Pakenham, *The Scramble for Africa 1876-1912* (London: Weidenfeld and Nicholson, 1991).

Specific

367-368. O. Ransford, *"Bid the Sickness Cease": Disease in the History of Black Africa* (London: John Murray, 1983).

369-370. P. Brent, *Black Nile: Mungo Park and the Search for the Niger* (London: Gordon Cremonesi, 1977).

370-371. Cartwright and Biddiss, 158–159.

371-374. O. Ransford, *David Livingstone: The Dark Interior* (New York: St. Martin's Press, 1978); M. Gelfand, *Livingstone the Doctor: His Life and Travels* (Oxford: Blackwell, 1957).

374-375 *Through the Dark Continent* (New York: Greenwood Press, 1969); R. Hall, *Stanley: An Adventurer Explored* (London: Collins, 1974); S. Newson-Smith, ed., *Quest: The Story of Stanley and Livingstone Told in Their Own Words* (London: Arlington, 1978); J. Bierman, *Dark Safari: The Life behind the Legend of Henry Morton Stanley* (New York: Knopf, 1990).

375-378. A. Hochschild, *King Leopold's Ghost: A Story of Greed, Terror, and Heroism in Colonial Africa* (London: Pan, 2002).

379-385. T. A. Nash, *Africa's Bane: The Tsetse Fly* (London: Collins, 1969); M. Lyons, *The Colonial Disease: A Social History of Sleeping Sickness in Northern Zaire, 1900–1940* (Cambridge: Cambridge University Press, 1992); M. Lyons, "African sleeping sickness: an historical review," *International Journal of STD & AIDS* 2 (1991): 20–25.

380-383. "Landmarks in trypanosome research," in *Trypanosomiasis and Leishmaniasis: Biology and Control*, ed. G. Hide et al. (London: CAB International, 1997), 1–37; J. J. McKelvey, Jr., *Man against Tsetse: Struggle for Africa* (Ithaca, N.Y.: Cornell University Press, 1973); de Kruif, chapter 9.

385-389. F. E. G. Cox, "History of human parasitology," *Clinical Microbiology Reviews* 15 (2002): 595–612; Foster; R. S. Desowitz, *New Guinea Tapeworms and Jewish Grandmothers: Tales of Parasites and People* (New York: Norton, 1987), chapter 6.

390. The 2015 Nobel Prize in Physiology or Medicine was awarded to William C. Campbell and Satoshi Ōmura " for their discoveries concerning a novel therapy against infections caused by roundworm parasites" Campbell and Ōmura discovered ivermectin.

387-390. T. V. Rajan, "The eye does not see what the mind does not know: the bacterium in the worm," *Perspectives in Biology and Medicine* 48 (2005): 31–41; R. Muller, "Onchocerciasis," in *The Wellcome Trust Illustrated History of Tropical Diseases*, ed. F. E. G. Cox (London: The Wellcome Trust, 1996), 304–309; K. Gustavsen et al., "Onchocerciasis in the Americas: from arrival to (near) elimination," *Parasites & Vectors* 4 (2011): 205; Centers for Disease Control and Prevention (CDC), "Progress toward elimination of onchocerciasis in the Americas—1993-2012," *MMWR. Morbidity and Mortality Weekly Report* 62 (2014): 405–408.

390-393. A. Tayeh, "Dracunculiasis," in Cox, 1996, 286–303; E. Ruiz-Tiben and D. R. Hopkins, "Dracunculiasis (Guinea worm disease) eradication," *Advances in Parasitology* 61 (2006): 275–309; A. R. Al-Awadi et al., "Guinea worm (Dracunculiasis) eradication: update on progress and endgame challenges," *Transactions of the Royal Society of Tropical Medicine and Hygiene* 108 (2014): 249–251.

393-395.   S. Howard, *Yellowjack, a History* (New York: Harcourt Brace, 1934); J. H. Powell, *Bring Out Your Dead: The Great Plague of Yellow Fever in Philadelphia in 1793* (Philadelphia: University of Pennsylvania Press, 1993); Oldstone, chapter 5; Franklin and Sutherland, 183–226; Altman, 129–158; de Kruif, chapter 11.

395-399,
403-405.   H. N. Simpson.

408-409.   P. J. Hotez and D. I. Pritchard, "," *Scientific American* 272 (1995): 68–74; P. J. Hotez et al., "Hookworm infection," *New England Journal of Medicine* 351 (2004): 799–807; T. L. Savitt and J. H. Young, ed., *Disease and Distinctiveness in the American South* (Knoxville: University of Tennessee Press, 1988); G. A. Schad and T. A. Nawalinski, "Historical introduction," in *Hookworm Infections*, ed. H. M. Gilles and P. A. J. Ball (Amsterdam: Elsevier, 1991), 1–14.

Chapter 16. Emerging and Reemerging Plagues

General

A. S. Khan, *The Next Pandemic: On the Front Lines Against Humankind's Gravest Dangers* (New York: Public Affairs, 2016).

Specific

411-412.   J. Lederberg, R. E. Shope, and S. C. Oaks, Jr., ed., *Emerging Infections: Microbial Threats to Health in the United States* (Washington: National Academy Press, 1992); M. E. Wilson, "Travel and the emergence of infectious diseases," *Emerging Infectious Diseases* 1 (1995): 39–46; ,"The emergence of new diseases," *American Scientist* 82 (1994): 52–60; L. Garrett, *The Coming Plague: Newly Emerging Diseases in a World Out of Balance* (New York: Farrar, Straus & Giroux, 1994); D. M. Morens et al., "The challenge of emerging and re-emerging infectious diseases," *Nature* 430 (2004): 242–249; V. R. Racaniello, "Emerging infectious diseases," *Journal of Clinical Investigation* 113 (2004): 796–798; R. Rosenberg, "Threat from emerging vectorborne viruses," *Emerging Infectious Diseases* 22 (2016): 910–911.

412-413.   K. .S. ,"Hantaviruses: four years after Four Corners," *Hospital Practice (1995)* 32 (1997): 93–94, 100, 103–108; S. Wrobel, "Serendipity, science, and a new hantavirus," *FASEB Journal* 9 (1995): 1247–1254; J. J. Núñez et al., "Hantavirus infections among overnight visitors to Yosemite National Park, California, USA, 2012," *Emerging Infectious Diseases* 20 (2014): 386–393; http://www.who.int/csr/don/27-may-2016-lassa-fever-nigeria/en/.

414-416.   P. Sampathkumar, "West Nile virus: epidemiology, clinical presentation, diagnosis, and prevention," *Mayo Clinic Proceedings* 78 (2003): 1137–1144; West Nile Virus. *Nature Medicine* 10 (12): 5101-5103. 2004; D. L. Morse, "West Nile virus—not a passing phenomenon," *New England Journal of Medicine* 348 (2003): 2173–2174; L. H. Gould and E. Fikrig, "West Nile virus: a growing concern?" *Journal of Clinical Investigation* 11 (2004): 1102–1107; J. J. Sejvar, "West Nile virus infection," in Scheld et al., 175–200.

416-424.   D. Freed, "The wrong man," *The Atlantic*, May 2010; J. Warrick, "FBI investigation of 2001 anthrax attacks concluded: US releases details," *Washington Post*, 20 February 2010; Brock; S. M. Blevins and M. S. Bronze, "Robert Koch and the 'golden age' of bacteriology," *International Journal of Infectious Diseases* 14 (2010): e744–e751; R. Dubos, *Pasteur and Modern Science* (Washington: ASM Press, 1998); Debré; Geison; A. Ullmann, "Pasteur-Koch: distinctive ways of thinking about infectious diseases," *Microbe* 2 (2007): 383–387; M. Moayeri and S. H. Leppla, "The roles of anthrax toxin in pathogenesis," *Current Opinion in Microbiology* 7 (2004): 19–24; L. Abrami et al., "Anthrax toxin: the long and winding road that leads to the kill," *Trends in Microbiology* 13 (2005): 72–78.

424-431. R. Rhodes, *Deadly Feasts: Tracking the Secrets of a Terrifying New Plague* (New York: Simon and Schuster, 1997); S. J. Collins et al., "Transmissible spongiform encephalopathies," *The Lancet* 363 (2004): 51–61; C. Musahl and A. Aguzzi, "Prions," in Lederburg, vol. 3, 809–823; s,"Transmission of prions," *Journal of Infectious Diseases* 186 (2002): S157–S165; ,"Balancing selection at the prion protein gene consistent with prehistoric kurulike epidemics," *Science* 300 (2003): 640–643; ,"Epidemiology. Tracking the fallout from mad cow disease," *Science* 289 (2000): 1452–1454; ,"Synthetic mammalian prions," *Science* 305 (2004): 673–676; B., "Detecting mad cow disease," *Scientific American* 291 (2004): 86–93; ,"Mad-cow disease in cattle and human beings," *American Scientist* 92 (2004): 334–341; S. B. Prusiner, ed., *Prion Biology and Diseases*, 2nd ed. (Cold Spring Harbor, N.Y.: Cold Spring Harbor Laboratory Press, 2004); S. B. Prusiner, *Madness and Memory: The Discovery of Prions—a New Biological Principle* (New Haven, Conn.: Yale University Press, 2014).

431-436. D. G. Bausch and A. Rojeck, "West Africa 2013: re-examining Ebola," in Scheld et al., 1–39; R. Fisher and L. Borio, "Ebola virus disease: therapeutic and potential preventative opportunities," in Scheld et al., 53–72; J. G. Breman et al., "Discovery and description of Ebola Zaire virus in 1976 and relevance to the West African epidemic during 2013-2016," *Journal of Infectious Diseases* 214 (2016): S93–S101; H. Feldmann et al., "Ebola virus: from discovery to vaccine," *Nature Reviews. Immunology* 3 (2003): 677–685; J. R. Spengler et al., "Perspectives on the West Africa Ebola virus disease outbreak, 2013-2016," *Emerging Infectious Diseases* 22 (2016): 956–963; R. Preston, "The Ebola wars," *The New Yorker*, 27 October 2014; K. Atkins et al., "Retrospective analysis of the 2014-2015 Ebola epidemic in Liberia," *American Journal of Tropical Medicine and Hygiene* 94 (2016): 833–839; C. Spencer, "Ebola isn't over yet," *New York Times*, 17 August 2015; K. A. Martins et al., "Ebola virus disease candidate vaccines under evaluation in clinical trials," *Expert Reviews of Vaccines* 15 (2016): 1101–1112; R. G. Swetha et al., "Ebolavirus Database: gene and protein information resource for ebolaviruses," *Advances in Bioinformatics* 2016 (2016): 1673284.

436-441. W. Slenczka, "Zika virus disease," in Scheld et al., 163–174; D. G. McNeil, Jr., *Zika: The Emerging Epidemic* (New York: Norton, 2016); J. C. Saiz et al., "Zika virus: the latest newcomer," *Frontiers in Microbiology* 7 (2016): 496; A. S. Fauci and D. M. Morens, "Zika virus in the Americas—yet another arbovirus threat," *New England Journal of Medicine* 374 (2016): 601–604; A. R. Plourde and E. M. Bloch, "A literature review of Zika virus," *Emerging Infectious Diseases* 22 (2016): 1185–1192; S. C. Weaver et al., "Zika virus: history, emergence, biology, and prospects for control," *Antiviral Research* 130 (2016): 69–80; A. A. Al-Qahtani et al., "Zika virus: a new pandemic threat," *Journal of Infection in Developing Countries* 10 (2016): 201–207; C. Chang et al., "The Zika outbreak of the 21st century," *Journal of Autoimmunity* 68 (2016): 1–13; J. Beck, "What to know about Zika virus," *The Atlantic*, 19 January 2015.

441-445. A. C. Steere et al., "The emergence of Lyme disease," *Journal of Clinical Investigation* 113 (2004): 1093–1101; M. Specter, "The Lyme wars," *The New Yorker*, 1 July 2013; G. Juckett, "Arthropod-borne diseases: the camper's uninvited guests," *Microbiology Spectrum* 3 (2015): IOL5-0001-2014; K. Tilly et al., "Biology of infection with *Borrelia burgdorferi*," *Infectious Disease Clinics of North America* 22 (2008): 217–234; J. Karow, "Battling Lyme disease," *Scientific American*, 5 July 2000; M. Lavelle, "Lyme disease surges north," *The Daily Climate*, 24 September 2014; N. A. Shadick et al., "A school-based intervention to increase Lyme disease preventive measures among elementary school-aged children," *Vector Borne and Zoonotic Diseases* 16 (2016): 507–515.

Bibliography

General Works on Disease and History

Abraham T. 2007. *Twenty First Century Plague: The Story of SARS*. Johns Hopkins University Press, Baltimore, MD.

Ackerknecht EH. 1965. *History and Geography of the Most Important Diseases*. Hafner, New York, NY.

Barker R. 1997. *And the Waters Turned to Blood*. Simon & Schuster, New York, NY.

Berenbaum MR. 1995. *Bugs in the System: Insects and Their Impact on Human Affairs*. Addison-Wesley, Boston, MA.

Bollet AJ. 1987. *Plagues & Poxes: The Rise and Fall of Epidemic Disease*. Demos, New York, NY.

Crawford DH. 2003. *The Invisible Enemy: A Natural History of Viruses*. Oxford University Press, Oxford, United Kingdom.

Crawford R. 1914. *Plague and Pestilence in Literature and Art*. Oxford University Press, Oxford, United Kingdom.

Crosby AW. 2004. *Ecological Imperialism: The Biological Expansion of Europe, 900-1900*. Cambridge University Press, Cambridge, United Kingdom.

Crosby AW, Jr. 1973. *The Columbian Exchange: Biological and Cultural Consequences of 1492*. Greenwood Press, New York, NY.

Dobson M. 2008. *Disease: The Extraordinary Stories behind History's Deadliest Killers*. Quercus, London, United Kingdom.

Goudsmit J. 2004. *Viral Fitness: The Next SARS and West Nile in the Making*. Oxford University Press, New York, NY.

Guillemin J. 1999. *Anthrax: The Investigation of a Deadly Outbreak*. University of California Press, Berkeley, CA.

Hammonds EM. 1999. *Childhood's Deadly Scourge: The Campaign to Control Diphtheria in New York City, 1880-1930*. Johns Hopkins Univerity Press, Baltimore, MD.

Hays JN. 1998. *The Burdens of Disease: Epidemics and Human Response in Western History*. Rutgers University Press, New Brunswick, NJ.

Hudson RP. 1983. *Disease and Its Control: The Shaping of Modern Thought*. Greenwood Press, New York, NY.

Karlen A. 1995. *Man and Microbes: Disease and Plagues in History and Modern Times*. Touchstone, New York, NY.

Kiple KF (ed). 1993. *The Cambridge World History of Human Disease*. Cambridge University Press, Cambridge, United Kingdom.

Knapp V. 1989. *Disease and Its Impact on Modern European History*. Edwin Mellen Press, Lewiston, NY.

Krause RM (ed). 2000. *Emerging Infections*. Academic Press, San Diego, CA.

Loudon I (ed). 1997. *Western Medicine: An Illustrated History*. Oxford University Press, Oxford, United Kingdom.

Markel H. 2004. *When Germs Travel: Six Major Epidemics That Have Invaded America and the Fears They Have Unleashed.* Vintage Books, New York, NY.

Porter R. 1997. *The Greatest Benefit to Mankind: A Medical History of Humanity from Antiquity to the Present.* HarperCollins, London, United Kingdom.

Quammen D. 2012. *Spillover: Animal Infections and the Next Human Pandemic.* Norton, New York, NY.

Rosenberg CE. 1992. *Explaining Epidemics: And Other Studies in the History of Medicine.* Cambridge University Press, Cambridge, United Kingdom.

Sankaran N. 2000. *Microbes and People: An A-Z of Microorganisms in Our Lives.* Oryx Press, Phoenix, AZ.

Singer C, Underwood EA. 1962. *A Short History of Medicine,* 2nd ed. Oxford University Press, Oxford, United Kingdom.

Spielman A, D'Antonio M. 2001. *Mosquito: A Natural History of Our Most Persistent and Deadly Foe.* Hyperion, New York, NY.

Stolley PD, Lasky T. 1995. *Investigating Disease Patterns: The Science of Epidemiology.* Scientific American Library/W.H. Freeman, New York, NY.

Tierno PM, Jr. 2004. *The Secret Life of Germs: What They Are, Why We Need Them, and How We Can Protect Ourselves against Them.* Simon & Schuster, New York, NY.

Watts S. 1997. *Epidemics and History: Disease, Power and Imperialism.* Yale University Press, New Haven, CT.

Wills C. 1996. *Yellow Fever, Black Goddess: The Coevolution of People and Plagues.* Helix Books, Cambridge, MA.

Winslow CEA. 1967. *The Conquest of Epidemic Disease: A Chapter in the History of Ideas.* Hafner Publishing, New York, NY.

Works Cited in the Text

Abbas AK, Lichtman AH, Pillai S. 2014. *Cellular and Molecular Immunology,* 8th ed. Saunders, Philadelphia, PA.

Abrami L, Reig N, van der Goot FG. 2005. Anthrax toxin: the long and winding road that leads to the kill. *Trends Microbiol* **13:**72–78.

Achtman M, Zurth K, Morelli G, Torrea G, Guiyoule A, Carniel E. 1999. *Yersinia pestis,* the cause of plague, is a recently emerged clone of *Yersinia pseudotuberculosis. Proc Natl Acad Sci U S A* **96:**14043–14048.

Aggleton P, O'Reilly K, Slutkin G, Davies P. 1994 Risking everything? Risk behavior, behavior change, and AIDS. *Science* **265:**341–345.

Al-Awadi AR, Al-Kuhlani A, Breman JG, Doumbo O, Eberhard ML, Guiguemde RT, Magnussen P, Molyneux DH, Nadim A. 2014. Guinea worm (Dracunculiasis) eradication: update on progress and endgame challenges. *Trans R Soc Trop Med Hyg* **108:**249–251.

Albert MR, Ostheimer KG, Breman JG. 2001. The last smallpox epidemic in Boston and the vaccination controversy, 1901-1903. *N Engl J Med* **344:**375–379.

Allen A. 2014. *The Fantastic Laboratory of Dr. Weigl: How Two Brave Scientists Battled Typhus and Sabotaged the Nazis.* Norton, New York, NY.

Al-Qahtani AA, Nazir N, Al-Anazi MR, Rubino S, Al-Ahdal MN. 2016. Zika virus: a new pandemic threat. *J Infect Dev Ctries* **10:**201–207.

Altman LK. 1998. *Who Goes First? The Story of Self-Experimentation in Medicine.* University of California Press, Berkeley, CA.

Anderson RM, May RM. 1985. Vaccination and herd immunity to infectious diseases. *Nature* **318:**323–329.

Arts EJ, Hazuda DJ. 2012. HIV-1 antiretroviral drug therapy. *Cold Spring Harb Perspect Med* **2:**a007161. doi:10.1101/cshperspect.a007161.

Askwith R. 13 September 1998. How aspirin turned hero. *Sunday Times,* London, United Kingdom.

Atkins KE, Pandey A, Wenzel NS, Skrip L, Yamin D, Nyenswah TG, Fallah M, Bawo L, Medlock J, Altice FL, Townsend J, Ndeffo-Mbah ML, Galvani AP. 2016. Retrospective analysis of the 2014-2015 Ebola epidemic in Liberia. *Am J Trop Med Hyg* **94:**833–839.

Atkinson S, Williams P. 2016. Yersinia virulence factors—a sophisticated arsenal for combating host defences. *F1000Res* **5:**1370. doi:10.12688/f1000research.8466.1.

Avery ME. 2000. *Gertrude B. Elion: 1918–1999.* Vol 78 of *Biographical Memoirs.* National Academy Press, Washington, DC.

Azman AS, Luquero FJ, Ciglenecki I, Grais RF, Sack DA, Lessler J. 2015. The impact of a one-dose versus two-dose oral cholera vaccine regimen in outbreak settings: a modeling study. *PLoS Med* **12:**e1001867. doi:10.1371/journal.pmed.1001867.

Badowski ME, Pérez SE, Biagi M, Littler JA. 2016. New antiretroviral treatment for HIV. *Infect Dis Ther* **5:**329–352.

Balter M. 2000. Epidemiology. Tracking the human fallout from 'mad cow disease.' *Science* **289:**1452–1454.

Barnes DS. 2000. Historical perspectives on the etiology of tuberculosis. *Microbes Infect* **2:**431–440.

Barnwell JW, Galinski MR. 1998. Invasion of vertebrate cells: erythrocytes, p 93–120. *In* Sherman IW (ed), *Malaria: Parasite Biology, Pathogenesis, and Protection.* ASM Press, Washington, DC.

Barry JM. 2004. *The Great Influenza: The Epic Story of the 1918 Pandemic.* Viking, New York, NY.

Barua D. 1992. History of cholera, p 1–36. *In* Barua D, Greenough WB, III (ed), *Cholera.* Plenum, New York, NY.

Bates B. 1992. *Bargaining for Life: A Social History of Tuberculosis, 1876–1938.* University of Pennsylvania Press, Philadelphia, PA.

Bausch DG, Rojeck A. 2016. West Africa 2013: re-examining Ebola, p 1–39. *In* Scheld WM, Hughes JM, Whitley RJ (ed), *Emerging Infections 10.* ASM Press, Washington, DC.

Baxby D. 1981. *Jenner's Smallpox Vaccine: The Riddle of Vaccinia Virus and Its Origin.* Heinemann, London, United Kingdom.

Bazin H. 2000. *The Eradication of Smallpox: Edward Jenner and the First and Only Eradication of a Human Infectious Disease.* Academic Press, San Diego, CA.

Beck J. 19 January 2015. What to know about Zika virus. *The Atlantic.*

Beckett DW. 1987. The striking hand of God: leprosy in history. *N Z Med J* 100:494–497.

Beer JJ. 1959. *Emergence of the German Dye Industry.* University of Illinois Press, Urbana, IL.

Bendiner E. 1989. Alexandre Yersin: pursuer of plague. *Hosp Pract (Off Ed)* **24:**121–128, 131–132, 135–138 passim.

Benedek TG, Erlen J. 1999. The scientific environment of the Tuskegee Study of Syphilis, 1920–1960. *Perspect Biol Med* **43:**1–30.

Berche P. 2001. The threat of smallpox and bioterrorism. *Trends Microbiol* **9:**15–18.

Bernstein SS. 1951. Smallpox and variolation; their historical significance in the American colonies. *J Mt Sinai Hosp N Y* **18:**228–244.

Bibel DJ. 1988. *Milestones in Immunology: A Historical Exploration.* Science Tech Publishers, Madison, WI.

Bierman J. 1990. *Dark Safari: The Life behind the Legend of Henry Morton Stanley.* Knopf, New York, NY.

Blevins SM, Bronze MS. 2010. Robert Koch and the 'golden age' of bacteriology. *Int J Infect Dis* **14:**e744–e751. doi:10.1016/j.ijid.2009.12.003.

Bloch M. 1989. *The Royal Touch: Sacred Monarchy and Scrofula in England and France.* Dorset, New York, NY.

Blower SM, McLean AR. 1994. Prophylactic vaccines, risk behavior change, and the probability of eradicating HIV in San Francisco. *Science* **265:**1451–1454.

Booth J. 9 January 2008. Woman in Los Angeles catches Black Death. *Los Angeles Times* Online, Los Angeles, CA.

Bordenave G. 2003. Louis Pasteur (1822-1895). *Microbes Infect* **5:**553–560.

Bratton TL. 1981. The identity of the plague of Justinian. (Part II). *Trans Stud Coll Physicians Phila* **3:**174–180.

Bray RS. 1996. *Armies of Pestilence: The Effects of Pandemics on History.* Butterworth, Cambridge, United Kingdom.

Breman JG, Heymann DL, Lloyd G, McCormick JB, Miatudila M, Murphy FA, Muyembé-Tamfun JJ, Piot P, Ruppol JF, Sureau P, van der Groen G, Johnson KM. 2016. Discovery and description of Ebola Zaire virus in 1976 and relevance to the West African epidemic during 2013-2016. *J Infect Dis* **214**(Suppl 3):S93–S101.

Brent P. 1977. *Black Nile: Mungo Park and the Search for the Niger.* Gordon Cremonesi, London, United Kingdom.

Britton WJ, Lockwood DN. 2004. Leprosy. *Lancet* **363:**1209–1219.

Brock TD. 1999. *Robert Koch: A Life in Medicine and Bacteriology.* ASM Press, Washington, DC.

Brophy PJ. 2002. Subversion of Schwann cells and the leper's bell. *Science* **296:**862–863.

Brown ED, Wright GD. 2016. Antibacterial drug discovery in the resistance era. *Nature* **529:**336–343.

Brown P. 2004. Mad-cow disease in cattle and human beings. *Am Sci* **92:**334–341.

Buckman R. 2003. *Human Wildlife: The Life That Lives on Us.* Johns Hopkins University Press, Baltimore, MD.

Burenhult G (ed). 1993. *The Illustrated History of Humankind,* 5 vols. HarperCollins, London, United Kingdom.

Byarugaba DK. 2009. Mechanisms of antimicrobial resistance, p 15–26. *In* Sosa AD, Byarugaba DK, Amábile-Cuevas CF, Hsueh PR, Kariuki S, Okeke IN (ed), *Antimicrobial Resistance in Developing Countries.* Springer, New York, NY.

Bynum H. 2012. *Spitting Blood: The History of Tuberculosis.* Oxford University Press, Oxford, United Kingdom.

Bynum WF, Overy C. 1998. *The Beast in the Mosquito: The Correspondence of Ronald Ross and Patrick Manson.* Editions Rodopi, Amsterdam, The Netherlands.

Caldwell M. 1988. *The Last Crusade: The War on Consumption, 1862–1954.* Atheneum, New York, NY.

Cameron CE, Lukehart SA. 2014. Current status of syphilis vaccine development: need, challenges, prospects. *Vaccine* **32:**1602–1609.**Carnathan DG, Wetzel KS, Yu J, Lee ST, Johnson BA, Paiardini M, Yan J, Morrow MP, Sardesai NY, Weiner DB, Ertl HC, Silvestri G.** 2015. Activated CD4⁺CCR5⁺ T cells in the rectum predict increased SIV acquisition in SIVGag/Tat-vaccinated rhesus macaques. *Proc Natl Acad Sci U S A* **112:**518–523.

Carniel E. 2000. Plague, p 654–661. *In* Lederberg J (ed), *Encyclopedia of Microbiology*, 2nd ed, vol 3. Academic Press, San Diego, CA.

Carrell JL. 2003. *The Speckled Monster: A Historical Tale of Battling Smallpox*. Dutton, New York, NY.

Cartwright FF. 1977. *A Social History of Medicine*. Longman, London, United Kingdom.

Cartwright FF, Biddiss M. 2000. *Disease and History*, 2nd ed. Sutton, Phoenix Mill, United Kingdom.

Celli A. 1933. *The History of Malaria on the Roman Campagna from Ancient Times*. Bale, London, United Kingdom.

Centers for Disease Control and Prevention (CDC). 2004. Fatal respiratory diphtheria in a U.S. traveler to Haiti—Pennsylvania, 2003. *MMWR Morb Mortal Wkly Rep* **52:**1285–1286. https://www.cdc.gov/mmwr/preview/mmwrhtml/mm5253a3.htm.

Centers for Disease Control and Prevention (CDC). 2013. Progress toward elimination of onchocerciasis in the Americas—1993-2012. *MMWR Morb Mortal Wkly Rep* **62:**405–408.

Chang C, Ortiz K, Ansari A, Gershwin ME. 2016. The Zika outbreak of the 21st century. *J Autoimmun* **68:**1–13.

Chien M, Morozova I, Shi S, Sheng H, Chen J, Gomez SM, Asamani G, Hill K, Nuara J, Feder M, Rineer J, Greenberg JJ, Steshenko V, Park SH, Zhao B, Teplitskaya E, Edwards JR, Pampou S, Georghiou A, Chou IC, Iannuccilli W, Ulz ME, Kim DH, Geringer-Sameth A, Goldsberry C, Morozov P, Fischer SG, Segal G, Qu X, Rzhetsky A, Zhang P, Cayanis E, De Jong PJ, Ju J, Kalachikov S, Shuman HA, Russo JJ. 2004. The genomic sequence of the accidental pathogen *Legionella pneumophila*. *Science* **305:**1966–1968.

Chung KT, Biggers CJ. 2001. Albert Léon Charles Calmette (1863-1933) and the antituberculous BCG vaccination. *Perspect Biol Med* **44:**379–389

Clark WR. 1995. *At War Within: The Double-Edged Sword of Immunity*. Oxford University Press, New York, NY.

Coale A. 1974. The history of the human population. *Sci Am* **231:**41–51.

Cole ST, Brosch R, Parkhill J, Garnier T, Churcher C, Harris D, Gordon SV, Eiglmeier K, Gas S, Barry CE, III, Tekaia F, Badcock K, Basham D, Brown D, Chillingworth T, Connor R, Davies R, Devlin K, Feltwell T, Gentles S, Hamlin N, Holroyd S, Hornsby T, Jagels K, Krogh A, McLean J, Moule S, Murphy L, Oliver K, Osborne J, Quail MA, Rajandream MA, Rogers J, Rutter S, Seeger K, Skelton J, Squares R, Squares S, Sulston JE, Taylor K, Whitehead S, Barrell BG. 1998. Deciphering the biology of *Mycobacterium tuberculosis* from the complete genome sequence. *Nature* **393:**537–544.

Collins SJ, Lawson VA, Masters CL. 2004. Transmissible spongiform encephalopathies. *Lancet* **363:**51–61.

Colwell RR. 1996. Global climate and infectious disease: the cholera paradigm. *Science* **274:**2025–2031.

Columbia University Press. 2001. Cecil John Rhodes. *In The Columbia Encyclopedia*, 6th ed. Columbia University Press, New York, NY.

Cook GC. 1996. The Asiatic cholera: an historical determinant of human genomic and social structure, p 18–53. *In* Drasar BS, Forrest BD (ed), *Cholera and the Ecology of* Vibrio cholerae. Chapman & Hall, London, United Kingdom.

Cooper GM, Temin RG, Sugden B (ed). 1995. *The DNA Provirus: Howard Temin's Scientific Legacy.* ASM Press, Washington, DC.

Cox FEG. 2002. History of human parasitology. *Clin Microbiol Rev* **15:**595–612.

Crosby AW. 2003. *America's Forgotten Pandemic: The Influenza of 1918*, 2nd ed. Cambridge University Press, Cambridge, United Kingdom.

Cunha CB, Cunha BA. 2006. Impact of plague on human history. *Infect Dis Clin N Am* **20:**253–272, viii.

Daniel T, Bates J, Dawnes K. 1984. History of tuberculosis, p 13–24. *In* Bloom BR (ed), *Tuberculosis: Pathogenesis, Protection and Control.* ASM Press, Washington, DC.

Daniel TM. 1997. *Captain of Death: The Story of Tuberculosis.* University of Rochester Press, Rochester, NY.

Daniel VS, Daniel TM. 1999. Old Testament biblical references to tuberculosis. *Clin Infect Dis* **29:**1557–1558.

Davis D, Gash-Kim TL, Heffernan EJ. 1998. Toxic shock syndrome: case report of a postpartum female and a literature review. *J Emerg Med* **16:**607–614.

Davis JC. 2004. *The Human Story: Our History, from the Stone Age to Today.* HarperCollins, New York, NY.

Debré P. 1998. *Louis Pasteur.* Johns Hopkins University Press, Baltimore, MD.

Deevey ES, Jr. 1960. The human population. *Sci Am* **203:**195–204.

Defoe D. 1969. *A Journal of the Plague Year.* Oxford University Press, New York, NY.

de Kruif P. 1954. *Microbe Hunters.* Harcourt Brace Jovanovich, San Diego, CA.

Desowitz RS. 1987. *New Guinea Tapeworms and Jewish Grandmothers: Tales of Parasites and People.* Norton, New York, NY.

Diamond J. 1997. *Guns, Germs, and Steel.* Norton, New York, NY.

Dixon CW. 1962. *Smallpox.* Churchill, London, United Kingdom.

Dobson AP, Carper ER. 1996. Infectious diseases and human population history. *Bioscience* **46:**115–126.

Dold C. February 1999. The cholera lesson. *Discover*, 71–75.

Dormandy T. 2000. *The White Death: A History of Tuberculosis.* New York University Press, New York, NY.

Drews J. 2000. Drug discovery: a historical perspective. *Science* **287:**1960–1964.

Dubos R. 1998. *Pasteur and Modern Science.* ASM Press, Washington, DC.

Dubos R, Dubos J. 1952. *The White Plague: Tuberculosis, Man, and Society.* Little, Brown, Boston, MA.

Duffy J. 1953. *Epidemics in Colonial America.* Louisiana State University Press, Baton Rouge, LA.

Dunbar R. 2004. *The Human Story: A New History of Mankind's Evolution.* Faber and Faber, London, United Kingdom.

Duthie MS, Gillis TP, Reed SG. 2011. Advances and hurdles on the way toward a leprosy vaccine. *Hum Vaccin* **7:**1172–1183.

Dwyer G, Levin SA, Buttel L. 1990. A simulation model of the population dynamics and evolution of myxomatosis. *Ecol Monogr* **60:**423–447.

Dye C, Gay N. 2003. Modeling the SARS epidemic. *Science* **300:**1884–1885.

Ebert D, Bull JJ. 2003. Challenging the trade-off model for the evolution of virulence: is virulence management feasible? *Trends Microbiol* **11:**15–20.

Echenberg M. 2011. *Africa in the Time of Cholera: A History of Pandemics from 1817 to the Present.* Cambridge University Press, New York, NY.

Elion GB. 1989. The purine path to chemotherapy. *Science* **244:**41–47.

Ewald PW. 1994. *Evolution of Infectious Disease.* Oxford University Press, New York, NY.

Eyler JM. 2003. Smallpox in history. *J Lab Clin Med* **142:**216–220.

Fan HY, Conner RF, Villarreal LP. 2013. *AIDS: Science and Society,* 7th ed. Jones and Bartlett, Sudbury, MA.

Farley J. 1992. *Bilharzia: A History of Imperial Tropical Medicine.* Cambridge University Press, Cambridge, United Kingdom.

Farmer P. 1999. *Infections and Inequalities: The Modern Plagues.* University of California Press, Berkeley, CA.

Farrell J. 1998. *Invisible Enemies: Stories of Infectious Disease.* Farrar, Straus & Giroux, New York, NY.

Fauci AS, Morens DM. 2016. Zika virus in the Americas—yet another arbovirus threat. *N Engl J Med* **374:**601–604.

Feldberg GD. 1995. *Disease and Class: Tuberculosis and the Shaping of Modern North American Society.* Rutgers University Press, New Brunswick, NJ.

Feldmann H, Jones S, Klenk HD, Schnittler HJ. 2003. Ebola virus: from discovery to vaccine. *Nat Rev Immunol* **3:**677–685.

Fenner F, Henderson DA, Arita I, Jezek Z, Ladnyi ID. 1988. *Smallpox and Its Eradication.* World Health Organization, Geneva, Switzerland.

Fenster JM. 2001. *Ether Day: The Strange Tale of America's Greatest Medical Discovery and the Haunted Men Who Made It.* HarperCollins, New York, NY.

Feodorova VA, Motin VL. 2012. Plague vaccines: current developments and future perspectives. *Emerg Microbes Infect* **1:**e36. doi:10.1038/emi.2012.34.

Feodorova VA, Sayapina LV, Motin VL. 2016. Assessment of live plague vaccine candidates. *Methods Mol Biol* **1403:**487–498.

Fiennes R. 1978. *Zoonoses and the Origins and Ecology of Human Disease.* Academic Press, London, United Kingdom.

Finer SE. 1952. *The Life and Times of Edwin Chadwick.* Methuen, London, United Kingdom.

Fisher R, Borio L. 2016. Ebola virus disease: therapeutic and potential preventative opportunities, p 53–72. *In* Scheld WM, Hughes JM, Whitley RJ (ed), *Emerging Infections 10.* ASM Press, Washington, DC.

Food and Drug Administration. 8 May 2015. FDA approves additional antibacterial treatment for plague. Food and Drug Administration, Silver Spring, MD. www.fda.gov/NewsEvents/Newsroom/PressAnnouncements/ucm446283.htm.

Foster WD. 1965. *A History of Parasitology.* Livingston, Edinburgh, United Kingdom.

Franklin J, Sutherland J. 1984. *Guinea Pig Doctors: The Drama of Medical Research through Self-Experimentation.* Morrow, New York, NY.

Freed D. May 2010. The wrong man. *The Atlantic.*

Gallo RC. 1987. The AIDS virus. *Sci Am* **256:**46–56.

Gardner WM (ed). 1915. *The British Coal Tar Industry.* Lippincott, Philadelphia, PA.

Garnier T, Eiglmeier K, Camus JC, Medina N, Mansoor H, Pryor M, Duthoy S, Grondin S, Lacroix C, Monsempe C, Simon S, Harris B, Atkin R, Doggett J, Mayes R, Keating L, Wheeler PR, Parkhill J, Barrell BG, Cole ST, Gordon SV, Hewinson RG. 2003. The complete genome sequence of *Mycobacterium bovis. Proc Natl Acad Sci U S A* **100:**7877–7882.

Garrett L. 1994. *The Coming Plague: Newly Emerging Diseases in a World Out of Balance.* Farrar, Straus & Giroux, New York, NY.

Geison GL. 1995. *The Private Science of Louis Pasteur.* Princeton University Press, Princeton, NJ.

Gelfand M. 1957. *Livingstone the Doctor: His Life and Travels.* Blackwell, Oxford, United Kingdom.

Gershon A. 2011. Vaccination against varicella and zoster: its development and progress, p 247–264. *In* Plotkin SA (ed), *History of Vaccine Development.* Springer, New York, NY.

Gill G, Burrell S, Brown J. 2001. Fear and frustration—the Liverpool cholera riots of 1832. *Lancet* **358:**233–237.

Glynn I, Glynn J. 2004. *The Life and Death of Smallpox.* Cambridge University Press, New York, NY.

Gottfried RS. 1983. *The Black Death: Natural and Human Disaster in Medieval Europe.* Free Press, New York, NY.

Goudsmit J. 1997. *Viral Sex: The Nature of AIDS.* Oxford University Press, New York, NY.

Gould LH, Fikrig E. 2004. West Nile virus: a growing concern? *J Clin Invest* **113:**1102–1107.

Granström M. 2011. The history of pertussis vaccination: from whole-cell to subunit vaccines, p 73–82. *In* Plotkin SA (ed), *History of Vaccine Development.* Springer, New York, NY.

Gregg CT. 1985. *Plague: An Ancient Disease in the Twentieth Century.* University of New Mexico Press, Albuquerque, NM.

Grmek M. 1989. *Diseases in the Ancient Greek World.* Johns Hopkins University Press, Baltimore, MD.

Grob GN. 2002. *The Deadly Truth: A History of Disease in America.* Harvard University Press, Cambridge, MA.

Gross L. 1995. How the plague bacillus and its transmission through fleas were discovered: reminiscences from my years at the Pasteur Institute in Paris. *Proc Natl Acad Sci U S A* **92:**7609–7611.

Gross L. 1996. How Charles Nicolle of the Pasteur Institute discovered that epidemic typhus is transmitted by lice: reminiscences from my years at the Pasteur Institute in Paris. *Proc Natl Acad Sci U S A* **93:**10539–10540.

Grundbacher FJ. 1992. Behring's discovery of diphtheria and tetanus antitoxins. *Immunol Today* **13:**188–189.

Grzybowski S, Allen EA. 1995. History and importance of scrofula. *Lancet* **346:**1472–1474.

Guilfoile PG. 2009. *Diphtheria.* Chelsea House, New York, NY.

Gussow Z. 1989. *Leprosy, Racism, and Public Health: Social Policy in Chronic Disease Control.* Westview, Boulder, CO.

Gustavsen K, Hopkins A, Sauerbrey M. 2011. Onchocerciasis in the Americas: from arrival to (near) elimination. *Parasit Vectors* **4:**205. doi:10.1186/1756-3305-4-205.

Haas F, Haas S. 1996. The origins of *Mycobacterium tuberculosis* and the notion of its contagiousness, p 3–20. *In* Rom WN, Garay SM (ed), *Tuberculosis.* Little, Brown, Boston, MA.

Hajjeh RA, Reingold A, Weil A, Shutt K, Schuchat A, Perkins BA. 1999. Toxic shock syndrome in the United States: surveillance update, 1979–1996. *Emerg Infect Dis* **5:**807–810.

Hall R. 1974. *Stanley: An Adventurer Explored.* Collins, London, United Kingdom.

Hamlin C. 1997. *Public Health and Social Justice in the Age of Chadwick: Britain, 1850-1854.* Cambridge University Press, Cambridge, United Kingdom.

Hammerschmidt DE. 2003. Bovine tuberculosis: still a world health problem. *J Lab Clin Med* **141:**359–360.

Hare R. 1970. *The Birth of Penicillin and the Disarming of Microbes.* Allen & Unwin, London, United Kingdom.

Harper KN, Zuckerman M, Armegelos GJ. 1 February 2014. Syphilis: then and now. *The Scientist.*

Harper KN, Zuckerman MK, Harper ML, Kingston JD, Armelagos GJ. 2011. The origin and antiquity of syphilis revisited: an appraisal of Old World pre-Columbian evidence for treponemal infection. *Am J Phys Anthropol* **146**(Suppl 53):99–133.

Harrison GA. 1978. *Mosquitoes, Malaria and Man: A History of the Hostilities since 1880.* Dutton, New York, NY.

Hempel S. 2007. *The Strange Case of the Broad Street Pump: John Snow and the Mystery of Cholera.* University of California Press, Berkeley, CA.

Henderson JS, McMaster PD, Kidd JG, Huggins C. 1971. *A Notable Career in Finding Out: Peyton Rous, 1879–1970.* Rockefeller University Press, New York, NY.

Herlihy D. 1997. *The Black Death and the Transformation of the West.* Harvard University Press, Cambridge, MA.

Hibbert C. 1982. *Africa Explored: Europeans in the Dark Continent, 1769-1889.* Allan Lane, London, United Kingdom.

Hill AV, Weatherall DJ. 1998. Host genetic factors in resistance to malaria, p 445–456. *In* Sherman IW (ed), *Malaria: Parasite Biology, Pathogenesis, and Protection.* ASM Press, Washington, DC.

Hilleman MR. 2011. The development of a live attenuated mumps virus vaccine in historic perspective and its role in the evolution of combined measles-mumps-rubella, p 207–218. *In* Plotkin SA (ed), *History of Vaccine Development.* Springer, New York, NY.

Ho EL, Lukehart SA. 2011. Syphilis: using modern approaches to understand an old disease. *J Clin Invest* **121:**4584–4592.

Hochschild A. 2002. *King Leopold's Ghost: A Story of Greed, Terror, and Heroism in Colonial Africa.* Pan, London, United Kingdom.

Hoffman SL, Vekemans J, Richie TL, Duffy PE. 2015. The march toward malaria vaccines. *Am J Prev Med* **49:**S319–S333. doi:10.1016/j.amepre.2015.09.011.

Holmes OW. 1909-1914. *The Contagiousness of Puerperal Fever.* P. F. Collier & Son, New York, NY.

Hopkins DR. 1983. *Princes and Peasants: Smallpox in History.* University of Chicago Press, Chicago, IL.

Hotchkiss RD. 2003. *Selman Abraham Waksman: 1888–1973.* Vol 83 of *Biographical Memoirs.* National Academy Press, Washington, DC.

Hotez PJ, Brooker S, Bethony JM, Bottazzi ME, Loukas A, Xiao S. 2004. Hookworm infection. *N Engl J Med* **351:**799–807.

Hotez PJ, Kamath A. 2009. Neglected tropical diseases in sub-Saharan Africa: review of their prevalence, distribution, and disease burden. *PLoS Negl Trop Dis* **3:**e412. doi:10.1371/journal.pntd.0000412.

Hotez PJ, Pritchard DI. 1995. Hookworm infection. *Sci Am* **272:**68–74.

Hotez PJ, Savioli L, Fenwick A. 2012. Neglected tropical diseases of the Middle East and North Africa: review of their prevalence, distribution, and opportunities for control. *PLoS Negl Trop Dis* **6:**e1475. doi:10.1371/journal.pntd.0001475.

Howard S. 1934. *Yellowjack, a History*. Harcourt Brace, New York, NY.

Hufthammer AK, Walløe L. 2013. Rats cannot have been intermediate hosts for *Yersinia pestis* during medieval plague epidemics in Northern Europe. *J Archaeol Sci* **40:**1752–1759.

Hulse EV. 1971. Joshua's curse and the abandonment of ancient Jericho. *Med Hist* **15:**376–386.

Hutcheon L, Hutcheon M. 1996. *Opera: Desire, Disease, Death*. University of Nebraska Press, Lincoln, NE.

Huxley E. 1975. *Florence Nightingale*. Weidenfeld and Nicholson, London, United Kingdom.

Inglesby TV, Dennis DT, Henderson DA, Bartlett JG, Ascher MS, Eitzen E, Fine AD, Friedlander AM, Hauer J, Koerner JF, Layton M, McDade J, Osterholm MT, O'Toole T, Parker G, Perl TM, Russell PK, Schoch-Spana M, Tonat K. 2000. Plague as a biological weapon: medical and public health management. *JAMA* **283:**2281–2290.

Ishizuka AS, Lyke KE, DeZure A, Berry AA, Richie TL, Mendoza FH, Enama ME, Gordon IJ, Chang LJ, Sarwar UN, Zephir KL, Holman LA, James ER, Billingsley PF, Gunasekera A, Chakravarty S, Manoj A, Li M, Ruben AJ, Li T, Eappen AG, Stafford RE, KC N, Murshedkar T, DeCederfelt H, Plummer SH, Hendel CS, Novik L, Costner PJ, Saunders JG, Laurens MB, Plowe CV, Flynn B, Whalen WR, Todd JP, Noor J, Rao S, Sierra-Davidson K, Lynn GM, Epstein JE, Kemp MA, Fahle GA, Mikolajczak SA, Fishbaugher M, Sack BK, Kappe SH, Davidson SA, Garver LS, Björkström NK, Nason MC, Graham BS, Roederer M, Sim BK, Hoffman SL, Ledgerwood JE, Seder RA. 2016. Protection against malaria at 1 year and immune correlates following PfSPZ vaccination. *Nat Med* **22:**614–623.

Jacobs CD. 2015. *Jonas Salk: A Life*. Oxford University Press, New York, NY.

Jenkins PN. 24 February 2014. Heroin addiction's fraught history. *The Atlantic*.

Jiménez MR. 2014. *Albert B. Sabin, 1906–1993*. National Academy of Sciences, Washington, DC. http://www.nasonline.org/publications/biographical-memoirs/memoir-pdfs/sabin-albert.pdf

Johanson D, Edgar B. 1996. *From Lucy to Language*. Simon and Schuster, New York, NY.

Johnson S. 2005. *The Ghost Map: The Story of London's Most Terrifying Epidemic—and How It Changed Science, Cities, and the Modern World*. Riverhead, New York, NY.

Jones DS. 2000. Syphilis, historical, p 538–544. *In* Lederberg J (ed), *Encyclopedia of Microbiology*, 2nd ed, vol 4. Academic Press, San Diego, CA.

Jones JH. 1981. *Bad Blood: The Tuskegee Syphilis Experiment*. Free Press, New York, NY.

Jones S, Martin R, Pilbeam D (ed). 1992. *The Cambridge Encyclopedia of Human Evolution*. Cambridge University Press, Cambridge, United Kingdom.

Juckett G. 2015. Arthropod-borne diseases: the camper's uninvited guests. *Microbiol Spectr* **3:**IOL5-0001-2014. doi:10.1128/microbiolspec.IOL5-0001-2014.

Kapuscinski R. 2002. *The Shadow of the Sun*. Vintage Books, New York, NY.

Karow J. 5 July 2000. Battling Lyme disease. *Scientific American*.

Kato-Maeda M, Small PM. 2001. User's guide to tuberculosis resources on the Internet. *Clin Infect Dis* **32:**1580–1588.

Katz SL. 2011. The history of measles virus and the development and utilization of measles virus vaccines, p 199–206. *In* Plotkin SA (ed), *History of Vaccine Development*. Springer, New York, NY.

Katz SL, Wilfert CM, Robbins FC. 2011. The role of tissue culture in vaccine development, p 145–149. *In* Plotkin SA (ed), *History of Vaccine Development*. Springer, New York, NY.

Kelly J. 2005. *The Great Mortality: An Intimate History of the Black Death, the Most Devastating Plague of All Time*. HarperCollins, New York, NY.

Khan AS. 2016. *The Next Pandemic: On the Front Lines Against Humankind's Gravest Dangers*. Public Affairs, New York, NY.

Kingston W. 2004. Streptomycin, Schatz v. Waksman, and the balance of credit for discovery. *J Hist Med Allied Sci* **59:**441–462.

Koch R. 1987. *Essays of Robert Koch*. Carter KC, trans. Greenwood Press, Westport, CT.

Koch R. 1999. The aetiology of tuberculosis, p 319–329. *In* Rothman DJ, Marcus S, Kiceluk SA (ed), *Medicine and Western Civilization*. Rutgers University Press, New Brunswick, NJ.

Kraut AM. 1995. Plagues and prejudice: nativism's construction of disease in the nineteenth- and twentieth-century New York City, p 65–90. *In* Rosner D (ed), *Hives of Sickness: Public Health and Epidemics in New York City*. Rutgers University Press, New Brunswick, NJ.

Larnick R, Ciochon R. 1996. The African emergence and early Asian dispersal of the genus *Homo*. *Am Sci* **84:**538–551.

Lastória JC, Abreu MA. 2014. Leprosy: a review of laboratory and therapeutic aspects—part 2. *An Bras Dermatol* **89:**389–401.

Lavelle M. 24 September 2014. Lyme disease surges north. *The Daily Climate*.

Lax E. 2004. *The Mold in Dr. Florey's Coat: The Story of the Penicillin Miracle*. Henry Holt, New York, NY.

Leakey M, Walker A. 1997. Early hominid fossils. *Sci Am* **276:**74–79.

Leakey R, Lewin R. 1992. *Origins Reconsidered: In Search of What Makes Us Human*. Doubleday, New York, NY.

Leavitt JW. 1996. *Typhoid Mary: Captive to the Public's Health*. Beacon Press, Boston, MA.

Lechevalier HA, Solotorovsky M. 1974. *Three Centuries of Microbiology*. Dover, New York, NY.

Lederberg J, Shope RD, Oaks SC, Jr (ed). 1992. *Emerging Infections: Microbial Threats to Health in the United States*. National Academy Press, Washington, DC.

Legname G, Baskakov IV, Nguyen HO, Riesner D, Cohen FE, DeArmond SJ, Prusiner SB. 2004. Synthetic mammalian prions. *Science* **305:**673–676.

Lenski RE, May RM. 1994. The evolution of virulence in parasites and pathogens: reconciliation between two competing hypotheses. *J Theor Biol* **169:**253–265.

Levins R, Awerbuch T, Brinkmann U, Eckardt I, Epstein P, Makhoul N, de Possas CA, Puccia C, Spielman A, Wilson AM. 1994. The emergence of new diseases. *Am Sci* **82:**52–60.

Levy SB. 1998. The challenge of antibiotic resistance. *Sci Am* **278:**46–53.

Levy SB, Marshall B. 2004. Antibacterial resistance worldwide: causes, challenges and responses. *Nat Med* **10**(12 Suppl):S122–S129.

Lewin R. 1999. *Human Evolution: An Illustrated Introduction*, 4th ed. Blackwell, London, United Kingdom.

Lewis K. 2013. Platforms for antibiotic discovery. *Nat Rev Drug Discov* **12:**371–387.

Ligon BL. 2006. Plague: a review of its history and potential as a biological weapon. *Semin Pediatr Infect Dis* **17:**161–170.

Lithgow KV, Cameron CE. 2017. Vaccine development for syphilis. *Expert Rev Vaccines* **16:**37–44.

Lockwood C, Stringer C. 2014. *The Human Story: Where We Come From and How We Evolved*. Natural History Museum, London, United Kingdom.

Luquero FJ, Sack DA. 2015. Effectiveness of oral cholera vaccine in Haiti. *Lancet Glob Health* **3:**e120–e121. doi:10.1016/S2214-109X(15)70015-X.

Lyons AS, Petrucelli RJ. 1987. *Medicine: An Illustrated History*. Abradale, New York, NY.

Lyons M. 1991. African sleeping sickness: an historical review. *Int J STD AIDS* **2** (Suppl 1):20–25.

Lyons M. 1992. *The Colonial Disease: A Social History of Sleeping Sickness in Northern Zaire, 1900–1940*. Cambridge University Press, Cambridge, United Kingdom.

Macfarlane G. 1979. *Howard Florey: The Making of a Great Scientist*. Oxford University Press, Oxford, United Kingdom.

Macfarlane G. 1984. *Alexander Fleming: The Man and the Myth*. Harvard University Press, Cambridge, MA.

Malthus T. 1927-1928. *An Essay on the Principle of Population*. Dent and Sons, London, United Kingdom.

Markell H. 1997. *Quarantine! East European Jewish Immigrants and the New York City Epidemics of 1892*. Johns Hopkins University Press, Baltimore, MD.

Marquardt M. 1951. *Paul Ehrlich*. Schuman, New York, NY.

Martins KA, Jahrling PB, Bavari S, Kuhn JH. 2016. Ebola virus disease candidate vaccines under evaluation in clinical trials. *Expert Rev Vaccines* **15:**1101–1112.

Maunder JW. 1983. The appreciation of lice. *Proc R Inst G B* **55:**1–31.

McClain CS. 1994. A new look at an old disease: smallpox and biotechnology. *Perspect Biol Med* **38:**624–639.

McEvedy C. 1988. The bubonic plague. *Sci Am* **258:**118–123.

McGeary J. 12 February 2001. Death stalks a continent. *Time*, 36–54. http://content.time.com/time/world/article/0,8599,2056158,00.html.

McKelvey JJ, Jr. 1973. *Man against Tsetse: Struggle for Africa*. Cornell University Press, Ithaca, NY.

McNeil DG, Jr. 2016. *Zika: The Emerging Epidemic*. Norton, New York, NY.

McNeill WH. 1976. *Plagues and Peoples*. Anchor Books, New York, NY.

Mead S, Stumpf MP, Whitfield J, Beck JA, Poulter M, Campbell T, Uphill JB, Goldstein D, Alpers M, Fisher EM, Collinge J. 2003. Balancing selection at the prion protein gene consistent with prehistoric kurulike epidemics. *Science* **300:**640–643.

Moayeri M, Leppla SH. 2004. The roles of anthrax toxin in pathogenesis. *Curr Opin Microbiol* **7:**19–24.

Mohammadi P, Desfarges S, Bartha I, Joos B, Zangger N, Muñoz M, Günthard HF, Beerenwinkel N, Telenti A, Ciuffi A. 2013. 24 hours in the life of HIV-1 in a T cell line. *PLoS Pathog* **9:**e1000361. doi:10.1371/journal.ppat.1003161.

Molotsky I. 21 March 1987. U.S. approves drug to prolong lives of AIDS patients. *New York Times*, New York, NY.

Moore J. 2004. The puzzling origins of AIDS. *Am Sci* **92:**540–549.

Morens DM, Folkers GK, Fauci AS. 2004. The challenge of emerging and re-emerging infectious diseases. *Nature* **430:**242–249.

Morse DL. 2003. West Nile virus—not a passing phenomenon. *N Engl J Med* **348:**2173–2174.

Muller R. 1996. Onchocerciasis, p 304–309. *In* Cox FEG (ed), *The Wellcome Trust Illustrated History of Tropical Diseases*. The Wellcome Trust, London, United Kingdom.

Münch R. 2003. Robert Koch. *Microbes Infect* **5:**69–74.

Murphy K, Weaver C. 2016. *Janeway's Immunobiology*, 9th ed. Garland, New York, NY.

Musahl C, Aguzzi A. 2000. Prions, p 809–823. *In* Lederberg J (ed), *Encyclopedia of Microbiology*, 2nd ed, vol 3. Academic Press, San Diego, CA.

Nash TA. 1992. *Africa's Bane: The Tsetse Fly*. Collins, London, United Kingdom.

Nesse RM, Williams GC. 1994. *Why We Get Sick: The New Science of Darwinian Medicine*. Random House, New York, NY.

Newson-Smith S (ed). 1978. *Quest: The Story of Stanley and Livingstone Told in Their Own Words*. Arlington, London, United Kingdom.

Nuland SB. 2003. *The Doctors' Plague: Germs, Childbed Fever, and the Strange Story of Ignac Semmelweis*. Norton, New York, NY.

Núñez JJ, Fritz CL, Knust B, Buttke D, Enge B, Novak MG, Kramer V, Osadebe L, Messenger S, Albariño CG, Ströher U, Niemela M, Amman BR, Wong D, Manning CR, Nichol ST, Rollin PE, Xia D, Watt JP, Vugia DJ, Yosemite Hantavirus Outbreak Investigation Team. 2014. Hantavirus infections among overnight visitors to Yosemite National Park, California, USA, 2012. *Emerg Infect Dis* **20:**386–393.

Nutton V. 2000. Portraits of science. Logic, learning, and experimental medicine. *Science* **295:**800–801.

Nye ER, Gibson ME. 1997. *Ronald Ross: Malariologist & Polymath, a Biography*. Macmillan, London, United Kingdom.

Offit PA. 2008. *Vaccinated: One Man's Quest to Defeat the World's Deadliest Diseases*. Smithsonian Books, Washington, DC.

Oldstone MB. 1998. *Viruses, Plagues, & History*. Oxford University Press, New York, NY.

Olsen PE, Hames CS, Benenson AS, Genovese EN. 1996. The Thucydides syndrome: Ebola déjà vu? (or Ebola reemergent?) *Emerg Infect Dis* **2:**155–156.

Oshinsky DM. 2005. *Polio: An American Story*. Oxford University Press, New York, NY.

O'Toole T. 1999. Smallpox: an attack scenario. *Emerg Infect Dis* **5:**540–546.

Ott K. 1996. *Fevered Lives: Tuberculosis in American Culture since 1870*. Harvard University Press, Cambridge, MA.

Pakenham T. 1991. *The Scramble for Africa 1876-1912*. Weidenfeld and Nicholson, London, United Kingdom.

Parish HG. 1968. *Victory with Vaccines: The Story of Immunization*. Livingston, Edinburgh, United Kingdom.

Parran T. 1937. *Shadow on the Land: Syphilis*. Reynal and Hitchcock, New York, NY.

Pead PJ. 2003. Benjamin Jesty: new light in the dawn of vaccination. *Lancet* **362:**2104–2109.

Pechous RD, Sivaraman V, Stasulli NM, Goldman WE. 2016. Pneumonic plague: the darker side of *Yersinia pestis*. *Trends Microbiol* **24:**190–197.

The People's Plague: Tuberculosis in America. 1995. PBS, Arlington, VA.

Pepin J. 2011. *The Origins of AIDS*. Cambridge University Press, Cambridge, United Kingdom.

Peterson RK. Fall 1995. Insects, disease and military history. *American Entomologist*, 147–160.

Pflieger AK, Khan AS. 1997. Hantaviruses: four years after four corners. *Hosp Pract (1995)* **32:**93–94, 100, 103–108.

Pier GB, Lyczak JB, Wetzler LM. 2004. *Immunology, Infection, and Immunity*. ASM Press, Washington, DC.

Plague confirmed as cause of park biologist's death. 17 November 2007. *San Diego Union-Tribune*, San Diego, CA.

Plourde AR, Bloch EM. 2016. A literature review of Zika virus. *Emerg Infect Dis* **22:**1185–1192.

Porter R (ed). 1996. *The Cambridge Illustrated History of Medicine*. Cambridge University Press, Cambridge, United Kingdom.

Powell JH. 1993. *Bring Out Your Dead: The Great Plague of Yellow Fever in Philadelphia in 1793*. University of Pennsylvania Press, Philadelphia, PA.

Prentice MB, Rahalison L. 2007. Plague. *Lancet* **369:**1196–1207.

Preston R. 2002. *The Demon in the Freezer: A True Story*. Random House, New York, NY.

Preston R. 27 October 2014. The Ebola wars. *The New Yorker*.

Pringle P. 12 June 2012. Notebooks shed light on antibiotic's contested discovery. *New York Times*, New York, NY.

Prusiner SB. 2004. Detecting mad cow disease. *Sci Am* **291:**86–93.

Prusiner SB (ed). 2004. *Prion Biology and Diseases*, 2nd ed. Cold Spring Harbor Laboratory Press, Cold Spring Harbor, NY.

Prusiner SB. 2014. *Madness and Memory: The Discovery of Prions—a New Biological Principle*. Yale University Press, New Haven, CT.

Quetel C. 1990. *The History of Syphilis*. Braddock J, Pike B, trans. Johns Hopkins University Press, Baltimore, MD.

Racaniello VR. 2004. Emerging infectious diseases. *J Clin Invest* **113:**796–798.

Rajan TV. 2005. The eye does not see what the mind does not know: the bacterium in the worm. *Perspect Biol Med* **48:**31–41.

Rambukkana A. 2004. How does *Mycobacterium leprae* target the peripheral nervous system? *Trends Microbiol* **8:**23–28.

Ransford O. 1978. *David Livingstone: The Dark Interior*. St. Martin's Press, New York, NY.

Ransford O. 1983. *"Bid the Sickness Cease": Disease in the History of Black Africa*. John Murray, London, United Kingdom.

Raoult D, Roux V. 1999. The body louse as a vector of reemerging human diseases. *Clin Infect Dis* **29:**888–911.

Razum O, Becher H, Kapaun A, Junghanss T. 2003. SARS, lay epidemiology, and fear. *Lancet* **361:**1739–1740.

Reichman LB, Hershfield ES (ed). 2000. *Tuberculosis: A Comprehensive International Approach*, 2nd ed. Marcel Dekker, New York, NY.

Rhodes R. 1997. *Deadly Feasts: Tracking the Secrets of a Terrifying New Plague*. Simon and Schuster, New York, NY.

Richardus JH, Oskam L. 2015. Protecting people against leprosy: chemoprophylaxis and immunoprophylaxis. *Clin Dermatol* **33:**19–25.

Richie TL, Billingsley PF, Sim BK, James ER, Chakravarty S, Epstein JE, Lyke KE, Mordmüller B, Alonso P, Duffy PE, Doumbo OK, Sauerwein RW, Tanner M, Abdulla S, Kremsner PG, Seder RA, Hoffman SL. 2015. Progress with *Plasmodium falciparum* sporozoite (PfSPZ)-based malaria vaccines. *Vaccine* **33:**7452–7461.

Riedel S. 2005. Plague: from natural disease to bioterrorism. *Proc (Bayl Univ Med Cent)* **18:**116–124.

Roberts LS, Janovy J, Jr. 2000. *Foundations of Parasitology*, 6th ed. McGraw-Hill, New York, NY.

Rosahn PD, Black-Schaffer B. 1943. Studies in syphilis: III. Mortality and morbidity findings in the Yale autopsy series. *Yale J Biol Med* **15:**587–602.

Rosebury T. 1971. *Microbes and Morals*. Viking, New York, NY.

Rosenberg CE. 1962. *The Cholera Years: The United States in 1832, 1849, and 1866*. University of Chicago Press, Chicago, IL.

Rosenberg R. 2016. Threat from emerging vectorborne viruses. *Emerg Infect Dis* **22:**910–911.

Ross R. 1923. *Memoirs*. John Murray, London, United Kingdom.

Ruiz-Tiben E, Hopkins DR. 2006. Dracunculiasis (Guinea worm disease) eradication. *Adv Parasitol* **61:**275–309.

Russell PF. 1955. *Man's Mastery of Malaria*. Oxford University Press, London, United Kingdom.

Saiz JC, Vázquez-Calvo Á, Blázquez AB, Merino-Ramos T, Escribano-Romero E, Martín-Acebes MA. 2016. Zika virus: the latest newcomer. *Front Microbiol* 7:496. doi:10.3389/fmicb.2016.00496.

Sallares R. 2002. *Malaria and Rome: A History of Malaria in Ancient Italy*. Oxford University Press, Oxford, United Kingdom.

Sampathkumar P. 2003. West Nile virus: epidemiology, clinical presentation, diagnosis, and prevention. *Mayo Clin Proc* **78:**1137–1143.

Savitt TL, Young JH (ed). 1988. *Disease and Distinctiveness in the American South*. University of Tennessee Press, Knoxville, TN.

Schad GA, Nawalinski TA. 1991. Historical introduction, p 1–14. *In* Gilles HM, Ball PAJ (ed), *Hookworm Infections*. Elsevier, Amsterdam, The Netherlands.

Schmid BV, Büntgen U, Easterday WR, Ginzler C, Walløe L, Bramanti B, Stenseth NC. 2015. Climate-driven introduction of the Black Death and successive plague reintroductions in Europe. *Proc Natl Acad Sci U S A* **112:**3020–3025.

Schuchat A, Fiebelkorn AP, Bellini W. 2016. Measles in the United States since the millennium: perils and progress in the postelimination era, p 131–142. *In* Scheld WM, Hughes JM, Whitley RJ (ed), *Emerging Infections 10*. ASM Press, Washington, DC.

Scott I. 1998. Heroin: a hundred-year habit. *History Today* **48:**6–8.

Scott S, Duncan CJ. 2001. *Biology of Plagues: Evidence from Historical Populations*. Cambridge University Press, Cambridge, United Kingdom.

Seder RA, Chang LJ, Enama ME, Zephir KL, Sarwar UN, Gordon IJ, Holman LA, James ER, Billingsley PF, Gunasekera A, Richman A, Chakravarty S, Manoj A, Velmurugan S, Li M, Ruben AJ, Li T, Eappen AG, Stafford RE, Plummer SH, Hendel CS, Novik L, Costner PJ, Mendoza FH, Saunders JG, Nason MC, Richardson JH, Murphy J, Davidson SA, Richie TL, Sedegah M, Sutamihardja A, Fahle GA, Lyke KE, Laurens MB, Roederer M, Tewari K, Epstein JE, Sim BK, Ledgerwood JE, Graham BS, Hoffman SL; VRC 312 Study Team. 2013. Protection against malaria by intravenous immunization with a nonreplicating sporozoite vaccine. *Science* **341:**1359–1365.

Sejvar JJ. 2016. West Nile virus infection, p 175–200. *In* Scheld WM, Hughes JM, Whitley RJ (ed), *Emerging Infections 10*. ASM Press, Washington, DC.

Sévère K, Rouzier V, Anglade SB, Bertil C, Joseph P, Deroncelay A, Mabou MM, Wright PF, Guillaume FD, Pape JW. 2016. Effectiveness of oral cholera vaccine in Haiti: 37-month follow-up. *Am J Trop Med Hyg* **94:**1136–1142.

Shadick NA, Zibit MJ, Nardone E, DeMaria A, Jr, Iannaccone CK, Cui J. 2016. A school-based intervention to increase Lyme disease preventive measures among elementary school-aged children. *Vector Borne Zoonotic Dis* **16:**507–515.

Sharp N. 3 December 2015. The return of syphilis. *The Atlantic.*

Sharp PM, Hahn BH. 2010. The evolution of HIV-1 and the origin of AIDS. *Philos Trans R Soc Lond B Biol Sci* **365:**2487–2494.

Sherman IW. 1998. A brief history of malaria and discovery of the parasite's life cycle, p 3–10. *In* Sherman IW (ed), *Malaria: Parasite Biology, Pathogenesis, and Protection.* ASM Press, Washington, DC.

Sherman IW. 2011. *Magic Bullets To Conquer Malaria: From Quinine to Qinghaosu.* ASM Press, Washington, DC.

Sherman IW. 2016. *Malaria Vaccines: The Continuing Quest.* World Scientific, London, United Kingdom.

Sherman IW (ed). *Great Scientists Speak Again II* (unpublished).

Shurkin JN. 1979. *The Invisible Fire: The Story of Mankind's Victory over the Ancient Scourge of Smallpox.* Putnam, New York, NY.

Silverstein AM. 2009. *A History of Immunology*, 2nd ed. Academic Press, San Diego, CA.

Simpson GG, Pittendrigh CS, Tiffany LH. 1957. *Life: An Introduction to Biology.* Harcourt Brace, San Francisco, CA.

Simpson HN. 1980. *Invisible Armies: The Impact of Disease on American History.* Bobbs-Merrill, Indianapolis, IN.

Slack P. 1985. *The Impact of Plague in Tudor and Stuart England.* Routledge Kegan & Paul, London, United Kingdom.

Slaughter FG. 1950. *Immortal Magyar: Semmelweis, Conqueror of Childhood Fever.* Schuman, New York, NY.

Slenczka W. 2016. Zika virus disease, p 163–174. *In* Scheld WM, Hughes JM, Whitley RJ (ed), *Emerging Infections 10.* ASM Press, Washington, DC.

Small H. 1998. *Florence Nightingale: Avenging Angel.* Constable, London, United Kingdom.

Small PM, Fujiwara PI. 2001. Management of tuberculosis in the United States. *N Engl J Med* **345:**189–200.

Smith FB. 1982. *Florence Nightingale: Reputation and Power.* St. Martin's Press, New York, NY.

Smith JS. 1990. *Patenting the Sun: Polio and the Salk Vaccine.* Morrow, New York, NY.

Snow J. 1855. *On The Mode of Communication of Cholera.* John Churchill, London, United Kingdom. http://www.ph.ucla.edu/epi/snow/snowbook.html.

Sompayrac L. 2015. *How the Immune System Works*, 5th ed. John Wiley & Sons, Chichester, United Kingdom.

Specter M. 1 July 2013. The Lyme wars. *The New Yorker.*

Spencer C. 17 August 2015. Ebola isn't over yet. *New York Times*, New York, NY.

Spengler JR, Ervin ED, Towner JS, Rollin PE, Nichol ST. 2016. Perspectives on West Africa Ebola virus disease outbreak, 2013-2016. *Emerg Infect Dis* **22:**956–963.

Stanley HM. 1969. *Through the Dark Continent.* Greenwood Press, New York, NY.

Steere AC, Coburn J, Glickstein L. 2004. The emergence of Lyme disease. *J Clin Invest* **113:**1093–1101.

Stenseth NC, Atshabar BB, Begon M, Belmain SR, Bertherat E, Carniel E, Gage KL, Leirs H, Rahalison L. 2008. Plague: past, present, and future. *PLoS Med* **5:**e3. doi:10.1371/journal. pmed.0050003.

Stine G. 2014. *AIDS Update 2014*. Pearson Benjamin Cummings, San Francisco, CA.

Strachey L. 2002. *Eminent Victorians*. Continuum, London, United Kingdom.

Stratmann L. 2003. *Chloroform: The Quest for Oblivion*. Sutton, Phoenix Mill, United Kingdom.

Stringer C, Andrews P. 2012. *The Complete World of Human Evolution*, 2nd ed. Thames and Hudson, London, United Kingdom.

Stringer C, Gamble C. 1993. *In Search of the Neanderthals*. Thames and Hudson, London, United Kingdom.

Sun L. 24 March 2016. TB cases increase in US for first time in 23 years. *Washington Post*, Washington, DC.

Swetha RG, Ramaiah S, Anbarasu A, Sekar K. 2016. Ebolavirus Database: gene and protein information resource for ebolaviruses. *Adv Bioinformatics* **2016:**1673284. doi:10.1155/2016/1673284.

Szybalski W. 1999. Maintenance of human-fed live lice in the laboratory and production of Weigl's exanthematic typhus vaccine, p 161–180. *In* Maramorosch K, Mahmood F (ed), *Maintenance of Human, Animal, and Plant Pathogen Vectors*. Science Publishers, Enfield, NH.

Tampa M, Sarbu I, Matei C, Benea V, Georgescu SR. 2014. Brief history of syphilis. *J Med Life* **7:**4–10.

Tattersall I. 1995. *The Last Neanderthal: The Rise, Success, and Mysterious Extinction of Our Closest Human Relatives*. Macmillan, New York NY.

Tattersall I. 2003. Once we were not alone. *Sci Am* **282:**56–62.

Tayeh A. 1996. Dracunculiasis, p 286–303. *In* Cox FEG (ed), *The Wellcome Trust Illustrated History of Tropical Diseases*. The Wellcome Trust, London, United Kingdom.

Taylor GM, Murphy E, Hopkins R, Rutland P, Chistov Y. 2007. First report of *Mycobacterium bovis* DNA in human remains from the Iron Age. *Microbiology* **153:**1243–1249.

Thatcher VS. 1984. *History of Anesthesia*. Garland, New York, NY.

Thompson M. 1949. *The Cry and the Covenant*. Doubleday, New York, NY.

Thucydides. *Peloponnesian Wars*, Book 2.

Tilly K, Rosa PA, Stewart PE. 2008. Biology of infection with *Borrelia burgdorferi. Infect Dis Clin North Am* **22:**217–234.

Torrey EF, Yolken RH. 2005. *Beasts of the Earth: Animals, Humans, and Disease*. Rutgers University Press, New Brunswick, NJ.

Trager W, Jensen JB. 1976. Human malaria parasites in continuous culture. *Science* **193:**673–675.

Tramont E. 2000. *Treponema pallidum* (syphilis), p 2474–2490. *In* Mandell J, Bennett J, Dolin R (ed), *Principles and Practice of Infectious Diseases*, 5th ed. Churchill Livingstone, New York, NY.

Travis AS. 1993. *The Rainbow Makers: The Origins of the Synthetic Dyestuffs Industry in Western Europe*. Lehigh University Press, Bethlehem, PA.

Trinkaus E, Shipman P. 1993. *The Neandertals: Changing the Image of Mankind*. Knopf, New York, NY.

Troutman J. 1994. The history of leprosy, p 11–25. *In* Hastings RC (ed), *Leprosy*, 2nd ed. Churchill Livingstone. Edinburgh, United Kingdom.

Tsang KW, Ho PL, Ooi GC, Yee WK, Wang T, Chan-Yeung M, Lam WK, Seto WH, Yam LY, Cheung TM, Wong PC, Lam B, Ip MS, Chan J, Yuen KY, Lai KN. 2003. A cluster of cases of severe acute respiratory syndrome in Hong Kong. *N Engl J Med* **348:**1977–1985.

Tuchman BW. 1978. *A Distant Mirror: The Calamitous 14th Century.* Ballantine, New York, NY.

Twigg G. 1985. *The Black Death: A Biological Reappraisal.* Schocken, New York, NY.

Ullmann A. 2007. Pasteur-Koch: distinctive ways of thinking about infectious diseases. *Microbe* **2:**383–387.

Ventola CL. 2015. The antibiotic resistance crisis. Part 1: causes and threats. *P T* **40:**277–283.

Ventola CL. 2015. The antibiotic resistance crisis. Part 2: management strategies and new agents. *P T* **40:**344–352.

Verma SK, Batra L, Tuteja U. 2016. A recombinant trivalent fusion protein Fi-LcrV-HSP70(II) augments humoral and cellular immune responses and imparts full protection against *Yersinia pestis.* *Front Microbiol* **7:**1053. doi:10.3389/fmicb.2016.01053.

Vickerman K. 1997. Landmarks in trypanosome research, p 1–37. *In* Hide G, Mottram JC, Coombs GH, Holmes PH (ed), *Trypanosomiasis and Leishmaniasis: Biology and Control.* CAB International, London, United Kingdom.

Vinten-Johansen P, Brody H, Paneth N, Rachman S, Rip M. 2003. *Cholera, Chloroform, and the Science of Medicine: A Life of John Snow.* Oxford University Press, New York, NY.

Waksman SA. 1964. *The Conquest of Tuberculosis.* University of California Press, Berkeley, CA.

Walking with Cavemen. 2003. BBC Video, London, United Kingdom.

Walsh C. 2003. *Antibiotics: Actions, Origins, Resistance.* ASM Press, Washington, DC.

Warrick J. 20 February 2010. FBI investigation of 2001 anthrax attacks concluded: US releases details. *Washington Post,* Washington, DC.

Weaver SC, Costa F, Garcia-Blanco MA, Ko AI, Ribeiro GS, Saade G, Shi PY, Vasilakis N. 2016. Zika virus: history, emergence, biology, and prospects for control. *Antiviral Res* **130:**69–80.

Weeks BS, Alcamo IE. 2009. *AIDS: The Biological Basis,* 5th ed. Jones and Bartlett, Sudbury, MA.

Weiss RA. 2001. The Leeuwenhoek Lecture 2001. Animal origins of human infectious disease. *Philos Trans R Soc Lond B Biol Sci* **356:**957–977.

Weissmann C, Enari M, Klöhn PC, Rossi D, Flechsig E. 2002. Transmission of prions. *J Infect Dis* **186**(Suppl 2)**:**S157–S165.

Weissmann G. 2005. Rats, lice, and Zinsser. *Emerging Infect Dis* **11:**492–496.

Williams G. 2010. *Angel of Death: The Story of Smallpox.* Palgrave MacMillan, Basingstoke, United Kingdom.

Willon P, Mason M. 17 October 2016. California Gov. Jerry Brown signs new vaccination law, one of nation's toughest. *Los Angeles Times,* Los Angeles, CA.

Wilson ME. 1995. Travel and the emergence of infectious diseases. *Emerg Infect Dis* **1:**39–46.

Winau F, Westphal O, Winau R. 2004. Paul Ehrlich—in search of the magic bullet. *Microbes Infect* **6:**786–789.

Winau F, Winau R. 2002. Emil von Behring and serum therapy. *Microbes Infect* **4:**185–188.

Wolfe RJ. 2001. *Tarnished Idol: William T. G. Morton and the Introduction of Surgical Anesthesia.* Norman Publishing, San Anselmo, CA.

Wood JW, Ferrell RJ, Dewitt-Aviña SN. 2003. The temporal dynamics of the fourteenth-century Black Death: new evidence from English ecclesiastical records *Hum Biol* **75:**427–448.

Woodruff HB. 2014. Selman A. Waksman, winner of the 1952 Nobel Prize for Physiology or Medicine. *Appl Environ Microbiol* **80:**2–8.

World Health Organization. 2015. *Global Tuberculosis Report 2016*. World Health Organization, Geneva, Switzerland. http://apps.who.int/iris/bitstream/10665/250441/1/9789241565394-eng.pdf?ua=1.

World Health Organization. 2016. Plague around the world, 2010–2015. *Wkly Epidemiol Rec* **91:**89–104.

Wrobel S. 1995. Serendipity, science, and a new hantavirus. *FASEB J* **9:**1247–1254.

Young D, Robertson B. 2001. Genomics: leprosy—a degenerative disease of the genome. *Curr Biol* **11:**R381–R383.

Zavis A. 19 August 2016. UN admits a role in cholera epidemic. *Los Angeles Times*, Los Angeles, CA.

Ziegler P. 1969. *The Black Death*. Harper, New York, NY.

Zimmer C. 1999. The stories behind the bones. *Science* **286:**1071, 1073–1074.

Zimmer C. 2007. *Smithsonian Intimate Guide to Human Origins*. Harper Perennial, New York, NY.

Zink AR, Grabner W, Nerlich AG. 2005. Molecular identification of human tuberculosis in recent and historic bone tissue samples: the role of molecular techniques for the study of historic tuberculosis. *Am J Phys Anthropol* **126:**32–47.

Zinsser H. 1935. *Rats, Lice and History*. Little, Brown, Boston, MA.

Index